Successes, Limitations, and Frontiers in Ecosystem Science

Springer

New York
Berlin
Heidelberg
Barcelona
Budapest
Hong Kong
London
Milan
Paris
Singapore
Tokyo

Michael L. Pace   Peter M. Groffman

Editors

Successes, Limitations, and Frontiers in Ecosystem Science

With 84 Illustrations

Springer

Michael L. Pace
Peter M. Groffman
Institute of Ecosystem Studies
Millbrook, NY 12545
USA

Cover photo by Molly Ahearn.

Library of Congress Cataloging-in-Publication Data
Successes, limitations, and frontiers in ecosystem science / [edited
  by] Michael L. Pace, Peter M. Groffman].
        p. cm.
     Papers from the Seventh Cary Conference held in Millbrook, N.Y.,
May 1997.
     Includes bibliographical references and index.
     ISBN 0-387-98476-3 (hardcover : alk. paper).—ISBN 0-387-98475-5
(softcover : alk. paper)
     1. Biotic communities—Research—Congresses. 2. Biotic
communities—Congresses. I. Pace, Michael L. II. Groffmann, Peter
M., 1958– . III. Cary Conference (7th : 1997 : Millbrook, N.Y.)
QH541.2.S92 1998
577.8′2—DC21   98-4683

Printed on acid-free paper.

Production coordinated by Chernow Editorial Services, Inc., and managed by Terry V.
Kornak; manufacturing supervised by Thomas King.
Typeset by Best-set Typesetter Ltd., Hong Kong.
Printed and bound by Maple-Vail Book Manufacturing Group, York, PA.
Printed in the United States of America.

9 8 7 6 5 4 3 2 1

ISBN 0-387-98476-3 Springer-Verlag New York Berlin Heidelberg   SPIN 10657930   (hardcover)
ISBN 0-387-98475-5 Springer-Verlag New York Berlin Heidelberg   SPIN 10657956   (softcover)

Acknowledgments

This book originated from the seventh Cary Conference held in May of 1997. We are indebted to all who worked so hard to make the conference successful. In particular, our steering committee of Mary Barber, Ingrid Burke, Anthony Janetos, William Lauenroth, Gene Likens, Gary Lovett, Pamela Matson, Judy Meyer, David Strayer, and Kathleen Weathers made valuable suggestions on topics, speakers, and participants, especially in the formative stages of planning for the conference. The National Center for Ecological Analysis and Synthesis (NCEAS) supported our efforts in hosting a workshop entitled "Establishing a Structure for the Synthesis and Integration of Progress in Ecosystem Science" in September of 1996. This allowed us, along with some of the conference speakers and participants, to explore formative issues in the development and practice of ecosystem science and to further develop topics and approaches to the 1997 Cary Conference.

We thank our authors who worked so hard to wrestle coherence from difficult and broad topics. Good humor prevailed despite our cajoling and deadlines; we are grateful for all their efforts. Conference participants provided critical and insightful reviews of manuscripts prior to the conference. The productive discussions that followed at the conference are evident in the pages within; they enabled us to produce this work efficiently. Steward Pickett, Jon Cole, and Moshe Shachak helped us shape the introductory and concluding chapters of the book.

We thank Timothy Wirth, Undersecretary of State for Global Affairs, for delivering the Conference keynote address, especially given the heavy demands of his schedule.

The Conference was supported by grants from the National Science Foundation, US Department of Energy, the National Aeronautics and Space Administration, and the Institute of Ecosystem Studies (IES). We thank all for their interest in ecosystem science and their support of the Cary Conference process.

The hard work of many made the Conference possible. The staff of the IES was crucial in attending to the needs of the Conference and the

Conferees both prior to the event and during the crucial days of the meeting. Although we cannot mention all, we would like to specifically thank the IES graduate students Jim Baxter, Mary Cadenasso, Greg Lewis, Eric Lilleskov, Robert Manson, and Bill Sobczak for their help with local transportation and daily logistics. Debbie Fargione and Marj Spoerri skillfully prepared many documents for the Conference and this book. Debbie Cosentino capably organized the complex travel arrangements and correspondence with participants and was involved from start to finish. Finally, Jan Mittan coordinated the Conference; without her ability, imagination and diligence standing behind us, the Conference would have been a headache. Instead, it was fun and productive.

Michael L. Pace
Peter M. Groffman

Contents

Contributors

Gregory P. Asner, Cooperative Institute for Research in Environmental Sciences and Environmental, Population, and Organismic Biology, University of Colorado, Boulder, CO 80309-0216, USA

James W. Baxter, Department of Ecology, Evolution, and Natural Resources, Rutgers University, New Brunswick, NJ 08903, USA

Denise L. Breitburg, Benedict Estuarine Research Center, Academy of Natural Sciences, St. Leonard, MD 20685, USA

Ingrid C. Burke, Natural Resources Ecology Laboratory, Colorado State University, Fort Collins, CO 80523, USA

Charles D. Canham, Institute of Ecosystem Studies, Millbrook, NY 12545, USA

Stephen R. Carpenter, Center for Limnology, University of Wisconsin, Madison, WI 53706, USA

Virginia H. Dale, Environmental Sciences Division, Oak Ridge National Laboratory, Oak Ridge, TN 37831-6083, USA

Meghan Q. Fellows, Pacific Estuarine Research Laboratory, San Diego State University, San Diego, CA 92182-4625, USA

Carl Folke, Natural Resources Management, Department of Systems Ecology, Stockholm University, S-106 91 Stockholm, and Beijer International Institute of Ecological Economics, The Royal Swedish Academy of Sciences, 05 Stockholm, Sweden

James H. Fownes, Department of Agronomy and Soils, University of Hawaii–Manoa, Honolulu, HI 96822, USA

Peter M. Groffman, Institute of Ecosystem Studies, Millbrook, NY 12545, USA

Colleen A. Hatfield, Benedict Estuarine Research Center, Academy of Natural Sciences, St. Leonard, MD 20685, USA

Lars O. Hedin, Section of Ecology and Systematics, Cornell University, Ithaca, NY 14853, USA

Robert W. Howarth, Cornell University, Ithaca, NY 14853, USA

Clive G. Jones, Institute of Ecosystem Studies, Millbrook, NY 12545, USA

W. Michael Kemp, Center for Environmental and Estuarine Studies, Horn Point Environmental Laboratory, University of Maryland, Cambridge, MD 21613, USA

Ann P. Kinzig, Department of Ecology, Princeton University, Princeton, NJ 08544-1003, USA

William K. Lauenroth, Department of Rangeland Ecosystem Science, Colorado State University, Fort Collins, CO 80523, USA

Gene E. Likens, Institute of Ecosystem Studies, Millbrook, NY 12545 USA

Gary M. Lovett, Institute of Ecosystem Studies, Millbrook, NY 12545, USA

Richard Lowrance, USDA Agricultural Research Service, Southeast Watershed Research Laboratory, Tifton, GA 31793, USA

James A. MacMahon, Department of Biology and Ecology Center, 5305 University Way, Utah State University, Logan, UT 84322-4400, USA

Pamela A. Matson, Department of Environmental Science, Policy, and Management, University of California, Berkeley, CA 94720-3110, USA

Jason Neff, Department of Biological Sciences, Stanford University, Stanford, CA 94305, USA

Michael L. Pace, Institute of Ecosystem Studies, Millbrook, NY 12545, USA

Eldor A. Paul, Department of Crop and Soil Sciences, Michigan State University, East Lansing, MI 48824, USA

Karen A. Poiani, The Nature Conservancy, Minneapolis, MN 55414, USA

Kurt S. Pregitzer, School of Forestry and Wood Products, Michigan Technological University, Houghton, MI 49931, USA

G. Philip Robertson, Department of Crop and Soil Sciences, W.K. Kellogg Biological Station, Michigan State University, Hickory Corners, MI 49060, USA

Steven W. Running, School of Forestry, University of Montana, Missoula, MT 59812, USA

Val H. Smith, Department of Systematics and Ecology, Environmental Studies Program, University of Kansas, Lawrence, KS 66045, USA

David Tilman, Department of Ecology, Evolution, and Behavior, University of Minnesota, St. Paul, MN 55108-6097, USA

Sally Trnka, San Diego State University, Pacific Estuarine Research Laboratory, San Diego, CA 92182-4625, USA

Peter M. Vitousek, Department of Biological Sciences, Stanford University, Stanford CA 94305, USA

Carl J. Walters, Fisheries Center, University of British Columbia, Vancouver, BC V6T IZ4, Canada

Kathleen C. Weathers, Institute of Ecosystem Studies, Millbrook, NY 12545, USA

Carol A. Wessman, Cooperative Institute for Research in Environmental Sciences and Environmental, Population, and Organismic Biology, University of Colorado, Boulder, CO 80309-0216, USA

Cathleen Wigand, Institute of Ecosystem Studies, Millbrook, NY 12545, USA

Donald R. Zak, School of Natural Resources and the Environment, University of Michigan, Ann Arbor, MI 48109-1115, USA

Joy B. Zedler, Department of Botany, University of Wisconsin, Madison, WI, 53706-1381, USA

Participants

Edward A. Ames, Mary Flagler Cary Charitable Trust, New York, NY 10168, USA

Juan J. Armesto, Universidad de Chile, Facultad de Ciencias, Departamento de Biologia, Laboratorio de Ecologia Forestal, Santiago, Chile, and Institute of Ecosystem Studies, Millbrook, NY 12545, USA

Mary C. Barber, Sustainable Biosphere Initiative, Ecological Society of America, Washington, DC 20036, USA

Barbara L. Bedford, Department of Natural Resources, Cornell University, Ithaca, NY 14853, USA

Janne Bengtsson, Department of Ecology and Environmental Research, Swedish University of Agricultural Sciences, Uppsala, Sweden

Alan R. Berkowitz, Institute of Ecosystem Studies, Millbrook, NY 12545, USA

Patrick J. Bohlen, Institute of Ecosystem Studies, Millbrook, NY 12545, USA

James T. Callahan, National Science Foundation, Arlington, VA 22230, USA

Nina F. Caraco, Institute of Ecosystem Studies, Millbrook, NY 12545, USA

Jonathan J. Cole, Institute of Ecosystem Studies, Millbrook, NY 12545, USA

Paul A. del Giorgio, Université du Québec à Montréal, Départment des Sciences Biologiques, Montréal, Québec H3C 3P8, Canada

Laurie E. Drinkwater, Rodale Institute, Kutztown, PA 19530, USA

Joan G. Ehrenfeld, Department of Ecology, Evolution, and Natural Resources, Cook College, Rutgers University, New Brunswick, NJ 08903, USA

Jerry W. Elwood, Office of Energy Research, US Department of Energy, Germantown, MD 20874, USA

Stuart E.G. Findlay, Institute of Ecosystem Studies, Millbrook, NY 12545, USA

Stuart G. Fisher, Department of Zoology, Arizona State University, Tempe, AZ 85287-1501, USA

Jerry F. Franklin, College of Forest Resources, University of Washington, Seattle, WA 98195, USA

Gretchen Long Glickman, Institute of Ecosystem Studies, Millbrook, NY 12545, USA

Arthur J. Gold, Natural Resources Science, University of Rhode Island, Kingston, RI 02881-0804, USA

Frank B. Golley, Institute of Ecology, University of Georgia, Athens, GA 30602, USA

James R. Gosz, Department of Biology, University of New Mexico, Albuquerque, NM 87131, USA

Nancy B. Grimm, Department of Zoology, Arizona State University, Tempe, AZ 85287-1501, USA

Robert O. Hall, Jr., Institute of Ecosystem Studies, Millbrook, NY 12545, USA

Graham P. Harris, CSIRO Land and Water, Canberra ACT 2601, Australia

Paul F. Hendrix, Institute of Ecology, University of Georgia, Athens, GA 30602-2602, USA

John E. Hobbie, The Marine Biological Laboratory, Ecosystems Center, Woods Hole, MA 02543, USA

Sarah E. Hobbie, Department of Biological Sciences, Stanford University, Stanford, CA 94305, USA

Vera Huszar, Institute of Ecosystem Studies, Millbrook, NY 12545, USA, and Universidade Federal de Rio de Janeiro, Brazil

Carol A. Johnston, Natural Resources Research Institute, University of Minnesota, Duluth, MN 55803, USA

Caroline A. Karp, Brown University, Center for Environmental Studies, Providence, RI 02912, USA

James R. Karr, Institute for Environmental Studies, University of Washington, Seattle, WA 98195-2200, USA

Elaine Matthews, Columbia University/NASA Goddard Institute for Space Studies, New York, NY 10025, USA

Judy L. Meyer, Institute of Ecology, University of Georgia, Athens, GA 30602-2602, USA

Richard S. Ostfeld, Institute of Ecosystem Studies, Millbrook, NY 12545, USA

Steward T.A. Pickett, Institute of Ecosystem Studies, Millbrook, NY 12545, USA

Isabel Reche, Institute of Ecosystem Studies, Millbrook, NY 12545, USA

O.J. Reichman, National Center for Ecological Analysis and Synthesis, Santa Barbara, CA 93101-3351, USA

William Robertson IV, The Andrew W. Mellon Foundation, New York, NY 10021, USA

Fabio Roland, Institute of Ecosystem Studies, Millbrook, NY 12545, USA and Universidade Federal de Juiz de Fora, Brazil

Osvaldo E. Sala, Departamento de Ecologia, Facultad de Agronomia, Universidade Buenos Aires, Buenos Aires 1417, Argentina

Moshe Snachak, The Jacob Blaustein Institute for Desert Research, Ben-Gurion University of the Negev, Israel 84990 and Institute of Ecosystem Studies, Millbrook, NY 12545, USA

David L. Strayer, Institute of Ecosystem Studies, Millbrook, NY 12545, USA

Sandy Tartowski, Institute of Ecosystem Studies, Millbrook, NY 12545, USA

Sara F. Tjossem, Department of Ecology, Evolution and Behavior, University of Minnesota, St. Paul, MN 55108, USA

Nico van Breemen, Department of Soil Science and Geology, Wageningen Agricultural University, 6700 AA Wageningen, The Netherlands

Diana Wall, Natural Resources Ecology Laboratory, Colorado State University, Fort Collins, CO 80523-1499, USA

Joseph S. Warner, Institute of Ecosystem Studies, Millbrook, NY 12545, USA

1
Needs and Concerns in Ecosystem Science

MICHAEL L. PACE AND PETER M. GROFFMAN

This book is about the status of ecosystem science. The papers within are organized into sections that explore successes, limitations, and frontiers in ecosystem research and are drawn from the seventh Cary Conference held in Millbrook, New York in May of 1997.

One purpose in focusing on ecosystem science is to foster among practitioners recognition of the common problems and current state of the discipline. We also hope to speak to a broader audience by describing advances and frontiers of the science. This book, therefore, is meant to be of interest to all ecologists and students as well as other environmental scientists interested in learning about ecosystem research.

We had several reasons for organizing a Cary Conference and subsequently a book around the theme of ecosystem research; these reasons are elaborated more thoroughly in the next section. Our rationale, however, can be summarized succinctly. First, all scientific endeavors need to take stock periodically. We ask here: what are some of the successes of the science?; what are some of the current limitations to advancement?; and, what don't we know and where do we need to go in the future? Second, we were interested in the motivations that underlie the development and exploration of ecosystem problems where both scientific and environmental concerns drive research. We wondered how this linkage between environmental problems and research affected the state and progress of ecosystem science. A third reason was our perception of a poor understanding by some ecologists of interests, questions, and concerns of work at the ecosystem scale. This view was clearly evident in discussions conducted at the fifth Cary Conference on linking species and ecosystems (Jones and Lawton 1995), reflecting the perceived split between ecosystem and population/community practitioners. Although there is no hope of dialogue with the willfully ignorant, the discipline needs to progress toward an integrated perspective. We hope this book will be a source for understanding that section of ecology concerned with ecosystems.

Needs and Concerns for Ecosystem Research

Ecosystem research is a core discipline in environmental science. Questions about widespread human-driven alteration of biogeochemical cycles, ecological complexity and biodiversity, and ecological response to climate change frequently focus on ecosystems. As a central and integrating environmental science, ecosystem studies have been highly successful. In the past thirty or so years, impressive discoveries have evolved from ecosystem research (Pomeroy and Alberts 1988) and some of these are described in the chapters of this volume. Given this recent history and the vitality of ongoing research, the discipline seems well poised to make important contributions in the future. Nevertheless, we believe ecosystem research faces important internal and external challenges. Within the discipline there is a stronger need for recognition and focus on core issues and problems motivated by theory, in addition to problems driven by environmental concerns. This need implies that a clear articulation of theory is required. Outside the discipline there is a clear need to better justify ecosystem science to other ecologists and environmental scientists as well as to the public.

As a "young," inherently multidisciplinary science, ecosystem ecology has had a tortuous, albeit dynamic, development. The strong links between environmental problems and ecosystem studies have attracted both positive attention and intense scrutiny to the discipline (McIntosh 1985; Hagen 1992). Arguably, environmental problems have driven much of the development in ecosystem studies, as for example, the problems of biomagnification of pesticides (Woodwell et al. 1967), eutrophication of aquatic systems (Schindler et al. 1972; Smith chapter 2), and acid deposition (Likens et al. 1972; Weathers and Lovett chapter 8). In these cases and others, ecosystem ecologists were among those initially detecting and diagnosing the causes, consequences, and cures for these problems. There has always been a strong interaction between ecosystem science and external pressures arising from human-driven changes in the environment. This situation is apt to continue.

Ecosystem ecology has also been strongly influenced by allied disciplines, by widening collaborations, and by the "style" of the science. Since the inception of ecosystem research, there have been conflicts as alluded to above between ecosystem and population ecologists over the theoretical underpinnings and practical value of ecosystem analysis (McIntosh 1985; O'Neill et al. 1986; Hagen 1992). The status of ecosystem ecology seems peripheral when viewed in the context of some popular textbooks (e.g. Ricklefs 1990; Krebs 1994), and one important text even denies the existence of ecosystem ecology (Begon et al. 1996). Nevertheless, there has been some softening of the divisions including recent attempts to integrate population and community studies with the variability and patterns of ecosystem processes (see examples in Jones and Lawton 1995). Ecosystem scientists have also built productive relationships both within the wider discipline of ecology and with geochemical and geophysical colleagues. The colloborations have evolved from the need to address questions related to the physical, chemical, and biological interactions in ecosystems. Collaborations have also extended to social scientists. The human enterprise in both its economic and social context are being integrated into ecosystem research. Ecosystem ecology is increasingly "big science" with large multi-investigator grants and multisite networks. This style of research is expensive. It also draws attention and occasional criticism from regulatory agencies and other scientists. In aggregate, these external pressures have shaped ecosystem science, and they compel ecosystem researchers to document their accomplishments and to articulate their vision of the science.

One barrier, however, to articulating a vision for ecosystem research is a lack of cohesion within the science. This is partly related to the multidisciplinary nature of ecosystem research. Ecosystem scientists are also microbiologists, hydrologists, biogeochemists, marine scientists, foresters, limnologists, agronomists, and a host of other disciplinary types. These additional specializations tend to divert or diffuse attention away from the common, core problems in ecosystem research. For example, presently there is no central journal that reports the major new results deriving from ecosystem research (although a new journal *Ecosystems* will begin publication in 1998). There is not even a small set of journals that could be cited. There is probably not agreement (or better, common awareness) among ecosystem scientists about central questions, concepts, and new developments in the discipline. There is also little evidence that the discipline is cohesive as a unified body of knowledge. What is the key textbook summarizing ecosystem research? There is not one but many, and these texts are almost always about specific types of ecosystems (e.g. Wetzel 1983; Day et al. 1988; Aber and Mellilo 1991; Mann and Lazier 1991; Mitsch and Gosselink 1993; Perry 1994; Valiela 1995). Nevertheless, a casual survey of these texts reveals discussion of many common topics that constitute the heart of ecosystem science—productivity, nutrient cycling, food-web interactions, fluxes within and across ecosystem boundaries, and the disturbance and recovery of systems.

There is also not wide agreement on theoretical issues in ecosystem research. Such concepts as disturbance, nutrient cycling, and mass-balance are widely used, however theoretical treatment of these subjects with explicit connection to field and experimental investigations remains limited. The development and application of general systems theory to ecosystem research was useful in providing a foundation for modeling studies (e.g. Odum 1983) but this body of theory has not coalesced as focal for the discipline. In addition, theoretical analyses of flow networks have advanced in recent years (Ulanowicz 1986). Although there have also been new analyses of the roles of material flow, cycling, and food-web structure in the broader context of stability theory (DeAngelis 1992), there is, still a disconnection between much of this theory and practice in ecosystem research. Part of the reason may be that the nature of theory (sensu Pickett et al. 1994) has not been widely discussed for ecosystems. A strong theoretical core, as for example, suggested by Reiners (1986), might allow more rapid progress and better sharing of questions and concepts across research sites and specialties.

Some of the external and internal factors that have influenced ecosystem ecology can be reconciled by focusing on the relationships between ecology and environmental problems. Many major advances in ecology have been driven by the emergence of new environmental problems that could not be solved with existing approaches and techniques. For example, ecosystem ecology emerged as a distinct discipline within ecology in the 1960s partly in response to biomagnification of pesticides and to eutrophication of aquatic ecosystems (Woodwell et al. 1967; Likens 1972). Landscape ecology emerged in the 1970s in response to the realization that many human activities have environmental effects over areas larger than individual communities or ecosystems, and the understanding that what we really manage in many cases is the variation and interactions of different ecological units within the landscape (Forman and Godron 1986). During the 1980s, regional- and global-scale concerns lead to the development of "earth system science," where ecology is integrated with such physical sciences as climatology and hydrology (Ojima et al. 1991). In the 1990s, recognition of the fundamental role that humans play in all ecological problems has led to the emergence of "integrated assessment," in which biological, physical, and social sciences come together (McDonnell and Pickett 1993; Groffman and Likens 1994). These changes in perspective and approach have influenced research directions. The relationships between environmental problems and progress in ecosystem ecology are highlighted in many of the discussions of successes in this book.

Successes, Limitations, and Frontiers

The motif for our assessment of ecosystem science is the analysis of successes, limitations, and frontiers. The book is organized accordingly by beginning with some notable successes in research (chapters 2 to 9). These chapters present a description of the successes, analyze conceptual and practical advances leading to and resulting from the work, and evaluate outcomes in the application of new knowledge to environmental problems. This latter criterion is an unusual one in an evaluation of science but is based on the connection between ecosystem research and environmental issues discussed above. The success stories are not uniform in the relative weight they place on description, analysis of scientific advances, and outcomes in application. Instead, chapters take different tacks. Some focus more on scientific aspects (Lowrance chapter 5; Robertson chapter 6) while some take the scientific accomplishments as a given and consider how and whether science impacted policy (Zedler et al. chapter 4). The role of human values and perceptions is considered central in chapter 3 by Dale. Interactions of science and environmental problems are clearly drawn in the chapters on eutrophication (Smith chapter 2) and acid deposition (Weathers and Lovett chapter 8). The conceptual underpinnings of ecosystem science are considered in relation to successes in restoration (McMahon chapter 9) and biogeochemical cycles (Burke et al. chapter 7). These strategies reflect the authors' attempts to tackle some very large topics. We hope the benefit to the reader is to encourage a consideration of the diversity of ways in which the science has advanced and influenced.

Not all has been success in ecosystem research. Many of the false starts and bad turns along the way are well described in historical analyses (McIntosh 1985; Hagen 1992; Golley 1993). We chose not to reiterate consideration of past failures but, instead, to focus on current limitations to the discipline. These are viewed in two chapters, one of which discusses intellectual limitations (Likens chapter 10), while the second considers problems in the relationship between scientists and managers (Walters chapter 11). The aggregate concern of these chapters is on some of the current practices and behaviors that haunt our efforts and limit advancement. These critical views provide both prognoses of our ills as well as visible pathways for improvement.

The final section of the book considers frontiers in research (chapters 13 to 19) and closes with our synthesis (chapter 20). The frontier chapters represent some but certainly not all of the critical areas; the focus is on what we don't know and what approaches and needs exist for advancement. These chapters offer a variety of issues requiring exploration. Frontier assessments are based on exploratory modeling (Vitousek et al. chapter 18), methodology and approaches (Carpenter chapter 12; Wessman and Asner chapter 14; Laurenroth et al.; chapter 16), and analyses of topical areas (Zak and Pregitzer chapter 15; Breitburg et al. chapter 17; Tilman chapter 19). Collectively, these chapters identify important questions and problems and indicate the excitement and challenge of the work ahead.

References

Aber, J.D., and J.M. Melillo. 1991. *Terrestrial ecosystems*. Saunders College Publications, Philadelphia, PA.

Begon, M., J.L. Harper, and C.R. Townsend. 1996. *Ecology: individuals, populations, and communities, 3rd edition*. Blackwell Science, Boston, MA.

Day, J.W. Jr., C.A.S. Hall, W.M. Kemp, and A. Yanez-Arancibia. 1988. *Estuarine ecology*. John Wiley & Sons, New York.

DeAngelis, D.L. 1992. *Dynamics of nutrient cycling and food webs*. Chapman & Hall, New York.

Forman, R.T.T., and M. Godron. 1986. *Landscape ecology*. John Wiley & Sons, New York.

Golley, F.B. 1993. *A history of the ecosystem concept in ecology: more than the sum of the parts.* Yale University Press, New Haven, CT.

Groffman, P.M., and G.E. Likens, eds. 1994. *Integrated regional models: interactions between humans and their environment.* Chapman & Hall, New York.

Hagen, J.B. 1992. *An entangled bank: The origins of ecosystem ecology.* Rutgers University Press, New Brunswick, NJ.

Jones, C.G., and J.H. Lawton, eds. 1995. *Linking species and ecosystems.* Chapman & Hall, New York.

Krebs, C.J. 1994. *Ecology: the experimental analysis of distribution and abundance.* Harper Collins College Publishers, New York.

Likens, G.E., ed. 1972. *Nutrients and eutrophication.* American Society of Limnology and Oceanography Special Symposium Volume 1. Allen Press, Lawrence, KS.

Likens, G.E., F.H. Bormann, and N.M. Johnson. 1972. Acid rain. *Environment* 14:33–40.

Mann, K.H., and J.R.N. Lazier. 1991. *Dynamics of marine ecosystems: biological-physical interactions in the oceans.* Blackwell Scientific Publications, Boston, MA.

McDonnell, M.J., and S.T.A. Pickett, eds. 1993. *Humans as components of ecosystems.* Springer-Verlag, New York.

McIntosh, R.P. 1985. *The background of ecology.* Cambridge University Press, New York.

Mitsch, W.J., and J.G. Gosselink. 1993. *Wetlands, 2nd edition.* Van Nostrand Reinhold, New York.

Odum, H.T. 1983. *Systems ecology: an introduction.* John Wiley & Sons, New York.

Ojima, D.S., T.G.F. Kittel, R. Rosswall, and B.H. Walker. 1991. Critical issues for understanding global change effects on terrestrial ecosystems. *Ecological Applications* 1:316–325.

O'Neill, R.V., D.L. DeAngelis, J.B. Waide, and T.F.H. Allen. 1986. *A hierarchical concept of ecosystems.* Princeton University Press, Princeton, NJ.

Perry, D.A. 1994. *Forest ecosystems.* Johns Hopkins University Press, Baltimore, MD.

Pickett, S.T.A., J. Kolasa, and C.G. Jones. 1994. *Ecological understanding.* Academic Press, San Diego, CA.

Pomeroy, L.R., and J.J. Alberts. 1988. *Concepts of ecosystem ecology.* Springer-Verlag, New York.

Reiners, W.A. 1986. Complementary models for ecosystems. *American Naturalist* 127:59–73.

Ricklefs, R.E. 1990. *Ecology.* W.H. Freeman, New York.

Schindler, D.W., G.J. Brunskill, S. Emerson, W.D. Broccker, and T.-H. Peng. 1972. Atmospheric carbon dioxide: its role in maintaining phytoplankton standing crops. *Science* 177:1192–1194.

Ulanowicz, R.E. 1986. *Growth and development: Ecosystems phenomenology.* Springer-Verlag, New York.

Valiela, I. 1995. *Marine ecological processes, 2nd edition.* Springer-Verlag, New York.

Wetzel, R.A. 1983. *Limnology, 2nd edition.* W.B. Saunders, Philadelphia, PA.

Woodwell, G.M., C.F. Wurster, and P.A. Isaacson. 1967. DDT residues in an east coast estuary: a case of biological concentration of a persistent insecticide. *Science* 156:821–824.

2
Cultural Eutrophication of Inland, Estuarine, and Coastal Waters

Val H. Smith

Summary

In the mid-1800s, when Justus von Liebig published his highly influential work relating soil fertility to the yields of agricultural crops, surface waters worldwide frequently were oligotrophic or poorly nourished. However, anthropogenic inputs of nutrients to surface waters have increased greatly during the past 150 years, and *eutrophication* is the process by which water bodies are made more well-nourished, or eutrophic, by an increase in their nutrient supply. We have learned as a consequence of cultural eutrophication that fresh and marine waters respond strongly to nitrogen and phosphorus inputs, and that this nutrient enrichment in turn can lead to highly undesirable changes in surface water quality. In this chapter, I review the process, the impacts, and the management of cultural eutrophication in inland, estuarine, and coastal marine waters. I also present two case studies (Lake Washington in Seattle, Washington, and Kaneohe Bay in Oahu, Hawaii) that highlight the study and subsequent control of eutrophication as a success in ecosystem science.

Introduction

Between 1840 and 1862, the German agricultural chemist Justus von Liebig published a series of books on the relationship between nutrients and plant production that has had a profound impact on the practice of agriculture. In these books, Leibig proposed that the yield of agricultural crops was proportional to soil fertility, and in one of his statements he concluded that "When a piece of land contains a certain amount of all the mineral constituents in equal quantity and in an available form, it becomes barren for any kind of plant when, by a series of crops, one of these constituents— as for example soluble silica—has been so far removed that the remaining quantity is no longer sufficient for a crop" (von Liebig 1855). In essence, Liebig proposed that the yield of a given species should be limited by the nutrient that was present in the least quantity in the environment relative to its demands for growth, and when phrased in this manner this statement has come to be known as Leibig's Law of the Minimum.

Liebig's law clearly has had a major impact on the practice of agronomy, but it also has provided a parallel conceptual foundation for the fields of terrestrial and aquatic ecology. As noted by Nixon et al. (1986), one of the oldest and most enduring metaphors in marine ecology is that of the sea as a farm. This analogy dates from the perceptive work of Brandt, who found that the abundance of plankton in freshwater lakes in the Holstein region of Germany was correlated with the amounts of nitrogen in the lake waters, and who extended Leibig's law to the oceans almost a century ago (Brandt 1899, 1901).

Attempts to categorize bodies of water by their relative productivity originated primarily in Europe, largely through the efforts of Einar Naumann in Sweden and August Theinemann in Germany (Hutchinson 1969, 1973; Carlson and Simpson 1995). Naumann (1919, 1929) first developed the framework that we now know as the *trophic state concept*. Through careful comparative regional studies, he concluded that phytoplankton production was determined primarily by the concentrations of nitrogen and phosphorus; that there were significant regional variations in productivity among lakes that correlated well with the geological characteristics of the lake's watershed; and that the algal productivity of the lake influenced the biology of the lake as a whole. As stressed in the historical review of Carlson and Simpson (1995), these conclusions are neither fuzzy speculation nor rigid dogmatism about how lakes should be classified. Rather, they are specific, testable statements about the connections that exist between the characteristics of a lake's watershed and its biology. As will be clear from the review below, Naumann's statements represent the genesis of our current ideas concerning nutrient loading and phosphorus-algal biomass relationships (Carlson and Simpson 1995).

Drawing from terminology developed by Weber (1907) for the classification of bogs, Naumann also developed the terms that we now use to categorize lakes by their nutrient supply (Hutchinson 1969, 1973). Lakes with relatively large supplies of nutrients are termed *eutrophic* (well nourished), and those having poor nutrient supplies are termed *oligotrophic* (poorly nourished). Lakes having intermediate nutrient supplies are termed *mesotrophic*. Eutrophication is the process by which bodies of water are made more eutrophic through an increase in their nutrient supply, and although these terms have most frequently been used to describe freshwater lakes and reservoirs, they can also be applied to flowing waters, estuaries, bays, sounds, and other partly enclosed marine waters (Edmondson 1995).

Just as nutrient enrichment has many profound effects on terrestrial ecosystems (Tamm 1991; Wedin and Tilman 1996), the eutrophication of aquatic ecosystems leads to numerous changes in ecosystem structure and function. Many of these changes can be highly undesirable from

TABLE 2.1. Major changes in ecosystem structure and function frequently associated with the eutrophication of aquatic ecosystems.

- Increased biomass of phytoplankton and suspended algae (Dillon and Rigler 1974; Jones and Bachmann 1976; Van Nieuwenhuyse and Jones 1996).
- Shifts in phytoplankton composition to bloom-forming species, many of which may be toxic or inedible (Paerl 1988; Cosper et al. 1989; Smith 1990a; Viviani 1992).
- Increased biomass of periphyton (Cattaneo 1987; Dodds et al. 1997).
- Increased biomass of attached marine macroalgae (Littler and Murray 1975; Baden et al. 1990; Rosenberg et al. 1990).
- Changes in vascular plant production, biomass, and species composition (Kemp et al. 1983; Stevenson 1988).
- Decreases in water column transparency (Dillon and Rigler 1975; Vollenweider 1992).
- Decreases in perceived aesthetic value of the body of water (Reckhow and Chapra 1983).
- Taste, odor, and water supply filtration problems (Welch and Lindell 1992).
- Depletion of deepwater oxygen concentrations (Reckhow and Chapra 1983; Håkanson and Wallin 1991; Vollenweider et al. 1992b).
- Increased fish production and harvest (Jones and Hoyer 1982; Nixon et al. 1986; Bachmann et al. 1996).
- Changes in consumer composition towards less desirable species (Kerr and Ryder 1992; Lenzi 1992).

the point of view of public and commercial use of the body of water. (Table 2.1).

As a result of the diverse impacts of eutrophication, a need to provide a practical solution to the problem began to arise in the late 1940s, and a successful ecosystem-based framework has subsequently been developed during the past five decades for the prevention and the management of cultural eutrophication. As will be seen below, our knowledge of eutrophication and its control has been developed through the complementary use of rigorous conceptual and theoretical foundations; careful comparative analyses of different bodies of water over both space and time; and controlled experimentation at three different levels of scale: (1) small-scale flask experiments such as bioassays; (2) medium-scale experiments such as mesocosms; and (3) whole-system manipulations. A rough time line for the development of our knowledge of eutrophication is presented in Table 2.2, which highlights some of the most pivotal events and developments in eutrophication research to approximately 1980. After that point, an explosion in the eutrophication literature occurred, which is summarized in more comprehensive overviews such Reckhow and Chapra (1983), Nixon et al. (1986), Ryding and Rast (1989), Harper (1992), Sutcliffe and Jones (1992), and Welch and Lindell (1992).

The Role of Nitrogen and Phosphorus as Limiting Nutrients

The concept of nutrient limitation of algal growth is the keystone of eutrophication research because it implies that (1) a single nutrient should be the primary limiting factor for algal growth in a given body of water; (2)

TABLE 2.2. Some pivotal events and developments in eutrophication.

- Development of the flush toilet.
- Principle of limiting nutrients (Leibig 1855).
- Links between nutrients and aquatic production (Brandt 1899, 1901; Johnstone 1908).
- First mass-balance for nitrogen in a marine system (Johnstone 1908).
- Trophic state classification system (Weber 1907; Theinemann 1921; Naumann 1929).
- Establishment of critical N and P levels for algal blooms in lakes (Sawyer 1947).
- Detection of *Oscillatoria rubescens* in Lake Washington (Edmondson et al. 1956).
- Redfield's (1958) study of elemental stoichiometry in natural waters.
- Mathematical models of nutrient conditions in coastal waters (Riley 1967).
- First OECD report on eutrophication (Vollenweider 1968).
- U.S. National Academy of Sciences report on eutrophication (National Academy of Sciences 1969).
- International symposium on nutrients and eutrophication (Likens 1972).
- Confirmation of the role of phosphorus in eutrophication (Schindler et al. 1972).
- Refinements of eutrophication models (Dillon and Rigler 1974, 1975; Jones and Bachmann 1976; Vollenweider 1976) and trophic state indices (Carlson 1977).
- Initial evidence that consumers modify the responses of aquatic ecosystem to eutrophication (Hrbácek et al. 1961; Shapiro 1979).
- Second OECD symposium on eutrophication (U.S. EPA 1981).

the observed algal growth in a given body of water should be proportional to the supply of this nutrient; and (3) the practical control of algal growth and of eutrophication in the body of water should involve restricting the inputs of this key nutrient to the system.

Despite a brief but intense controversy concerning the relative roles of carbon (C), nitrogen (N), and phosphorus (P) in eutrophication (cf. Likens 1972; Schindler 1985a, 1990), N and P are generally considered to be the two primary limiting nutrients for algae and vascular plants in aquatic ecosystems because both are frequently in short supply relative to their cellular demands for growth (Hutchinson 1973). As is true for terrestrial plants, this conclusion follows directly from Liebig's Law of the Minimum, which has in fact had a long history in plankton ecology (De Baar 1994). Because N and P are the two primary growth-limiting nutrients in aquatic ecosystems, and because the biogeochemical cycles of these two important elements are closely interrelated (Schindler 1985b; Smith 1993), this review will consider the roles of both nutrients in the process, impacts, and management of eutrophication.

Because the inputs of N and P to natural waters can vary dramatically, the question of the relative degrees of N- versus P-limitation of algal growth has been discussed extensively since the early studies of Brandt (1899, 1901). Although evidence for nutrient limitation can be derived from four different levels of inquiry (Hecky and Kilham 1988), most of our knowledge of the relative importance of N versus P as growth-limiting nutrients is based primarily on direct evidence from bioassays (in which the response of algal growth to N or P is evaluated by making additions of one or both nutrients in small volume containers), or from inferred evidence based on elemental ratios. The extensive algal bioassay programs for the assessment of nutrient limitation undertaken in the 1970s (e.g. Chiaudani and Vighi 1974; Miller et al. 1974; Forsberg 1981) led to the development of objective guidelines for the assessment of N- versus P-limitation. Physiological indicators that indicate nutrient limitation and can be rapidly measured on samples of phytoplankton have also proven to be extremely valuable as well. St. Amand et al. (1989) concluded that physiological indicators are rapid, can be run more frequently than bioassays, and lead to the same conclusions as nutrient-enrichment bioassays.

In the absence of direct bioassays or physiological measurements, however, elemental ratios frequently are used to infer nutrient limitation. For example, Pearsall (1932) used inorganic N/P ratios as an indicator of N- versus P-limitation of algal growth in lakes of the English Lake District, and the value of using elemental ratios as an objective indicator of nutrient limitation was emphasized by Redfield (1958). In this provocative synthesis, Redfield demonstrated strong effects of the biota on the chemistry of natural marine waters, and proposed that the nutrient content of living algal cells could be characterized on average by a molar ratio of $106C:16N:1P$ ($40C:7N:1P$ by weight, the Redfield ratio). By extension from Liebig's Law of the Minimum, algae that experience an N/P-supply ratio less than the Redfield ratio should be limited by N. Conversely, algae experiencing an N/P supply ratio greater than the Redfield ratio should be limited by P, and the Redfield ratio has since been used extensively in the empirical assessment of nutrient-limitation and dynamics in marine waters (Howarth et al. 1996; Downing 1997; Tyrrell and Law 1997).

However, the phytoplankton is composed of many species, each of which may have their own optimum N/P ratio for growth (for example, see lists in Smith 1982 and in Hecky and Kilham 1988). A single N/P ratio (such as the Redfield ratio) for nutrient-limitation assessment thus may not be entirely appropriate, and the empirical use of variable elemental ratios in the nutrient-limitation assessment of freshwaters is illustrated well by the study of Sakamoto (1966). Sakamoto concluded from careful comparative studies of Japanese lakes that N-limitation of summer algal biomass was most likely when the epilimnetic total N/total P (TN/TP) ratio was less than 10:1 by weight; that summer algal biomass was limited by P when the epilimnetic TN/TP ratio was greater than 17; and that either N or P may be limiting at intermediate TN/TP ratios. These criteria were confirmed independently by Forsberg (1981).

Nutrient-Loading Models and Their Application

The inputs of biologically available N and P to bodies of water are primarily derived from direct atmospheric deposition, fluvial inputs, and groundwater flows. The sum of these three external inputs is termed the *external load*.

Both the absolute and the relative external loads of N and P differ significantly among different bodies of water, however, and the loading of N and P to single bodies of water also may change over time. To predict the effects of these two nutrients on receiving waters, it is necessary to be able to predict how water column nutrients will vary as the external N and P inputs are changed.

The second vital development in eutrophication research was the development of quantitative relationships between external nutrient inputs and the resulting water column concentrations of nutrients in the body of water itself. This major advance came about as a result of the successful application of the mass-balance concept to aquatic ecosystems. Surprisingly, the first annual mass-balance for an aquatic ecosystem may have been made by Johnstone (1908), who developed a mass-balance for N in the North Sea. It was almost six decades later before the mass-balance method was applied to chemical substances in lakes (see Edmondson 1961; Reckhow and Chapra 1983), and this approach apparently was not quantitatively applied to eutrophication modelling on a broad basis until the Organization for Economic Cooperation and Development (OECD) report of Vollenweider (1968). In the thirty years since the publication of Vollenweider (1968), a rigorous quantitative framework has been developed to predict the responses of freshwater lakes and reservoirs to eutrophication. Central to this framework is the calculation of a mass-budget for both N and P, and the calculation of a similar budget for water. It is interesting to note, however, that the utility of the mass-balance approach was not immediately appreciated even by Thienemann, who attended an early presentation by E.A. Thomas of P budgets for Swiss lakes, and who remarked to Vollenweider that this "had nothing to do with limnology" (Vollenweider, personal communication).

P-Loading Models for Lakes

Because of the prevalence of P-limitation in lakes, most current lake management frameworks focus primarily on the control of P-loading (e.g. OECD 1982; Reckhow and Chapra 1983; Heyman and Lundgren 1988; Ryding and Rast 1989; Sas 1989; Foy et al. 1996). Numerous P-loading models of varying structure currently can be found in the literature (e.g. Dillon and Rigler 1975; Canfield and Bachmann 1981; OECD 1982; Salas and Martino 1991). Although a detailed overview of these models is beyond the scope of this chapter, their derivation and use is comprehensively summarized in Reckhow and Chapra (1983).

In lakes, nutrient losses in the outflowing water and nutrient retention caused by sedimention from the water column correlate strongly with the hydraulic characteristics of the body of water, and these terms are typically estimated empirically from the results of the lake's water budget (cf. Reckhow and Chapra 1983 for an excellent historical overview). In

Vollenweider's most recent P-loading models (OECD 1982), P retention and in-lake TP concentrations are estimated from an empirical expression involving the hydraulic residence time of water in the lake (t_w, yr):

$$TP = 1.55 \left[P_{in} / \left(1 + \sqrt{t_w} \right) \right]^{0.82} \tag{1}$$

in which TP ($mg\,m^{-3}$) is the mean annual concentration of total P in the water column of the lake, and P_{in} is the mean annual concentration of total phosphorus in the inflow. Equation 2.1 and many of its variants have been applied very successfully to many lakes worldwide, and other models are also now available for shallow river-run reservoirs that have a strong directional flow of water (Ernst et al. 1994; Kennedy 1995). However, these models do not necessarily work well for all systems (Nürnberg 1984; Sas 1989), and we require a more detailed knowledge of the factors that govern net P retention if we are to predict more precisely the changes in water column concentrations that will occur when the external loading is altered.

N-Loading Models for Lakes

Although N is also an important limiting nutrient for algal growth in freshwaters (see below), relatively few models for mean annual TN (TN, $mg\,m^{-3}$) currently exist. Notable exceptions, however, can be found in Kratzer (1979), Bachmann (1981, 1984), Canfield and Bachmann (1981), Baker et al. (1981, 1985), OECD (1982), and Reckhow (1988), all of which present input-output models analogous to equation 1 for TN in lakes. These models are very relevant to any assessment of the current or future N status of lakes.

As is true of P-loading models (equation 1), current N-loading models rely on estimates of hydraulic residence time or its inverse, hydraulic flushing rate. In addition, a model term is required either for apparent N attenuation (σ_N) or retention (R_N). The estimation of the latter two terms is somewhat problematical, however. Although N attenuation and retention are strongly influenced by hydraulic residence time in a manner that is generally analogous to that for P, there is a gaseous phase in the N cycle that is not present in the P cycle. Although N in the water column can be lost either to the sediments via sinking or through channelized outflow, N also may accumulate in the water column because of biological N fixation (Howarth et al. 1988a,b), or may be lost to the atmosphere via microbial denitrification (Seitzinger 1988). Some advances are being made, however. For example, Jensen et al. (1990) has used a direct statistical approach to predict TN concentrations in lakes from N-loading and hydrology, and developed regression models relating in-lake concentrations of total N to inflow TN in shallow Danish lakes. More recently, Hellström (1996) has made a detailed empirical analysis of N dynamics in lakes that accounts for direct external loading, N-fixation inputs, and both fluvial and denitrification losses. Howarth et al. (1996) also have explored trends in N retention over a wide range of hydraulic conditions that are representative of lakes, reservoirs, streams, and rivers in a typical North Atlantic basin.

Loading Models for Marine Systems

Although Schindler (1977) stressed the relevance of eutrophication studies in lakes to the estuarine environment, nutrient-loading models for estuarine and coastal marine are still uncommon. One reason for the current lack of such models may be in part that it is extremely difficult to obtain good nutrient budgets for such large systems (Vollenweider 1992); it is hard to find comparable coastal marine systems with well-defined boundaries (Wallin and Håkanson 1991); and there are numerous complex factors that influence the subsequent dispersal, sedimentation, and recirculation of nutrient inputs to coastal areas (Håkanson and Wallin 1991). Although empirical models relating nutrients to ecosystem responses for estuaries (O'Connor 1981), for fjords (Aure and Stigebrandt 1989), and for coastal areas (Håkanson 1994) can be found in the literature, apparently few authors have sought to develop or to test models to predict water column nutrient concentrations from external nutrient loading, hydrology, and the morphometric characteristics of the receiving marine body of water.

An exception is the analysis of Lee and Jones (1981), who examined the applicability of the OECD modelling approach to the Potomac estuary and upper Chesapeake Bay. These authors noted that two of the most significant questions about applying the OECD approach to estuaries were how determine an appropriate mean depth and an appropriate hydraulic residence time for the different regions of an estuary. However, Lee and Jones (1981) focussed more on the general consistency between OECD-predicted values for algal biomass and transparency than on the ability of the OECD model to predict water column TP concentrations in three regions of the Potomac estuary. It is noteworthy that in a recent workshop on nutrient assessment (U.S. EPA 1996), the authors of the state of the science section on models for estuaries and coastal waters mentioned only three large-scale simulation models (ie. WASP5; the Tidal Prism Model, and CE-QUAL-ICM), and they urged the development of simpler empirical models for marine waters that are patterned after the Vollenweider/OECD approach.

Effects of Nutrient Loading on Aquatic Ecosystems

Just as the effects of fertilizing a terrestrial field are most easily perceived as an increase in plant growth, the effects of increased N and P supplies are typically manifested as increases in the population densities of algae and plants. The third key to our understanding and to the practical control of eutrophication has been the development of models that link nutrient concentrations in bodies of water to critical aspects of water quality that are perceived by the general public as being important and worthy of preservation. These critical aspects, which have been termed "quality variables of concern" by Reckhow and Chapra (1983), may sometimes vary from one body of water to another. For example, the primary water quality aspects of value to someone living on or making recreational use of a lake shore may differ from someone using the lake either for recreational or commercial fishing.

Perhaps the most important and commonly cited quality variable of concern is the accumulation of algal biomass, which is readily perceived by the public. As will be seen below, there is a remarkable unity in the response of algal biomass to eutrophication in every aquatic ecosystem that has been studied, and a demonstrated success in reducing algal biomass to publicly acceptable levels has been a common feature of all successful eutrophication control efforts. An additional quality variable of concern that is commonly of value to the public is water transparency, and the differences in N, P, and these two quality variables of concern in bodies of water in four different trophic states are summarized in Table 2.3. It is of great interest to note that the values of these parameters are somewhat more conservative in marine systems than in lakes, and that this difference probably reflects a somewhat greater sensitivity by the public to the perceived degradation of marine waters. Because of space limitations for this

TABLE 2.3. Average characteristics of freshwater[1] and coastal marine waters[2] of four different trophic states.[3]

	Lakes			
Trophic state	TN, $mg\,m^{-3}$	TP, $mg\,m^{-3}$	chl a, $mg\,m^{-3}$	SD, m
Oligotrophic	<350	<10	<3.5	>4
Mesotrophic	350–650	10–30	3.5–9	2–4
Eutrophic	650–1,200	30–100	9–25	1–2
Hypertrophic	>1,200	>100	>25	<1

	Marine			
Trophic state	TN, $mg\,m^{-3}$	TIN, $mg\,m^{-3}$	chl a, $mg\,m^{-3}$	SD, m
Oligotrophic	<260	<10	<1	>6
Mesotrophic	260–350	10–30	1–3	3–6
Eutrophic	350–400	30–40	3–5	1.5–3
Hypertrophic	>400	>40	>5	<1.5

1. Nürnberg 1996.
2. Håkanson 1994.
3. The terms oligotrophic, mesotrophic, and eutrophic correspond to systems receiving low, intermediate, and high inputs of nutrients. Hypertrophic is the term used for systems receiving greatly excessive nutrient inputs. Note: TN = total nitrogen; TP = total phosphorus; chl a = cholorophyll a; and SD = Secchi disk transparency.

review, algal growth will be the primary focus in the discussion of responses to eutrophication below; the reader is referred to key references in Table 2.1 for other quality variables of concern in both freshwater and marine systems.

Effects of N and P on Phytoplankton Biomass in Lakes and Reservoirs

The preponderance of evidence suggests that the growth of phytoplankton in a majority of lakes and reservoirs worldwide is limited by P (Schindler 1977). However, nitrogen limitation of phytoplankton biomass has also been widely observed in lakes (Bachmann 1981; Walker 1984; Reckhow 1988; Smith 1990a), some of which are of very low fertility (Morris and Lewis 1988; Goldman and Jassby 1990; Downing and McCauley 1992). In particular, there is a strong trend towards increasing N-limitation with eutrophication (Miller et al. 1974; Forsberg 1977; Ahl 1975, 1979; OECD 1982; White 1983; Downing and McCauley 1992). An increase in the probability of N-limitation with increasing trophic state can even be seen in Antarctic lakes (Priddle et al. 1986).

Although a general relationship between nutrients and algal production was apparently first noted by Brandt almost a century ago (see Table 2.2), the development of the first quantitative criteria for eutrophication risk in lakes was made by Sawyer (1947). Based on his comparative studies of Wisconsin lakes, Sawyer concluded that if spring lakewater N and P concentrations exceeded critical levels ($>300 \, \text{mg m}^{-3}$ total inorganic N and $>10 \, \text{mg m}^{-3}$ phosphate P), nuisance algal blooms could be expected during the summer months. Sawyer's numerical criteria were strongly supported by the landmark OECD report of Vollenweider (1968), and the linkages between nutrient loading, lake trophic state, and the summer biomass of phytoplankton were subsequently quantified by a wide variety of workers worldwide (Sakamoto 1966; Lund 1970; Edmondson 1972; Shannon and Brezonik 1972; Dillon and Rigler 1974; Ahl 1975; Jones and Bachmann 1976; Vollenweider 1976; Carlson 1977).

As can be seen in Figure 2.1a,b, there is a very strong correlation between the summer concentration of TP and the summer mean algal biomass (as indicated by concentrations of the photosynthetic pigment chlorophyll a) in freshwater lakes and reservoirs. This relationship is strongly hyperbolic (Straskraba 1985; McCauley et al. 1989; Prairie et al. 1989; Jones and Knowlton 1993), with a tendency towards decreasing algal yields at concentrations of TP $> 100 \, \text{mg m}^{-3}$. Although this apparent plateau could reflect potential light-limitation at very high levels of algal biomass, it is very evident from the TN/TP ratios in the lakes shown in Figure 2.1a that N-limitation may also play a very strong role in reducing algal yields in these hypertrophic lakes. In human-made reservoirs containing high levels of

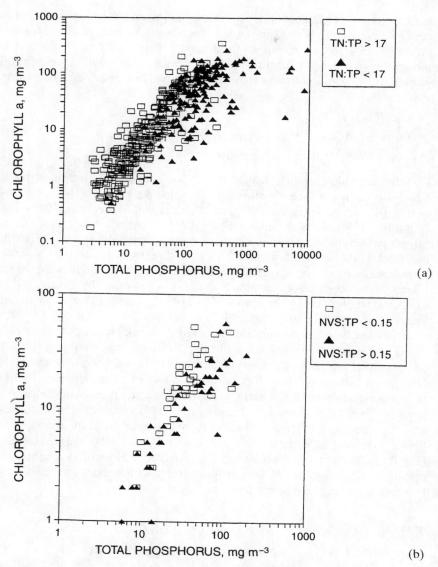

FIGURE 2.1. (a) Relationship between growing season mean chlorophyll *a*, total phosphorus, and total nitrogen in freshwater lakes (data from Smith 1982 and Prairie et al. 1989). The data are coded as P-limited lakes (TN:TP > 17 by mass) and N− or N + P-limited lakes (TN:TP < 17 by mass) according to the empirical criteria of Sakamoto (1966). (b) Relationship between growing season mean chlorophyll *a*, total phosphorus, and non-volatile solids in Missouri reservoirs (data from Jones and Knowlton 1993). The data are coded as low- to moderate-turbidity reservoirs (NVS:TP < 0.15) and high turbidity reservoirs (NVS:TP > 0.15) according to the empirical criteria of Jones and Knowlton (1993).

suspended soils, N-limitation is often less evident, and the presence of these non-volative solids (NVS) can strongly reduce the amount of algal biomass produced at a given concentration of total P (Figure 2.1b). This effect is thought to stem largely from the effects of NVS on light-limitation of algal growth in the water column (Hoyer and Jones 1983; Jones and Knowlton 1993).

Effects of N and P on Attached Algae (Periphyton) in Lakes

Periphyton, which are the algae that grow on submerged aquatic surfaces, have also been consistently shown to be sensitive indicators of increased nutrient inputs to lakes (Collins and Weber 1978; Sladeckova 1979). For example, Goldman and deAmezaga (1975) found a strong response of littoral periphyton to urban development in the watershed of Lake Tahoe, and Loeb (1986) concluded that the increased growth of attached algae was a sensitive indicator of eutrophication of the lake. Additional evidence for effects of eutrophication on attached algae is provided by Ennis (1975), Loeb and Reuter (1981), Stevenson et al. (1985), Hansson (1988), Kann and Falter (1989), and Jacoby et al. (1991).

As in the case of phytoplankton, much of the work on periphyton growth in lakes has focussed on P as a limiting nutrient. For example, in their comparative analyses of periphyton growth in lakes, Cattaneo and Kalff (1980) and Cattaneo (1987) focussed on the role of P, and demonstrated a strong correlation between periphyton biomass and epilimnetic concentrations of TP in lakes. Similarly, Jacoby et al. (1991) presented an empirical model predicting periphyton biomass from concentrations of soluble reactive P in lakes. In contrast to Cattaneo and Kalff (1980) and Hansson (1988), Lalonde and Downing (1991) observed a unimodal response of epiphytic algal biomass to lake trophic state, with maximum periphyton biomass occurring at intermediate fertility levels.

Effects of N and P on Periphyton and Suspended Algal Biomass in Flowing Waters

As was noted over a decade ago by Welch (1978), the effects of increased nutrient inputs unfortunately remain much less well understood for attached and suspended algae in streams than for phytoplankton. However, phosphorus enrichment studies by Stockner and Shortreed (1978), Elwood et al. (1981), Bothwell (1985, 1989), and Peterson et al. (1985) have demonstrated P-limitation of periphyton biomass in situ, and experimental studies of artificial laboratory channels (e.g. Krewer and Holm 1982; Horner et al. 1983) have also clearly shown that the periphyton biomass in flowing waters is frequently dependent on P availability. McGarrigle (1993) concluded that mean annual streamwater concentrations of dissolved inorganic P $<47 \, \text{mg m}^{-3}$ in Irish streams were necessary to avoid excessive growth of benthic algae.

Many other investigators have also provided evidence for N-limitation, or joint N- and P-limitation, of attached algal growth in flowing waters (Stockner and Shortreed 1978; Crawford 1979; Gregory 1980; Marcus 1980; Triska et al. 1983; Jones et al. 1984; Grimm and Fisher 1986; Burton et al. 1991). For example, both Grimm and Fisher (1986) and Pringle (1987) noted significant effects of N/P ratios on stream periphyton growth. The N/ P ratio criteria that very closely approximate those developed by Sakamoto (1966) for lakes were also used to assess potential N- or P-limitation of algal growth in streams by Schanz and Juon (1983), who examined nutrient limitation in periphyton in the river Rhine and eight of its tributaries. In a recent study of limiting nutrients in sub-alpine, forest, agricultural, and urban Australian streams, Chessman et al. (1992) also concluded that nutrient limitation of periphyton growth was widespread, and that the identity of the limiting nutrient could be inferred from streamwater total inorganic nitrogen (TIN)/soluble reactive phosphorus (SRP) ratios. Valuable additional literature on nutrient limitation of periphyton in streams is summarized by Tate (1990) and by Grimm and Fisher (1986).

As is true of periphyton in lakes, the biomass of both attached algae in streams has been found to be strongly influenced by concentrations of TP in the water (Dodds et al. 1997). The relationship between periphyton biomass and TP appears to be strongly curvilinear in streams (Figure 2.2a), and data from the lakes studied by Cattaneo (1987) fit surprisingly well within this general trend (however, see Lalonde and Downing 1991). The data in Figure 2.2b further suggest that the influence of N-limitation on algal biomass may be even more evident in streams than in lakes, and Dodds et al. (1997) have proposed a quantitative framework for predicting and managing the effects of N- and P-loading on the biomass of stream periphyton. A somewhat different modelling approach based on instream concentrations of dissolved nutrients has also been proposed by Welch et al. (1989).

Suspended algae may also be found in streamwater, and although the factors governing their development is much more poorly understood than for lake phytoplankton (Soballe and Kimmel 1987; Reynolds 1988), these suspended algae probably originate from the growth of periphyton in the streambed (Swanson and Bachmann 1976). As is the case for periphyton, the relationship between suspended algal biomass and TP in streams is strongly curvilinear (Figure 2.3), but the suspended algal biomass at a given level of TP is much lower than the equivalent biomass of phytoplankton in lakes (Van Nieuwenhuyse and Jones 1996).

Thus, in streams as well as lakes, the biomass of algae suspended in the water column is more strongly dependent upon nutrient availability than on hydrology (Basu and Pick 1996). As in lakes, the primary limiting nutrient

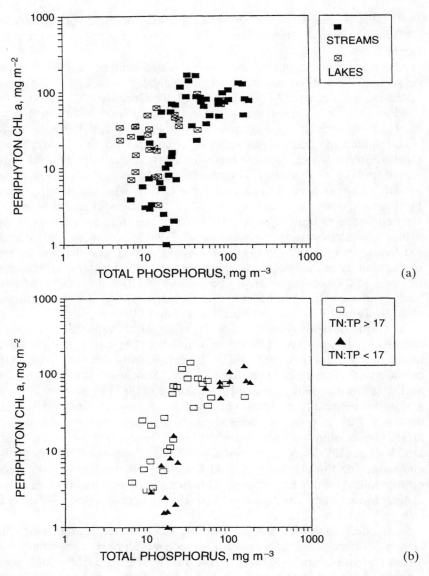

FIGURE 2.2. (a) Relationship between periphyton chlorophyll *a* and total phosphorus in freshwater streams and lakes. Lake data are from Cattaneo (1987). (b) Relationship between periphyton chlorophyll *a*, total phosphorus, and total nitrogen in freshwater streams. The data are coded as P-limited streams (TN:TP > 17 by mass) and N- or N + P-limited streams (TN:TP < 17 by mass) according to the empirical criteria of Sakamoto (1966).

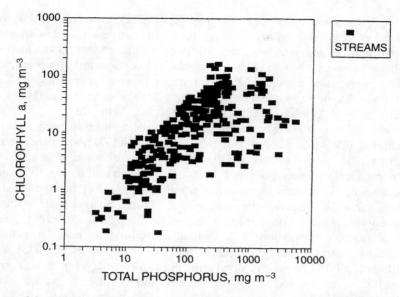

FIGURE 2.3. Relationship between suspended chlorophyll *a* and total phosphorus in freshwater streams (data from Van Nieuwenhuyse and Jones 1996).

appears to be P; however, there is some evidence for an effect of N as well. Although data for TN were not available from the study of Van Nieuwenhuyse and Jones (1996), a joint effect of TN and TP on suspended chlorophyll was found in an earlier study by Jones et al. (1984). This pattern would be expected from the interactive effects of N and P on periphyton biomass that are evident in Figure 2.2b above.

Effects of N and P on Estuarine and Coastal Marine Ecosystems

In their excellent review of nutrients and marine ecosystems, Nixon et al. (1986) pointed out that it was the development of quantitative methods for plankton and water chemistry in the latter part of the 19th century that allowed Brandt (1901) and others to make the first correlations between N availability and plankton concentrations in coastal and open ocean waters. Soon afterwards, Johnstone (1908) developed a mass-balance for N in the North Sea, and this focus on N as a primary limiting nutrient for phytoplankton growth in estuarine and coastal marine waters has continued for the past century.

As in lakes and rivers, an increase in the supply of nutrients can have profound effects on receiving water quality in both estuarine and coastal marine ecosystems and reviews of many of the diverse problems associated with marine eutrophication and its practical control can be found in Nixon

(1990), Valiela (1992), Vollenweider et al. (1992a), Howarth (1993), and Malone et al. (1996). Many of these problems are associated with algal blooms, and the nutrient-limitation status of marine phytoplankton in situ has been assessed extensively based on the use of bioassays, physiological measurements, and elemental stoichiometry (for general reviews see Carpenter and Capone 1983; Hecky and Kilham 1988; Howarth et al. 1988a, b; Seitzinger 1988; Vitousek and Howarth 1991; Howarth 1993).

Studies of the nutrient-limitation status of coastal marine environments by Smayda (1974), Goldman (1976), Yentsch et al. (1977), and others have supported the general conclusion that N-limitation is common. In addition, both empirical evidence (Boynton et al. 1982; Nixon and Pilson 1983) and experimental mesocosm fertilization studies (Figure 2.4) have suggested that the biomass and production of phytoplankton in many coastal systems is strongly correlated with external N inputs (see also Howarth 1993). These studies support suggestions that N-loading reductions may be needed to control coastal marine eutrophication (e.g. Goldman et al. 1973), and Wallin et al. (1992) have developed eutrophication models for coastal areas of the Baltic Sea that are based on N-loading and on total N concentrations in the water column.

However, in their review of the evidence for a general N-limitation of marine ecosystems, Hecky and Kilham (1988) concluded that the evidence in support of N-limitation in salt water was fundamentally weaker than the evidence for P-limitation in freshwater. For example, in some semi-

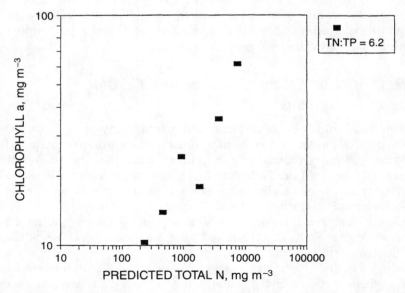

FIGURE 2.4. Relationship between observed mean chlorophyll *a* concentrations and predicted water column total nitrogen concentrations in the experimental mesocosms (MERL tanks) of Nixon et al. (1984).

enclosed systems for which complete nutrient budgets could be calculated, P rather than N was found to be limiting (e.g., Smith 1984). Evidence for P-limitation has also been found in the Peel-Harvey estuary, Australia (McComb and Humphries 1992); Florida Bay (Fourqurean et al. 1993); a small tidal river tributary to the Chesapeake Bay (Correll 1981); and in the Chesapeake Bay proper following the implementation of point-source controls (Magnien et al. 1992). It is probable that the degree of N- versus P-limitation of algal productivity in any given coastal marine ecosystem is dependent on the N/P-supply ratio to its waters, which will vary with the land use and human population density of nearby terrestrial ecosystems and with the local balance between fluvial, advective, and atmospheric nutrient inputs (see Caraco 1995; Howarth et al. 1995, 1996; Downing 1997).

Evidence for a strong nutrient dependence of algal productivity in marine waters suggests that a modelling approach similar to that used in lakes (e.g. OECD 1982; Reckhow and Chapra 1983) could potentially be applied to the management of eutrophication in estuarine and coastal marine ecosystems as well. This hypothesis is supported by the strong empirical correlations between phytoplankton biomass and phosphorus observed by Ketchum (1969), and the applicability of such a general modelling framework to marine waters has been considered by Vollenweider (1992).

The general importance of both N and P as limiting factors for marine phytoplankton is further supported by the results of a comparative analysis of data from marine systems located worldwide. As can be seen in Figure 2.5, the strong correlation between seasonal mean water column TP concentrations and seasonal mean chlorophyll a in marine systems is strikingly similar to that presented earlier for lakes in Figure 2.1. The strength of the relationship in Figure 2.5 strongly suggests that many of the general methods and tools used in freshwater eutrophication should indeed be applicable to the control and management of algal blooms in estuarine and coastal marine systems. Several aspects of the OECD (1982) approach in fact have been used successfully in a recent study of eutrophication in the northern Adriatic Sea (cf. Giovanardi and Tromellini 1992; Vollenweider et al. 1992b) and an analysis of a much more extensive dataset than that shown here is currently underway by J. Meeuwig of McGill University (Meeuwig et al. 1997). I predict that these empirical approaches can be successfully used in the future to develop predictive tools for other such marine quality variables of concern as the biomass and cover of attached marine macroalgae, vascular plants (e.g. sea grasses), and corals.

Effects of Eutrophication on Phytoplankton Community Structure

The enrichment of surface waters with N and P not only causes increases in total phytoplankton biomass, but also causes pronounced shifts in algal community structure. The effects of the chemical composition of lake water

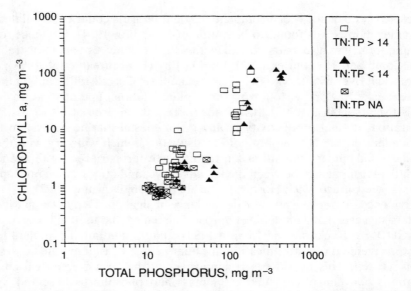

FIGURE 2.5. Relationship between mean chlorophyll *a*, total phosphorus, and total nitrogen in estuarine and coastal marine waters. The data are coded empirically as potentially P-limited systems (TN:TP > 14 by mass) and potentially N- or N + P-limited systems (TN:TP < 14 by mass). Note: NA indicates cases for which total nitrogen data were not available; in almost all of these cases, however, the original sources of these data provided objective evidence for primary P-limitation (Fourqurean et al. 1993; Vollenweider et al. 1992b).

on phytoplankton have been appreciated since before the turn of the century. For example, Apstein (1896, as cited in Hutchinson 1967) suggested the existence of a relationship between N availability and late summer blooms of cyanobacteria. Unfortunately, however, we still have only a rudimentary understanding of the factors that determine phytoplankton species composition and succession. Clearly, many potential factors may be involved (Reynolds 1984, 1987, 1997; Harris 1986; Sommer et al. 1986; Paerl 1988), but whole-lake fertilization experiments (Schindler 1977; Flett et al. 1980; Findlay et al. 1994) have clearly confirmed that changes in the nutrient loading to lakes can have profound effects on phytoplankton community structure.

One of the most consistent effects of eutrophication on lakes is an increase in the frequency and intensity of summer blooms of cyanobacteria (Vollenweider 1968; Pick and Lean 1987; Duarte et al. 1992; Carpenter et al. 1997). Cyanobacterial blooms have been associated worldwide with nuisance water quality conditions such as surface scums, filter clogging, taste and odor problems, and summer fishkills (Reynolds and Walsby 1975; Horne 1979; Gregor and Rast 1981; Smith 1990a). Cyanobacteria also frequently produce potent toxins (Carmichael 1991), and are frequently considered to be a poor food resource for aquatic consumers (Kohl and Lampert 1991). Dominance by cyanobacteria in lake phytoplankton thus typically creates undesirable water quality conditions, and also can potentially decrease the efficiency with which primary production is transferred up the food web through herbivores to higher trophic levels. As has been noted by Horne (1979), the management of lakes to avoid blooms of nuisance N-fixing blue-green algae is a complex and problematic issue.

The dramatic increase that occurs in both the absolute and the relative biomass of cyanobacteria that accompanies the eutrophication of lakes is illustrated in Figure 2.6. One possible explanation for this strong directional shift in phytoplankton species composition comes from resource-ratio theory (Tilman 1982; Grover 1997). This theory suggests that the supplies of N, P, and silicon (SI), and the availability of light in the water column are important determinants of algal species composition. An important implication of resource-ratio theory for eutrophication management is its prediction that N-limitation (low N/P-supply ratios) should favor dominance by cyanobacteria. In fresh waters, cyanobacteria are apparently better N competitors than other phytoplankton taxa (Tilman et al. 1982), and thus would be expected to dominate under conditions of N-limitation, as evidenced by low N/P-supply ratios (Schindler 1977; Flett et al. 1980; Findlay et al. 1994), or by low N/P ratios in the water column (Smith 1983, 1986; Smith et al. 1994). Because low N/P ratios typically accompany cultural eutrophication

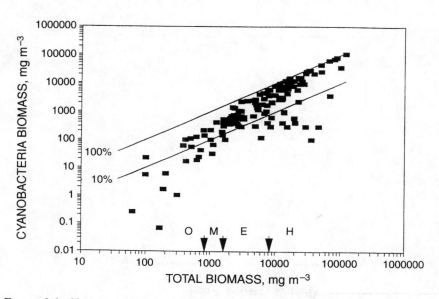

FIGURE 2.6. Changes in the absolute and relative biomass of cyanobacteria in freshwater lakes. The approximate boundaries for the four trophic state categories (i.e. oligotrophic, mesotrophic, eutrophic, and hypertrophic) are from Rott (1980).

(cf. Downing and McCauley 1992), this trend towards N-limitation provides one potential explanation for the increasing cyanobacterial dominance in lakes seen in Figure 2.6 (however, see Jensen et al. 1994).

Species shifts and blooms of undesirable phytoplankton species also are typical of nutrient-enriched estuaries and coastal marine waters (Paerl 1988), and such blooms have become increasingly common worldwide during the past ten years (Cosper et al. 1989; Nixon 1990; Burkholder et al. 1992; Anderson 1994). Although the causes of these sudden shifts towards nuisance taxa are not yet well understood, it is probable that changes in nutrient-loading or nutrient-supply ratios to these waters again may in part be responsible. For example, Officer and Ryther (1980) have concluded that the enrichment of coastal areas by S-poor domestic effluents may cause localized Si-limitation and lead to shifts from diatoms to dominance by other algal taxa. Sommer (1991) has in fact found evidence that shifts in Si/N ratios may be responsible for shifts in the relative dominance of marine phytoplankton. There is also evidence that N/P supply ratios may influence phytoplankton species composition in some low salinity environments. For example, Niemi (1979) has implicated low N/P ratios as a factor contributing to blooms of nitrogen-fixing cyanobacteria in the brackish Baltic Sea, and I concluded in a recent comparative analysis of data from the freshwater Potomac estuary (cf. Limno-Tech 1991) that low N/P ratios strongly favor cyanobacterial dominance in this system as well (Figure 2.7).

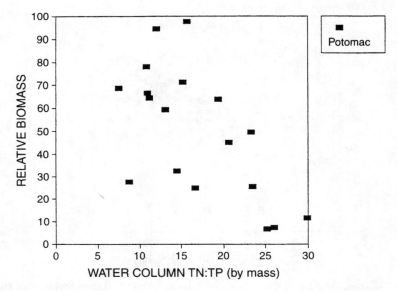

FIGURE 2.7. Relationship between the summer mean relative biomass of cyanobacteria and summer mean TN/TP ratios in the freshwater Potomac estuary (data from Limno-Tech 1991).

Case Study: Lake Washington, Seattle, Washington

The eutrophication and recovery of Lake Washington provides an excellent example of the sensitive responses of freshwater lakes to increases in external nutrient loading, and their subsequent recovery following external nutrient-loading control. Lake Washington is a large lake (mean depth 32.9 meters; surface area 87.6km^2) that lies parallel to the Puget Sound and adjacent to the city of Seattle, Washington. Seattle grew rapidly following its settlement in 1851, and by 1922 there were thirty outfalls containing a combination of raw sewage and storm runoff from Seattle, serving 50,000 people (Edmondson 1991). These inputs were of concern to the public because Lake Washington was for many years a source of drinking water, and in 1936 a major sewerage system was built to divert sewage into Puget Sound from the developed Lake Washington side of the hills separating the lake from the Sound. However, this was only a partial solution because further development proceeded, and by the 1950s the lake was receiving nutrient-rich discharge from eleven wastewater treatment plants employing secondary treatment. At its peak just before diversion began in 1963, twenty million gallons of treated sewage effluent were discharged into the lake per day.

Lake Washington thus went through two major periods of sewage pollution, the first of which ended in 1936, and the second of which ended in 1968 when all secondary effluent was finally diverted from the lake. This second phase concluded a unique series of events that began with the sudden appearance of a worrisome species of phytoplankton in the lake detected by limnologists in the summer of 1955, causing a rapid escalation of public concern and debate, and culminating in an historic public vote in 1958 to divert the remaining secondary effluent from the lake into Puget Sound (Edmondson 1991, 1994).

The environmental signal that led to this historic action was the observation by Dr. G.C. Anderson of the cyanobacterium *Oscillatoria rubescens* in the lakewater on June 15, 1955. This species was already well known as an indicator of eutrophication and deteriorating water quality in European lakes (Hasler 1947). Dr. W.T. Edmondson of the University of Washington had been monitoring the limnology of the lake for the prior five years, and immediately expressed strong concern that deterioration of water quality in Lake Washington was imminent (Edmondson 1979). Because the scope of this problem was so large, Edmondson immediately sought funding for a basic research program on the lake, and initially regarded the project as a whole-lake study of the effects of fertilization with nutrients. Funding for this work first came from the National Institutes of Health, and then from the newly formed National Science Foundation, which has provided essentially continuous funding for research on Lake Washington for almost forty years (Edmondson 1991). Funding from the Andrew W. Mellon Foundation has enabled the work to continue since 1986 (Edmondson, personal communication).

Because of his understanding of basic limnology and his empirical knowledge of the European experience with eutrophication, Edmondson also was able to make a series of very accurate predictions. He wrote in February of 1957 to an advisory committee to the mayor of Seattle, "If fertilization continues, the series of changes already started can be expected to continue. Within a few years, we can expect to have serious scum and odor nuisances ... I would expect distinct trouble here within five years, although isolated occurrences might come earlier." (Edmondson 1979). Moreover, he was able to link the problem (the appearance of *Oscillatoria rubescens* and the potential for nuisance summer algal blooms) directly to inputs of secondary sewage effluent to the lake. It was his opinion that if the sewage were diverted from the lake within a few years, there would be no further trouble with algal blooms in the future; implicit in this was the idea that the proportion of cyanobacteria in the phytoplankton would decline following sewage diversion, and thus that the water quality of the lake would improve perceptively (Edmondson 1979).

The ensuing public debate was lively and not without controversy, and has strong parallels with current political debates about regulation and the proper role of government in people's lives (see Edmondson 1991). However, Edmondson's warnings were persuasive and led to a public vote in 1958 that established the Municipality of Metropolitan Seattle (METRO) and charged it with the responsibility of solving sewerage problems both in the metropolitan area and in Lake Washington (Edmondson 1979). As a result of this vote, sewage diversion from the lake began before really serious deterioration of water quality began. Nonetheless, between the time of the vote and the first phase of the diversion began in February 1963, highly visible increases in summer phytoplankton biomass occurred and the transparency of the lake water was cut in half from ca. 2 meters in 1955 to ca. 1 meter in 1963. As predicted by Edmondson, however, the lake responded quickly and sensitively to the reduction of external nutrient loading with declines in both total P concentrations and algal biomass. As can be seen in Figure 2.8, the pattern of eutrophication and recovery was similar to and predictable from the general relationships between total phosphorus and chlorophyll that were shown earlier in Figure 2.1. Moreover, the quality of the phytoplankton improved dramatically as well, with rapid declines in the relative biomass of nuisance cyanobacteria (Figure 2.9).

The water quality of Lake Washington has remained very acceptable during the thirty years since sewage diversion was completed, and in fact now resembles the quality seen in the early 1930s. Moreover, the Lake Washington situation produced ecosystem-level information that has been valuable in at least two ways. First, the sensitivity of the response of the lake both to reductions in external nutrient loading and to changes in its biological structure (see Edmondson and Abella 1988) has increased our scientific understanding of the factors that influence the ability of a lake to respond to restorative actions. In addition, the public action that established

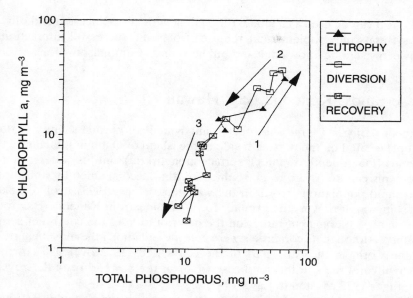

FIGURE 2.8. Response of summer mean chlorophyll *a* to changes in summer mean lakewater total phosphorus concentrations during (1) the eutrophication phase (1950–1963); (2) the diversion phase (1963–1968); and (3) the recovery phase (1968–1980) of Lake Washington, Seattle, WA.

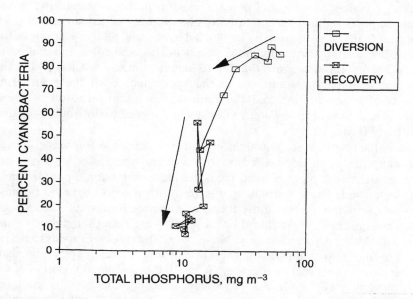

FIGURE 2.9. Responses in the summer mean relative biomass of cyanobacteria to changes in summer mean lakewater total phosphorus concentrations during the diversion and recovery phases of Lake Washington, Seattle, WA.

METRO can be used as a model for dealing with eutrophication and other ecosystem-scale problems that require strong and successful interactions between scientific knowledge and public values (Edmondson 1991).

Case Study: Kaneohe Bay, Hawaii

Kaneohe Bay is a large subtropical embayment of normal salinity (33 to 35 ppt) located on the northeast side of the island of Oahu in the Hawaiian Islands (mean depth 6 meters; surface area 46 km^2), and its water quality problems can be traced to eutrophication that accompanied urbanization and rapid population growth in the watershed (Laws 1993). Unlike Lake Washington, however, the primary focus of concern in Kaneohe Bay was not on algal blooms but rather on the deterioration of the coral reef community. Although its coral reefs were considered among the most beautiful in the world in the early part of the century, by 1972 over 99% of the original coral reefs in the southeast sector of the bay had been destroyed (Maragos 1972 as cited in Laws 1993).

The case study of eutrophication and its reversal in Kaneohe Bay has been well summarized by Laws (1993), who noted that until roughly 1940 the Kaneohe Bay watershed was used primarily for agricultural production, and the local population was only about 5,000 inhabitants. Rapid changes from the cultivation of the native taro root to commercial production of coffee, bananas, sugar cane, pineapples, and perhaps oranges occurred; more important, however, the quality and quantity of stream runoff from the land to the sea was strongly influenced by the use of large sections of the watershed for the grazing of cattle, sheep, horses, and goats. In addition to these factors, rapid urbanization led to increases in soil erosion and in the amount of impervious surfaces in the watershed. Both the smothering effects of sedimentation and the reduction in salinity caused by these increasing inputs of freshwater to the bay undoubtedly contributed to coral mortality (Laws 1993).

As in the case of Lake Washington, however, Kaneohe Bay was also used for many years as a receptacle first for raw sewage and later for secondary sewage treatment plant effluent from the increasing human population in the watershed. These nutrient inputs were of profound concern because hermatypic corals such as those found in Kaneohe Bay contain the symbiotic populations of algae that they require for their continued growth and survival. Although they are also filter-feeders, these corals appear to derive most of their nutrition from the excretion products of their symbiotic algae rather than from direct planktivory (Laws 1993). The eutrophication of the Bay by sewage effluent had three primary effects on these corals. First, the clarity of the overlying water was greatly reduced by the growth of phytoplankton algae, and light availability for the coral's symbiotic algae was reduced. Second, the sewage inputs favored the growth of sponges and

TABLE 2.4. Effects of sewage diversion on water quality in Kaneohe Bay, Oahu, Hawaii.[1]

Parameter	Southeast sector		Remainder of bay	
	Pre-diversion	Post-diversion	Pre-diversion	Post-diversion
chl a, $\mathrm{mg\,m^{-3}}$	2.0	1.1	0.67	0.53
Secchi depth, m	4.9	6.4	7.1	7.9
TIN, $\mathrm{mg\,m^{-3}}$	26	10	23	11
SRP, $\mathrm{mg\,m^{-3}}$	14	5	6.5	2.5

[1] Note: TIN is the concentration of total inorganic nitrogen, and SRP is the concentration of dissolved inorganic phosphorus in the water column.

zoanithids, which overgrew many of the coral reefs. Finally, eutrophication stimulated the growth of the attached bubble alga *Dictyosphaera cavernosa*, which may envelop and then kill a coral head (Laws 1993).

Concerns for the reefs led to the diversion of wastewater effluent from the bay between December of 1977 and June of 1978. Because the hydraulic retention time of the bay is only about two weeks, it was expected that water column nutrient concentrations would drop rapidly following sewage diversion, and that both the water quality and the growth of coral reefs would improve dramatically (Laws 1993). As can be seen in Table 2.4, dissolved inorganic nitrogen and phosphorus concentrations declined, and the water quality of the bay indeed improved rapidly. Moreover, in the southeastern sector (which was most strongly influenced by the sewage inputs), live coral coverage increased from ca. zero in 1971 to 20% by 1990 (Laws 1993). This represents a major and relatively rapid recovery from the effects of eutrophication by this important marine ecosystem, but Laws (1993) also notes that careful planning and control of land use in the watershed will be necessary to provide further protection of the reefs as a recreational resource and tourist attraction.

Key Elements of Success

The eutrophication of inland, estuarine, and coastal marine waters leads to a wide variety of consequences that are considered to be undesirable by the public (see Tables 2.1 and 2.3). Figures 1 to 9 reveal that there is nonetheless a remarkable unity in the response of aquatic ecosystems to nutrient enrichment: algal growth and primary productivity in all of these systems is strongly influenced by the P-supply. The two case studies presented in this chapter demonstrate the general sensitivity of freshwater and marine aquatic ecosystems to increases in their external nutrient supplies, and also confirm the use of external nutrient-loading control as the cornerstone of cultural eutrophication management.

It is important to consider here whether there have been identifiable elements that have made eutrophication research such a success in ecosys-

TABLE 2.5. Keys to success in eutrophication research and management.

- The responses to eutrophication were very similar across all aquatic systems.
- The responses to eutrophication were easily perceived and understood by the public.
- The resources affected by eutrophication were highly valued by the public; their degradation was uniformly disliked; their restoration was broadly supported; and the public were willing to put in time, energy, and money into their restoration and protection.
- There were few monetary or societal benefits to the degradation of these resources.
- The responses to eutrophication were in most cases inherently reversible.
- The generation times of the aquatic organisms involved were rapid relative to both ecological and political time frames, and there was therefore rapid user satisfaction and perception of the success (or lack of success) of control measures.
- Willingness of local, state, and federal agencies and private organizations to fund eutrophication research at all levels of complexity (laboratory to whole systems).
- Willingness of experts from a wide variety of disciplines to collaborate in understanding and solving the problem.
- The models used in eutrophication management were grounded in a strong conceptual framework; were typically noncomplex (both conceptually and empirically), and thus could be used and understood both by the public and by experts directly involved in eutrophication management; did not require expensive or complicated computing tools; and were portable and relatively insensitive to geographical location.
- Monitoring played a vital role both in the detection of the initial responses of these ecosystems to eutrophication, and in the detection of their responses to restoration measures.

tem science, because these elements may help to provide a guide in the solution of other equally important problems in ecosystem ecology. I suggest that there indeed have been a number of key elements to our success in eutrophication research (Table 2.5). Clearly, there must always be a strong and effective dialog between scientists and the managers who implement science to solve real-world problems (for an interesting and provocative viewpoint on this issue, see Cullen 1990). However, I conclude that the primary key to success is the essential dialog that must occur between scientists and the public who ultimately must support (and fund) our efforts to protect and conserve natural ecosystems. Edmondson (1991) has emphasized, "if you explain it well enough, people will do the right thing," and his extremely thoughtful reflections on the uses of ecology in Lake Washington and beyond should be required reading by all ecosystem ecologists.

Where Do We Go from Here?

Despite these clear successes in ecosystem science, there is much more eutrophication research to be done in the future. For example, Sas (1989) emphasized that the inflowing concentration of P is not the only factor that determines lakewater TP concentrations, even at steady state, and also stressed that that the TP concentration in the lake itself is not the only factor that determines the growth of algae in the system. I extend the admonition of Sas (1989) here to include N as well. To further improve our success in controlling eutrophication, it is very important that we more fully understand the ways in which key non-nutrient factors such as the system's morphometry and hydrology (Biggs and Close 1989; Monbet 1992; Håkanson 1994) and biological structure (Moss 1990; Carpenter and Kitchell 1993; Mazumder 1994; Lacroix et al. 1996; Sarnelle 1996) modify the physical, chemical, and biological expressions of eutrophication in receiving bodies of water. It is also equally important to evaluate more fully the relative importance of diffuse and internal sources of nutrient loading, which may strongly influence the rate of recovery of a body of water following the implementation of external nutrient-loading controls (e.g. Nürnberg 1984; Sas 1989; Carvalho et al. 1995).

It also must be recognized both by the public and by managers that the new trophic state to which a body of water can be improved will always be restricted by historical and geological constraints. For example, many lakes and reservoirs are naturally mesotrophic or eutrophic because they are imbedded in a nutrient-rich landscape, and they cannot be transformed into pristine, crystal-clear systems despite the best management efforts (Anderson 1995). These limitations to eutrophication management are evident from the presence of distinct ecoregions in the natural landscape (Omernik et al. 1991). Additionally, the trophic state of many freshwater and marine systems may be dependent more upon nutrient loading from nonpoint sources than on point sources such as domestic sewage effluent or industrial inputs. Many of the success stories in lake eutrophication have involved reductions in point source nutrient inputs, which are much more easily regulated. However, many inland, estuarine, and coastal waters receive a majority of their nutrient inputs from nonpoint sources that are much less easily managed and controlled. To manage these systems effectively, we need to develop more precise models that link land use to water quality and water-management decisions (e.g., Summer et al. 1990; Scarborough and Peters 1996; Soranno et al. 1996; Carpenter et al. 1998).

In many aquatic ecosystems the appearance of nuisance taxa such as cyanobacteria, toxic dinoflagellates, or bubble algae may be of primary concern to users of the body of water. However, in other systems the primary source of concern may be water clarity; the biomass or yield of other taxa such as finfish or shellfish; or measures of ecosystem degradation such as coral reef dieoffs or deepwater deoxygenation. It is, therefore, very important to use models that relate these key site-specific quality variables of concern to nutrient loading when considering the types and extent of external nutrient controls that will be used in water quality management.

In the case of lakes, nutrient control strategies for eutrophication management typically focus on P removal or wastewater diversion, and not on N removal. As was evident in the case study of Lake Washington, this strategy has had a strong history of success in lakes and reservoirs (Smith and Shapiro 1981; OECD 1982; Vismarra et al. 1985; Cooke et al. 1986), and as has been noted by Sas (1989), eutrophied lakes tend to recover reluctantly but steadily once the P supply has been decreased. To the extent to which N/P-supply ratios influence cyanobacterial dominance in lakes, the focus on P control and wastewater diversion has had the additional effect of increasing N/P-supply ratios and thus favoring taxa other than cyanobacteria in lake phytoplankton. That a reduction in P, and not N, was the best management strategy for controlling algal blooms in the river Rhine was also similarly concluded by van Steveninck et al. (1992).

The serious eutrophication problems that we now see developing in our estuaries and coastal areas serve to emphasize that much more work is needed in the future if we are to maintain and preserve desirable water quality in these vitally important ecosystems. Several authors have concluded that inputs of both N and P need to be restricted to control marine eutrophication (Howarth 1993). The marine case study presented suggests that a strategy of dual N- and P-loading reductions via sewage diversion has worked reasonably well in Kaneohe Bay, Hawaii. Similarly, dramatic improvements have occurred during the past ten years in Hillsborough Bay, a highly impacted subdivision of Tampa Bay, Florida (Johansson and Lewis 1992). Recent declines in chlorophyll *a* and improvements in water clarity and dissolved oxygen concentrations in the bay have been attributed to reductions in N- and P-loading that followed the implementation of advanced wastewater treatment and the restriction of nutrient inputs from fertilizer industries. Coincidentally with these improvements in the water quality, seagrass and attached macroalgae have revegetated shallow areas around the bay that had previously been barren for several decades (Johansson and Lewis 1992).

In other systems such as the freshwater Potomac River estuary, an N-control strategy has been proposed to control nuisance algal growth. However, because this system is susceptible to blooms of cyanobacteria, it has been argued that joint N- and P-control will be necessary to maintain N/P-loading ratios high enough to avoid further increasing cyanobacterial dominance in the estuary (Limno-Tech 1991). The lower Neuse River estuary (North Carolina) is also thought to be N-limited, and blooms of such nuisance phytoplankton taxa as cyanobacteria are becoming increasingly frequent (Paerl et al. 1995). Nitrogen-loading control measures have been proposed for this watershed as well, and concerns have been raised that an N-control policy could potentially result in low N/P-supply ratios, and further increase the probability of cyanobacterial blooms in this system (Paerl, personal communication).

The Chesapeake Bay is the largest estuary in the United States, and Fisher et al. (1992) have concluded that controlling eutrophication in the bay and its subestuaries would require basin-specific management practices for both N and P. D'Elia and Boynton (1997) have recently argued that biological nutrient control (which removes both N and P) has indeed been a highly successful approach in the Chesapeake Bay watershed. This eco-system not only may provide an important benchmark test of this new biological nutrient removal technology for the control of estuarine eutrophication, but also may provide a large-ecosystem nutrient manipulation experiment similar to the whole-system manipulation studies made by Schindler and others (cf. Carpenter et al. 1995).

Much more research is needed on the relative effects of N-, P-, and N+P-removal strategies in a wide variety of systems, however, to determine which of these three strategies is the most successful and most cost-effective policy for eutrophication control in estuarine and coastal marine waters. For example, if the supply of P is confirmed to be a factor in the formation of toxic marine dinoflagellate blooms (Smayda 1997), then the implemention of nutrient control policies that focus solely on N removal may not be a desirable strategy for the control of nuisance phytoplankton growth associated with coastal marine eutrophication.

The management and control of cultural eutrophication is clearly a strong success story in ecosystem science. However, there is a great deal that is left to be done. As was stressed in a national nutrient assessment workshop recently held by the United States Environmental Protection Agency (U.S. EPA 1996), it is vitally important that we take the general tools that have been developed for lakes and reservoirs and begin to modify and to apply these vital tools to the management of eutrophication in flowing waters, wetlands, estuaries, and coastal marine waters. Moreover, other human alterations of the Earth's ecosystems are substantial and growing yearly (Vitousek et al. 1997), and it is my hope that many of the lessons learned from eutrophication research are transportable and can ultimately be applied to the solution of many other globally important issues as well. Happily, there is evidence that the approaches used so successfully to address the problem of cultural eutrophication can also be used in the solution of other ecosystem-scale problems such as acidification, heavy metal inputs, and radioisotope pollution (Håkanson and Peters 1995).

Acknowledgments. This chapter is dedicated with very deep gratitude to Dr. W.T. Edmondson, who has played a central role in the study, under-standing, and management of eutrophication, and who has been a much valued friend and colleague for almost two decades. I also wish to thank my mentor, Joseph Shapiro, who guided my studies and gave me a profound appreciation for the Hutchinson legacy; David Schindler, who kindly shared with me data from his landmark studies of eutrophication at the ELA; and Richard Vollenweider, who has provided warm support for my research on eutrophication, and shared with me a very personal history of limnology. Some of the data presented here were kindly provided by W.T. Edmondson, J.R. Jones, M. Knowlton, and A. Cattaneo. W. Baillargeon assisted in the compilation of many of the data shown in Figure 2.2. I also thank Mike Pace and reviewers at the Cary Conference for helping to improve this chapter.

References

Ahl, T. 1975. Effects of man-induced and natural loading of phosphorus and nitrogen on the large Swedish lakes. *Verhandlungen der Internationale Vereinigung für Theoretische und Angewandte Limnologie* 19:1125–1132.

Ahl, T. 1979. Natural and human effects on trophic evolution. *Archiv für Hydrobiologie Beihefte Ergebnisse der Limnologie* 13:259–277.

Anderson, D.M. 1994. Red tides. *Scientific American* 271:62–68.

Anderson, N.J. 1995. Naturally eutrophic lakes: reality, myth or myopia? *Trends in Ecology and Evolution* 10:137–138.

Apstein, C. 1896. *Das Süsswasserplankton. Methode und Resultate der quantitativen Untersuchung.* Kiel.

Aure, J., and A. Stigebrandt. 1989. On the influence of topographic factors upon the oxygen consumption rate in sill basins of fjords. *Estuarine and Coastal Shelf Science* 28:59–69.

Bachmann, R.W. 1981. Prediction of total nitrogen in lakes and reservoirs. Pages 320–324 in *Restoration of lakes and inland waters.* U.S. EPA 440/5-81-010, Washington, DC.

Bachmann, R.W. 1984. Calculation of phosphorus and nitrogen loadings to natural and artificial lakes. *Verhandlungen der Internationale Vereinigung für Theoretische und Angewandte Limnologie* 22:239–243.

Bachmann, R.W., B.L. Jones, D.D. Fox, M. Hoyer, L.A. Bull, and D.E. Canfield, Jr. 1996. Relationship between trophic state indicators and fish in Florida (U.S.A.) lakes. *Canadian Journal of Fisheries and Aquatic Sciences* 53:842–855.

Baden, S.P., L.O. Loo, L. Pihl, and R. Rosenberg. 1990. Effects of eutrophication on benthic communities including fish: Swedish west coast. *Ambio* 19:113–122.

Baker, L.A., P.L. Brezonik, and C.R. Kratzer. 1981. *Nutrient loading-trophic state relationships in Florida lakes.* Florida Water Resources Research Center Publication 56, University of Florida, Gainesville.

Baker, L.A., P.L. Brezonik, and C.R. Kratzer. 1985. Nutrient loading models for Florida lakes. Pages 253–258 in *Lake and reservoir management: practical implications.* North American Lake Management Society, Merrifield, VA.

Basu, B.K., and F.R. Pick. 1996. Factors regulating phytoplankton and zooplankton biomass in temperate rivers. *Limnology and Oceanography* 41:1572–1577.

Biggs, B.J.F., and M.E. Close. 1989. Periphyton biomass dynamics in gravel bed rivers: the relative effects of flows and nutrients. *Freshwater Biology* 22:209–231.

Bothwell, M.L. 1985. Phosphorus limitation of lotic periphyton growth rates: an intersite comparison using continuous-flow troughs (Thompson River system, British Columbia). *Limnology and Oceanography* 30:527–54.

Bothwell, M.L. 1989. Phosphorus-limited growth dynamics of lotic periphytic diatom communities: areal biomass and cellular growth rate responses. *Canadian Journal of Fisheries and Aquatic Sciences* 46:1293–1301.

Boynton, W.R., W.M. Kemp, and C.W. Keefe. 1982. A comparative analysis of nutrients and other factors influencing estuarine phytoplankton production. Pages 93–109 in V.S. Kennedy, ed., *Estuarine comparisons.* Academic Press, New York.

Brandt, K. 1899. Uber das Stoffwechsel In Meere. *Wissenschaftliche Meere-suchunger, Abt. Kiel Band* 4:213–230.

Brandt, K. 1901. Life in the ocean. Pages 493–506 in *Annual report of the Board of Regents of the Smithsonian Institution for the year ending June 30, 1900.* Washington, DC.

Burkholder, J.M., K.M. Mason, and H.B. Glasgow. 1992. New "phantom" dinoflagellate is the causative agent of major estuarine fish kills. *Nature* 358:407–410.

Burton, T.M., M.P. Oemke, and J.M. Molloy. 1991. Contrasting effects of nitrogen and phosphorus additions on epilithic algae in a hardwater and a softwater stream in Northern Michigan. *Verhandlungen der Internationale Vereinigung für Theoretische und Angewandte Limnologie* 24:1644–1653.

Canfield, D.E., Jr. and R.W. Bachmann. 1981. Prediction of total phosphorus concentrations, chlorophyll *a*, and Secchi depths in natural and artificial lakes. *Canadian Journal of Fisheries and Aquatic Sciences* 38:414–423.

Caraco, N. 1995. Influence of human populations on P transfers to aquatic systems: a regional scale study using large rivers. Pages 235–244 in H. Tiessen, ed., *Phosphorus in the global environment: transfers, cycles and management.* SCOPE 54. John Wiley & Sons, Chichester, UK.

Carlson, R.E. 1977. A trophic state index for lakes. *Limnology and Oceanography* 22:361–369.

Carlson, R.E., and J.T. Simpson. 1995. Trophic state. Pages 73–92 in *A coordinator's guide to volunteer lake monitoring methods.* North American Lake Management Society, Madison, WI.

Carmichael, W.W. 1991. Toxic freshwater blue-green algae (cyanobacteria): an overlooked health threat. *Health and Environment Digest* 5:1–4.

Carpenter, E.J., and D.G. Capone. 1983. *Nitrogen in the marine environment.* Academic Press, New York.

Carpenter, S.R., and J.F. Kitchell, eds. 1993. *The trophic cascade in lakes.* Cambridge University Press, Cambridge, UK.

Carpenter, S.R., S.W. Chisholm, C.J. Krebs, D.W. Schindler, and R.F. Wright. 1995. Ecosystem experiments. *Science* 269:324–327.

Carpenter, S.R., D. Bolgrien, R.C. Lathrop, C.A. Stowe, T. Reed, and M.A. Wilson. 1998. Ecological and economic analysis of lake eutrophication by nonpoint pollution. *Australian Journal of Ecology* 23:68–79.

Carvalho, L., M. Beklioglu, and B. Moss. 1995. Changes in a deep lake following sewage diversion—a challenge to the orthodoxy of external phosphorus control as a restoration strategy? *Freshwater Biology* 34:399–410.

Cattaneo, A. 1987. Periphyton in lakes of different trophy. *Canadian Journal of Fisheries and Aquatic Sciences* 44:296–303.

Cattaneo, A., and J. Kalff. 1980. The relative contribution of aquatic macrophytes and their epiphytes to the production of macrophyte beds. *Limnology and Oceanography* 25:280–289.

Chessman, B.C., P.E. Hutton, and J.M. Burch. 1992. Limiting nutrients for periphyton growth in sub-alpine, forest, agricultural and urban streams. *Freshwater Biology* 28:349–361.

Chiaudani, G., and M. Vighi. 1974. The N:P ratio and tests with *Selenastrum* to predict eutrophication in lakes. *Water Research* 8:1063–1069.

Collins, G.B., and C.I. Weber. 1978. Phycoperiphyton (algae) as indicators of water quality. *Transactions of the American Microscopical Society* 97:36–43.

Cooke, G.D., E.B. Welch, S. Peterson, and P. Newroth. 1986. *Lake and reservoir restoration*. Butterworth, Boston, MA.

Correll, D.L. 1981. Eutrophication trends in the water quality of the Rhode River (1971–1978). Pages 425–435 in B.J. Nielson and L.E. Cronin, eds. *Estuaries and nutrients*. Humana Press, Clifton, NJ.

Cosper, E.M., V.M. Bricelj, and E.J. Carpenter, eds. 1989. *Novel phytoplankton blooms: causes and impacts of recurrent brown tides and other unusual blooms*. Springer-Verlag, New York.

Crawford, J.K. 1979. *Limiting nutrients in streams: a case study of Pine Creek, Pennsylvania*. Ph.D. thesis. Pennsylvania State University, University Park.

Cullen, P. 1990. The turbulent boundary between water science and water management. *Freshwater Biology* 24:201–209.

DeBaar, H.J.W. 1994. von Liebig's law of the minimum and plankton ecology (1899–1991). *Progress in Oceanography* 33:347–386.

D'Elia, C.F., and W.R. Boynton. 1997. A case history of nutrient control in a Chesapeake Bay tributary: indications of successful policy development and implementations. Page 138 in *Abstracts of the 1997 Annual Meeting, American Society of Limnology and Oceanography*.

Dillon, P.J., and F.H. Rigler. 1974. The phosphorus-chlorophyll relationship in lakes. *Limnology and Oceanography* 19:767–773.

Dillon, P.J., and F.H. Rigler. 1975. A simple method for predicting the capacity of a lake for development based on lake trophic state. *Journal of the Fisheries Research Board of Canada* 31:1518–1531.

Dodds, W.K., V.H. Smith, and B. Zander. 1997. Controlling excess benthic chlorophyll levels in streams: a general approach and application to the Clark Fork River. *Water Research* 31:1738–1750.

Downing, J.A. 1997. Marine nitrogen: phosphorus stoichiometry and the global N:P cycle. *Biogeochemistry* 37:237–252.

Downing, J.A., and E. McCauley. 1992. The nitrogen: phosphorus relationship in lakes. *Limnology and Oceanography* 37:936–945.

Duarte, C., S. Agustí, and D.E. Canfield, Jr. 1992. Patterns in phytoplankton community structure in Florida lakes. *Limnology and Oceanography* 37:155–161.

Edmondson, W.T. 1961. Changes in Lake Washington following an increase in the nutrient income. *Verhandlungen der Internationale Vereinigung für Theoretische und Angewandte Limnologie* 14:167–175.

Edmondson, W.T. 1972. Nutrients and phytoplankton in Lake Washington. Pages 172–193 in G.E. Likens, ed. *Nutrients and eutrophication*. Special Symposium 1, American Society of Limnology and Oceanography. Allen Press, Lawrence, KS.

Edmondson, W.T. 1979. Lake Washington and the predictability of limnological events. *Archiv für Hydrobiologie Beihefte Ergebnisse der Limnologie* 13:234–241.

Edmondson, W.T. 1991. *The uses of ecology: Lake Washington and beyond*. University of Washington Press, Seattle.

Edmondson. W.T. 1994. Sixty years of Lake Washington: a curriculum vitae. *Lake Reservoir Management* 10:75–84.

Edmondson, W.T. 1995. Eutrophication. Pages 697–703 in *Encyclopedia of Environmental Biology, Volume 1*. Academic Press, New York.

Edmondson, W.T., and S.E.B. Abella. 1988. Unplanned biomanipulation in Lake Washington. *Limnologica* (Berlin) 19:73–79.

Edmondson, W.T., G.C. Anderson, and D.R. Peterson. 1956. Artificial eutrophication of Lake Washington. *Limnology and Oceanography* 1:47–53.

Elwood, J.W., J.D. Newbold, A.F. Trimble and R.W. Stark. 1981. The limiting role of phosphorus in a woodland stream ecosystem: effects of P enrichment on leaf decomposition and primary producers. *Ecology* 62:146–158.

Ennis, G.L. 1975. Distribution and abundance of benthic algae along phosphate gradients in Kootenay Lake, British Columbia. *Verhandlungen der Internationale Vereinigung für Theoretische und Angewandte Limnologie* 19:562–570.

Ernst, M.R., W. Frossard, and J.L. Mancini. 1994. Two eutrophication models make the grade. *Water Environment and Technology* 10:15–16.

Findlay, D.L., R.L. Hecky, L.L. Hendzel, M.P. Stainton, and G.W. Regehr. 1994. Relationship between N_2-fixation and heterocyst abundance and its relevance to the nitrogen budget of Lake 227. *Canadian Journal of Fisheries and Aquatic Sciences* 51:2254–2266.

Fisher, T.R., E.R. Peele, J.W. Ammerman, and J.W. Harding, Jr. 1992. Nutrient limitation of phytoplankton in Chesapeake Bay. *Marine Ecology Progress Series* 82:51–63.

Flett, R.J., D.W. Schindler, R.D. Hamilton, and N.E.R. Campbell. 1980. Nitrogen fixation in Canadian Precambrian Shield lakes. *Canadian Journal of Fisheries and Aquatic Sciences* 37:494–505.

Forsberg, C. 1977. Nitrogen as a growth factor in fresh water. *Marine Ecology Progress Series* 8:275–290.

Forsberg, C. 1981. Present knowledge on limiting nutrients. Page 37 in *Restoration of lakes and inland waters*. U.S. EPA 440/5-81-010, Washington, DC.

Fourqurean, J.W., R.D. Jones, and J.C. Zieman. 1993. Processes influencing water column nutrient characteristics and phosphorus limitation of phytoplankton biomass in Florida Bay, FL, USA: inferences from spatial distributions. *Estuarine and Coastal Shelf Science* 36:295–314.

Foy, R.H., C.E. Gibson, and T. Champ. 1996. The effectiveness of restricting phosphorus loading from sewage treatment works as a means of controlling eutrophication in Irish lakes. Pages 134–152 in P.S. Giller and A.A. Miller, eds. *Disturbance and recovery of ecological systems*. Royal Irish Academy, Dublin.

Giovanardi, F., and E. Tromellini. 1992. Statistical assessment of trophic conditions. Application of the OECD methodology to the marine environment. Pages 211–233 in R.A. Vollenweider, R. Marchetti, and R. Viviani, eds. *Marine coastal eutrophication. The response of marine transitional systems to human impact: problems and perspectives for restoration*. Science of the Total Environment Supplement 1992. Elsevier Scientific, Amsterdam, The Netherlands.

Goldman, C.R., and E. deAmezaga. 1975. Primary productivity in the littoral zone of Lake Tahoe, California–Nevada. *Symposia Biologica Hungarica* 15:49–62.

Goldman, C.R., and A. Jassby. 1990. Spring mixing and annual primary production at Lake Tahoe, California–Nevada. *Verhandlungen der Internationale Vereinigung für Theoretische und Angewandte Limnologie* 24:504.

Goldman, J., K. Tenore, and D. Stanley. 1973. Inorganic nitrogen removal from waste water: effect on phytoplankton growth in coastal marine waters. *Science* 80:955–956.

Goldman, J.C. 1976. Identification of nitrogen as a growth-limiting nutrient in wastewaters and coastal marine waters through continuous culture algal assays. *Water Research* 10:97–104.

Gregor, D.J., and W. Rast. 1981. Benefits and problems of eutrophication control. Pages 166–171 in *Restoration of lakes and inland waters*. U.S. EPA 440/5-81-010, Washington, DC.

Gregory, S.V. 1980. Effects of light, nutrients and grazing on periphyton communities in streams. Ph.D. dissertation, Oregon State University, Corvallis.

Grimm, N.B., and S.G. Fisher. 1986. Nitrogen limitation in a Sonoran Desert stream. *Journal of the North American Benthological Society* 5:2–15.

Grover, J.P. 1997. *Resource competition*. Chapman & Hall, London.

Håkanson, L. 1994. A review of effect-dose-sensitivity models for aquatic ecosystems. *Internationale Revue der Gesamten Hydrobiologie* 79:621–667.

Håkanson, L., and R.H. Peters. 1995. *Predictive limnology: methods for predictive modelling*. SPB Academic Publishing, Amsterdam, The Netherlands.

Håkanson, L., and M. Wallin. 1991. An outline of ecometric analysis to establish load diagrams for nutrients/eutrophication. *Environmetrics* 2:49–68.

Hansson, L.-A. 1988. Effects of competitive interactions on the biomass development of planktonic and periphytic algae in lakes. *Limnology and Oceanography* 33:121–128.

Harper, D. 1992. *Eutrophication of freshwaters: principles, problems, restoration*. Chapman & Hall, New York.

Harris, G.P. 1986. *Phytoplankton ecology: structure, function and fluctuation*. Chapman & Hall, New York.

Hasler, A.D. 1947. Eutrophication of lakes by domestic sewage. *Ecology* 28:383–395.

Hecky, R.E., and P. Kilham. 1988. Nutrient limitation of phytoplankton in freshwater and marine environments: a review of recent evidence on the effects of enrichment. *Limnology and Oceanography* 33:796–822.

Hellström, T. 1996. An empirical study of nitrogen dynamics in lakes. *Water Environment Research* 68:55–65.

Heyman, U., and A. Lundgren. 1988. Phytoplankton biomass and production in relation to phosphorus. Some conclusions from field studies. *Hydrobiologia* 170:211–227.

Horne, A.J. 1979. Management of lakes containing N_2-fixing blue-green algae. *Archiv für Hydrobiologie Beihefte Ergebnisse der Limnologie* 13:133–144.

Horner, R.R, E.B. Welch, and R.B. Veenstra. 1983. Development of nuisance periphytic algae in laboratory streams in relation to enrichment and velocity. Pages 121–134 in R.G. Wetzel, ed. *Periphyton of Freshwater Ecosystems*. W. Junk, The Hague, The Netherlands.

Howarth, R.W. 1993. The role of nutrients in coastal waters. Pages 177–202 in *Managing wastewater in coastal urban areas*. Report from the National Research Council Committee on Wastewater Management for Coastal Urban Areas. National Academy Press, Washington, DC.

Howarth, R.W., H. Jensen, R. Marino, and H. Postma. 1995. Transport to and processing of P in near-shore and oceanic waters. Pages 323–345 in H. Tiessen, ed. *Phosphorus in the global environment: transfers, Cycles and Management*. SCOPE 54. John Wiley & Sons, Chichester, UK.

Howarth, R.W., G. Billen, D. Swaney, A. Townsend, N. Jaworski, K. Lajtha, et al. 1996. Regional nitrogen budgets and riverine N & P fluxes for the drainages to the North Atlantic Ocean: natural and human influences. *Biogeochemistry* 35:75–139.

Howarth, R.W., R. Marino, and J.J. Cole. 1988a. Nitrogen fixation in freshwater, estuarine, and marine ecosystems. 2. Biogeochemical controls. *Limnology and Oceanography* 33:688–701.

Howarth, R.W., R. Marino, J. Lane, and J.J. Cole. 1988b. Nitrogen fixation in freshwater, estuarine, and marine ecosystems. 1. Rates and importance. *Limnology and Oceanography* 33:669–687.

Hoyer, M.V., and J.R. Jones. 1983. Factors affecting the relation between phosphorus and chlorophyll *a* in midwestern reservoirs. *Canadian Journal of Fisheries and Aquatic Sciences* 40:192–199.

Hrbácek, J., M. Dvoráková, M. Korínek, and L. Procházková. 1961. Demonstration of the effect of the fish stock on the species composition of zooplankton and the intensity of metabolism of the whole plankton association. *Verhandlungen der Internationale Vereinigung für Theoretische und Angewandte Limnologie* 14:192–195.

Hutchinson, G.E. 1967. *A treatise on limnology. Vol. II. Introduction to lake biology and the limnoplankton.* John Wiley & Sons, New York.

Hutchinson, G.E. 1969. Eutrophication, past and present. Pages 17–26 in *Eutrophication: causes, consequences, correctives*. National Academy of Sciences, Washington, DC.

Hutchinson, G.E. 1973. Eutrophication. The scientific background of a contemporary practical problem. *American Scientist* 61:269–279.

Jacoby, J.M., D.D. Bouchard, and C.R. Patmont. 1991. Response of periphyton to nutrient enrichment in Lake Whelan, WA. *Lake and Reservoir Management* 7:33–43.

Jensen, J.P., P. Kristensen, and E. Jeppesen. 1990. Relationships between nitrogen loading and in-lake nitrogen concentrations in shallow Danish lakes. *Verhandlungen der Internationale Vereinigung für Theoretische und Angewandte Limnologie* 24:201–204.

Jensen, J.P., E. Jeppesen, K. Olrik, and P. Kristensen. 1994. Impact of nutrients and physical factors on the shift from cyanobacterial to chlorophyte dominance in shallow Danish lakes. *Canadian Journal of Fisheries and Aquatic Sciences* 51:1692–1699.

Johansson, J.O.R., and R.R. Lewis, III. 1992. Recent improvements in water quality and biological indicators in Hillsborough Bay, a highly impacted subdivision of Tampa Bay, Florida, USA. Pages 1199–1216 in R.A. Vollenweider, R. Marchetti, and R. Viviani, eds. *Marine coastal eutrophication. The response of marine transitional systems to human impact: problems and perspectives for restoration.* Science of the Total Environment Supplement 1992. Elsevier Scientific, Amsterdam, The Netherlands.

Jones, J.R., and R.W. Bachmann. 1976. Prediction of phosphorus and chlorophyll levels in lakes. *Journal of the Water Pollution Control Federation* 48:2176–2182.

Jones, J.R., and M.V. Hoyer. 1982. Sportfish harvest predicted by summer chlorophyll-*a* concentration in midwestern lakes and reservoirs. *Transactions of the American Fisheries Society* 111:176–179.

Jones, J.R., and M.F. Knowlton. 1993. Limnology of Missouri reservoirs: an analysis of regional patterns. *Lake and Reservoir Management* 8:17–30.

Jones, J.R., M.M. Smart and J.N. Burroughs. 1984. Factors related to algal biomass in Missouri Ozark streams. *Verhandlungen der Internationale Vereinigung für Theoretische und Angewandte Limnologie* 22:1867–1875.

Johnstone, J. 1908, 1977. *Conditions of life in the sea.* Cambridge University Press. Reprinted by Arno Press, New York.

Kann, J., and C.M. Falter. 1989. Periphyton as indicators of enrichment in Lake Pend Oreille, Idaho. *Lake and Reservoir Management* 5:39–48.

Kemp, W.M., R.R. Twilley, J.C. Stephenson, W.R. Boynton, and J.C. Means. 1983. The decline of submerged vascular plants in upper Chesapeake Bay: summary of results concerning possible causes. *Journal of the Marine Technology Society* 17:78–85.

Kennedy, R.H. 1995. *Application of the BATHTUB model to selected southeastern reservoirs.* Techical Report EL-95-15, U.S. Army Engineer Waterways Experimental Station, Vicksburg, MS.

Kerr, S.R., and R.A. Ryder. 1992. Effects of cultural eutrophication on coastal marine fisheries: a comparative approach. Pages 599–614 in R.A. Vollenweider, R. Marchetti, and R. Viviani, eds. *Marine coastal eutrophication. The response of marine transitional systems to human impact: problems and perspectives for restoration.* Science of the Total Environment Supplement 1992. Elsevier Scientific, Amsterdam, The Netherlands.

Ketchum, B. 1969. Eutrophication of estuaries. Pages 17–209 in *Eutrophication: Causes, consequences, correctives.* National Academy of Sciences, Washington, DC.

Kohl, J.-G., and W. Lampert, eds. 1991. Interactions between zooplankton and blue-green algae (Cyanobacteria). *Internationale Revue der Gesamten Hydrobiologie* 76:1–88.

Kratzer, C.R. 1979. Application of input-output models to Florida lakes. M.S. thesis, University of Florida, Gainesville.

Krewer, J.A., and H.W. Holm. 1982. The phosphorus-chlorophyll *a* relationship in periphytic communities in a controlled ecosystem. *Hydrobiologia* 94:173–176.

Lacroix, G., F. Lescher-Montoué, and R. Pourriot. 1996. Trophic interactions, nutrient supply, and the structure of freshwater pelagic food webs. Pages 162–179 in M.E. Hochberg, J. Clobert, and R. Barbault, eds. *Aspects of the genesis and maintenance of biological diversity.* Oxford University Press, Oxford, UK.

Lalonde, S., and J.A. Downing. 1991. Epiphyton biomass is related to lake trophic status, depth, and macrophyte architecture. *Canadian Journal of Fisheries and Aquatic Sciences* 48:2285–2291.

Laws, E.A. 1993. *Aquatic pollution, 3rd. edition.* John Wiley & Sons, New York.

Lee, G.F., and R.A. Jones. 1981. Application of the OECD eutrophication modeling approach to estuaries. Pages 549–568 in B.J. Nielson and L.E. Cronin, eds. *Estuaries and nutrients.* Humana Press, Clifton, NJ.

Lenzi, M. 1992. Experiences for the management of Orbetello Lagoon: eutrophication and fishing. Pages 1189–98 in R.A. Vollenweider, R. Marchetti, and R. Viviani, eds. *Marine coastal eutrophication. The response of marine transitional systems to human impact: problems and perspectives for restoration.* Science of the Total Environment Supplement 1992. Elsevier Scientific, Amsterdam, The Netherlands.

Likens, G.E., ed. 1972. *Nutrients and eutrophication.* Special Symposium 1, American Society of Limnology and Oceanography. Allen Press, Lawrence, KS.

Limno-Tech. 1991. *Evaluation of nitrogen removal eutrophication risk for the freshwater Potomac estuary.* Limno-Tech, Inc. Final report prepared for the Metropolitan Washington Council of Governments, Washington, DC.

Littler, M.M., and S.N. Murray. 1975. Impact of sewage on the distribution, abundance and community structure of rocky intertidal macro-organisms. *Marine Biology* 30:277–291.

Loeb, S.L. 1986. Algal biofouling of oligotrophic Lake Tahoe: causal factors affecting algal production. Pages 159–173 in L.V. Evans and K.D. Hoagland, eds. *Algal biofouling. Elsevier*, Amsterdam, The Netherlands.

Loeb, S.L., and J.E. Reuter. 1981. The epilithic periphyton community: a five-lake comparative study of community productivity, nitrogen metabolism and depth-distribution of standing crop. *Verhandlungen der Internationale Vereinigung für Theoretische und Angewandte Limnologie* 21:346–352.

Lund, J.W.G. 1970. Primary production. *Water Treatment and Examination* 19:332–358.

Magnien, R.E., R.M. Summer, and K.G. Sellner. 1992. External sources, internal nutrient pools, and phytoplankton production in Chesapeake Bay. *Estuaries* 15:497–516.

Malone, T.C., A. Màlej, and N. Smodlaka. 1996. Trends in land-use, water quality and fisheries: a comparison of the Northern Adriatic Sea and the Chesapeake Bay. *Periodicum Biologorum* 98:137–148.

Maragos, J.E. 1972. A study of the ecology of Hawaiian reet corals Ph.D. dissertation, University of Hawaii 290 p.

Marcus, M.D. 1980. Periphytic community response to chronic nutrient enrichment by a reservoir discharge. *Ecology* 61:387–399.

Mazumder, A. 1994. Phosphorus-chlorophyll relationships under contrasting herbivory and thermal stratification: predictions and patterns. *Canadian Journal of Fisheries and Aquatic Sciences* 51:401–407.

McCauley, E., J.A. Downing, and S. Watson. 1989. Sigmoid relationships between nutrients and chlorophyll among lakes. *Canadian Journal of Fisheries and Aquatic Sciences* 46:1171–1175.

McComb, A.J., and R. Humphries. 1992. Loss of nutrients from catchments and their ecological impacts in the Peel-Harvey estuarine system, Western Australia. *Estuaries* 15:529–537.

McGarrigle, M.L. 1993. Aspects of river eutrophication in Ireland. *Annals of Limnology* 29:355–364.

Meeuwig, J., R.H. Peters, and J. Rasmussen. 1997. Patterns in the fog: a cross-system comparison of chlorophyll *a*: nutrient relations in estuaries and lakes. Page 237 in *Abstracts of the 1997 Annual Meeting*, American Society of Limnology and Oceanography.

Miller, W.E., T.E. Maloney, and J.C. Greene. 1974. Algal productivity in 49 lakes as determined by algal assays. *Water Research* 8:667–679.

Monbet, Y. 1992. Control of phytoplankton biomass in estuaries: a comparative analysis of microtidal and macrotidal estuaries. *Estuaries* 15:563–571.

Morris, D.P., and W.M. Lewis, Jr. 1988. Phytoplankton nutrient limitation in Colorado mountain lakes. *Freshwater Biology* 20:315–327.

Moss, B. 1990. Engineering and biological approaches to the restoration from eutrophication of shallow lakes in which aquatic plant communities are important components. *Hydrobiologia* 200:367–378.

National Academy of Sciences. 1969. *Eutrophication: causes, consequences, correctives*. National Academy of Sciences, Washington, DC.

Naumann, E. 1919. Några synpunkter angående limnoplanktons ökologi med särskild hänsyn till fytoplankton. *Svensk Botanisk Tidskrift* 13:129–163.

Naumann, E. 1929. The scope and chief problems of regional limnology. *Internationale Revue der Gesamten Hydrobiologie* 21:423.

Niemi, A. 1979. Blue-green algal blooms and N:P ratio in the Baltic Sea. *Acta Botanica Fennica* 110:57–61.

Nixon, S.W. 1990. Marine eutrophication: a growing international problem. *Ambio* 19:101.

Nixon, S.W., and M.E.Q. Pilson. 1983. Nitrogen in estuarine and coastal marine ecosystems. Pages 565–648 in E.J. Carpenter and D.C. Capone, eds. *Nitrogen in the marine environment*. Academic Press, New York.

Nixon, S.W., M.E.Q. Pilson, C.A. Oviatt, P. Donaghay, B. Sullivan, and S. Seitzinger. 1984. Eutrophication of a coastal marine ecosystem—an experimental study using the MERL microcosms. Pages 105–135 in M.J.R. Fasham, ed. *Flows of energy and materials in marine ecosystems*. Plenum, New York.

Nixon, S.W., C.A. Oviatt, J. Frithsen, and B. Sullivan. 1986. Nutrients and the productivity of estuarine and coastal marine ecosystems. *Journal of the Limnological Society of South Africa* 12:43–71.

Nürnberg, G.K. 1984. The prediction of internal phosphorus load in lakes with anoxic hypolimnia. *Limnology and Oceanography* 29:111–124.

Nürnberg, G.K. 1996. Trophic state of clear and colored, soft- and hardwater lakes with special consideration of nutrients, anoxia, phytoplankton and fish. *Lake and Reservoir Management* 12:432–447.

O'Connor, D.J. 1981. Modeling of eutrophication in estuaries. Pages 183–223 in B.J. Nielson and L.E. Cronin, eds. *Estuaries and nutrients*. Humana Press, Clifton, NJ.

OECD. 1982. *Eutrophication of waters: monitoring, assessment and control*. Organisation for Economic and Cooperative Development, Paris, France.

Officer, C.B., and J.H. Ryther. 1980. The possible importance of silicon in marine eutrophication. *Marine Ecology Progress Series* 3:83–91.

Omernik, J.M., C.M. Rohm, R.A. Lillie, and N. Mesner. 1991. Usefulness of natural regions for lake management: analysis of variation among lakes in northwestern Wisconsin, U.S.A. *Environmental Management* 15:2881–293.

Paerl, H.W. 1988. Nuisance phytoplankton blooms in coastal, estuarine, and inland waters. *Limnology and Oceanography* 33:823–847.

Paerl, H.W., M.A. Mallin, C.A. Donahue, M. Go, and B.L. Peierls. 1995. *Nitrogen loading sources and eutrophication of the Neuse River estuary, NC: direct and indirect roles of atmospheric deposition*. University of North Carolina Water Resources Research Institute Report 291, Raleigh.

Pearsall, W.H. 1932. Phytoplankton in English lakes. II. The composition of the phytoplankton in relation to dissolved substances. *Journal of Ecology* 20:241–262.

Peterson, B.J., J.E. Hobbie, A.E. Hershey, M.A. Lock, T.E. Ford, J.R. Vestal, et al. 1985. Transformation of a tundra stream from heterotrophy to autotrophy by addition of phosphorus. *Science* 229:1383–1386.

Pick, F.R., and D.R.S. Lean. 1987. The role of macronutrients (C,N,P) in controlling cyanobacterial dominance in temperate lakes. *New Zealand Journal of Marine and Freshwater Research* 21:425–434.

Prairie, Y.T., C.M. Duarte, and J. Kalff. 1989. Unifying nutrient-chlorophyll relationships in lakes. *Canadian Journal of Fisheries and Aquatic Sciences* 46:1176–1182.

Priddle, J., I. Hawes, J.C. Ellis-Evans, and T.J. Smith. 1986. Antarctic aquatic ecosystems as habitats for phytoplankton. *Biological Reviews* 61:199–238.

Pringle, C.M. 1987. Effects of water and substratum nutrient supplies on lotic periphyton growth: an integrated bioassay. *Canadian Journal of Fisheries and Aquatic Sciences* 4:619–629.

Pringle, C.M., P. Paaby-Hansen, P.D. Vaux, and C.R. Goldman. 1986. *In situ* nutrient assays of periphyton growth in a lowland Costa Rica stream. *Hydrobiologia* 134:207–213.

Reckhow, K.H. 1988. Empirical models for trophic state in southeastern U.S. lakes and reservoirs. *Water Resources Bulletin* 24:723–734.

Reckhow, K.H. and S.C. Chapra, 1983. *Engineering Approaches for Lake Management. Vol. 1: data analysis and empirical modeling.* Butterworth, Boston, MA.

Redfield, A.C. 1958. The biological control of chemical factors in the environment. *American Scientist* 46:205–222.

Reynolds, C.S. 1984. *The ecology of freshwater phytoplankton.* Cambridge, New York.

Reynolds, C.S. 1987. Cyanobacterial water-blooms. *Advances in Botanical Research* 13:67–143.

Reynolds, C.S. 1988. Potamoplankton: paradigms, paradoxes and prognoses. Pages 285–311 in F.E. Round, ed. *Algae and the aquatic environment.* Biopress, London, UK.

Reynolds, C.S. 1997. Vegetation processes in the pelagic: a model for ecosystem theory. *Excellence in Ecology, volume 9.* Ecology Institute, Luhe, Germany.

Reynolds, C.S., and A.E. Walsby. 1975. Water blooms. *Biological Reviews* 50:437–481.

Riley, G.A. 1967. Mathematical model of nutrient conditions in coastal waters. *Bulletin of the Bingham Oceanographic Colloquia* 19(2):72–80.

Rosenberg, R., R. Elmgren, S. Fleischer, P. Jonsson, G. Persson, and H. Dahlin. 1990. *Marine eutrophication case studies in Sweden.* Ambio 19:102–108.

Rott, E. 1980. Some results from phytoplankton counting intercalibrations. *Schweizerische Zeitschrift für Hydrologie* 43:34–62.

Ryding, S-O., and W. Rast. 1989. *The control of eutrophication of lakes and reservoirs. Man and the Biosphere, volume 1.* UNESCO, Paris, France.

Ryther, J.H., and W.M. Dunstan. 1971. Nitrogen, phosphorus, and eutrophication in the coastal marine environment. *Science* 171:1008–1013.

Sakamoto, M., 1966. Primary production by phytoplankton community in some Japanese lakes and its dependence on lake depth. *Archiv für Hydrobiologie* 62:1–28.

Salas, H.J., and P. Martino. 1991. A simplified phosphorus trophic state model for warm-water tropical lakes. *Water Research* 25:341–350.

Sarnelle, O. 1996. Predicting the outcome of trophic manipulation in lakes—a comment on Harris (1994). *Freshwater Biology* 35:339–342.

Sawyer, C.N. 1947. Fertilization of lakes by urban drainage. *Journal of the New England Water Works Association* 61:109–127.

Sas, H. 1989. *Lake restoration by reduction of nutrient loading: expectations, experiences, extrapolations.* Academia Verlag, Richarz, St. Augustin.

Scarborough, G.C., and R.H. Peters. 1996. Predictability of phophorus load, hydrological load and lake total phosphorus concentration. *Lake and Reservoir Management* 12:420–431.

Schanz, F., and H. Juon. 1983. Two different methods of evaluating nutrient limitations of periphyton bioassays, using water from the River Rhine and eight of its tributaries. *Hydrobiologia* 102:187–195.

Schindler, D.W. 1977. Evolution of phosphorus limitation in lakes. *Science* 195:260–262.

Schindler, D.W. 1981. Studies of eutrophication in lakes and their relevance to the estuarine environment. Pages 71–82 in B.J. Nielson and L.E. Cronin, eds. *Estuaries and nutrients*. Humana Press, Clifton, NJ.

Schindler, D.W. 1985a. The coupling of elemental cycles by organisms: evidence from whole-lake chemical perturbations. Pages 225–250 in W. Stumm, ed. *Chemical processes in lakes*. John Wiley & Sons, New York.

Schindler, D.W. 1985b. Interrelationships between the cycles of elements in freshwater ecosystems. Pages 113–123 in G.E. Likens, ed. *Some perspectives of the major biogeochemical cycles, SCOPE 17*. John Wiley & Sons, New York.

Schindler, D.W. 1990. Experimental perturbations of whole lakes as tests of hypothesis concerning ecosystem structure and function. *Oikos* 57:25–41.

Schindler, D.W., G.J. Brunskill, S. Emerson, W.D. Broecker, and T.-H. Peng. 1972. Atmospheric carbon dioxide: its role in maintaining phytoplankton standing crops. *Science* 177:1192–1194.

Seitzinger, S.P. 1988. Denitrification in freshwater and coastal marine ecosystems: ecological and geochemical significance. *Limnology and Oceanography* 33:702–724.

Shannon, E.E., and P.L. Brezonik. 1972. Eutrophication analysis: a multivariate approach. *Environmental Engineering Division ASCE* 98:(SA1):37–57.

Shapiro, J. 1979. The need for more biology in lake restoration. Pages 161–167 in *Lake restoration*. U.S. EPA 440/5-79-001, Washington, DC.

Sladeckova, A. 1979. Periphyton assays *in situ*. Pages 205–209 in P. Marvan, S. Pribil, and O. Lhotsky, eds. *Algal assays and monitoring eutrophication*. E. Schweizerbart'sche Verlagsbuchhandlungen, Stuttgart, Germany.

Smayda, T.J. 1974. Bioassay of the growth potential of the surface water of lower Narragansett Bay over an annual cycle using the diatom *Thalassiosira pseudonana* (oceanic clone, 13-1). *Limnology and Oceanography* 19:889–901.

Smayda, T.J. 1997. Is phosphorus a significant nutrient factor in harmful algal blooms? Page 306 in *Abstracts of the 1997 Annual Meeting*, American Society of Limnology and Oceanography.

Smith, S.V. 1984. Phosphorus versus nitrogen limitation in the marine environment. *Limnology and Oceanography* 29:1149–1160.

Smith, V.H. 1982. The nitrogen and phosphorus dependence of algal biomass in lakes: an empirical and theoretical analysis. *Limnology and Oceanography* 27:1101–1112.

Smith, V.H. 1983. Low nitrogen to phosphorus ratios favor dominance by blue-green algae in lake phytoplankton. *Science* 221:669–671.

Smith, V.H. 1986. Light and nutrient effects on the relative biomass of blue-green algae in lake phytoplankton. *Canadian Journal of Fisheries and Aquatic Sciences* 43:148–153.

Smith, V.H. 1990a. Phytoplankton responses to eutrophication in inland waters. Pages 231–249 in I. Akatsuka, ed. *Introduction to applied phycology*. SPB Academic Publishing, Amsterdam, The Netherlands.

Smith, V.H. 1990b. Nitrogen, phosphorus, and nitrogen fixation in lacustrine and estuarine ecosystems. *Limnology and Oceanography* 35:1852–1859.

Smith, V.H. 1993. Effects of nitrogen and phosphorus ratios on nitrogen fixation in agricultural and pastoral ecosystems. *Biogeochemistry* 18:19–35.

Smith, V.H., and J. Shapiro, 1981. Chlorophyll-phosphorus relations in individual lakes. Their importance to lake restoration strategies. *Environmental Science and Technology* 15:444–451.

Smith, V.H., V.J. Bierman, Jr., B.L. Jones, and K.E. Havens. 1994. Historical trends in the Lake Okeechobee ecosystem. IV. Nitrogen: phosphorus ratios, cyano-bacterial biomass, and nitrogen fixation potential. *Archiv für Hydrobiologie Monographische Beitrage* 107:69–86.

Soballe, D.M., and B.L. Kimmel. 1987. A large-scale comparison of factors influencing phytoplankton abundance in rivers, lakes, and impoundments. *Ecology* 68:1943–1954.

Sommer, U. 1991. Comparative nutrient status and competitive interactions of two Antarctic diatoms (*Corethron criophilum* and *Thalassiosira antarctica*). *Journal of Plankton Research* 13:61–75.

Sommer, U., Z.M. Gliwicz, W. Lampert, and A. Duncan. 1986. The PEG-model of seasonal succession of planktonic events in fresh waters. *Archiv für Hydrobiologie* 106:433–471.

Soranno, P.A., S.L. Hubler, S.R. Carpenter, and R.C. Lathrop. 1996. Phosphorus loads to surface waters: a simple model to account for spatial pattern of land use. *Ecological Applications* 6:865–878.

St. Amand, A.L., P.A. Soranno, S.R. Carpenter, and J.J. Elser. 1989. Algal nutrient deficiency: growth bioassays versus physiological indicators. *Lake and Reservoir Management* 5:27–36.

Stevenson, J.C. 1988. Comparative ecology of submersed grass beds in freshwater, estuarine, and marine environments. *Limnology and Oceanography* 33:867–893.

Stevenson, R.J., R. Singer, D.A. Roberts, and C.W. Boylen. 1985. Patterns of benthic algal abundance with depth, trophic status, and acidity in poorly buffered New Hampshire lakes. *Canadian Journal of Fisheries and Aquatic Sciences* 42:1501–1512.

Stockner, J.G., and K.R.S. Shortreed. 1978. Enhancement of autotrophic production by nutrient addition in a coastal rainforest stream on Vancouver Island. *Journal of the Fisheries Research Board of Canada* 35:28–34.

Straskraba, M. 1985. Managing of eutrophication by means of ecotechnology and mathematical modelling. Pages 79–90 in R. Vismarra, R. Marforio, V. Mezzanotte, and S. Cernuschi, eds. *Lake pollution and recovery*. Proceedings of the International Congress, European Water Pollution Control Association, Associazione Nazionale di Ingegneria Sanitaria, Milan, Italy.

Summer, R.M., C.V. Alonso, and R.A. Young. 1990. Modeling linked watershed and lake processes for water quality management decisions. *Journal of Environmental Quality* 19:421–427.

Sutcliffe, D.W., and J.G. Jones, eds. 1992. *Eutrophication: research and application to water supply*. Freshwater Biological Association, Ambleside, UK.

Swanson, C.D., and R.W. Bachmann. 1976. A model of algal exports in some Iowa streams. *Ecology* 57:1076–1080.

Tamm, C.O. 1991. *Nitrogen in terrestrial ecosystems. Questions of productivity, vegetational changes, and ecosystem stability*. Springer-Verlag, New York.

Tate, C.M. 1990. Patterns and controls of nitrogen in tallgrass prairie streams. *Ecology* 71:2007–2018.

Theinemann, A. 1921. Seetypen. *Naturwissenschaften* 18:1–3.

Tilman, D. 1982. *Resource competition and community structure*. Princeton Monographs in Population Biolgy 17. Princeton University Press, Princeton, NJ.

Tilman, D., S.S. Kilham, and P. Kilham. 1982. Phytoplankton community ecology: the role of limiting nutrients. *Annual Review of Ecology and Systematics* 13:349–373.

Triska, F.J., V.C. Kennedy, R.J. Avanzino, and B.N. Reilly. 1983. Effect of simulated canopy cover on regulation of nitrate uptake and primary production by natural periphyton assemblages. Pages 129–159 in T.D. Fontaine III and S.M. Bartell, eds. *Dynamics of lotic ecosystems*. Ann Arbor Science, Ann Arbor, MI.

Tyrrell, T., and C.S. Law. 1997. Low nitrate: phosphate ratios in the global ocean. *Nature* 387:793–796.

U.S. EPA. 1981. *Restoration of inland lakes and waters*. EPA 440/5-81-010, Washington, DC.

U.S. EPA. 1996. *Proceedings of the national nutrient assessment workshop, December 4–6, 1995*. U.S. EPA 822-R-96-004, Washington, DC.

Valiela, I. 1992. Coupling of watersheds and coastal waters: an introduction to the dedicated issue. *Estuaries* 15:429–430.

Van Nieuwenhuyse, E.E., and J.R. Jones. 1996. Phosphorus-chlorophyll relationship in temperate streams and its variation with stream catchment area. *Canadian Journal of Fisheries and Aquatic Sciences* 53:99–105.

van Steveninck, E.D. de Ruyter, W. Admiraal, L. Breebaart, G.M.J. Tubbing, and B. van Zanten. 1992. Plankton in the River Rhine: structural and functional changes observed during downstream transport. *Journal of Plankton Research* 14:1351–1368.

Vismarra, R., R. Marforio, V. Mezzanotte, and S. Cernuschi, eds. 1985. *Lake pollution and recovery*. Proceedings of the International Congress, European Water Pollution Control Association, Associazione Nazionale di Ingegneria Sanitaria, Milan, Italy.

Vitousek, P.M., H.A. Mooney, J. Lubchenko, and J.M. Melillo. 1997. Human dominance of Earth's ecosystems. *Science* 277:494–499.

Vitousek, P.M., and R.W. Howarth. 1991. Nitrogen limitation on land and in the sea: how can it occur? *Biogeochemistry* 13:87–115.

Viviani, R. 1992. Eutrophication, marine biotoxins, human health. Pages 631–662 in R.A. Vollenweider, R. Marchetti, and R. Viviani, eds. *Marine coastal eutrophication. The response of marine transitional systems to human impact: problems and perspectives for restoration*. Science of the Total Environment Supplement 1992. Elsevier Scientific, Amsterdam, The Netherlands.

Vollenweider, R.A. 1968. *Scientific fundamentals of lake and stream eutrophication, with particular reference to phosphorus and nitrogen as eutrophication factors*. Technical Report DAS/DSI/68.27. OECD, Paris, France.

Vollenweider, R.A. 1976. Advances in defining critical loading levels for phosphorus in lake eutrophication. *Memorie del Istituto Italiano di Idrobiologia* 33:53–83.

Vollenweider, R.A. 1992. Coastal marine eutrophication: Principles and control. Pages 1–20 in R.A. Vollenweider, R. Marchetti, and R. Viviani, eds. *Marine coastal eutrophication. The response of marine transitional systems to human impact: problems and perspectives for restoration*. Science of the Total Environment Supplement 1992. Elsevier Scientific, Amsterdam, The Netherlands.

Vollenweider, R.A., R. Marchetti, and R. Viviani, eds. 1992a. Marine coastal eutrophication. The response of marine transitional systems to human impact: *problems and perspectives for restoration*. Science of the Total Environment Supplement 1992. Elsevier Scientific, Amsterdam, The Netherlands.

Vollenweider, R.A., A. Rinaldi, and G. Montanari. 1992b. Eutrophication, structure and dynamics of a marine coastal system: results of a ten-year monitoring along the Emilia-Romagna coast (Northwest Adriatic Sea). Pages 63–105 in R.A. Vollenweider, R. Marchetti, and R. Viviani, eds. *Marine coastal eutrophication. The response of marine transitional systems to human impact: problems and perspectives for restoration*. Science of the Total Environment Supplement 1992. Elsevier Scientific, Amsterdam, The Netherlands.

von Liebig, J. 1855. Principles of agricultural chemistry with special reference to the late researches made in England. Reprinted in L.R. Pomeroy, ed. *Cycles of essential elements*. Benchmark Papers in Ecology, vol. 1, pages 11–28. Dowden, Hutchinson, and Ross, UK.

Walker, W.W., Jr. 1984. Empirical prediction of chlorophyll in reservoirs. Pages 292–297 in *Lake and reservoir management*. U.S. EPA 440/5-84-001, Washington, DC.

Wallin, M., and L. Håkanson. 1991. The importance of inherent properties of coastal areas. *Marine Pollution Bulletin* 22:381–388.

Wallin, M., L. Håkanson, and J. Persson. 1992. Load models for nutrients in coastal areas, especially from fish farms (in Swedish with English summary). *Nordiska ministerrådet* 1992:502.

Weber, C.A. 1907. Aufbau und vegetation der Moore Norddeutschlands. Botanische Jahrbuch 40. *Beiblatt zo den botanischen Jahrbuchern*. 90:19–34.

Wedin, D.A., and D. Tilman. 1996. Influence of nitrogen loading and species composition on the carbon balance of grasslands. *Science* 274:1720–1723.

Welch, E.B., R.R. Horner, and C.R. Patmont. 1989. Prediction of nuisance periphytic biomass: a management approach. *Water Research* 23:401–405.

Welch, E.B., and T. Lindell. 1992. *Ecological effects of wastewater: applied limnology and pollutant effects, 2nd edition*. Chapman & Hall, New York.

Welch, J. 1978. *The impact of inorganic phosphates in the environment*. Final report. Office of Toxic Substances. U.S. EPA, Washington, DC.

White, E. 1983. Lake eutrophication in New Zealand—a comparison with other countries of the Organisation for Economic Co-operation and Development. *New Zealand Journal of Marine and Freshwater Research* 17:437–444.

Yentsch, C.M., C.S. Yentsch, and L.R. Strube. 1977. Variations in ammonium enhancement, an indication of nitrogen deficiency in New England coastal phytoplankton populations. *Journal of Marine Research* 35:537–555.

3
Managing Forests as Ecosystems: A Success Story or a Challenge Ahead?

VIRGINIA H. DALE

Summary

To manage forests as ecosystems, the many values they hold for different users must be recognized, and they must be used so that those assets are not destroyed. Important ecosystem features of forests include nutrient cycling, habitat, succession, and water quality. Over time, the ways in which humans value forests have changed as forest uses have altered and as forests have declined in size and quality. Both ecosystem science and forest ecology have developed approaches that are useful to manage forests to retain their value. A historical perspective shows how changes in ecology, legislation, and technology have resulted in modern forest-management practices. However, current forest practices are still a decade or so behind present ecosystem science. Ecologists have done a good job of transferring their theories and approaches to the forest-management classroom but have done a poor job of translating these concepts into practice. Thus, the future for ecosystem management requires a closer linkage between ecologists and other disciplines. For example, the changing ways in which humans value forests are the primary determinant of forest-management policies. Therefore, if ecologists are to understand how ecosystem science can influence these policies, they must work closely with social scientists trained to assess human values.

There is as yet no ethic dealing with man's relation to land and to the animals and plants which grow upon it.

Aldo Leopold, 1949

Introduction

Almost fifty years after Leopold's writings, a forest ethic is still not in place. Such an ethic would serve as a guide for managing forest resources under diverse conditions. In the same sense that social ethics include a responsibility to "do the right thing," a forest ethic would set the standards used by both private and public landowners in protecting the diverse values of forests. Leopold's viewpoint clearly defines the role of ecologists: They can convey to forest managers the ecological principles upon which such a standard is built.

This chapter considers the benefits imparted to forest management by using ecological information. First, the management goals are set forth by discussing forest attributes that people value. The implementation of forest practices is influenced not only by changes in ecological understanding but also by legislation and technological developments. Therefore, ecological, legislative, and technological changes relevant to forest management are considered together. Next, ecological principles used in forest management to retain the valued attributes are discussed. Finally, future challenges for managing forests as sustainable ecosystems are proposed. Throughout this discussion, impacts that forestry itself has had upon ecological thought and research are mentioned.

The practice of forestry has been greatly altered in recent years by the combined effects of changes in legislation, increasing environmental sensibility of politically active citizens, and improved scientific understanding of ecosystems. These changes have occurred throughout the world but are most dramatically reflected in renewed forestry schemes in the temperate forests of the United States and Europe. As a result, the vast majority of the examples in this chapter come from the United States. However, it is important to note that developing countries that hold a large fraction of the world's forest are, at best, still struggling to move away from conventional approaches to forestry. Unfortunately, discussion of this struggle is beyond the scope of this chapter.

Attributes of Forests That People Value

Managing forests as ecosystems requires an awareness of the diverse attributes of forests and how they relate to human values. The benefits from forests are many. Forests serve as watersheds that provide purified water for rivers, streams, and municipal use as well as being important habitats for wildlife. Dry forests have widely spaced trees interspersed with grass and shrubs that serve as grazing lands. Forests offer many recreational opportunities, including hiking, camping, hunting, and fishing. Some people value forests for their beauty and the spiritual experience they provide. Trees also supply the raw materials for paper, building supplies, furniture, and pharmaceutical supplies. Finally, forests are a critical part of the global carbon cycle, for trees store large amounts of carbon in their tissue and, in mass, provide the largest storage site for terrestrial carbon (Post et al. 1990). When the carbon cycle is out of balance, it can instigate global changes in both temperature and precipitation.

Although an economic benefit relates to some of these forest attributes, considering only the monetary aspects of the forest is unsatisfactory. Even in the 1940s, Leopold discussed the difficulty of putting a value on forests other than in an economic sense. He argued that many of the benefits arising from the natural interconnectedness of the environment are not appreciated until they are lost or at risk. Indeed, gradual and largely unnoticed loss of forest-ecosystem components frequently occurs until a threshold is reached at which point these losses impact upon human values. As soon as a threshold is reached, people cry out for a return to a system that provides some of the benefits that are now in jeopardy. The possibility of going beyond the threshold has been made extremely clear in a science fiction book entitled *The Word for World is Forest* by Ursula LaGuin. In this novel, the forest is so valuable that it actually becomes the name of the planet. During the course of the book, interplanetary expansion and the need for wood products decimates the forest and diminishes the life that is dependent upon the diverse attributes of the forest. This novel may not be far from one future path along which the Earth could proceed. To avoid this possibility, ecologists argue for a long-term perspective on the use and value of forest resources.

Too often, however, the short-term impact of these gradual losses of forest attributes is not obvious. People can substitute one organism for another, or may synthesize a pharmaceutical product that formerly came from a forest plant or animal. But lost ecosystem properties, aesthetic qualities, or spiritual attributes may be more difficult to mimic. Moreover, forest ecosystems provide and protect water quality, water yield, and habitats that support other flora and fauna. For example, the fungus-induced demise of American chestnut trees in the early twentieth century resulted in not only the loss of timber but also seeds, which are an important food source for forest fauna. In addition, it changed nutrient turnover from a system dominated by chestnut trees (which have high nutrient turnover rate of the leaves and slow nutrient turnover of woody tissue) to one just the opposite (the chestnut oaks that largely replaced the chestnut trees have slow nutrient turnover in leaves, and the wood decays rapidly). Because this replacement caused such a complex alteration in ecosystem properties, the full extent of these changes may never be known.

Forest-management policies reflect the changing values that people have for forests. Thus, it is important to determine how the ways that humans value forests have changed over time. Throughout history, forests have served as sources of building materials and fuel and as habitats for game. As forests became scarce in Europe, commercial forestry arose as a means to ensure a sufficient supply of wood and wildlife. One goal of forest management is to maintain sufficient quality, quantity, and diversity of wood products for commercial uses. Tree diversity is an important aspect of management because the qualities of individual tree species affect their use. For example, in the past Port Orford cedar was preferred for venetian blinds and ship masts because of its strength when cut in long sections.

Presently, oak is preferred for floors because of its hardness. Douglas fir and pine make excellent structural material for buildings because they grow rapidly yet are strong.

At any one time, the forest industry seeks to have sufficient wood of appropriate size, species, and quality available for each of its mills distributed across the country. About 130 years ago, wood became a source of pulp for paper. Since that time, many tree plantations have been established to meet the need for pulp. Recently, the value of maintaining species habitats and forest health (e.g., resistance from fire and insect outbreaks) has been reemphasized. Great strides have been made in recognizing the ecological values of forests and in communicating to public and private owners how to preserve these values (Franklin 1989; Salwasser 1990; Kessler et al. 1992; National Research Council 1992; Swanson and Franklin 1992). Yet there are still gaps. For example, the public does not do a good job of integrating the many attributes of forests, nor have ecosystem scientists adequately demonstrated the integration process (Naiman in press). Also, because no single stand maintains all these human values simultaneously, a coordinated effort between federal and state land holders, private forest companies, citizens, and citizen groups is necessary to maintain an appropriate balance of forest resources within a region.

Adaptive management of forest systems recently arose as a way to preserve diverse values amid complex forest ownerships. This approach monitors how environmental decisions affect forests by tracking such ecosystem attributes as trophic dynamics, succession, and nutrient cycling (Christensen et al. 1996). The management is adaptive in the sense that the lessons learned from past actions and their effects are used to guide future actions.

Monitoring forest attributes provides a challenge in itself. The USDA Forest Service in coordination with the Environmental Protection Agency (EPA) has monitored attributes meant to reflect the diverse values of forests, including the ecological perspective [as measured by EPA's Environmental Monitoring and Assessment Program (EMAP) see Table 3.1] (Lewis and Conkling 1994). Despite the fact that these efforts were meant to cover a comprehensive range of ecological attributes, animal conditions in forests were largely ignored. Even so, the attributes included are too numerous and technical for the general public to grasp a basic understanding of them.

Therefore, how these ecological attributes relate to human values must also be determined. Russell et al. (submitted) are in the process of examining how human values relate to the EMAP forest metrics. They have identified dimensions that map both the ecological indicators and human values (Table 3.1). These dimensions are both understandable to the public and focus on forest-management goals. Research is currently underway to find out if these dimensions capture the public's values and also to query trained ecologists to determine if the mapping strategy is appropriate.

TABLE 3.1. Example indicators, dimensions, and values for forest systems.

Indicators	Dimensions	Values
Soil acidity	Forest type	Aesthetics
Foliar chemistry	Tree size	Recreation
Soil class	Visible plant damage	Cultural importance
Crown measures	Fragmentation	Religious importance
Leaf area	Visibility	Consumptive use
Branch evaluations	Recreation use	Services of nature
Plant diversity	Extractive use	
Ozone bioindicators		
Lichen diversity		
Visible damage		
Vegetation structure		

Results from interviews to date suggest that even people who have never been in a forest understand these dimensions and feel that they relate to forest values. However, the exact set and definition of such a set of dimensions will probably vary by community and by a particular communities' relations to forests. (For instance, how much of the community's economy depends on forest products? Do forests play a role in the religious beliefs of the community?) In addition, the importance of specific dimensions may vary over time as changes in ecological understanding, legislation, and technology occur.

The Influence of Changes in Ecological Understanding, Legislation, and Technology on Forest Management

To understand a science it is necessary to know its history.

August Compte (1798–1857)

The ecological, legislative, and technological context in which forests are grown and used is illustrated in Figure 3.1. By considering changes in these arenas, a better understanding of the forces that influence forest use and the effects of those forces on forest ecosystems can be gained.

Ecological Advances

Great advances have been made this century in the ecological understanding of how forests grow and interact with their environment. The roots of this understanding go back to the introduction of plant geography between 1765 and 1812, which explored the relationship between the spatial location of a forest and environmental gradients. In 1840, Liebig presented what is now know as the Law of the Minimum, which states that the growth and distribution of a species is dependent on the factor most limiting in the environment. [This theory was later modified by Shelford (1913) and Good

(1931, 1953) to acknowledge that various limitations act in concert with each another]. The first ecological textbook was authored by Warming in 1895. Merriam's concept of life zones (1894, 1898) set the stage for considering environmental factors responsible for the elevational and latitudinal distribution of species. Shortly thereafter, work by Cowles (1911) and Clements (1916) led to an interpretation of succession as the gradual changes in species composition and site conditions that occur in the aftermath of a disturbance. Gleason's (1926) perceptions on succession were contrary to the individualistic notions of Clements and led Gleason to introduce the idea of plant associations, which formed the basis for the concept of the *ecosystem*, a term coined by Tansley (1939). The concept of environmental gradients and their effect on plant distribution was demonstrated by Whittaker (1956).

The International Biological Program (IBP) marked an intense focus on biome properties that stretched from 1960 to 1974 (e.g., van Dobben and Lowe-Connell 1975; O'Neill et al. 1975; Levin 1976; Waring 1980). Experimental forests were established and monitored that demonstrated the intricate ecological relationships within an entire watershed (Bormann and Likens 1981). The large amount of data collected and analyzed during this period was complemented by a set of ecosystem models that were developed and exercised to assess the major components and interactions of

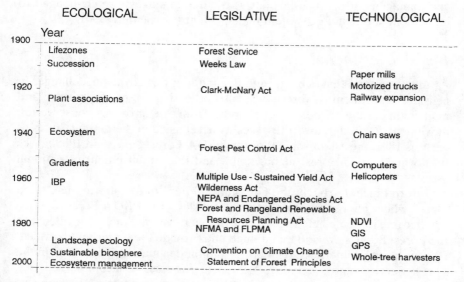

Year	ECOLOGICAL	LEGISLATIVE	TECHNOLOGICAL
1900			
	Lifezones	Forest Service	
	Succession	Weeks Law	
			Paper mills
1920		Clark-McNary Act	Motorized trucks
	Plant associations		Railway expansion
1940	Ecosystem		Chain saws
		Forest Pest Control Act	
	Gradients		Computers
1960	IBP	Multiple Use - Sustained Yield Act	Helicopters
		Wilderness Act	
		NEPA and Endangered Species Act	
		Forest and Rangeland Renewable	
1980		Resources Planning Act	NDVI
		NFMA and FLPMA	GIS
	Landscape ecology		GPS
	Sustainable biosphere	Convention on Climate Change	Whole-tree harvesters
2000	Ecosystem management	Statement of Forest Principles	

FIGURE 3.1. Landmarks in U.S. forestry. Acronyms are as follows: IBP (International Biological Program) NEPA (National Environmental Protection Act), NFMA (National Forest Management Act), FLPMA (Federal Land Policy and Management Act), NDVI (normalized-difference vegetation index), GIS (geographic information system), and GPS (global positioning system).

ecological systems. The discipline of systems ecology arose from the under-standing that resulted from these efforts. Forest-simulation models were developed that considered various factors affecting forest growth and de-velopment (Botkin et al. 1972). The IBP also stimulated research in evolu-tionary ecology and plant demography.

During the past two decades several new directions in ecological analysis have arisen. Landscape ecology, for example, focuses on the spatial rela-tions of sites and considers the size, shape, and location of ecosystems (Foreman and Godron 1986; Turner 1989; Turner and Gardner 1991). This perspective forces ecologists to consider more than just natural systems. The large spatial context of landscape ecology requires an acknowledge-ment of the role that anthropogenic influences have also had on shaping the landscape.

More recently, the concept of a sustainable biosphere (Lubchenco et al. 1991) has led ecologists to consider ways in which human actions can result in the long-term viability of an ecosystem and the planet. The more general need to protect the habitats of species that may become at risk led to the development of *ecosystem management* as a tool to consider the diversity of resources within a forest (FEMAT 1993; Slocombe 1993; Christensen et al. 1996). The basic concept of this approach is to maintain the critical func-tional aspects of the forest and riparian systems. These practices, now being adopted by the USDA Forest Service, lead to a holistic view of forest values and management practices. Today's resource managers are challenged to implement successful ecosystem management (Carpenter 1996).

Legislative Changes

At the same time that ecological understanding was advancing, changes in legislation pertinent to forests was also occurring. Steps to preserve and manage federally owned forests arose from the growing conservation movement. Although forest reserves were set aside in the United States as early as 1891, the establishment of the USDA Forest Service in 1905 marks the formal start of forest management and the establishment of national forests.

Gifford Pinchot brought the German traditions of forestry to the United States, where he was instrumental in establishing the USDA Forest Service and became its first chief. The multiple forest uses that Pinchot instilled into the Forest Service are still maintained and provide the underlying justifica-tion for having wilderness, recreation, and timber-management areas in the Forest Service lands of today. Yet, in many cases, noncommercial tree species are not recognized as valuable ecological members of the forest community, nor has there been long-term promotion of the nonmonetary value of forests (e.g. building soils).

During the twentieth century, a steady stream of legislation provided the means to establish and manage federally owned forests. The Weeks Law of

1911 authorized the purchase of forests that serve as watersheds for important rivers. In 1924, the Clarke-McNary Act expanded forest management to include fire protection and permitted the purchase of land for timber production. The Forest Pest Control Act of 1947 provided federal cooperation for states and private landowners to manage insect pests and diseases. The Multiple Use-Sustained Yield Act was passed in 1960 to require national forests to produce a sustained yield of timber while protecting other forest resources (reflecting some of Gifford Pinchot's goals for forest management). In 1964, the Wilderness Act was passed, which preserves wilderness areas within national forests where no roads or buildings may be built and timber harvest is prohibited. Because of numerous controversies that arose over the use of clear-cutting, the Forest and Rangeland Renewable Resources Planning Act of 1974 established procedures for the review and management of forests. In 1976, the National Forest Management Act (NFMA) and the Federal Land Policy and Management Act (FLPMA) further defined land and resource planning on the National Forests (Wilkinson and Anderson 1987) and focused attention on the need to manage riparian areas (Gregory 1997).

Two generally applicable laws have greatly influenced forest-management practices on federal lands. The National Environmental Protection Act (NEPA), passed in 1970, requires formal review before changes can be made on federal lands. The Endangered Species Act, passed in 1973, requires protection of threatened and endangered species. Legal actions to protect the red-cockaded woodpecker and the spotted owl have dictated forest-management practices in forests containing those species. In some locations, the forest managers must retain the forest structure necessary for these rare and endangered birds. Although these legal actions only pertain to sites where rare species occur, they have implications for forest management in general. In the case of the forests in the U.S. Pacific Northwest, a process has emerged that bases the management plan on socioeconomic values as well as ecosystem attributes of the forest (FEMAT 1993). This process involves deliberation between the interested parties and consideration of the spatial locations of forests of particular interest for specific values (Shannon 1997). For example, some locations are critical for timber harvest because of their location relative to sawmills. Other forests may have the old trees and large enough area to be important habitat for spotted owls.

The same conservation movement that spawned protection and management of federally owned forests also promoted responsible management of state and private forest lands. Between 1890 and 1917, state forest agencies were established, forest research and educational institutions were put in place, and pest-management programs were organized (Irland 1982). Beginning in the late 1940s and continuing to the present day, increased housing and business development have threatened some suburban forests. But a new conservation movement of the 1960s sparked an emphasis on open space. The Land and Water Conservation Fund and the beautification programs established under Lyndon Johnson's administration brought a new interest in forest acquisition for recreation and conservation. Beginning with the Oregon Forest Practices Act of 1972, some states have implemented laws regulating forest practices on state and private lands that provide some riparian protection as well as a diversity of criteria for forest management that varies from state to state (Gregory 1997). However, the adoption of a state forest-protection law is currently under debate in other states (e.g., Tennessee).

Today there is a conflict between the rights of private landowners and federal and state laws, epitomized by the Wise Use movement. Some private landowners feel that their management should be entirely open-ended, yet federal and state regulations may restrict some actions (e.g., in the case of wetlands). Again, this conflict is an arena in which ecological knowledge can contribute to development of a land-use ethic that would guide the standards used by landowners in protecting the diverse values of forests.

International trade practices and regulations also affect forest practices within different countries. For example, the concern for using wood that has been harvested from sustainable plantations rather than from old growth has lead for a need to define and certify sustainable-forestry practices. Also, forests and biodiversity issues received much attention as a part of the 1992 United Nations Conference on the Environment and Development (UNCED), which resulted in a number of instruments, including the Conventions on Climate Change and Biological Diversity and the Statement of Forest Principles. These agreements reflect the concerns for the sustainable use and management of biodiversity in forests in addition to bringing international pressure for the development of a means of determining the diversity of forests and estimating potential effects of specific management policies (Stork et al. 1997).

Technological Changes

Technological advances have influenced those laws that could be put in place and those aspects of ecological systems that could be analyzed. In the early 1900s, trees were cut by hand, horses or mules removed the logs from the forest, and rivers often provided long-distance transport (Karamanski 1989). Today's logging operations, however, rely on machines. Paper mills established by the 1920s greatly increased the demand for wood, particularly from small trees. During the 1920s, motorized trucks became common in the forests in addition to the railway expansion, which also had direct and indirect effects on forests. Trains could be used to transport trees from the forest to the mill, and trains also allowed food products to be shipped from the Midwest, permitting the less profitable agricultural lands in the eastern United States to revert to forests. However, in the process of establishing the railway system many trees were harvested and used for railway ties. All of these technological advances expanded the area cleared and increased the number of trees harvested from a site.

The chain saw may be the invention that most influenced the amount of wood taken from a site. The chain saw was invented in 1926 but did not become prevalent in the United States until the 1940s (Hall 1977). The chain saw not only allowed more wood to be cut in a set time but also permitted more trees to be cut within a stand. Thus, not only was timber more easily harvested, but more branches and cull trees were cut for firewood or other purposes. Thus, chain saws bring about a "cleaner" forest but one with less dead wood, which serves as animal habitat and is part of the nutrient-cycling and water-purification processes. Although there was great concern that the chain saw would replace jobs, more people are now employed in the timber industry than were at the turn of the century. Also, chain saws have made the work of logging much more efficient and safe.

The economic value of chain saws was made clear in a recent survey of farmers in Rondonia, Brazil (Dale and Pedlowski 1992). In 1991, none of the eighty-nine farmers we surveyed owned a chain saw. Shortly thereafter, however, the successful candidate for governor of the state gave away chain saws as part of his election campaign. The effect on forests of the rapid introduction of chain saws into the region is just beginning to be observed. The major tool for observing these changes is satellite imagery.

Satellite imagery in now commonly used in the planning process and to monitor changes in forest resources over time. The development of the normalized-difference vegetation index (NDVI) that relates reflectance, as measured in the image, to forest conditions was the breakthrough that made remote imagery a useful tool (Tucker 1979). The use of global positioning sysytems (GPS) now allows accurate recording of a location on the ground and therefore augments the value of remote imagery. Satellite imagery is particularly important for international forestry. As late as 1990 (Dale 1990), we still did not know how much of the Earth's surface was covered by forests. Also, the extent of large-scale natural disturbances and anthropogenic clearing was not known. For example, a large fire burned millions of hectares of land in southeast Asia but the area was not even known to be burning until it was identified in a satellite image.

The use of helicopters and small planes in forests became a part of some forest operations in the 1970s. Aircraft are used to distribute seeds, control fires, perform controlled burns, spray herbicides, spread fertilizers, and occasionally to harvest trees from otherwise inaccessible sites. Because of their high cost, helicopters and planes are not routinely used except for the cost-effective applications of herbicides and fertilizers.

The development of computers has also changed the logging industry. The control room in modern paper mills contains sophisticated computers to monitor and alert the operators of all aspects of the papermaking process. Computers are also used to decide how to cut a tree to maximize the value of the wood and minimize waste. Computers allow for the development and use of detailed simulation models of forest development. These models project situations in which sets of species grow and compete for resources. Geographic information systems (GIS) are the hardware and software combination that allows spatially explicit information to be integrated into a map of site conditions. The combination of simulation models with GIS technology allows the spatial arrangement of trees to be analyzed. The models have been used both for improving understanding of forest interactions and for planning harvest and selection regimes. As mentioned before, this spatial integration of information is critical to the adaptive management policy now being espoused for forest management (FEMAT 1993).

Machines for whole-tree harvesting are routinely used in Finland and Sweden to harvest trees in such a fashion that the soil and roots are not disturbed, minimizing impacts on soil runoff and nutrient cycling (Figure 3.2). Furthermore, the trees can be cut and lifted with these machines so that damage to adjacent trees is negligible, retaining the economic value of those trees, while preserving the forest as a habitat for animals.

Overall, some technologies (e.g. chain saws) have resulted in cleaner forests that are less likely to support such ecosystem characteristics as habitat and water purification. Whole-tree harvesters may be the best example of a technology that retains many of the ecosystem features of forests.

FIGURE 3.2. The machine used for whole-tree harvesting removes the above-ground portion of the tree with minimum impact to the soil or surrounding vegetation.

Integrating the Landmarks in Forest Management

Taken together, these ecological, legislative, and technological advances during the past two centuries allow foresters to implement a diversity of management practices. But just because the means are available does not always imply that the practice will be used. For example, helicopter logging now allows trees to be removed from slopes more steep and inaccessible than was previously possible, but helicopters are generally not cost effective and are rarely used for harvesting. In addition to technological advances, awareness of some basic ecological concepts as well as legislative restrictions frame today's management practices. A review of the ecological concepts taught to foresters and their management implications clarifies how ecological knowledge has helped forest managers understand the diversity of forces that determine which species grow where, how forests develop, and how disturbances affect forests. These concepts were derived by examining forestry textbooks (e.g., the text by Oliver and Larsen 1996):

Gradient concept: Environmental gradients are partially responsible for the distribution of species, ecosystems, and biomes. Use: Sometimes silvicultural practices take advantage of the ecological understanding of gradients and encourage planting or maintenance of species mixes and densities appropriate for a site and a region. Often, however, these natural gradients are ignored, and monocultures of trees are planted over large areas.

Limitation concept: Leibig (1840) introduced the concept that the growth and distribution of species depend on the most limiting environmental factor. This Law of the Minimum is now understood to imply that most factors provide both upper and lower limits and frequently factors act in concert. Use: Forest management for sustained yield depends upon this concept. For example, thinning regimes are put in place to minimize competition for scarce resources and thus to maximum tree volume for a given stand age.

Disturbance: Such disturbances such as fire, windthrows, and insect outbreaks, are a natural part of many forests and affect forest structure, composition, and function. Use: Forestry, more than any other area of resource extraction, manages for the long term (with rotation times ranging from ten to 110 years). Thus, foresters must recognize the potential for a disturbance to occur within a management cycle. Practices are put in place to either manage the disturbance (e.g., by fire breaks) or manage the system so that it is less susceptible to a disturbance (e.g., by reducing the amount of flammable litter or thinning trees so they have less probability to be attacked by beetles).

Succession: Succession is the gradual replacement over time of one species by another that results in an increase in biomass in forests, a change in species-life-history characteristics, a change in light levels, a switch from high to low net primary production, and an alteration of the storage of nutrients from the soil to the standing biomass. Use: Knowledge of successional processes pervades forest management. Some practices create initial stages of succession by spreading seeds, planting seedlings, or leaving seed trees in place. The overall goal of silviculture is to artificially move stands through seral stages by such activities as thinning. Harvesting is scheduled to maximize profitability of timber yield by cutting trees prior to their peak in net primary production. Harvesting trees at a time when much of the nutrients are still being cycled within the soil and leaving branches onsite reduce the loss of nutrients from the site.

Watershed concept: Forests cleanse water and reduce runoff. Changes in upland forests affect riparian systems. Use: Some management practices are designed to enhance water quality and quantity, reduce soil erosion, and maintain the riparian system. Trees are not cut on steep slopes or adjacent to rivers. Downed branches are left onsite, and the litter layer is disturbed as little as possible. Stream-bank corridors of trees are left in place, and logging roads are maintained by providing proper drainage that reduces the chances of landslides into streams.

Structure concept: Many ecological attributes can be linked to structural elements of the forest (e.g., size of both live and dead trees, and canopy configuration). Therefore, many functional and compositional features of the forest can be preserved by maintaining structural features. Use: Ecosystem management is largely built upon maintaining or establishing structural features. This approach suggests ways that forest products can be harvested, while structural features, such as large or downed trees, can be left in place.

Recent Trends in Ecosystem Science

The major advances in ecosystem science in recent decades have been built upon these concepts. In the 1960s, forest management operated under the belief that actions to promote timber production are not necessarily good for other forest-related values. However, ecosystem science has demonstrated that structural complexity is an important feature of natural forest ecosystems and that complexity is created naturally through disturbances. Because these disturbances occur on a scale larger than a stand, a landscape view was adopted. Traditional practices of clear-cutting even-aged stands do not recognize the importance of structure or disturbance in shaping a forest. Major questions for ecologists that accept these concepts are, therefore, how much structural complexity is needed and how can management activities (e.g. logging) be arranged in time and space to promote these features. Finally, it is important to know the relationships between forest stands and nonforest sites as well as the relative importance of those relationships.

This review shows that many forest practices are built upon ecological concepts and vice versa. Much of what ecologists know about forests has been learned by working with foresters. Even so, a gap remains between present ecological understanding and forest-management practices. For example, this gap is evident in forest-management practices that are frequently focused on timber harvesting in small parcels, instead of considering multiple uses of the entire landscape. Unfortunately, some recent textbooks also focus on management for timber at a stand level rather than promoting ecosystem-management practices (e.g. Oliver and Larsen 1996).

However, new forestry approaches are also being developed to promote practices that will enhance ecosystem properties (Kohm and Franklin 1997). For example, long rotations and variable-retention harvesting will both maintain ecologically important structural features of forests while allowing harvest to continue (Franklin et al. 1997). Alternatively, previously unproductive agricultural lands can now grow fiber for energy with intensive short rotations (e.g. ten years). Proposals for these management practices take many forms, but three basic ecological impacts are (1) a return to the size structure more typical of natural forests, (2) consideration of long-term and broad-scale implications of forest management, and (3) the perspective of the forest as an interacting system.

The public understanding of ecosystem processes is also deficient. The public typically associates ecology with the environmental movement and does not understand such critical aspects of ecosystems as feedbacks, changes over time and space, or the concept of a system itself. For example, Naiman (in press) points out that the public does not understand how water affects land processes and vice versa.

Challenges Ahead for Managing Forests as Sustainable Ecosystems

The position statement posed by the editors of this volume is that the transfer of ecological knowledge into forestry practices has been a success story. However, I would have to respond to this statement in a qualified manner. Today, many forestry practices have developed from ecological studies, but it is also clear that forest practices are not based on the most recent ecosystem science. Rather, forestry is a success story for ecosystem science in the sense that there is an established process for moving ecological theory into practice in our national forests as well as in the private forest industry. However, this process takes time and is largely based on personal interchange and experiences. An example of this transfer is given by looking at the authors of the recent book on *Creating a Forestry for the 21st Century* (Komb and Franklin 1997). Many of those authors are from the research branch of the USDA Forest Service. A gap exists between the practicing foresters and the research foresters but because some members of these two groups are in the same organization and professional societies, this gap is not as wide as it is in other fields.

Finally, a wide difference exists in application of forest-protection laws. For example, in the United States, some state laws protect streams, whereas other states have no such protection. The challenge for ecosystem scientists is not just to publish their papers, but also to present their new ideas and understanding in a way that is useful and applicable for resource managers.

As I close this paper, I am reminded of the words said by my father, who was a researcher, a surgeon, and somewhat of a poet:

Now I hear the closing music. I see the autumn leaves. I remember the be-all and end-all of the surgeon. I see the green clad figures converge upon the still form beneath the brilliant overheads. . . . I balance the anguish of the bereaved against the happiness of the salvaged success. I bow beneath my failure and search for surer judgement, wider knowledge, and greater skill. Finally, supported by education, backed by training, and borne upward by my association with you [my colleagues], the reason for research and the end-point for experience is revealed. I see the patient who relies upon us. (Dale 1981)

These words make me wonder if we as ecologists recognize the ultimate use of our research. Do we see the natural-resource manager as a "patient" who needs the understanding and tools we develop? Do we recognize the environment as the be-all and the end-all of our efforts? I believe the answer is both yes and no. Natural-resource management is much improved by the many contributions of ecologists. Yet, too often, we as ecologists do not consider the end use of our work. We may be involved in investigating interesting scientific hypotheses but may not consider the management implications of our research. Often, the working forester or land manager is most familiar with pertinent land-management issues. Yet, ecologists discuss their research with managers too infrequently. Therefore, more contact is needed between these two endeavors. Addressing this issue strikes at the balance between directed and undirected research programs. Much of the applied forestry research is funded by the USDA Forest Service and private timber companies that have a clear mandate to meet current management regulations. Although some of the more innovative forestry approaches have arisen out of basic research endeavors, it may be years before the knowledge is applied. Programs directed at resolving particular forestry issues may provide the best balance between basic and applied research. For example, the Forest Ecosystem Management Team (FEMAT) took on the task of defining ecosystem management in such a way as to make use of the latest ecological thinking (FEMAT 1993).

Products from our research efforts need to be evaluated in the context of reaching the relevant audience. Social scientists can help ecologists identify the characteristics of the interested parties and those forest values that are important to them. Scientific papers may elegantly lay out results, but these results may not be attainable by land managers. Too many times our products are immersed in jargon, complicated analysis or theory, with the management implication being unclear. A recent analysis of tools used by environmental decision makers shows that the results of research efforts need to be available to managers and not just described in papers (Dale and English 1998). Thus, the challenges ahead for scientists are to find ways to deliver research results to the forest manager.

Irland (1982) has summarized future trends in forest use: Suburban sprawl will continue. The values that the public attributes to forests will change, with many people recognizing multiple uses of forests and others appreciating only wood products. Demand for wood and forest products will increase with a corresponding rise in prices that will add to the existing pressures to harvest forests. Speculative booms in rural land will diminish the size and increase the turnover of privately owned forest parcels. Reduced amounts of private land will be available for hunting, fishing, and hiking. Mechanization of the forest industry will reduce the number of jobs.

Up to now, for some small landowners forest management for timber in the United States has reflected the lack of wood shortage. Future trends will exert pressures to manage forest lands more efficiently. Presently, ineffective thinning, sloppy harvesting, and inappropriate clear-cutting have led to loss of productivity. Of the 21% of U.S. land that is managed as commercial forest (195 million hectares or 480 million acres), only 8% was seeded or planted between 1950 and 1978 (Oliver and Larsen 1996). The rest was allowed to regenerate naturally. Rapid turnover in land ownership, land fragmentation, and land-use conversion result in loss of forests or erode the value of forests for wildlife habitat or as watersheds. Owners of large amounts of land, on the other hand, have been increasing productivity significantly through improved genetics, fertilization, planting, and more efficient harvesting.

Ecological science offers a long-term and broad-spatial perspective that is essential to sustainable management of forest resources. The long view instills a sense of responsibility for the forest resources. Forestry companies that own rather than their lands tend to cut and replant the forest in such a way that promotes long-term benefits. When land management is performed for the long term, the ecological interactions are an integral aspect of the changes in the forest. Yet even in the short term, forest resources can be appropriately managed based on a knowledge of succession, effects of gradients, environmental limits, and the roles of disturbance and watershed properties. It is the responsibility of science to educate land managers about how these principles come into play in forest management, and it is the responsibility of the public to challenge land managers to develop an ethic that can preserve the ecosystem features of the forests. Ethical resource management built upon sound ecological principles is the key to appropriate forest stewardship.

66   V.H. Dale

Acknowledgements. The author appreciates comments on an earlier draft of the paper from Juan Armesto, Jan Bengtsson, Charles Canham, Jerry Franklin, Robin Graham, Peter Groffman, Erik Lilleskov, Gary Lovett, and Richard Ostfeld. Donald Zak suggested some useful references. Thanks also to Fred and Linda O'Hara for an excellent job of editing the draft manuscript. Oak Ridge National Laboratory is managed by Lockheed Martin Energy Research, Corp. for the Department of Energy under contract DE-AC05-96OR22464. This is Environmental Sciences Division publication number 4684.

References

Bormann, F.H., and G.E. Likens. 1981. *Pattern and process in a forested ecosystem.* Springer-Verlag, New York.

Botkin, D.B., J.F. Janak, and J.R. Wallis. 1972. Some ecological consequences of a computer model of forest growth. *Ecology* 60:849–872.

Carpenter, R.A. 1996. Ecology should apply to ecosystem management. *Ecological Applications* 6:1373–1377.

Christensen, N.L., A.M., Bartuska, J.H. Brown, S.R. Carpenter, C. D'Antonio, R. Francis, et al. 1996. The report of the Ecological Society of America committee on the scientific basis for ecosystem management. *Ecological Applications* 6:665–691.

Clements, F.E. 1916. *Plant succession: an analysis of the development of vegetation.* Carnegie Institution of Washington Publication 242: Carnegie Institution of Washington, Washington, DC.

Cowles, H.C. 1911. The causes of vegetation cycles. *Botanical Gazette* 51:161–183.

Dale, V.H. 1990. Strategy for monitoring the effects of land use change on atmospheric CO_2 concentrations. Pages 422–431 in Proceedings of *Global natural resource monitoring and assessments: preparing for the 21st century*, Venice, Italy, September 25–30, 1989. American Society for Photogrammetry and Remote Sensing, Bethesda, MD.

Dale, V.H., and M.R. English. 1999. Tools to aid environmental decision making. Springer-Verlag, New York.

Dale, V.H., and Pedlowski, M.A. 1992. Farming the forests. *Forum for Applied Research and Public Policy* 7:20–21.

Dale, W.A. 1981. *Then and now.* Presentation to the International Society of Surgery, Dallas, TX.

FEMAT. 1993. *Forest ecosystem management: an ecological, economic and social assessment.* U.S. Government Printing Office, Washington, DC.

Foreman, R.T.T., and M. Godron. 1986. *Landscape ecology.* John Wiley & Sons, New York.

Franklin, J.F. 1989. Toward a new forestry. *American Forestry* 11:37–44.

Franklin, J.F., D.R. Berg, D.A. Thornburgh, and J.C. Tappeiner. 1997. Alternative silvicultural approaches to timber harvesting: variable retention harvest systems. Pages 111–140 in K.A. Kohm, and J.F. Franklin, eds. *Creating a forestry for the 21st century.* Island Press, Washington, DC.

Gleason, H.A. 1926. The individualistic concept of the plant association. *Bulletin of the Torrey Botanical Club* 53:1–20.

Good, R.E. 1931. A theory of plant geography. *The New Phytologist* 30:149–203.

Good, R.E. 1953. *The geography of the flowering plants, 2nd edition.* Longmans, Green, and Co., New York.

Gregory, S.V. 1997. Riparian management in the 21st century. Pages 69–86 in K.A. Kohm and J.F. Franklin, eds. *Creating a forestry for the 21st century.* Island Press, Washington, DC.

Hall, W. 1977. *Barnacles Parp's chain saw guide.* Rodale Press, Emmaus, PA.

Irland, L.C. 1982. *Wildlands and woodlots: the story of New England's forests.* University Press of New England, Hanover, NH.

Karamanski, T.J. 1989. *Deep woods frontier: a history of logging in northern Michigan.* Wayne State University Press, Detroit, MI.

Kessler, W.B., H. Salwasser, C.W. Cartwright, Jr., and J. Caplan. 1992. New perspectives for sustainable natural resource management. *Ecological Applications* 2:221–225.

Kohm, K.A., and J.F. Franklin, eds. 1997. *Creating a forestry for the 21st century.* Island Press, Washington, DC.

Leopold, A. 1949. *A Sand County Almanac.* Oxford University Press, New York.

LeGuin, U.K. 1972. *The Word for World is Forest.* Berkley Publishing Corporation, New York.

Levin, S.A., ed. 1976. *Ecological theory and ecosystem models.* The Institute of Ecology, Athens, GA.

Lewis, T.E., and B.L. Conkling. 1994. *Forest health monitoring: southeast loblolly/ shortleaf pine demonstration interim report.* EPA Project Summary, Report Number EPA/620/SR-94/006, Washington, DC.

Liebig, H. 1840. *Chemistry in its agriculture and physiology.* Taylor and Walton, London, U.K.

Lubchenco, J.A., A.M. Olson, L.B. Brubaker, S.R. Carpenter, M.M. Holland, S.P. Hubbell et al. 1991. The sustainable biosphere initiative: An ecological research agenda. *Ecology* 72:371–412.

Merriam, C.H. 1894. Laws of temperature control of the geographic distribution of animals and plants. *National Geography* 6:229–238.

Merriam, C.H. 1898. Life zones and crop zones of the United States. United States *Department of Agriculture Biological Survey Division Bulletin* 10:9–79.

Naiman, R. 1998. Watersheds, landscapes, and society. *Landscape Ecology* (in press).

National Research Council. 1992. *Conserving biodiversity: a research agenda for development agencies.* National Academy Press, Washington, DC.

Oliver, C.D., and B.C. Larsen. 1996. Forest stand dynamics. John Wiley & Sons, New York.

O'Neill, R.V., W.F. Harris, B.S. Ausmus, and D.E. Reichle. 1975. A theoretical basis for ecosystem analysis with particular reference to element cycling. Pages 28–40 in F.G. Howell, J.B. Gentry, and M.H. Smith, eds. *Mineral cycling in southeastern ecosystems.* ERDA-CONF-740513. U.S. Energy Research and Development Administration, Washington, DC.

Post, W.M., T.-H. Peng, W. Emanuel, A.W. King, V.H. Dale, and D.L. DeAngelis. 1990. The global carbon cycle. *American Scientist* 78:310–326.

Russell, C., V.H. Dale, M. Hadley, M. Kane, and R. Gregory. 1998 Applying multi-attribute utility techniques to environmental valuation: a forest ecosystem study. Vanderbilt Institute of Public Policy Working Paper.

Salwasser, H. 1990. Gaining perspective: forestry for the future. *Journal of Forestry* 88:32–38.

Shannon, M.A. and A.R. Antypas. 1997. Open institutions: uncertanty and ambiguity in 21st-century forestry. Pages 437–446 in K.A. Kohm, and J.F. Franklin, eds. *Creating a forestry for the 21st century.* Island Press, Washington, DC.

Shelford, V.E. 1913. *Animal communities in temperate America.* University of Chicago Press, Chicago, IL.

Slocombe, D.S. 1993. Implementing ecosystem-based management. *BioScience* 43:612–622.

Stork, N.J., T.J.B. Boyle, V. Dale, H. Eeley, B. Finegan, M. Lawes, et al. 1997. *Criteria and indicators for assessing the sustainability of forest management: Conservation of biodiversity.* Center for International Forestry, Bogor, Indonesia.

Swanson, F.J., and J.F. Franklin. 1992. New forestry principles from ecosystem analysis of Pacific Northwest forests. *Ecological Applications* 2:262–274.

Tansley, A.G. 1939. The plant community and the ecosystem. *Journal of Ecology* 27:513–530.

Tucker, C.J. 1979. Red and photographic infrared linear combinations monitoring vegetation. *Remote Sensing of Environment* 8:127–150.

Turner, M.G. 1989. Landscape ecology: the effect of pattern on process. *Annual Review of Ecology and Systematics* 20:171–197.

Turner, M.G., and R.H. Gardner. 1991. *Quantitative methods in landscape ecology: the analysis and interpreation of landscape heterogeniety.* Springer-Verlag, New York.

van Dobben, W.H., and R.H. Lowe-Connell, eds. 1975. *Unifying concepts in ecology.* W. Junk, The Hague, The Netherlands.

Waring, R.H., ed. 1980. *Forests: fresh perspectives from ecosystem analysis.* The fortieth annual biology colloquium. Oregon State University Press, Corvallis, OR.

Warming, J.E. 1895. *Plantesamfund, Gruntraek af den Okologiske Plantegeografi.* Philipsen, Copenhagen, Denmark.

Wilkinson, C.F. and H.M. Anderson, 1987. Land and Resource Planning in the National Forests. Washington, D.C. Island Press.

Whittaker, R.H. 1956. Vegetation of the Great Smoky Mountains. *Ecological Monographs* 26:1–80.

USDA Forest Service. 1992. *Ecosystem management of the national forests and grasslands.* Memorandum 1330–1. USDA Forest Service, Washington, DC.

4
Wastelands to Wetlands: Links Between Habitat Protection and Ecosystem Science

Joy B. Zedler, Meghan Q. Fellows, and Sally Trnka

Wetlands are a vital element in the biosphere, but they are disappearing and being degraded rapidly. Until recently, they have been regarded largely as nuisances to be drained, cleared, filled, or inundated. Now we have begun to realize that in their natural state wetlands produce numerous benefits for society, benefits which are either irreplaceable if lost or can only be replaced at great expense.

Secretary of Interior Bruce Babbitt

Summary

The use of science in protecting the functions of wetland ecosystems has been viewed as a success story. We looked for evidence that scientific efforts affected two important actions: (1) the National Wildlife Refuge System was expanded substantially in the 1930s, including many wetlands, (2) legislation in 1972 protected the "waters of the United States" from dredging and filling. The first action was catalyzed by the droughts and catastrophic declines of waterfowl populations during the 1930s. A few key scientists were positioned to acquire and protect wetlands, making choices based on scientific opinions about waterfowl populations. In the 1960s, ecosystem science played a stronger role, with research on freshwater wetlands revealing water quality improvement functions, and studies of tidal marshes linking coastal fisheries to high productivity in salt marshes. Wetland protection followed from strong statements made about these links, even though the science base was limited. The methods by which wetland scientists influenced policy are obscure, because few scientists record such activities. We provide a few case studies of how ecologists have translated science into management action.

The effect of wetland protection on science and wetland science on ecosystem science are also viewed as successes. Hundreds of papers on wetland functioning have appeared since 1972, and the science now supports more than one journal.

Introduction

Public attitudes about wetlands have undergone an enormous transformation from being considered wastelands to gaining acceptance as highly valued ecosystems. Our task was to explore the role of wetland science in the metamorphosis. Today, wetlands (see Box 4.1) are regarded widely as conducting three sets of functions that serve society: habitat or food-chain-support functions, water quality improvement, and hydrologic functions, such as flood abatement and shoreline stabilization. These values are recognized by a large segment of the public, and legislation passed in the 1970s protects the nation's wetlands from many detrimental activities. A large amount of research has been undertaken to elucidate wetland ecosystem functions, and a science of wetlands has emerged. This chapter explores the relationship between the recognition of wetland values, the passage of wetland-protection legislation, and the development of wetland science.

Box 4.1
Terminology

Before the term "wet land" (later wetland) became widely used, individual habitats were referred to by name (e.g. marsh, swamp, bog, fen, mire). The inclusive term was used widely in a National Wildlife Federation campaign in 1955, but there was no clear definition until protective legislation made it essential. The USDI Fish and Wildlife Service published the following definition with its 1979 classification scheme:

Wetlands are lands transitional between terrestrial and aquatic systems where the water table is usually at or near the surface or the land is covered by shallow water. For purposes of this classification wetlands must have one or more of the following three attributes: (1) at least periodically, the land supports predominantly hydrophytes; (2) the substrate is predominantly undrained hydric soil; and (3) the substrate is nonsoil and is saturated with water or covered by shallow water at some time during the growing season of each year (Cowardin et al. 1979).

A regulatory definition of wetlands was developed for use by the U.S. Army Corps of Engineers and the Environmental Protection Agency (EPA). It is more restrictive and is often referred to as the definition for jurisdictional wetlands (i.e. those for which permits were needed before dredging or filling could be undertaken under Section 404 of the Clean Water Act):

Those areas that are inundated or saturated by surface or goundwater at a frequency and duration sufficient to support, and that under normal circumstances do support, a prevalence of vegetation typically adapted for life in saturated soil conditions. Wetlands generally include swamps, marshes, bogs, and similar areas (EPA, 40 CFR 230.0 and CE, 33 CFR 328.3)

In 1993, Congress asked the EPA to have the National Academy of Science define wetlands to help settle an ongoing debate. The scientific community reached consensus on the following definition:

A wetland is an ecosystem that depends on constant or recurrent, shallow inundation or saturation at or near the surface of the substrate. The minimum essential characteristics of a wetland are recurrent, sustained inundation or saturation at or near the surface and the presence of physical, chemical, and biological features reflective of recurrent, sustained inundation or saturation. Common diagnostic features of wetlands are hydric soils and hydrophytic vegetation. These features will be present except where specific physio-chemical, biotic, or anthropogenic factors have removed them or prevented their development (NRC 1995).

Debate continues, but the focus has shifted toward the specific criteria that should be used to delineate wetlands, the degree to which there should be regional standards, and the exact functions (processes) and values (functions of benefit to society) that different wetlands provide.

The connotation of the word wetland has not always been positive. Some have equated the term with wasteland, an association that has persisted. William Patrick Jr. (1994) recalls a 1980 (approximately) submission of a scientific proposal on the movement of nutrients from uplands to wetlands, which was then described as a study of waste-lands by a federal agency scientist. Others have confused wetlands with ponds, assuming that they must always have standing water. The term "wetland" indicates a system with a dual nature, neither terrestrial nor aquatic. "Wetlands are uniquely both, but are rarely treated as such" (Williams 1990). Jon Kusler et al. (1994) recently called attention to the confusion over wetland hydrology and stated in clear terms that wetland water levels fluctuate.

A major question today is, how wet is wet? (EDF and WWF 1992). Legislation before congress (HR1330, SB851) would require that jurisdictional wetlands have water at the soil surface for twenty-one consecutive days during the growing season. The current standard is that the rooting zone be saturated for >12.5% of the growing season. It has been estimated that the proposed change would exclude protection for at least 50% of the nation's jurisdictional wetlands (EDF and WWF 1992).

We began by reviewing the roots of wetland protection, that is, the conservation movement. Until wetlands became identified with specific values, there was little distinction between wetland and upland conservation efforts. Wetlands and uplands alike were degraded as European settlers moved west across North America, but the waters of the United States, and eventually the associated wetlands, were singled out for protection. Scientists played a role in the process (see Box 4.2), but too few of their activities have been recorded for a thorough evaluation of the methods of information transfer.

Box 4.2
How individuals have protected wetlands

Ordinary citizens and scientists have influenced wetland protection by:

Writing a book. By 1908, Okefenokee Swamp was threatened by major logging. In 1918, a small, local group organized as the Okefenokee Preservation Society. Their mission was to "secure its reservation and preservation for public, educational, scientific and recreational uses." In 1926, McQueen and Mizell, a local lawyer and lumberman, published *History of the Okefenokee Swamp*, which initiated public interest and newspaper coverage. In 1931, the Conservation of Wildlife Resources committee was sent to the site by the U.S. Senate. In 1937, Okefenokee National Wildlife Refuge was established. In 1974, it was given higher protection as a wilderness area (Thomas 1976).

Buying the land. In 1959, The Great Swamp of New Jersey was threatened by New York Port Authority plans for a jetport. Citizens formed "The Great Swamp Committee of the North American Wildlife Federation." They pooled money and purchased or donated land. In 1960, 3,000 acres were given to the U.S. Fish and Wildlife Service, creating the Great Swamp National Wildlife Refuge (Thomas 1976).

Organizing, publicizing, and advising. Charles Wharton, a biology professor at Georgia State University documented habitat functions, estimated their dollar value, and recorded the efforts of many individuals involved in protecting Georgia's Alcovy Swamp (Wharton 1973). The editor of Georgia's Game and Fish Commission publication spoke against the project at a public hearing and initiated a debate with the town newspaper editor. The subsequent Commission magazine featured "an exposé on channelization." Flyers and bumper stickers appeared in opposition of channelization. The Georgia Natural Areas Council showed officials of the Soil Conservation Service (SCS) the problems of channelization—spoil banks, dead trees, and disrupted fish habitat. The Georgia Conservancy was organized, and in 1969 published a report on the Alcovy Swamp; they also asked Governor Maddox to study the issue, and he gave his approval for the Conservancy to "assault the Washington bulwarks."

Congressman Blackburn, who represented a district downstream of the Alcovy (which would receive sediment from the proposed dredging), took a stand against the project and wrote articles for The Wilderness Society and *Field and Stream* magazine. It became clear that the beneficiaries of swamp drainage would be double-dipping in federal funds: They would receive aid for drainage in order to bring land into production; they would also receive aid for keeping land out of production by putting it in the soil bank. Momentum grew, and the congressman arranged for eight Georgia Conservancy members to meet with the SCS Director in Washington, D.C., but to little avail. At the Department of Interior, however, there were more receptive ears, as the wetland drainage efforts of SCS conflicted with waterfowl conservation efforts from the Duck Stamp program. An August 6, 1969, a letter to the Secretary of Interior recommended that the channelization plan be reevaluated. The SCS agreed in September and restudy began. It was a turning point and a reprieve for the Alcovy River (Wharton 1973).

Drafting legislation. Three ecologists played key roles in drafting the 1972 Water Pollution Control Act (Likens, personal communication). Shortly after completing his Ph.D. under R.H. Whittaker, Walter Westman was selected to be a "Congressional Fellow." Funds for this position had been raised by faculty and friends at Cornell University. Westman was assigned to the Senate Committee on Public Works, chaired by Senator Jennings Randolph. George Woodwell and Gene Likens were members of an advisory panel, whose task was to advise on legislation for clean water and other environmental issues. The bill was drafted by committee staff, including Westman, the Minority Counsel Tom Jorling (who had a master's degree in ecology), and Majority Counsel Leon Billings. This team, with its technical expertise, drafted some very strong terminology (e.g. "The objective of this Act is to restore and maintain the chemical, physical, and biological integrity of the Nation's waters. In order to achieve this objective it is hereby declared that . . . it is the national goal that the discharge of pollutants into the navigable waters be eliminated by 1985 . . . ").

Our review of the role of wetland science in achieving wetland protection was guided by alternative hypotheses, that wetland-protection legislation came about because of wetland science or that legislation drove the science. To uncover the cause-effect relationships, we emphasized the timing of different efforts: the syntheses of scientific research, the publication of papers on wetlands, the abundance of popular writings on wetlands, legislative acts, and newspaper accounts of wetlands.

American Attitudes About Wetlands

The first Europeans to arrive in North America settled near river-fed, tidally flushed ecosystems, which are called estuaries. Although this gave them access both to drinking water and navigable waters, it also introduced them to several problems presented by estuarine "real estate." Farming was difficult in wet soils, flooding was often a threat, and people were subject to the problems of mosquitoes, snakes, alligators (Patrick 1994), and "swamp fever," that is, typhoid, diphtheria, and cholera (Siry 1983). Various corrective measures were undertaken, such as drainage to make wetlands cultivable, dredging to control flooding, and filling to raise the land. Reclaiming estuaries (i.e. draining the wet lands) was considered a sign of progress in the late 18th century (Siry 1983). Dredging and filling became common and eventually these practices were encouraged in formal legislation. Collectively, the Swampland Acts of 1849, 1850, and 1860 gave lands that were unfit for cultivation to fifteen states for implementation of flood control measures. The result was further drainage and conversion to upland. In the swamps and bottomland forests of the southeastern United States, logging was an equally widespread agent of destruction (Thomas 1976).

Out of this destruction, however, came the beginnings of a land ethic. Writers began to record their concerns that our natural resources were being ravaged and that our natural heritage was being lost (Table 4.1). Ecologists (e.g. Meine 1988) credit Aldo Leopold for clarifying the message: "A land ethic, then, reflects the existence of an ecological conscience, and this in turn reflects a conviction of individual responsibility for the health of the land" (Leopold 1949). As one historian put it, "Throughout the rise of conservation as a national policy in the United States, it is *individuals* who have led the way by awakening the American people to the need for caring for the land, and whose writings and political efforts have fostered effective government policies" (Strong 1971).

Wetlands as Habitat

Among the earliest organized attempts to change the nation's habits were sportsmen. Fishing enthusiasts were concerned about aquatic habitat degradation as early as the 1840s, and fish hatchery research was underway by the 1850s in an attempt to replenish declining fish populations. Among the influential spokespersons was Massachusetts biologist, Louis Agassiz, who

TABLE 4.1. Early concerns about lost habitat.

1851:	Henry David Thoreau said . . . "in Wildness is the preservation of the World." Nash (1967) identified this as a new concept to Americans.
1864:	George Perkins Marsh published "Man and Nature; or, Physical Geography as Modified by Human Action"—a politician and naturalist who recognized the destructiveness of human activities (Strong 1971).
1872:	Yellowstone became a National Park—the first large-scale wilderness preservation in the public interest (Nash 1967). However, supporters of the bill before congress justified the park on the basis of its "uselessness to civilization," rather than its value as wilderness (Nash 1967).
1870s–early 1920s:	John Muir wrote essays and books in defense of wilderness; he challenged and opposed the "use" of wild lands (Nash 1967). In 1897, he and Gifford Pinchot clashed over the issue of sheep grazing in forest reserves (Nash 1967).
1913:	Gifford Pinchot, former Chief Forester, said to the U.S. House of Representatives in a hearing concerning the Hetch Hetchy dam site: "Now, the fundamental principle of the whole conservation policy is that of use, to take every part of the land and its resources and put it to that use in which it will best serve the most people. . . ." (Nash 1970).
1920s:	Post-war attitudes indicated that nature was valued for outdoor recreation, but not as wilderness per se; emphasis was still on use (Nash 1967).
1934:	A President-appointed Committee on Wildlife Restoration (Thomas Beck, J.N. Darling, and Aldo Leopold) reported Bureau of Biological Survey data, leading to a national program of waterfowl restoration (Eisenhower and Chew 1935).
1930s and 1940s:	Aldo Leopold wrote articles on the need to respect nature, the land ethic, and the need to develop an ecological conscience (Leopold 1949; Meine 1988; Flader and Callicott 1991). He helped form the Wilderness Society and, in addition to his many writings on game and habitat conservation, he criticized the straightening of rivers and the drainage of wetlands.

helped anglers influence the formation of the U.S. Fish Commission in 1971 (Siry 1983). Simultaneously, hunters blamed the smaller duck populations on declining wetland habitat (Milton Weller, personal communication). Duck-hunting clubs formed and began buying up lands around Chesapeake Bay during the 1840s (Siry 1983). Milton Weller (personal communication) dates the recognition that birds require wetlands throughout their nesting and wintering ranges to bird migration studies published around the turn of the 20th century. Scientists played a role, especially those who functioned both as naturalists and hunters or fishermen. Those who held positions in federal agencies no doubt had many opportunities to influence policy. Even though their science was descriptive, they suggested cause-effect relationships that indicated the need to protect wild lands.

The value of wetlands for ducks was made clear by the legendary droughts of the early 1930s, with visible reductions in both wetland areas and duck populations. The droughts prompted duck hunters to band together to raise money to rebuild waterfowl populations; during 1937 the group became Ducks Unlimited, Inc. Although the droughts were catastrophic, there was a silver lining, as property values dropped to levels that made land acquisition feasible. In 1934, the Bureau of Biological Survey had over $6 million for purchase of migratory bird refuges through the sale of duck stamps (Eisenhower and Chew 1935). The "wetland acquisition programs of the 1930s . . . were the result of the combined pressures of the duck-hunting lobby and a public consensus that migratory birds were an integral part of the wild heritage of America" (Williams 1990). Nongovernmental organizations played an early role in protection efforts.

From 1930 to 1932 on the Atlantic Coast, a massive eelgrass (*Zostera marina*) decline had occurred, with visible losses of shallow water vegetation from North Carolina to New England (Cottam 1934). There were two reasons that the public took note of this plant—it was widely used as packing and insulating material, and it was known to support geese. Clarence Cottam (1934), of the U.S. Bureau of Biological Survey, reported that the loss of shallow water vegetation caused an alarming decrease in sea brant and a less precipitous decline in Canada geese along the Atlantic Coast. He analyzed the contents of brant guts before and after eelgrass declines and showed their dependency on plant foods. Whether or not this work led to management initiatives is unclear, but it demonstrates an early interest in cause-effect analysis of wetland-waterfowl dynamics.

During 1955, The National Wildlife Federation's theme for National Wildlife Week was "Save America's Wetlands," and it drew attention to the losses of U.S. wetlands since settlement: "Of the more than one hundred and twenty million acres of marshes and swamps originally lying within our boundaries, less than a fourth remain fit for use by waterfowl and other marsh life. Most of the loss has been through artificial drainage." (Errington 1957). Also in 1955, the USDI Fish and Wildlife Service published Circular 39, which reported the value of freshwater wetlands for waterfowl, presented a national inventory with twenty wetland types, and mapped the nation's wetlands according to their value to waterfowl. The documentation was qualitative, but it represented an important stage in the development of wetland science, namely, the assessment of the resource. In 1977, Bellrose presented quantitative data relating the density of ducks to the density of ponds, stating unequivocally that "The abundance of ponds in the Prairie Pothole Region is the most important single factor regulating the production of mallards and no doubt other dabbling species" (Bellrose 1979).

Coastal Wetlands as a Food Base for Coastal Fisheries

The emphasis on protecting coastal wetlands came later and had a stronger ecosystem science basis. While efforts were being made to purchase inland wetlands to support waterfowl, aquatic ecologists were beginning to transform the concept of ecosystem, ultimately linking coastal fish production to the growth of tidal marsh grasses. Two limnologists had an early impact on these developments; Stephen A. Forbes' 1887 paper "The Lake as Microcosm" and Raymond Lindemann's 1942 paper "The Trophic-Dynamic Aspects of Ecology" shifted ecological attention away from the more straightforward relationships between species and habitat types and toward the concept (and quantification) of energy flow. Food webs were seen as linking top-level consumers not only to areas where they fed but also to foods that might be produced in one location, exported, and then used in another. Several ecologists sought to quantify the flow of energy through food webs, and decomposers and detritus became recognized as important ecosystem components (Odum 1959; Darnell 1961). Among these ecologists was John Teal, who calculated that 45% of the carbon fixed by Georgia salt marsh grasses was exported to coastal waters "before the marsh consumers have a chance to use it and in so doing permit the estuaries to support an abundance of animals" (Teal 1962). As Teal (personal communication) is quick to point out, he never intended for the final sentence of his paper to be so widely quoted; furthermore, the 45% value was obtained by subtraction, not measurement.

The detritus-export concept or "outwelling hypothesis" (Odum 1971) formed the basis of many subsequent research projects e.g. stable-isotope-ratio studies to detect the signature of marsh vegetation in the detritus of coastal waters (Haines 1977), and evaluations of the import-export patterns of many estuaries. But criticism did not gain momentum until well after there was widespread acceptance of the concept that commercial fish production was strongly dependent on coastal wetlands (Joe Larson, personal communication; Odum 1961). The National Estuary Protection Act of 1968 and the Coastal Zone Management Act of 1972 were subsequent protection actions. Much later, Scott Nixon (1980) evaluated the literature on outwelling and concluded that:

There may be an export of organic carbon from many tidal marshes, and the export may provide a carbon supplement equivalent to a significant fraction of the open water primary production in many areas of the South. It is even possible that this carbon may contribute measurably to the standing crop of organic carbon in the water at any one time. But it does not appear to result in any greater production of finfish or shellfish than is found in other coastal areas without salt marsh organic supplements.

He further chastises estuarine scientists for not being more critical of their own work and for overstating the case in order to help protect marshes. "The scientist," he said, "must remain skeptical" (Nixon 1980). A case with some parallels concerns Massachusetts' inland wetlands, where groundwater recharge was advanced as a reason for protection. The concept led to 1965 legislation for freshwater wetlands; later, most Massachusetts freshwater wetlands were shown to be discharge, rather than recharge, systems (Larson, personal communication).

Wetlands as Filters That Improve Water Quality

The main protective legislation for the Nation's wetlands is not a habitat- or wildlife-protection law but an effort to provide clean water. The Federal Water Pollution Control act of 1972, later amended and named the Clean Water Act, established that dredge and fill operations required a permit in order to protect the waters of the United States. Several ecologists played a role in drafting this legislation (see Box 4.2). Although initially interpreted as being just the deeper, navigable waters, jurisdiction was extended to shallow waters by a 1975 court case (Table 4.2).

TABLE 4.2. Legislative actions and agency policies affecting wetlands.[1]

1871	U.S. Fish Commission created. First federal attempt to stop natural resource depletion.
1890	Rivers and Harbors Act gave the U.S. Army Corps of Engineers authority to regulate dredging and filling activities affecting navigable waters.
1902	Reclamation Act. Led to a survey of "wastelands" that could be reclaimed for agriculture (M. Weller, personal communication)
1934	Migratory Bird Hunting Stamp Act. Duck Stamps to be purchased by hunters; revenues for acquiring land or easements on duck habitat.
1934	Fish and Wildlife Coordination Act. U.S. Army Corps of Engineers must consult the Bureau of Fisheries and the Biological Survey about any plans to impound rivers. Established foundation for water quality control and need for biological integrity.
1937	Federal Aid to Wildlife Restoration Act. States given grants to acquire, restore, and maintain wildlife areas.
1940	U.S. Fish and Wildlife Service (US FWS) created.
1940–	USDA assisted landowners in draining wetlands under the Agricultural Conservation Program. For Minnesota, North Dakota, and South Dakota, this ended in 1962 if wildlife were threatened. Remainder ended in 1977 with President Jimmy Carter's order.
1946	Amendment to the Fish and Wildlife Coordination Act allowed lands acquired for flood control to be used in fish and wildlife propagation.
1958	Amendment to the Fish and Wildlife Coordination Act gave fish and wildlife concerns equal weight in plans for water-resource projects (dams, flood control, irrigation, generating plants).
1961	Wetlands Loan Act. Interest-free loans for wetland acquisitions and easements.
1963	Massachusetts became the first state to protect wetlands.
1964	Land and Water Conservation Fund Bill. Allowed fees to be charged in national parks and revenues to be used in further park acquisitions.
1965	Water Quality Act required states to adopt and enforce water quality standards.
1966	Clean Waters Restoration Act provided funds for wastewater treatment facilities and a study of estuarine water quality.
1967	Fish and Wildlife Coordination Act.
1968	Land and Water Conservation Fund Act.
1968	National Estuary Protection Act. Required surveys to identify marshlands to protect; established goal of acquisition and restoration.
1969	National Environmental Policy Act (NEPA). Required environmental impact analysis of federal projects.

TABLE 4.2. *Continued*

1972	Federal Water Pollution Control Act. Required permits to discharge dredge and fill material into navigable waters.
1972	Coastal Zone Management Act.
1973	Endangered Species Act.
1973	Flood Disaster Protection Act.
1974	Federal Aid to Wildlife Restoration Act.
1975	Federal court expanded the definition of navigable waters to mean "waters of the United States."
1977	Clean Water Act. Amended the Federal Water Pollution Control Act. Section 404 regulates disposal of dredged or fill material in the waters of the United States.
1977	President Carter signed Executive Order 11990 (Protection of Wetlands), requiring federal agencies to minimize wetland loss and protect wetland values and Executive Order 11988 (Floodplain Management) requiring agencies to avoid activities in floodplains.
1978	Council on Environmental Quality defined mitigation alternatives under NEPA (i.e. avoid, minimize, rectify, reduce, compensate) (Dennison and Berry 1993).
1985	Food Security Act of 1985. Eliminated subsidies to farmers who converted wetlands to agriculture.
1986	Emergency Wetlands Resources Act of 1986 (PL 99-645). Congress directed preparation of National Wetlands Priority Conservation Plan.
1988	President George Bush established the "No net loss" policy.
1989	US FWS published the National Wetlands Priority Conservation Plan. Acquisition to be prioritized by degree of public benefit, representation of rare or declining wetland types within an ecoregion, and subject to further loss or degradation.
1990	Food, Agriculture, Conservation and Trade Act established the Wetland Reserve Program: Landowners may volunteer to be paid to restore previously drained wetlands. Amended the "Swampbuster" provisions in the Food Security Act of 1985; denied benefits to farmers who planted crops on wetlands converted after December 23, 1985, or who converted wetlands to agriculture after Nov. 28, 1990 (USDI 1994).
1993	President Bill Clinton endorsed the "No net loss" policy.
1996	Secretary of Interior Bruce Babbitt made the Cowardin et al. (1979) classification system the federal standard for classifying wetlands.

[1] Sources: Siry 1983; Williams 1990; USDI 1994; Kusler and Opheim 1996.

Wetlands do affect water quality, so we are unsure why the 1972 legislation was not more inclusive. Presumably, there was early recognition that eliminating wetlands degraded water quality downstream by releasing sediments. The ability of wetlands to remove nutrients and contaminants would have been much less obvious. Wharton (1970) quoted a 1967 water quality report that attributed improvements in streamwater chemistry to adjacent swamps, but the concept of wetlands as "kidneys" was just beginning to develop. Elsewhere, in Europe, wetlands were being constructed and harnessed for wastewater treatment, a practice that later caught on in the United States (Sloey et al. 1978; Brix 1994; Kadlec 1994). There are now many constructed wetlands that are used to treat wastewater (EPA 1993) in addition to many books on the subject.

Following a general understanding that wetlands help improve water quality, there were many attempts to quantify the link. Most of this research began in the late 1970s (Nixon and Lee 1986; dates of references in Godfrey et al. 1985; Mitsch 1994; Reddy and D'Angelo 1994). Developing the link between wetlands and water quality improvement requires analysis of uptake, storage and release of materials (i.e. a mass-balance approach). Such work is not only difficult, it is expensive. There are still relatively few natural wetlands that have been thoroughly analyzed for their water quality improvement potential. We conclude that comparisons (water quality before and after wetland loss and in areas with and without wetlands) led decision-makers to extend protection to wetlands and that ecosystem science (mass-balance approaches) grew out of the need to quantify the relationship and apply the knowledge in wastewater treatment.

Critical Activities of Scientists

There is little doubt that key individuals played major roles in protecting wetlands and in developing ecosystem science (see Box 4.2). In addition to direct efforts to protect specific wetlands, perhaps the most important policy-influencing activity that wetland scientists undertook was the listing of wetland values. In the early 1900s, McCallie (1911) listed one use for wetlands: potential agriculture. Now, there are dozens of papers that describe the values of wetlands. We summarized nine versions of functions that benefit society (Table 4.3), dating back to Shaw and Fredine's (1956) circular on waterfowl and wildlife habitat value. The differences among lists are in the details; most recognize direct economic uses, as well as habitat, productivity, hydrologic, and water quality values.

The next step was assessing the dollar value of wetland functions. The first to put dollar values on wetland functions was Charles Wharton, a biology professor at Georgia State University. In *The Southern River Swamp* (Wharton 1970), he directly challenged the proposal to drain 4,327 acres of Alcovy River swampland. As an alternative, he recommended retaining the swamps as open space for use in education and recreation, protection of water resources, and support of wildlife, claiming an annual value of over $7,000,000 for Alcovy River Swamp (Table 4.4). His report was plainly written, but it carried the weight of 104 citations from the scientific literature, and it helped establish the swamp as a reserve (Thomas 1976). Later, Gosselink et al. (1974) published an estimate for tidal marshes; summing the contributions of wetlands as habitat, food production, wastewater treatment, and flood protection, they suggested that an acre of tidal marsh was worth $83,000. Lugo and Brinson (1978) reported other estimates ranging from $33,175 to $118,000 per acre. There are obvious pitfalls to the dollar-value approach, as many commercial uses of property could generate more revenues in the short term. The lack of a single, far-reaching method for assessing the value of wetlands is a problem, although H.T. Odum (1971) has proposed solar energy as the basic unit of comparison.

An interview with Professor Wharton suggested that behind-the-scenes activities have been even more important to wetland protection efforts and perhaps indicative of much of the unwritten history of wetland protection. Like Aldo Leopold, Wharton had a passion for the land, access to scientific facts, and an ability to present strong arguments in nontechnical language. Dr. Wharton developed a number of political contacts and took the time to show them his favorite wetlands. He persuaded the Georgia governor to establish a Natural Areas Council, with half of the members to be scientists. He then chaired the council, which advised the governor on natural resource protection (Wharton, personal communication).

Research reviews and compilations further helped make wetland science accessible to policy and decision-makers. Coastal wetland scientists were first to summarize and publicize their rapidly growing knowledge. A 1959 conference on salt marshes eventually grew into a national scientific society (Estuarine Research Federation) with its own journal (*Estuaries*, called *Chesapeake Science* from 1960 to 1977). Broader research review efforts began to appear in the 1970s. Darnell et al. (1976) reviewed 1,023 references and summarized the types of physical and chemical alterations to wetlands resulting from drainage, mining, impoundment, canalization, dredging, and bank or shoreline construction. Negative impacts were ranked from highest to lowest as: direct habitat loss, addition of suspended solids, and modification of water levels and flow regimes (Darnell et al. 1976). In 1978, a National Symposium on Wetlands was held (Greeson et al. 1979) to discuss several categories of wetland values: food chain, habitat, hydrologic and hydraulic, water quality maintenance, and use (harvest and heritage). Kusler (1978) presented a list of eight areas of research needed to implement wetland protection and management initiatives: functions of each type of wetland and their value to society, evaluation methods, impact assessment, hazard evaluation, cost-benefit evaluation of using upland vs. wetland sites for development, mapping of wetland boundaries, characteristics of individual wetlands, and data on functions of individual wetlands. After the workshop, fifty-six scientists met to address 233 questions provided by resource-management agencies. Important questions were: What rapid assessment techniques would indicate primary productivity, carrying capacity, and pollution effects? How should wetland boundaries be determined? Is there a minimum size for a valuable wetland? And how long does it take for constructed wetlands to mature and provide "full" value?

Related reviews were undertaken for riparian ecosystems, which are sometimes considered wetlands (Lowrance, chapter 6): Johnson and McCormick (1979) evaluated strategies for protection and management of floodplain wetlands and other riparian ecosystems for the USDA Forest Service. They reviewed seventeen papers on characteristics of floodplain

TABLE 4.3. Comparison of wetland values listed by various authors.

Function	Shaw and Fredine 1956	Wharton 1970	Greeson, Clark and Clark 1979	Wentz 1981	Adamus 1983	Sather and Smith 1984	Feierabend and Zelazny 1987	Mitsch and Gosselink 1993	Kusler, et al. 1994
Socioeconomic									
"Active" recreation	X	X			X	X			
"Passive" recreation		X			X	X			
Aesthetic appeal		X			X			X	
Agriculture	X								
Commercial fishing	X		X		X	X		X	X
Crops—rice, cranberries, hay, etc.	X							X	X
Education	X	X							
Firebreaks	X							X	
Peat fuel	X							X	
Pelts—mink, nutria, etc.	X							X	X
Timber	X	X			X			X	X
Habitat									
Biodiversity	X	X		X		X		X	X
Habitat—fish/shellfish	X		X	X	X	X	X	X	X
Habitat—invertebrates			X						X
Habitat—plants		X							X
Habitat—rare species						X		X	X
Habitat—waterfowl	X	X	X	X		X		X	X
Habitat—wildlife	X	X	X		X	X	X		X
Transitional zone habitat					X		X	X	

Productivity
- Food web
- Inorganic importers
- Organic exporters
- Primary productivity
- Secondary productivity

Hydrology
- Aquifer discharge
- Aquifer recharge
- Flood control
- Groundwater storage
- Shore anchoring
- Storm energy abatement
- Water conservation
- Watershed protection/maintenance

Water Quality
- Decomposition
- Global nutrient cycles
- Nutrient removal—burial
- Nutrient Removal—denitrification
- Toxic substance removal
- Wastewater treatment
- Water quality improvement

TABLE 4.4. Estimated economic values of the Alcovy
River Swamp ecosystem.[1]

Education (underestimated)	$1,320,917
Public use (@$3)	4,080,000
Groundwater storage and recharge	228,014
Water quality protection	1,013,232
Productivity (underestimated; most is lumber)	546,940
Total	$7,189,103

[1] From Wharton 1970.

wetlands and other riparian ecosystems, nineteen papers on their values, and eighteen papers on their management, including position papers by the U.S. Fish and Wildlife Service (US FWS), the Soil Conservation Service (SCS), and the Bureau of Land Management, all of which had jurisdiction over significant national wetlands. Their summary called for a National Riparian Program, a recommendation that was also made by the National Research Council (NRC) in 1995, but which has not yet been fulfilled.

The debate over wetland values continued into the 1980s. Sather and Smith (1984) summarized several literature reviews as background for a National Wetland Values Assessment Workshop held in Alexandria, VA. Wetland values were considered in five categories: (1) hydrology, (2) water quality, (3) food chain, (4) habitat, and (5) socioeconomic values. For each, they listed specific values and indicated how well authenticated they were. Also in the 1980s, the FWS published a series of wetland community profiles (Table 4.5), which further publicized the diversity of wetland types and summarized the status of understanding of many of the Nation's wetland types.

TABLE 4.5. Community profiles published by United States Fish and Wildlife Service.

Date	Author	Title[1]
1979	Peterson, C., N. Peterson	. . intertidal flats of North Carolina
1980	Ward, G. Jr., N. Armstrong	. . Matagorda Bay, Texas: Its hydrography, ecology, and fishery resources
1981	Bahr, L., W. Lanier	. . intertidal oyster reefs of the south Atlantic Coast
1982	Gallaway, B., G. Lewbel	. . petroleum platforms in the northwestern Gulf of Mexico
1982	Nixon, Scott W.	. . New England high salt marshes
1982	Odum, W. et al.	. . the mangroves of south Florida
1982	Sharitz, R., J. Gibbons	. . southeastern shrub bogs (pocosins) and Carolina bays
1982	Wharton, C.	. . bottomland hardwood swamps of the Southeast
1982	Whitlach, Robert B.	. . New England tidal flats
1982	Zedler, J.	. . southern California coastal salt marshes
1982	Zieman, J.	. . the seagrasses of south Florida
1983	Josselyn, M.	. . San Francisco Bay tidal marshes

TABLE 4.5. *Continued*

Date	Author	Title[1]
1983	Seliskar, D., J. Gallagher	. . tidal marshes of the Pacific Northwest coast
1984	Gosselink, J.	. . delta marshes of coastal Louisiana
1984	Hobbie, J.	. . tundra ponds of the Arctic coastal plain
1984	Jaap, W.	. . the south Florida coral reefs
1984	Odum, W.E.	. . tidal freshwater marshes of the United States east coast
1984	Phillips, R.C.	. . eelgrass meadows in the Pacific Northwest
1984	Simenstad, C.	. . estuarine channels of the Pacific Northwest coast
1984	Stout, J.	. . irregularly flooded salt marshes of the northeastern Gulf of Mexico
1984	Thayer, G. et al.	. . eelgrass meadows of the Atlantic coast
1984	Wiedemann, A.	. . Pacific Northwest coastal sand dunes
1985	Foster, M., D. Schiel	. . giant kelp forests in California
1986	Eckbald, J.	. . pools 11–13 of the upper Mississippi River
1986	Herdendorf, C. et al.	. . Lake St. Clair wetlands
1986	Jahn, L., R. Anderson	. . pools 19 and 20, upper Mississippi River
1986	Teal, J.	. . regularly flooded salt marshes of New England
1987	Armstrong, N.	. . open-bay bottoms of Texas
1987	Damman, A., T. French	. . peat bogs of the glaciated northeastern United States
1987	Glaser, P.	. . patterned boreal peatlands of northern Minnesota
1987	Herdendorf, C.	. . the coastal marshes of western Lake Erie
1987	Zedler, P.	. . southern California vernal pools
1988	Hay, M., J. Sutherland	. . rubble structures of the South Atlantic Bight
1988	Nichols, F., M. Pamatmat	. . the soft-bottom benthos of San Francisco Bay
1988	Ohmart, R. et al.	. . the lower Colorado River from Davis Dam to the Mexico–United States international boundary
1989	Faber, P. et al.	. . riparian habitats of the southern California coastal region
1989	Herbold, B., P. Moyle	. . the Sacramento–San Joaquin Delta
1989	Kantrud, H. et al.	Prairie basin wetlands of the Dakotas
1989	Laderman, A.D.	. . Atlantic white cedar wetlands
1989	Minshall, G. et al.	. . stream and riparian habitats of the Great Basin region
1989	Vince, S. et al.	. . hydric hammocks
1989	Zieman, J., R. Zieman	. . the seagrass meadows of the west coast of Florida
1990	Wiegert, R., B. Freeman	Tidal salt marshes of the southeastern Atlantic Coast
1993	Golet, F.	. . red maple swamps in the glaciated northeast
1995	Bellis, V., J. Keough	. . maritime forests of the Southern Atlantic Coast

[1] Most begin with: The ecology of.

In the absence of a wetland research funding program within the National Science Foundation, much of the wetland science has taken place in federal laboratories. The U.S. Army Corps of Engineers initiated research to assist in delineating wetlands, and in 1985, the EPA established a wetlands research program to evaluate wetland functions and wetland restoration capabilities (Zedler and Kentula 1985). Academic scientists have received funding from the National Sea Grant Program for work in estuaries and coastal marshes, and from other programs within the National Science Foundation. There is still very little direct evaluation of wetland boundaries and identifying characters, and there is still too little information on reliable indicators of wetland functions.

A current thrust of the Army Corps of Engineers is the development of a hydrogeomorphic approach to the classification of wetland functions (Brinson 1993). The goal is to provide a rapid method of assessing the effects of a project on wetland functions. Biological functions are related to geomorphic setting (i.e. where the wetland is located in the landscape), water source and its transport (i.e. fed by precipitation, surface flow, or groundwater), and hydrodynamics (i.e. direction and strength of water movement). To ensure that regulatory agencies protect wetland functions they need reliable indicators that are easily identified. Not all wetlands are critical for all of the widely used list of functions (habitat, water quality improvement, and hydrologic functions); for example, small vernal pools play little role in groundwater recharge, because they overlie an impermeable clay pan. Using Brinson's (1993) approach, attributes of the wetland in question are compared to reference systems that have the same hydrogeomorphology, for which values have been identified. The Army Corps of Engineers is in the process of describing reference systems and associated values, in preparation for implementing this new approach to functional assessment.

Scientists are, of course, free to develop research programs that are proactive and responsive to agency needs (see Box 4.3). The main constraint on such efforts is funding, and it may take many years to accumulate the information needed and influence wetland management.

Box 4.3
A contemporary case: changing attitudes about wastewater discharge

Few scientists bother to write down their management-advising activities, in part because it is hard to know when one's advice is being heard, who listened and took action, which advisory activities were most influential, and what decisions were influenced. One of us (Zedler) has worked for twenty years to change attitudes about southern California coastal wetlands on at least four major issues, that is, to gain recognition that the region's wetlands function differently from those on the Atlantic and Gulf coasts, to explain the difficulties in restoring wetlands to levels that will support endangered species, to specify that mitigators not be allowed to damage natural wetlands on the promise that they will provide functionally equivalent replacements, and to make dischargers of wastewater aware that fresh water—independent of the materials it carries—can also have negative impacts on coastal ecosystems by lowering water and soil salinity. Suggesting that freshwater discharge could harm downstream wetlands runs counter to dogma that estuaries are brackish-water systems. This issue and the research behind it provide insight into how wetland science can change management attitudes.

During the 1970s, few people considered that the southern California coastal wetlands might be sensitive to augmented freshwater inflows. There were few data on estuarine salinity and no major dischargers of nonsaline water. Municipal effluent was treated and discharged to the ocean, and the effects of various sewage spills on coastal wetlands went unmonitored. Regulatory agencies were aware of the problems of nutrients and contaminants in wastewater but there were no reports of a separate effect of salinity dilution. Thus, regulators were concerned about water quality, but not water quantity. If water could be treated sufficiently to lower nutrient and contaminant levels below EPA standards, it was considered suitable for discharge to local streams.

The desire to discharge wastewater to coastal streams is very strong in southern California, because urbanization is occurring well inland of existing ocean outfalls and building longer pipes to release wastewater into the ocean would be very costly. If wastewater facilities could discharge treated effluent directly into coastal rivers, just as in other parts of the world, not only would the cost of piping be saved, but seasonally dry streambeds could be "revived." The terms "live stream discharge" and "stream enhancement" caught the attention of decision-makers early on; they promised midwestern-style streams that would match our midwestern-style green lawns and agriculture, all of which depend on imported water. San Diego imports about 90% of its potable water from the Colorado River and northern California. What would happen if substantial quantities of imported water were discharged year-round to coastal streams?

The catastrophic floods of January and February 1980 first made me aware of the substantial effects that excess freshwater inflows could have on southern California's saline wetlands. One wetland (Tijuana Estuary) flooded and retained its salt marsh character, while another (San Diego River Marsh) flooded and converted to a cattail marsh (Zedler 1981, 1985). The differences in the two systems were that (a) the San Diego River's hydroperiod was prolonged for months by reservoir drawdown, releasing freshwater well beyond the normal rainy season, and (b) the water was confined to a narrow flood control channel and salt marsh, so that reservoir discharges continuously

Box 4.3 *Continued*

inundated the native vegetation. The large, obvious change in the salt marsh was shown to relate to the prolonged dilution of soil salinity (Beare 1984, Beare and Zedler 1987). This catalyzed other studies of the effects of low salinity on salt marshes (Zedler and Beare 1986; Brenchley-Jackson 1992).

Aware of proposals to begin discharging imported water, I asked the EPA to sponsor an evaluation of natural and augmented river flows. They did; Robert Koenigs helped review the literature and William Magdych developed a simple model to predict salinity changes with different discharge regimes. The model suggested that the San Diego River Estuary would show a significant drop in salinity if discharges exceeded 5 MGD (Zedler et al. 1984a, b). Working directly with an agency put the findings in the hands of the regulators. In this case, the EPA funds were passed through the San Diego Association of Governments (SANDAG), which provided access to local decision-makers. Public presentations to SANDAG board members (e.g. mayors, county supervisors, and so forth) required a simple message. Slides showing the San Diego River in flood and the salt marsh before and after flooding got the attention of agency staff members, who began to accept the message that they should be cautious about "live stream discharges." Nevertheless, a proposal by the Padre Dam Water District to discharge 4 MGD to San Diego River was approved in 1988.

In the mid 1980s, the EPA, the State Department, and the International Boundary and Water Commission began planning new wastewater treatment facilities at the US–Mexico border, which is just upstream of Tijuana Estuary, a National Estuarine Research Reserve. Part of this estuary is a National Wildlife Refuge for endangered species. Staff members from the EPA asked me how much discharge (i.e. salinity reduction) Tijuana Estuary could tolerate, indicating that regulators had become aware of potential detrimental impacts. An SDSU colleague, Dr. Charles Cooper, was a member of the Tijuana Estuary Management Authority; he understood well the implications of our research and helped carry the message to the planning process. I remember how delighted I was when a young EPA staff member introduced himself and explained to me that wastewater discharges could potentially impact the plants and animals of Tijuana Estuary by lowering the salinity—I knew then that our message had been heard!

Subsequently, we began to study the effects of urban runoff and showed that (1) street run-off and wastewater spills lower salt marsh soil salinity (Zedler et al. 1990); (2) exotic plants can invade the salt marsh when salinity is lowered (Kuhn and Zedler 1997); and (3) invasive exotics can outcompete native plants (Callaway and Zedler in press). Our evidence that salt marsh vegetation might change seems to have had little impact, however, because there is no economic impact. More effective has been the observation that natural flooding kills many invertebrates and some fishes (Onuf 1987). The potential for artificially prolonged streamflows to kill animals led me to solicit master of science research on a high-profile species, the California halibut. Baczkowski (1992) conducted a series of experiments and concluded that nursery-size halibut prefer seawater salinity to brackish water and that the tiniest halibut are most negatively affected by lowered salinity.

Another study, funded by California Sea Grant, explored an alternative pulsed-discharge concept. The idea was to give regulators options other than preventing discharges. Reasoning that treated effluent could be held in wastewater wetlands and discharged in pulses, I hypothesized that the constructed freshwater wetland would behave like an Atlantic Coast tidal freshwater marsh and be a highly effective polishing system. This was shown to be true (Busnardo et al. 1992, Sinicrope et al. 1992) and pulsed discharges were posed as a win-win alternative (Zedler et al. 1994).

There is still much to be done before guidelines can be provided for maximum inflows during the normal dry season. Presently under development is a salinity-dilution model for Tijuana Estuary by Kamman, Goodwin, and Sobey (Philip Williams and Association, and University of California Berkeley). This effort is the most costly to date, requiring high-precision surveys of the estuarine channels and salt marsh topography, continuous monitoring of water levels and water salinities, and intermittent sampling of water salinity throughout the estuary. A groundwater component has recently been added (K. Thorbjarnarson, SDSU, in progress). Still to be explored are impacts on the most sensitive estuarine species (e.g. macroinvertebrates). Yet, the potential of a negative impact has influenced decisions: current policy is not to discharge any treated effluent to Tijuana River; instead, the treated water will be returned to Mexico for reuse or discharged to an ocean outfall that is under construction.

Stream enhancement is still in the plans for other coastal rivers. During 1988, the Regional Water Quality Control Board listed ten streams that could receive treated wastewater, up to 10 MGD in one case, 15 MGD in three, 20 MGD in two, and 30 MGD in three cases. Tijuana Estuary was held to 12.5 MGD, in keeping with our report to EPA and SANDAG (Zedler et al. 1994). Elsewhere in the region, Irvine Ranch Water District recently proposed to begin discharging 5 MGD to Upper Newport Bay. The City of Newport Beach opposed this inflow and a major controversy developed. The alternative of

Box 4.3 *Continued*

piping excess effluent to an adjacent watershed for reuse is preferable, but more costly. Key actors in this interaction have been the City Attorney and a citizen action group. Both sought scientific advice and used it in organized testimony before the Regional Water Quality Control Board. The Water District also hired consultants, so the decision-makers were barraged with information from experts on both sides of the issue. After lengthy testimony, the Board approved the discharge with a requirement that impacts be monitored. The result has been an improved monitoring program for Upper Newport Bay. Potentially, there will be funding for salinity modeling and tests of sensitive species. The ultimate result could be an adaptive management approach, with the discharger indicating discharge regimes that can be used in experimental microcosms, and the response of native species indicating appropriate discharge regimes.

It is difficult to know which actions caused agencies to consider the role of salinity dilution along with preexisting concerns about eutrophication and contamination. Scientific publications appeared well after decisions were made, so oral presentations and letters were essential in transmitting the knowledge. A notice from the Regional Water Quality Control Board indicated that they changed their Stream Enhancement document in response to my letter; it is rare to see such acknowledgment. Our "gray literature" (e.g., Zedler 1981; Zedler et al. 1984a, b) was probably also critical to the process, because draft reports were available when the information was needed. The involvement of master's students (P. Beare, S. Baczkowski, M. Busnardo, and T. Sinicrope) was essential, because they were able to address specific problems in a short period of time. A long-term, adaptive research program was no doubt a key feature—that is, one that focuses on the needs of users, uses management plans to suggest studies (e.g. consequences of alternative management actions), brings the results to bear on the issue, and pursues further research based on those responses.

Educating the Public About Wetlands

Many scientists have influenced policy by writing for the average American. Through their descriptions of natural history and their translations of wetland science (Table 4.6), they no doubt have influenced public sentiment and actions. The 1960s and 70s were peak decades for such contributions to the literature, but at least two major tributes to wetlands in North America

TABLE 4.6. Popular writings that describe the environment (especially wetlands) in a favorable light.

1926	McQueen and Mizell	*History of Okefenokee Swamp*—The account by a country lawyer and a lumberman of a place they hoped to have set aside as a federal sanctuary. "This magnificent Swamp, one of the very largest, if not the largest, in the entire United States, is truly a nature-lover's paradise" (p. 9).
1949	Leopold	*Sand County Almanac*—Early statement of the land ethic.
1957	Errington, P.	*Of Men and Marshes*—Personal account of hunting and studying marshes of the upper Midwest; shows love of and respect for wetlands.
1951	Carson	*The Sea Around Us*—An oceanography book for nonscientists. Prepublication excerpts appeared in *New Yorker Magazine.* The book was on the *New York Times* bestseller list for eighty-one weeks (Siry 1983).
1961	Odum, E.	Article in *The Conservationist* described the high productivity of tidal wetlands in Georgia; made the link between marshes and fisheries.
1962	Carson	*Silent Spring*—Early call for environmental activism; one mention of estuaries: "The inshore waters—the bays, the sounds, the river estuaries, the tidal marshes—form an ecological unit of the utmost importance . . . were they no longer habitable these seafoods would disappear from our tables" (p. 149).
1963	Udall	*The Quiet Crisis*: "Lumbering . . . during most of the nineteenth century . . . did more than anything else to awaken us to the fallacy of the Myth of Superabundance" (p. 99).
1964	Teal and Teal	*Portrait of an Island*: "The marsh is of inestimable value to man just as it is" (p. 125).
1966	Niering, W.	*The Life of The Marsh*—A picture book describing the habitats and organisms of North American wetlands; promotes values in a chapter called "Wetlands or Wastelands?" "In most cases filling has caused an unnecessary and irreversible loss" (p. 163).
1966–1972		McGraw-Hill published twelve additional books in "*The Life of . . .*" series, indicating public interest in habitats.
1967	Nash, R.	*Wilderness and the American Mind*
1969	Teal and Teal	*Life and Death of a Salt Marsh*
1970	Amos, W.	*The Infinite River*—Poetic prose about a hypothetical river in the North Atlantic, its animals and plants, along the course from headwaters to the ocean. Chapter on "The Intruders" describes human impacts.
1976	Thomas	*The Swamp* (picture book)
1976	Warner	*Beautiful Swimmers*—A heartwarming story of blue crab fishermen and their dependence on Chesapeake Bay;

TABLE 4.6. *Continued*

		problems with eutrophication are noted; recognized importance of salt marsh for estuarine productivity.
1991	Niering	*Wetlands of North America*—Beautifully illustrated description of wetlands.
1994	Kusler	*Scientific American* article about the fluctuating nature of water levels in wetlands and why policy-makers should not expect wetlands to be wet all the time.

(Niering 1991) and Australia (McComb and Lake 1990) appeared in recent years. Although it is difficult to measure the impact of such publications, their intent usually is clear. For example, a beautiful booklet on woodlands, wetlands, and wildlife in New York City parks "is part of a public–private initiative aimed at increasing environmental awareness and developing a citizen stewardship ethic for the city's natural resources" (Urban Forest and Education Program 1994).

Scientists have played many other roles in the debates over the value and definition of wetlands. Hardest to document are their discussions, letter writing activities, public speeches, and connections with influential people. Yet we know from our own experience (Box 4.3) that it is not enough to publish sound ecosystem science. The information has to be conveyed to the right people at the right time and in a form in which it can be heard. Citizen activist groups often play that intermediary role of "science translator." Thomas (1976) recounts a familiar series of events surrounding the Okefenokee swamp and the Great Swamp of New Jersey (Box 4.2).

Scientific knowledge is not equal around the country (Table 4.7), and the roles that scientists play may well differ in response to how long wetland functions have been studied, how many scientists are working on wetlands in a particular region, and the network of citizen activists. In southern California, the knowledge base on wetland functioning is relatively limited, but the citizen action network is very well organized. Nearly every coastal wetland has its own "friends" group, and the groups act collectively in many umbrella organizations (e.g. Save California Wetlands lists dozens of subgroups). Where there are no landmark papers linking wetland areas to significant societal values, scientists still can be asked to testify at hearings, write letters expressing their science-based opinions, attend meetings to offer advice, or serve as technical advisors on specific projects or issues.

Some scientists play multiple roles as researchers, writers for the public, and policy-makers. Eugene Odum's paper in *The Conservationist* translated the findings of several Georgia salt marsh scientists into plain language: "When the average citizen looks at the vast green marshes of the southeast . . . he is likely to regard them as wastelands because he sees no

TABLE 4.7. Geographic distribution of papers cited by Nixon and Lee (1986).[1]

Region	Number of papers cited
Alaska	31
Pacific Coast	39
Gulf and South Atlantic	161
North Atlantic	92
North Central–Great Lakes	38
Interior: Midcentral	25
Total	386

[1] Their review of literature on wetlands and water quality emphasizes papers from 1977 on.

direct use by man, when, as a matter of fact, he is looking at an important source of sea food!" (Odum 1961). John Teal functioned as a scientist, a writer for laypersons (Teal and Teal 1964, 1969), and also a decision-maker. At one time, he was a member of the Conservation Committee of the town of Falmouth, MA, which had permitting authority for wetlands (J. Teal, personal communication). Charles Wharton (personal communication) chaired a natural resources advisory council for the state of Georgia and once led Jimmy Carter on a personal tour of Altamaha River wetlands. Several personality attributes would seem to be important for individual scientists to influence decisions (Box 4.4).

Box 4.4
Attributes for effectiveness in the policy arena

Individual scientists who made a difference in the wetland-protection arena had many traits that helped them influence decision-makers, among them being: persistence; the ability to state complex concepts simply (i.e. the ability to "see the forest" in spite of all the trees); willingness to be in the public eye; and being viewed as trustworthy by their listeners, i.e. having a solid standing in science. Eugene Odum exemplifies these attributes. His activities in the spotlight (e.g. working with state legislators, testifying at hearings, working on environmental committees, writing articles for the public) and behind the scenes (e.g. helping the University of Georgia's Ecology Club prepare and distribute flyers, bumper stickers, and lapel buttons urging the citizenry to "Save Our Marshes," as well as creating newspaper cartoons that depicted marsh destruction in an unfavorable light) all led the State of Georgia to develop its 1970

> ## Box 4.4 *Continued*
>
> Coastal Marsh Protection Act. A major threat (a proposal to strip mine the marshes for phosphate) and an organized coalition were viewed as essential to accomplish long-range protection goals (E. Odum, personal communication).

The effect of scientific activities on policy is not easily documented, however, and the recollections of scientists who watched the decision-making process and knew what science was being used must be relied upon. Joe Larson (personal communication) recalls that in 1963 the first wetland regulations in the world were passed by the Massachusetts legislature because the state's coastal communities were persuaded that the loss of marshes was largely responsible for reduction of finfish harvest. This inference was drawn from articles from the scientific community (Larson personal communication). The Massachusetts legislation was years ahead of federal action. However, in Connecticut, the SCS played a major role in the state's wetland program. The SCS held a series of workshops on wetland functions and values, and a wetland identification program based on soils and soil maps was initiated. "The rise of wetland regulation came out of an informed public. . . . ; it is not surprising that it should have started in New England, because we have such a long tradition of local involvement in government. . . . the town meeting is a mini-legislature" (Joe Larson, personal communication).

Still other scientists have catalyzed significant actions as government employees or advisors to agencies. Allan Hirsch worked with the EPA and US FWS, and started the National Wetlands Inventory. Hank Sather worked with the US FWS, SCS, and the Bureau of Land Management to achieve interagency cooperation on wetland initiatives and the National Wetland Inventory (Sather, personal communication). John Clark headed the Conservation Foundation and got the National Wetland Technical Council started. This ad hoc group of wetland scientists independently advised federal agencies on wetland issues; it was chaired by Joe Larson, who organized scientists in each region of the United States to evaluate the status of knowledge of wetland functions (Zedler et al. 1985; Brown and Lugo 1986; Sather and Low 1986; van der Valk and Hall 1986; van der Valk 1989). Jon Kusler, a rare individual with both scientific and legal credentials, created the Association of State Wetland Managers and helped them convene annual meetings with scientific presentations on a variety of such issues as mitigation, hydrology, and urban wetlands.

The message that wetlands are diverse, dynamic, interesting, and worthy of protection has been carried to the public and decision-makers in many formats. In a variety of educational films for the classroom and television, wetlands are portrayed as fascinating habitats. Even television ads by major oil companies portray wetland wildlife as worthy of protection and habitat enhancement.

Contemporary Attitudes and Actions

A three-year study of American attitudes toward wildlife and natural habitats (Kellert 1979) showed that wetlands were valued widely for wildlife. The national, random sample produced 3,107 responses, of which 57% disagreed and 39% agreed with the statement "I approve of building on marshes that ducks and other nonendangered wildlife if the marshes are needed FOR HOUSING DEVELOPMENT." Williams (1990) dates the present desire to protect wetlands to the early 1970s and offers these three reasons: (1) massive modifications to wetlands by the U.S. Army Corps of Engineers, with no consideration of ecological impacts; (2) increased public awareness of the beauty and valuable functions of coastal wetlands, and (3) concern for inland wetlands, which were being lost to urbanization and agriculture and damaged by dumping of toxic wastes.

To characterize recent attitude shifts, we conducted a library search of the indices of four major newspapers (*New York Times*, *Los Angeles Times*, *Washington Post*, and *The Wall Street Journal*) to assess interest in, and alternative attitudes about, wetlands and estuaries. For articles published in 1986 and later, The National Newspaper Index was searched on computer for the keywords, "wetland*" and "estuar*" (* allows alternative endings). For articles published before 1986, the annual or biennial indexes for each newspaper were used (those for the NY Times were published by the New York Times Company, NY; those for other papers by University Microfilms International, Ann Arbor). No associated terms were searched unless they were cross-referenced under wetland* or estuar*. A total of 855 articles was found; only summaries of each article appeared in these index volumes.

Summaries of articles about a wetland or estuary were read and classified as being either (1) "pro," (i.e. reporting something favorable about wetlands); (2) "con," (i.e. having a negative attitude about wetlands); (3) "debate," (i.e. articles announcing hearings, arguments between politicians, or letters to the editor); or (4) "info," (i.e. articles announcing hearings or providing information on events, both for and against wetlands (Table 4.8). About half the articles found were read and classified by two of us (Trnka and Fellows) in order to reduce bias; the remainder were read by one of that team.

The New York Times had the longest record (dating to 1961) and the largest number of articles of interest, with nearly 500 in the past thirty years (Figure 4.1). The term *wetland* first appeared in the index for the New York Times in 1967, and articles about wetlands and estuaries increased steadily

TABLE 4.8. Articles on wetlands and estuaries in four major newspapers.[1]

	LAT	NYT	WP	WSJ
Number classified as "pro"	30	205	43	11
Number classified as "con"	25	37	15	12
Number classified as "debate"	58	118	44	25
Number classified as "info"	60	132	29	11
Total number of articles	173	492	131	59

1. LAT = Los Angeles Times, NYT = New York Times, WP = Washington Post, WSJ = Wall Street Journal). Article summaries were classified in four categories (see text).

thereafter, with articles falling in the "pro" category substantially outnumbering those in the "con" category in the first decade (1967 to 77; Figure 4.1). The Los Angeles Times, Washington Post, and The Wall Street Journal, each reported far fewer articles of interest, with totals of 173, 131, and fifty-nine respectively (Table 4.8). Of the 855 articles, only about 10% (eighty-nine) were considered "con" (Table 4.8). Of those classified as "pro," 18% appeared before 1970, 40% appeared in 1971 to 80, 22% in 1981 to 1990, and 20% after 1990. The New York Times not only had the largest number of articles on wetlands, it also published the majority (71%) of all articles classified as "pro." The New York Times data suggest that interest in wetlands has increased steadily.

A heightened awareness of wetland values among the public has not always led to wise decision-making, however. For example, between 1962 and 1971, the U.S. Army Corps of Engineers spent some $30 million to dredge the Kissimmee River and drain 12,000 to 14,000 hectare (ha) of wetlands; the 166-km river became a 90-km canal 10 m deep and 100 m wide (Koebel 1995). Major wetland loss continued through the 1980s. Williams (1990) attributed the 250,000 hectare per year loss rate to the lack of interagency coordination and the lack of a comprehensive management approach. Wetlands were protected through either land-use or water quality legislation, but the functioning of wetlands and whole watersheds was not. So, it was not surprising that the straightening of the Kissimmee River failed to provide the desired flood protection. Because channelization also had devastating effects on aquatic and terrestrial wildlife, the channelization process is being reversed. Weirs are being used to force flows back into the historic river channel in one of the most ambitious wetland restoration projects in the nation.

Overall, progress toward wetland protection, management, and restoration has been encouraging but slow (see Table 4.2). The National Park System was established under Abraham Lincoln, but it included only 22 million acres by 1940 (Strong 1971). There was little further acquisition until the 1960s when Stuart Udall became Secretary of the Interior and influenced federal policy on habitat protection and clean water. The 1964 Land and Water Conservation Fund Bill helped parks become self-

sustaining by allowing fees to be charged and revenues used for land purchase. The 1964 Wilderness Bill was further evidence of Udall's influence in
setting aside land as habitat. In 1968, Congress passed the National Estuary
Protection Act, which identifies, protects, and restores estuaries. George
Bush initiated (and Bill Clinton reaffirmed) a national policy that there

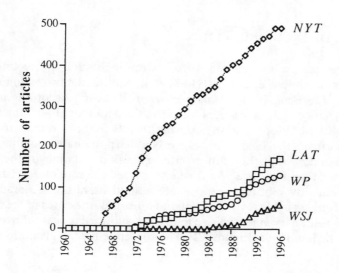

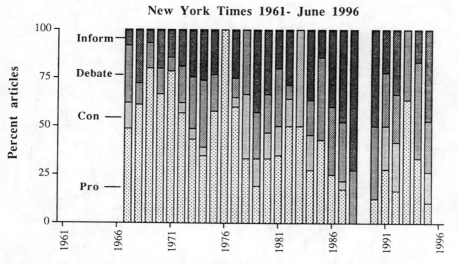

FIGURE 4.1. Articles published in four newspapers (NYT = New York Times, LAT
= Los Angeles Times, WP = Washington Post, WSJ = Wall Street Journal) indicate
increasing interest in wetlands (top graph) and changing attitudes about wetlands
(bottom graph).

should be no overall net loss in wetland acreage and function and that the quality and quantity of wetlands should be increased. During the 1990s, there has been substantial interest in wetland restoration. The 1990 Wetlands Reserve Program specified a goal of restoring a million acres of wetlands, and the National Research Council (1992) has recommended restoring 10 million acres by the year 2010.

The Growth of Wetland Science

Wetland science caught fire in the 1970s, after legislation protecting wetlands was already in place. In 1988, Hook et al. published a two-volume set of papers on *The Ecology and Management of Wetlands*. According to Williams (1990), some 2050 references were cited in Hook et al., of which <5% were published before 1960, 8% during the 1960s, 38% during the 1970s, and 48% during the 1980s. This suggests a sharp increase beginning in the 1970s, assuming that the authors cited papers in relation to their availability, rather than their age. We further explored the rate of publishing wetland studies by plotting citations of wetland literature. Adamus (1983) cited 970 papers in his review of wetland functions; the dates of these publications suggest an exponential rise in numbers toward a peak of over 100 per year in the late 1970s (Figure 4.2). The subsequent drop in citations

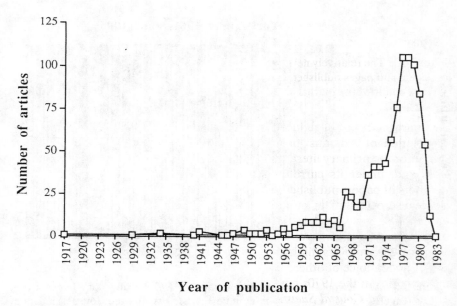

FIGURE 4.2. Over 1,000 citations in Adamus (1983) indicate a late-1970s increase in publications on wetland functions.

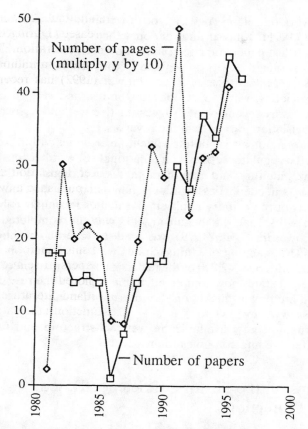

FIGURE 4.3. The relatively new journal, *Wetlands*, increased substantially in number of papers and pages published in the late 1980s and 1990s. Data provided by Doug Wilcox, editor of the journal since 1988.

is in part the result of reduced access to papers being published during the completion of Adamus' project. The journal *Wetlands,* now a principal outlet for the primary literature on wetlands (Figure 4.3) was established in 1981 and, under its current editor, has shown steady expansion in the numbers of papers published. Some of the increase is real growth, but some is because of a shift in publishing outlets, as scientists began to perceive *Wetlands* as a suitable destination for articles that might otherwise have been submitted elsewhere. That wetland science can support a journal of its own is evidence of its maturity.

The appearance of other new journals indicates that the burst of publishing activities in the 1970s has been sustained through the 1980s and 1990s. *Estuarine and Coastal Shelf Science* (called *Estuarine and Coastal Marine Science* from 1973 to 1980) specializes in coastal systems, including wetlands; *Aquatic Botany* (1975 ff.) grew out of National Science Foundation (NSF) funding for work on eelgrass beds, a type of wetland; *Wetlands Ecology and Management* (1991 ff.) has an international focus; *Ecological Engineering* (1992 ff.) publishes papers on constructed wetlands, and *Restoration Ecology* (1993 ff.) attracted papers on wetland restoration. A new journal, *Mangroves and Salt Marshes* focuses on coastal wetlands. These journals, their sponsoring societies, wetland ecology textbooks, and college courses in wetland ecology are recent events, suggesting that wetland science grew out of wetland legislation, rather than the reverse.

Wetland protection also generated scientific activities within governmental agencies. To regulate wetlands on a national scale, it became necessary to characterize, identify, and map them, that is, a National Wetland Inventory (NWI) was needed. Classification schemes ensure that individuals in different parts of the country apply similar names to similar habitats. The classification work that was begun in 1953 was not completed until the system of Cowardin et al. (1979) was made a federal standard in 1996 (Table 4.9). The mapping of United States wetlands began about twenty years ago and, with over 42,000 of the necessary 54,000 maps needed for the lower forty-eight states now completed. Some 18,000 NWI maps have been digitized and are now available on the Internet (see the NWI home page, URL address: www.nwi.fws.gov).

The information now available on wetland structure and functioning clearly is extensive and continues to grow.

Reflections on the History of Science-Based Wetland Protection

Based on the post-1972 increase in literature, it seems to us that wetland science followed, more than it drove, wetland protection. Another line of reasoning supports this contention: If science were the limiting factor in bringing about wetland protection, then it should be easier to conserve wetlands today than in the 1960s. However, it has been difficult just to retain the level of protection achieved in the 1970s. The 1990s have seen repeated efforts to undo much of the protection that developed before the publishing boom. In 1991, a national controversy arose over wetland definitions. Vice President Dan Quayle's Committee on Competitiveness proposed restricting wetland protection by changing the definition to exclude the drier wetland types. The EPA received approximately 80,000 public comments about the proposed change, indicating its highly controversial nature. The changes were not made, and the status quo was sustained.

If science were the critical factor, protection should have been strengthened by now, given so much information on wetland functions (no studies suggest that wetlands lack valuable functions). Instead, the plethora of comments sent to the EPA and the information explosion seem to have confused decision-makers (see Box 4.1). In 1993, Congress asked the EPA

TABLE 4.9. Scientific efforts that contributed substantially to wetland protection.

1953	Martin et al. (1953) published a wetland classification scheme with twenty types; it was used for the first inventory.
1956	First Wetland Inventory and Circular 39.
1959	Salt marsh conference, Sapelo, GA. Salt marsh recognized as an ecosystem. Teal presented his energy flow model, with a statement that 46% of total primary production was transformed by marsh consumers. "This means the salt marsh is producing and exporting enough energy to support a larger community than that living on the marsh" Ragotzkie, et al. 1959).
1971	An intergovernmental convention met in Ramsar, Iran, and developed an international treaty (the Ramsar Convention) that laid the groundwork for protecting the wetlands of global significance to waterbirds. The United States did not join the Ramsar Convention until 1985 (US FWS 1989).
1975	National Wetland Inventory established.
1976	EPA published "Impacts of construction activities in wetlands of the United States" by Darnell et al. (1976).
1978	National Symposium on Wetlands, Lake Buena Vista, Florida, and publication of the proceedings, *Wetlands functions and values: The state of our understanding* (Greeson, et al. 1979).
1979	Johnson, R.R., and J.F. McCormick. 1978. *Strategies for protection and management of floodplain wetlands and other riparian ecosystems.* USDA Forest Service, Washington, DC.
1979	Clark and Clark scientists' report, Lake Buena Vista.
1979	Cowardin et al. (1979) classification scheme was published, recognizing the variety of wetlands; Adamus (1983) vol. 1, p. 2 states that there are 210,240 distinguishable types in the classification system.
1980	Society of Wetland Scientists formed.
1981	Brinson, et al. (1981a,b) *Riparian ecosystems* published with 525 references. Discusses functions and attributes of riparian ecosystems; includes a chapter on the value of riparian ecosystems.
1982	U.S. Army Corps of Engineers established the wetland research program to refine wetland delineation, quantify wetland values for use in permit evaluation, and develop techniques for wetland restoration. (Office of Technology Assessment 1984).
1983	Adamus published *A Method of Wetland Functional Assessment* for Federal Highways recognizing many functions—Adamus listed twelve and cited >1,000 refs).
1983	National Wetland Values Assessment Workshop (Sather and Stuber 1984)
1984	The US FWS published *An overview of major wetland functions and values* (Sather and Smith 1994).
1984	Office of Technology Assessment published *Wetlands: Their Use and Regulation.*
1985	EPA Wetlands Research Plan (Zedler and Kentula 1985).
1987	The National Wildlife Federation published a *Status Report on Our Nation's Wetlands* (Feierabend and Zelazny 1987).

to have the National Academy of Sciences (NAS) study the issue of wetland definition. A national panel of scientists was formed and they deliberated for two years before reaching consensus. In May of 1995, their book was published (NRC 1995). If scientific input were limiting, Congress should have implemented the recommendations. The most that can be said is that the report forestalled an even greater attack on the Clean Water Act (John Meagher, EPA Wetlands Division, personal communication). Still sitting before the Legislature are the Hayes Bill (HR1330), and its companion Senate Bill (SB851), which threaten to define many wetlands out of existence. If enacted into law, the "wetness standard" would increase, thereby excluding some 50% of the jurisdictional wetlands from protection (EDF and WWF 1992). If enacted, all wetlands in the United States would have to be classified under a new function/value-based system with developers allowed to dredge or fill "type C" wetlands (i.e. those with fewer readily identifiable functions) without a Section 404 permit. A great deal of science was brought to bear on the issue, but the resulting actions were minimal (see Box 4.1).

Is the effect of an information explosion primarily one of confusion? Before the 1970s, there were fewer individuals to serve as spokespersons, smaller groups to consult to reach consensus, less peer review, and less literature to comprehend. What has changed to a large extent is the technical level of the debates over wetland protection. Both the questions and answers are highly complex, and few scientists can simplify information while still remaining true to its content. At the same time, few policy-makers can take the time to understand the details. And there are more voices to be heard. Controversies over wetland delineation, assessment techniques, mapping units, water quality improvement functions, mitigation agreements, and restoration procedures can easily occupy teams of scientists. In local decision-making arenas, it is not uncommon for opponents to hire separate sets of experts to testify for and against a proposed action (see Box 4.3). One team may offer convincing testimony that a wetland will remove over 90% of the nitrogen in wastewater so that a discharge to a coastal water body should be permitted, while another may argue effectively that phosphorus is the limiting nutrient downstream and the discharge should be prevented (Box 4.3). Sometimes the objective may be to obfuscate, rather than elucidate, the facts so that the basis for decisions-making shifts away from science, leaving economic and political forces to prevail.

The complexity of science and the conflicting testimonies of experts suggest that decision-makers need an intermediary body to identify areas of scientific consensus and develop science-based recommendations. Scientists have difficulty providing testimony that is both simple and correct; their data sets are too complex and broad generalities are too hard to support unequivocally. And there is never enough data to answer all questions. We suspect that citizen action groups (Table 4.10) are more often responsible for improved management of habitat than individual scientists, and they are increasingly skilled at soliciting and using scientific knowledge. We believe that scientists should play a stronger role, along the lines of a mediation panel.

TABLE 4.10. Organizations that work to protect, manage, and restore wetlands and other wild lands.[1]

Association of State Wetland Managers
Ducks Unlimited
Environmental Law Institute
Environmental Defense Fund
Friends of the Earth
National Audubon Society
National Wildlife Federation
Natural Resources Defense Council
Sierra Club
The Conservation Fund
The Nature Conservancy
The Wilderness Society
World Wildlife Fund

[1] See Appendix B in Kusler and Opheim 1996.

At the national level, the NAS and its working arm, the NRC, develop scientific consensus for selected issues. The staff, board, and commission members of the NRC identify issues requiring scientific advice or respond to requests from Congress or government agencies to assist in resolving a particular scientific issue. (One of the NRC's most widely read reports (NRC 1992) came about because a staff member, Sheila David, attended a conference on ecosystem restoration and became so interested that she proposed a study on restoration of aquatic ecosystems and lessons that could be learned from any successes or failures.)

An NRC committee to study an issue is selected from a list compiled by the staff with recommendations being made by members of the NAS, NRC board and commission members, and others in the scientific community. The NRC attempts to achieve a balance of expertise on its committee as well as a variety of viewpoints, and a balance of composition by gender, age, ethnicity, and geographic location. Final committees are therefore made up of a subset of scientists from throughout the United States and elsewhere, with the final filter being the candidate's availability for a one to three year commitment. These committees have tackled the issues of aquatic ecosystem restoration (NRC 1992), wetland delineation (NRC 1995), and presently the functions of riparian zones and strategies for management. Such a process does not ensure scientific consensus on all aspects of a debate, but it can lead to a clear statement of the status of the science and a document that summarizes the areas in which there is scientific agreement. When controversy remains, the reports point to critical areas needing further research.

We recommend that a similar consensus-building mechanism be developed to handle regional issues, so that more of the environmental conflicts can be explored in a nonpolitical arena. We further recommend that

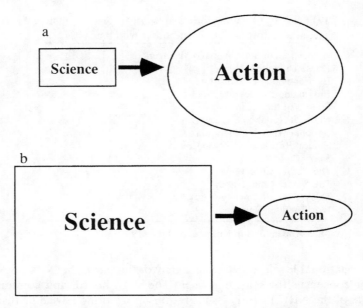

FIGURE 4.4. Linkages between wetland science and habitat protection. a. Compared to later protection efforts, the science available in the 1930s was used to take major actions (a greatly expanded National Wildlife Refuge System). b. In the 1990s, a great deal of science is available to be brought to bear on management issues, but the standards of evidence have increased, and it takes much more science to bring about even minor management actions (see Box 4.3).

decision-makers and environmental activists both look to science for information and that scientists be more aware of the need to simplify complex information and provide it in usable formats. Finally, we suggest that scientists become more directly involved in wetland management.

The Role of Wetland Studies in Ecosystem Science

Consistent with other success stories in this book, we list several contributions of wetland studies to the development of ecosystem science. Our list is based on discussions at the Cary Conference on how wetland research advanced the field, rather than a thorough literature review.

Studies of wetlands leading up to and following the Clean Water Act necessarily took an ecosystem view and landscape perspective, because wetlands were recognized as a significant part of the watershed and the water cycle. The focus on water quality improvement functions required whole-system and whole-watershed studies. Ewel and Odum's (1978) whole-ecosystem analysis of the wastewater-treatment capability of individual cypress domes stood out as a bold new approach that emphasized inputs, transformations, and outputs of intact natural wetlands, as well as replicated mesocosm-scale experimental treatments. The concept of manipulating wetlands to achieve improved wastewater treatment was popularized, and the field of ecological engineering (ecotechnology) developed (Mitsch and Jørgensen 1989). Mesocosms have since been used widely in ecosystem studies (Beyers and Odum 1993).

There was early recognition that ecosystems provided many kinds of services (such as nutrient removal), not just production of consumables (such as timber). An emphasis on belowground processes and microbial activity was essential for studies of nutrient retention, denitrification, and nutrient transformations. At the finest scale, chemical transformations at the anaerobic-aerobic interface were explored, giving attention to the importance of understanding microbes and microzones in ecosystem processes (Pomeroy and Wiegert 1981).

Wetland studies bridged work at the microzone scale with that at the landscape scale, following recognition that microbes played critical roles in wetlands that collected nutrients from the entire watershed. A landscape approach was essential, and the position of a wetland in the watershed was recognized as an important variable in the system's effectiveness in nutrient removal (Johnston et al. 1990). Today we have a hydrogeomorphic approach to classifying wetland functions that emphasizes landscape position (Brinson 1993), and attempts are being made across the nation to relate local ecosystem functions to habitat and community structure, hydrology, and geomorphology.

New methods also grew out of wetland research. Efforts to identify the base of food webs were unsuccessful using traditional stomach-content analyses, because plant materials were rapidly digested. Early on, stable-isotope techniques were developed to track the carbon of primary producers through food chains (Haines 1977), and subsequent multiple stable-isotope analyses have had widespread use in a variety of ecosystem studies.

Interest in global change led to studies of wetlands as sources vs. sinks for carbon. Peat was recognized as a major carbon sink, and methane production as an important process in greenhouse warming. Because wetlands cover a tiny fraction of the globe, wetland processes might not be expected to affect global cycles of elements; the role of wetlands is disproportionate to their area (for a review of our understanding of global cycles, see the chapter by Burke et al. chapter 7).

Because the Clean Water Act requires that damage to wetlands be mitigated, many wetland restoration efforts have been undertaken, and studies of the shortcomings of various restoration attempts are expanding our knowledge of what limits wetland ecosystem functioning (see the March 1996 special feature on mitigation in *Ecological Applications,* Zedler 1996, and MacMahon chapter 9). Mitigation projects offer unique opportunities for large-scale habitat creation and manipulation exercises that can support outstanding scientific work (see the December 1994 special issue on the Des Plaines River Wetlands Demonstration Project in *Ecological Engineering*). The most ambitious wetland restoration projects have pushed ecosystem modelers toward extremely ambitious goals, from linking nutrient inflows and fisheries within the watershed of Chesapeake Bay to predicting the effects of water level controls on the biota of south Florida, including the entire Everglades (Davis and Ogden 1994) and the Kissimmee River (*Restoration Ecology* Special Issue, September 1995).

For the future, we predict continued success in advancing the science of ecosystems. We also predict further success in bringing wetland science to bear on habitat conservation efforts, especially with regard to endangered species and habitat conservation planning. A large percentage of the nation's imperiled plants and animals depend on wetlands for at least part of their life cycle. With more than 50% of the wetland habitat gone in the conterminous United States (Dahl 1990), there is much to learn about species-wetland dependencies and wetland-landscape relationships.

Acknowledgments. We thank Stuart Findlay, Judy Meyer, Eugene Odum, Karen Poiani, and Cathy Wigand for their thoughtful comments on the draft manuscript. We are indebted to Peter Groffman and Michael Pace for inviting the contribution and for their enthusiastic encouragement during its preparation.

References

Adamus, P.R. 1983. *A method for wetland functional assessment.* U.S. Department of Transportation Report FHWA-1P-82–83, Vol. I–II. Washington, DC.

Amos, W.H. 1970. *The infinite river: a biologist's vision of the world of water.* Random House-Ballantine, NY.

Baczkowski, S.L. 1992. *The effects of a decreased salinity on juvenile California halibut,* Paralichthys californicus. M.S. thesis, San Diego State University, CA.

Beare, P.A. 1984. *Salinity tolerance in cattails* (Typha domingensis Pers.): *explanations for invasion and persistence in a coastal salt marsh.* M.S. thesis, San Diego State University, CA.

Beare, P.A., and J.B. Zedler. 1987. Cattail invasion and persistence in a coastal salt marsh: the role of salinity. *Estuaries* 10:165–170.

Bellrose, F.C. 1979. Waterfowl and wetlands—an integrated review. Pages 1–16 in T.A. Bookhout, ed. *Proceedings of the 1977 Symposium of the North Central Section of The Wildlife Society.* The Wildlife Society, Madison, WI.

Beyers, R.J., and H.T. Odum. 1993. *Ecological microcosms.* Springer-Verlag, New York.

Brenchley-Jackson, J. 1992. *Factors controlling the vegetation dynamics of* Spartina foliosa *and* Salicornia virginica *in California's coastal wetlands.* Ph.D. dissertation, University of California, Davis.

Brinson, M.M. 1993. *A hydrogeomorphic classification for wetlands*. Waterways Experiment Station Technical Report WRP-DE-4, U.S. Army Corps of Engineers, Washington, DC.

Brinson, M.M., B.L. Swift, R.C. Plantico, and J.S. Barclay. 1981a. *Riparian ecosystems: their ecology and status*. Eastern Energy and Land Use Team, National Water Resources Analysis Group, US FWS, FWS/OBS-81/17. Kearneysville, WV.

Brinson, M., B. Swift, R. Plantico, and J. Barclay. 1981b. *Riparian ecosystems: their ecology and status*. US FWS FWS/OBS-81/17, Washington, DC.

Brix, H. 1994. Constructed wetlands for municipal wastewater treatment in Europe. Pages 325–333 in W.J. Mitsch, ed. *Global wetlands: Old World and New*. Elsevier Science B.V., Amsterdam, The Netherlands.

Brown, S., and A.E., Lugo, eds. 1986. *Caribbean regional wetland functions*. Proceedings of a Workshop, October 28–29, 1986. Institute of Tropical Forestry, Rio Piedras, Puerto Rico.

Busnardo, M.J., R.M. Gersberg, R. Langis, T.L. Sinicrope, and J.B. Zedler. 1992. Nitrogen and phosphorus removal by wetland mesocosms subjected to different hydroperiods. *Ecological Engineering* 1:287–307.

Callaway J.C., J.B. Zedler, and D.L. Ross. 1997. Using tidal salt marsh mesocosms to aid wetland restoration. *Restoration Ecology* 5:135–146.

Callaway, J.C., and J.B. Zedler. In press. Interactions between a salt marsh native perennial (*Salicornia virginica*) and an exotic annual (*Polypogon monspeliensis*) under varied salinity and hydroperiod. *Wetlands Ecology and Management*.

Carson, R. 1951. *The sea around us*. Oxford University Press, New York.

Carson, R. 1962. *Silent spring*. Houghton Miffin, Boston, MA.

Clark, J., and J. Clark, eds. 1979. *Scientists' Report*. The National Symposium on Wetlands, Lake Buena Vista, Florida. November 6–9 1978. National Wetlands Technical Council, Washington, DC.

Cottam, C. 1934. *The eel-grass shortage in relation to waterfowl*. Transactions of the Twentieth American Game Conference. American Game Association, Washington, DC.

Cowardin, L.M., V. Carter, F.C. Golet, and E.T. LaRoe. 1979. *Classification of wetland and deepwater habitats of the United States*. U.S. Fish and Wildlife Service. FWS/OBS-79/31.

Dahl, T. 1990. *Wetlands losses in the United States, 1780's to 1980's*. U.S. Fish and Wildlife Service, Washington, DC.

Darnell, R.M. 1961. Trophic spectrum of an estuarine community, based on studies of Lake Pontchartrain, Louisiana. *Ecology* 42:553–568.

Darnell, R.M., W.E. Pequegnat, B.M. James, F.J. Benson, and R.A. Defenbaugh. 1976. *Impacts of construction activities in wetlands of the United States*. U.S. EPA Office of Research and Development, Corvallis Environmental Research Laboratory, Corvallis, OR.

Davis, S.M., and J.C. Ogden, eds. 1994. *Everglades, the ecosystem and its restoration*. St. Lucie Press, Delray Beach, FL.

Dennison, M.S., and J.F. Berry. 1993. *Wetlands: guide to science, law, and technology*. Noyes Publications, Park Ridge, NJ.

Eisenhower, M.S., and A.P. Chew, eds. 1935. *Yearbook of agriculture 1935*. U.S. Government Printing Office, Washington, DC.

108   J.B. Zedler, M.Q. Fellows, and S. Trnka

EDF, and WWF. 1992. *How wet is a wetland? The impacts of the proposed revisions to the federal wetlands delineation manual.* New York, NY and Washington, DC.

EPA. 1993. *Constructed wetlands for wastewater treatment and wildlife habitat.* EPA832-R-93–005. Washington, DC.

Errington, P.L. 1957. *Of men and marshes.* Macmillan, New York.

Ewel, K.C., and H.T. Odum. 1978. Cypress swamps for nutrient removal and wastewater recycling. Pages 181–198 in M.B. Wanielista and W.W. Eckenfelder, Jr., eds. *Advances in water and wastewater treatment biological nutrient removal.* Ann Arbor Scientific Publications, Inc., MI.

Feierabend, S.J., and J.M. Zelazny. 1987. *Status report on our nation's wetlands.* National Wildlife Federation, Washington, DC.

Flader, S.L., and J.B. Callicott. eds. 1991. *The river of the mother of God and other essays by Aldo Leopold.* University of Wisconsin Press, WI.

Forbes, S.A. 1887. The lake as a microcosm. *Illinois Natural History Survey Bulletin* 15:537–550.

Godfrey, P.J., E.R. Kaynor, S. Pelczarski, and J. Benforado. 1985. *Ecological considerations in wetlands treatment of municipal wastewaters.* Van Nostrand Reinhold, New York.

Gosselink, J.G., E.P. Odum, and R.M. Pope. 1974. *The value of the tidal marsh.* Publ. No. LSU-SG-74–03, Center for Wetland Resources, Louisiana State University, Baton Rouge.

Greeson, J.E., J.E. Clark, and J.R. Clark. 1979. *Wetland functions and values: State of our understanding.* Proceedings of the National Symposium on Wetlands held in Disneyworld Village, Lake Buena Vista, Florida. November 7–10, 1978. American Water Resources Association. Minneapolis, MN.

Haines, E.B. 1977. The origins of detritus in Georgia salt marsh estuaries. *Oikos* 29:254–260.

Hook, D.D., W.H. McKee, Jr., H.K. Smith, J. Gregory, V.G. Burrell, Jr., M.R. DeVoe, et al. 1988. *The ecology and management of wetlands, Vol. 1–2.* Croom Helm, London, UK.

Johnston, C.A., N.E. Detenbeck, and G.J. Niemi. 1990. The cumulative effect of wetlands on stream water quality and quantity. A landscape approach. *Biogeochemistry* 10:105–141.

Johnson, R.R., and J.F. McCormick. 1979. *Strategies for protection and management of floodplain wetlands and other riparian ecosystems.* USDA Forest Service GTR-WO-12. Washington, DC.

Kadlec, R.H. 1994. Wetlands for water polishing: free water surface wetlands. Pages 335–349 in W.J. Mitsch, ed. *Global wetlands: Old World and New.* Elsevier Science B.V., Amsterdam, The Netherlands.

Kellert, S.R. 1979. *Public attitudes toward critical wildlife and natural habitat issues.* USDI FWS, Washington, DC.

Koebel, J.W. Jr. 1995. An historical perspective on the Kissimmee River restoration project. *Restoration Ecology* 3:149–159.

Kuhn, N., and J.B. Zedler. 1997. Differential effects of salinity and soil saturation on native and exotic plants of a coastal salt marsh. *Estuaries* 20:391–403.

Kusler, J. 1978. Wetland protection: is science meeting the challenge? Pages 31–42 in P.E. Greeson, J.R. Clark, and J.E. Clark, eds. 1979. *Wetland functions and values: The state of our understanding.* Proceedings of the National Symposium on Wetlands. American Water Resources Association, Minneapolis, MN.

Kusler, J.A., W.J. Mitsch, and J.S. Larson. 1994. Wetlands. *Scientific American* January:64–70.

Kusler, J.A., and T. Opheim. 1996. *Our national wetland heritage: a protection guide, 2nd ed.* Environmental Law Institute, Washington, DC.

Leopold, A. 1949. *A Sand County almanac, and sketches here and there.* Oxford University Press, New York.

Lindeman, R.L. 1942. The trophic-dynamic concept of ecology. *Ecology* 23:399–418.

Lugo, A., and M. Brinson. 1978. Calculations of the value of salt water wetlands. Pages 120–130 in P.E. Greeson, J.R. Clark, and J.E. Clark, eds. 1979. *Wetland functions and values: the state of our understanding.* Proceedings of the National Symposium on Wetlands. American Water Resources Association, Minneapolis, MN.

Martin, A.C., N. Hotchkiss, F.M. Uhler, and W.S. Bourn. 1953. *Classification of wetlands of the United States.* USDI Fish and Wildlife Service, Special Scientific Report—Wildlife No. 20.

McCallie, S.W. 1911. *A preliminary report on drainage reclamation in Georgia.* Georgia Geological Survey Bulletin, Atlanta.

McComb, A.J., and P.S. Lake. 1990. *Australian wetlands.* Collins/Angus & Robertson Publishers, North Ryde, New South Wales.

McQueen, A.S., and H. Mizell. 1926. *History of Okefenokee Swamp.* Press of Jacobs and Co. of Georgia and Florida, Folkston, GA.

Meine, C. 1988. *Aldo Leopold: his life and work.* University of Wisconsin Press, WI.

Mitsch, W.J. ed. 1994. *Global wetlands: Old World and New.* Elsevier Science B.V., Amsterdam, The Netherlands.

Mitsch, W.J., and S.E. Jørgensen. 1989. *Ecological engineering: an introduction to ecotechnology.* John Wiley & Sons, New York.

Mitsch, W.J., and J.G. Gosselink. 1993. *Wetlands,* Second Edition. Van Nostrand Reinhold, New York.

Nash, R. 1967. *Wilderness and the American mind.* Yale University Press, New Haven, CT.

Nash, R. 1968. *The American environment: readings in the history of conservation.* Addison-Wesley, Reading, MA.

Nash, R., ed. 1970. *The call of the wild.* George Braziller, New York.

National Research Council. 1992. Restoration of Aquatic Ecosystems: *Science, Technology, and Public Policy.* National Academy Press, Washington, D.C.

Niering, W.A. 1966. *The life of the marsh: the North American wetlands.* McGraw-Hill Book Company: New York.

Niering, W.A. 1991. *Wetlands of North America.* Thomasson-Grant, Charlottesville, VA.

Nixon, S.W. 1980. Between coastal marshes and coastal waters. A review of twenty years of speculation and research on the role of salt marshes in estuarine productivity and water chemistry. Pages 437–525 in P. Hamilton and K.B. Macdonald, eds. *Estuarine and wetland processes.* Plenum Publishing Corp., New York.

Nixon, S.W., and V. Lee. 1986. *Wetlands and water quality: a regional review of recent research in the United States on the role of freshwater and saltwater wetlands as sources, sinks, and transformers of nitrogen, phosphorus and various heavy metals.* U.S. Army Corps of Engineers, Waterways Experiment Station, Vicksburg, MS.

NRC. 1992. *Restoration of aquatic ecosystems: science, technology, and public policy.* National Academy Press, Washington, DC.

NRC. 1995. *Wetlands characteristics and boundaries.* National Academy Press, Washington, DC.

Odum, E.P. 1959. *Fundamentals of ecology.* W.B. Saunders, Philadelphia, PA.

Odum, E.P. 1961. The role of tidal marshes in estuarine production. State of New York Conservation Department. *The Conservationist*, June–July 1961:12–35.

Odum, E.P. 1971. *Fundamentals of ecology, 3rd edition.* W.B. Saunders, Philadelphia, PA.

Odum, E.P. and A.E. Smalley. 1959. *Comparison of population energy flow of a herbivorous and a deposit-feeding invertebrate in a salt marsh ecosystem.* Proccedings of the National Academy of Science 45:617–622.

Odum, H.T. 1971. *Environment, power and society.* Wiley Interscience. New York.

Onuf, C.P. 1987. *The ecology of Mugu Lagoon: an estuarine profile.* U.S. Fish and Wildlife. Service, Biological Report 85(7.15) Washington, DC.

Office of Technical Assessment. 1984. *Wetlands: their use and regulation.* OTA-O-206. U.S. Congress, Office of Technology Assessment. Washington, DC.

Patrick, W.H., Jr. 1994. From wastelands to wetlands. *Journal of Environmental Quality* 23:892–896.

Pomeroy, L.R., and R.G. Wiegert. eds. 1981. *The ecology of a salt marsh.* Springer-Verlag, New York.

Ragotzkie, R.A., L.R. Pomeroy, J.M. Teal, and D.C. Scott, eds. 1959. *Proceedings: salt marsh conference held at the Marine Institute of the University of Georgia, Sapelo Island, Georgia, March 25–28, 1958.* Marine Institute, University of Georgia. Athens, GA.

Reddy, K.R., and E.M. D'Angelo. 1994. Soil processes regulating water quality in wetlands. Pages 309–324 in W.J. Mitsch, ed. *Global wetlands: Old World and New.* Elsevier Science B.V., Amsterdam, The Netherlands.

Sather, J.H., and J. Low. 1986. *Great Basin desert and montane regional wetland functions.* Proceedings of a Workshop held at Logan Utah. February 27–28, 1986.

Sather, J.H., and R.D. Smith. 1984. *An overview of major wetland functions and values.* Western Energy and Land Use Team, Division of Biological Services, US FWS, FWS/OBS-84/18. Washington, DC.

Sather, J.H., and P.J.R. Stuber. 1984. *Proceedings of the National Wetland Assessment Workshop 1983.* US FWS Western Energy and Land Use Team. FWS/OBS-84/12. Washington, DC.

Shaw, S.P., and C.G. Fredine. 1956. *Wetlands of the United States: their extent and their value to waterfowl and other wildlife.* Circular 39. USFWS USDI. Washington, DC.

Sinicrope, T.L., R. Langis, R.M. Gersberg, M.J. Busnardo, and J.B. Zedler. 1992. Metal removal by wetland mesocosms subjected to different hydroperiods. *Ecological Engineering* 1:309–322.

Siry, J.V. 1983. *Marshes of the ocean shore: development of an ecological ethic.* Texas A&M University Press, College Station.

Sloey, W.E., F.L. Spangler, and C.W. Fetter, Jr. 1978. Management of freshwater wetlands for nutrient assimilation. Pages 321–340 in R.E. Good, D.F. Whigham, and R.L. Simpson, eds. *Freshwater wetlands: ecological processes and management potential.* Academic Press, New York.

Strong, D.H. 1971. *The conservationists*. Addison-Wesley Pub.

Teal, J.M. 1962. Energy flow in the salt marsh ecosystem of Georgia. *Ecology* 43:614–624.

Teal, J., and M. Teal. 1969. *Life and death of a salt marsh*. Little, Brown, Boston, MA.

Teal, M., and J. Teal. 1964. *Portrait of an island*. Atheneum, New York.

Thomas, B. 1976. *The swamp*. W.W. Norton & Co. Inc. New York.

Udall, S.L. 1963. *The quiet crisis*. Avon Books, New York.

Urban Forest and Education Program. 1994. *Woodlands, wetlands & wildlife: a guide to the natural areas of New York City parks*. City of New York Parks and Recreation and City Parks Foundation, New York.

US FWS (Fish and Wildlife Service). 1989. *National wetlands priority conservation plan*. USDI Fish and Wildlife Service. Washington, DC.

USDI (Department of Interior). 1994. *The impact of federal programs on wetlands, Vol. II, A report to Congress by the Secretary of the Interior*. U.S. Dept. of the Interior, Washington, DC.

van der Valk, A., ed. 1989. *Northern prairie wetlands*. Iowa State University Press, Ames.

van der Valk, A., and J. Hall. 1986. *Alaska: regional wetland functions*. Proceedings of a Workshop held at Anchorage, AK. May 28–29, 1986.

Warner, W.W. 1976. *Beautiful swimmers: watermen, crabs, and the Chesapeake Bay*. Little, Brown, Boston, MA.

Wentz, W.A. *Wetland values and management*. U.S. Department of Fish and Wildlife, Washington, D.C.

Wharton, C.H. 1970. *Southern river swamp*. Georgia State University, Atlanta.

Wharton, C.H. 1973. The Alcovy River. Pages 15–25 in *Citizens' Advisory Committee on Environmental Quality. Citizens make the difference: case studies of environmental action*. Citizens' Advisory Committee on Environmental Quality, Washington, DC.

Williams, M. 1990. *Protection and retrospection. Wetlands: a threatened landscape*. Special Publications Series, Institute of British Geographers. Basil Blackwell, Inc., London, England.

Zedler, J.B. 1981. The San Diego River marsh before/after the 1980 flood. *Environment Southwest* 495:20–22.

Zedler, J.B. 1985. Wastewater input to coastal wetlands: management concerns. Pages 127–133 in Godfrey et al., eds. *Ecological considerations in wetlands treatment of municipal wastewaters*. Van Nostrand Reinhold, New York.

Zedler, J.B. 1996. Ecological issues in wetland mitigation: an introduction to the forum. *Ecological Applications* 6:33–37.

Zedler, J.B., and P.A. Beare. 1986. Temporal variability of salt marsh vegetation: the role of low-salinity gaps and environmental stress. Pages 295–306 in D. Wolfe, ed. *Estuarine Variability*. Academic Press, New York.

Zedler, J.B., M. Busnardo, T. Sinicrope, R. Langis, R. Gersberg, and S. Baczkowski. 1994. Pulsed-discharge wastewater wetlands: the potential for solving multiple problems by varying hydroperiod. Pages 363–368 in W.J. Mitsch, ed. *Global wetlands: Old World and New*. Elsevier, Amsterdam, The Netherlands.

Zedler, J.B., T. Huffman, and M. Josselyn. 1985. *Pacific regional wetland functions*. Proceedings of a Workshop held at Mill Valley, CA. April 14–16, 1985.

Zedler, J.B., and M.E. Kentula. 1985. *Wetlands research plan*. Environmental Research Laboratory, Office of Research and Development, U.S. Environmental Protection Agency, Corvallis, OR.

Zedler, J.B., R. Koenigs, and W. Magdych. 1984a. *Freshwater release and southern California coastal wetlands: management plan for the beneficial use of treated wastewater in the Tijuana River and San Diego River estuaries*. San Diego Association of Governments, San Diego, CA.

Zedler, J.B., R. Koenigs, and W. Magdych. 1984b. *Freshwater release and southern California coastal wetlands; Report 1, Streamflow for the San Diego and Tijuana Rivers; Report 2, Review of salinity effects and predictions of estuarine responses to lowered salinity*. San Diego Association of Governments, San Diego.

Zedler, J.B., C.S. Nordby, and T. Griswold. 1990. *Linkages: among estuarine habitats and with the watershed*. Technical Memorandum under contract number NA89AA-D-CZ043. National Oceanic and Atmospheric Administration, National Ocean Service, Marine and Estuarine Management Division. Washington, DC.

5
Riparian Forest Ecosystems as Filters for Nonpoint-Source Pollution

RICHARD LOWRANCE

Summary

Riparian (streamside) ecosystems have been the subject of ecosystem research for about the last twenty years. Ecosystem research has shown that riparian areas are especially effective controllers of nitrogen and sediment movement to streams and other water bodies. Nitrogen control is primarily the result of biotic processes. Sediment control is primarily caused by physical processes that are enhanced by such biotic interactions as enhanced infiltration and leaf litter at the soil surface. Riparian areas are increasingly being used as a landscape- and watershed-management technique to reduce the risk of pollution and to create and preserve healthy stream ecosystems. The management practices recommended for riparian areas are based on ecosystem research. Riparian ecosystem policies and management are a success story for the application of ecosystem science to real-world problems. Riparian ecosystem research has been used to develop generalized management practices for USDA action agencies such as the Natural Resources Conservation Service and the Forest Service. Ecosystem-management recommendations for riparian areas have been incorporated into such programs as the Coastal Zone Management Act Reauthorization and the Chesapeake Bay Program. The USDA has incorporated riparian policies into the Conservation Reserve Program (CRP) reauthorized in the 1996 Farm Bill. The CRP allows farmers to voluntarily enroll such environmentally sensitive lands as riparian ecosystems because of the high environmental benefits associated with riparian areas compared to other lands. The success of riparian ecosystem research in shaping riparian policy and management results from multiple factors including: agreement among major studies on the effectiveness of riparian ecosystems for nonpoint-source pollution control; flexibility of management practices; and use of riparian ecosystems for multiple functions.

Introduction

Common sense tells us that healthy streams depend on healthy streamside or riparian areas. Riparian ecosystems are adjacent to aquatic ecosystems, and are often ecotones or transition zones with distinct vegetation and soil characteristics resulting from their landscape position. In most humid and subhumid regions, natural riparian ecosystems are native forests that are adapted to unregulated hydrologic regimes. In the past two decades, research on riparian forest ecosystems has shown that they can help stabilize streambanks, control nonpoint-source pollution, and control the environment of aquatic organisms. The past two decades of research on riparian ecosystems has been translated into myriad programs to provide protection, enhancement, and restoration of riparian ecosystems in forest, urban, and agricultural environments.

Riparian ecosystems are of special interest because they provide multiple benefits and are important as components of landscape- and watershed-scale management because of these multiple benefits. Research on riparian ecosystems has typically focused on one of three areas: (1) riparian ecosystems as habitat for terrestrial and wetland organisms; (2) riparian ecosystems as controllers of aquatic ecosystems; or (3) riparian ecosystems as processors of exogenous inputs and regulators of output. Management recommendations for riparian ecosystems have typically integrated pollution control and aquatic ecosystem functions.

In this chapter, I will first provide a brief overview of how water quality and nonpoint-source pollution control programs have stimulated interest in riparian forest ecosystems. Next, I will provide an overview of the role of ecosystem and watershed research in the development of riparian forest buffers as a nonpoint-source pollution control practice. Third, I will review the programs for riparian forest buffer restoration, enhancement, and preservation that are based on our knowledge of these ecosystems. Finally, I will summarize my perspective on the successes of riparian ecosystem research and present a set of ecosystem management research topics that should be addressed by ongoing or future research.

Riparian Ecosystems and Clean Water

The Federal Water Pollution Control Act Amendments of 1972, the Federal Clean Water Act amendments of 1978, and subsequent amendments allow states, tribes, and other jurisdictions to set their own water quality standards but require that beneficial uses of water must at a minimum provide for "the protection and propagation of fish, shellfish, and wildlife" and provide for "recreation in and on the water." Section 305(b) of the Clean Water Act (CWA) requires that states biennially survey their water quality for attainment of the so-called "fishable and swimmable" goals of the act and report the results to the EPA. As pollutant loadings from point sources have decreased, the proportion of water quality impairment attributable to nonpoint-source pollution has steadily increased (US EPA 1994). Nonpoint-source pollution has many definitions. Technically, the term nonpoint-source is defined to mean any source of water pollution that does not meet the legal definition of "point source" (US EPA 1993). Nonpoint-source pollution is caused by rainfall or snowmelt moving over or through the ground. As the water moves, it carries natural and anthropogenic pollutants, finally depositing them in rivers, streams, lakes, wetlands, coastal waters, and groundwater. Hydrologic modification is also considered a form of nonpoint-source pollution (US EPA 1993).

The EPA reports that of surveyed rivers, 36% are impaired (do not meet water quality standards), and that most of the impairment is the result of nonpoint-sources (US EPA 1994). Agriculture is the primary source of impairment, being associated with 60% of the reported problems (Figure 5.1). The leading pollutants are bacteria, silt, and nutrients (US EPA 1994). Although the surveyed segments only account for 17% of the nation's streams and rivers, over 224,000 miles of impaired streams were identified in the survey (Figure 5.1). Agriculture impacts the majority of the impaired streams through nonpoint-source pollution.

Concern over agricultural nonpoint-source pollution is not new, however. Considerable research was conducted on the causes, consequences, and control of agricultural nonpoint-source pollution in response to passage of the Water Pollution Control Act Amendments of 1972 (Chesters and Shierow 1985). By 1978 when the CWA was passed, literally hundreds of projects were completed or ongoing to determine the causes and consequences of agricultural nonpoint-source pollution (Chesters and Shierow 1985). Although research activity diminished in the early 1980s, the 1982 305(b) report documented that agriculture was the most pervasive source of pollution in every region of the country (Myers et al. 1985). By 1985, the administrator of the EPA stated that "nonpoint-source pollution occupies a place near the top of the EPA agenda" (Thomas 1985). Perhaps the many years of concern with nonpoint-source pollution and the apparent seriousness of the problem in the most recent water quality assessments (US EPA 1994) indicates that new approaches for control are necessary. Most practices to control nonpoint-source pollution are designed to reduce edge-of-field loadings. Such practices as fertilizer and manure management, conservation tillage, terraces, and integrated pest management are implemented to reduce the loadings of nutrients, pesticides, or sediments moving to natural waters from agriculture. Certain types of relatively simple buffer systems such as grass waterways and sediment detention basins have been used to help control nonpoint-source pollution. These practices rely on the physical attributes of the system to detain sediment and sediment-borne pollutants. Until the 1990s, the deliberate use of complex natural, managed, or created ecosystems as buffers was rare.

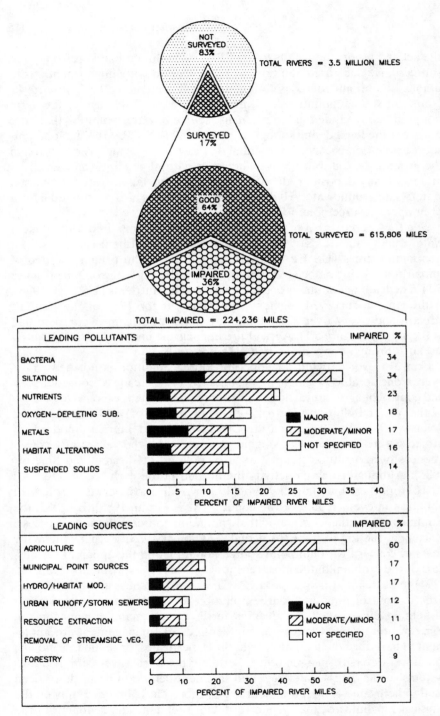

FIGURE 5.1. Results of biennial survey of nations rivers (US EPA 1994). Based on 1994 section 305(B) reports submitted by states, tribes, territories, commissions, and the District of Columbia.

116

Interest in buffer systems has increased because of the recognized need to evaluate water quality on a watershed scale. Concepts embodied in the CWA, especially the requirement for states to develop and implement total maximum daily load (TMDL) plans for target watersheds, make it essential that the control of nonpoint-source pollution be evaluated at the watershed scale (US EPA 1991). Section 303(d) of the CWA establishes the TMDL process to provide for water quality-based controls when technology-based controls are inadequate to achieve state water quality standards. The TMDL establishes the allowable loadings for an impaired body of water. Particularly important to the use of riparian forest buffers as a best management practice (BMP), "EPA recognizes that it is appropriate to use the TMDL process to establish control measures for quantifiable non-chemical parameters that are preventing the attainment of water quality standards" (US EPA 1991). When these nonchemical parameters are such factors as stream temperature, channel morphology, or habitat alteration, riparian forest ecosystems are relevant to both chemical and nonchemical sources of impairment. Use of such complex ecosystems as riparian forests and wetlands as buffers is compatible with evaluation and control of nonpoint-source pollution at the watershed scale.

Riparian Ecosystems and Water Quality

Early research on riparian ecosystems in agricultural watersheds was only partially motivated by concerns over nonpoint-source pollution. Beginning in the late 1970s, watershed science, agricultural water pollution research, and ecosystem ecology all influenced the study of riparian forest buffer systems. These three approaches can be characterized by the position of the observer. Watershed science, especially the study of gaged watersheds, allowed us to stand at the bottom of the watershed looking uphill and examine the integrated response of the entire system. Soil science and porous media transport (i.e. agrichemical transport) allowed us to stand in the field and look downhill at the fate and transport of pollutants. Ecosystem science allowed us to stand in the riparian ecosystem and look both ways: uphill at the inputs, the processing, and the ecosystem-level effects of those inputs; and downhill at the effect of the riparian ecosystems on the stream.

Little River Watershed: Investigations of Riparian Ecosystems in Agriculture

Little River Watershed (LRW) is a USDA-ARS experimental watershed in the Gulf-Atlantic Coastal Plain near Tifton, Georgia (Figure 5.2) established to understand hydrology and water quality of southeastern Coastal Plain watersheds. Land use in the LRW is about 60% agriculture and

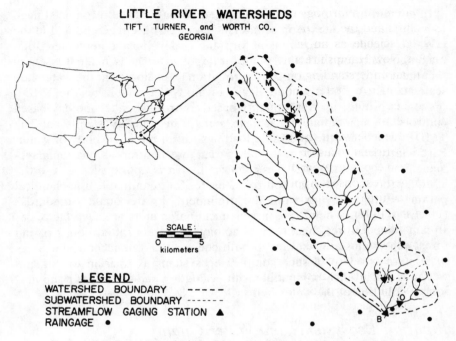

FIGURE 5.2. Location of Little River Watershed in the Coastal Plain.

depends on large inputs of Nitrogen (N) and Phosphorus (P) fertilizer. Although N budgets for agricultural lands of LRW were not being done in the early 1970s, it was well known that uptake of N by harvested crops accounted for 50% or less of the N applied to the crop (Frissel 1977). Therefore, there was abundant N available for transport from the watershed. Yet, the water quality data being collected for streamflow and rainfall indicated that there was less nitrate-N leaving the watershed in streamflow than entered in rainfall (Yates and Sheridan 1983). It appeared that the watershed acted as a sink for N. In other regions, agriculture was known to be a major source of nitrate to surface waters through discharges from tile drains and movement to shallow groundwater (Kohl et al. 1971; Chichester 1976). Did these coastal plain fields behave in a fundamentally different way than fields that were sources of N pollution? Did the entire watershed behave in a fundamentally different way? Studies from small field-sized drainages in the LRW showed that most of the N was transported from fields as nitrate in shallow groundwater (Jackson 1973), yet very little N left the LRW in the nitrate form in streamflow. Clearly something was occurring in the watershed to reduce nitrate transport.

These limited findings about nitrate budgets for LRW helped expand the horizons from that of watershed hydrology to a broader field of watershed science. This is an important distinction because concurrent with much of the watershed hydrology being done in agricultural watersheds, ecosystem ecologists and the USDA Forest Service (USDAFS) were developing watershed science as an integrative discipline in such places as Hubbard Brook, New Hampshire and Coweeta, North Carolina. The approach was to use an understanding of nutrient-cycling processes within forest ecosystems to gain a better understanding of the integrated response of the watershed—both water quantity and quality (Borman and Likens 1967; Likens et al. 1977; Swank and Douglas 1977). A cross-fertilization of ideas was needed between the agricultural watershed/water pollution research with its original focus on water management and the forested watershed/ ecosystem ecology research with its focus on fundamental understanding of watershed responses. Applying the techniques used in forested watersheds to agricultural watersheds would obviously require a number of modifications. Input/output budgets for agricultural watersheds would require accurate estimates of fertilizer and N-fixation inputs as well as precipitation and atmospheric deposition. We expected that nutrient cycling in agroecosystems would control the outputs of nutrients and the watershed response. We also expected that the spatial configuration of land use in the LRW would affect the watershed response. When viewed from the air, many such coastal plain watersheds as LRW were easily divisible into two systems—an upland agricultural source area and a riparian forest receiving area where the vegetation was mostly regenerated pines and hardwoods

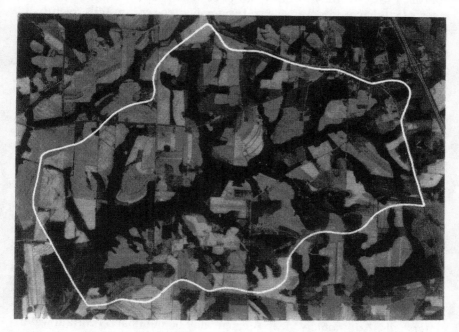

FIGURE 5.3. Aerial photograph of a subwatershed of Little River.

(Figure 5.3). In most cases, the riparian forests formed a continuous buffer along the stream. Soil maps showed that the land-use configuration was clearly a response to soil suitability for agriculture. Many of the riparian forests were developed on wetland soils and because of the natural low-gradient channel of Little River and its tributaries, there was little opportunity for enhanced drainage of the wetland soils for agriculture. Even when drained, most of the wetland soils were unsuitable for agriculture without irrigation and extra inputs of lime and fertilizer because of low-moisture-holding capacity and high acidity. Clearly, farmers were farming the best agricultural soils and leaving the poorer agricultural soils for natural forest regeneration.

Ecologists and watershed scientists collaborated on a proposal to the National Science Foundation (NSF) in 1978 entitled "Nutrient Cycling in an Agricultural Watershed Ecosystem." The specific objectives of the project, which started in October of 1978 were: (1) to understand internal nutrient-cycling processes and input/output relationships of selected field-scale management systems; (2) to understand the internal nutrient-cycling processes and input/output relationships for the riparian forests; and (3) understand the dynamics of nutrient transport from the entire LRW and its gaged subwatersheds and relate these dynamics to the riparian ecosystems.

Research outlined in this proposal was organized around the watershed approach for the study of nutrient cycles. The proposal cited five advantages of studying properly selected watershed ecosystems (Bormann and Likens 1967):

1. Watersheds represent natural, easily definable ecosystems.
2. Properly selected and gaged watersheds allow evaluation of deep seepage and erosional nutrient losses.
3. The watershed approach makes the calculation of mineral budgets for an entire ecosystem possible.
4. This approach allows evaluation of land-water interactions in the context of various land-management policies.
5. Internal nutrient-cycling processes and the role of individual ecosystems in larger biosphere processes can be examined.

The overall objectives and approaches of the proposed research recognized and sought to deal with complications resulting from the complexity and variability of watersheds dominated by intensive agricultural management systems (Lowrance et al. 1985). Among these differences were:

1. Watersheds can be defined by topographic boundaries, but many field-sized agroecosystems may exist within the topographic watershed.
2. Irrigation and artificial drainage can alter hydrology of agricultural watersheds. Additionally, deep seepage may be minimal on certain watersheds but these watersheds may not adequately represent areas where deep seepage occurs.

3. Nutrient budgets are more difficult to calculate because of the inputs and outputs of nutrients, energy, and materials associated with agriculture. External factors (e.g. prices and other market variables affect both inputs and outputs).
4. Individual crop fields exhibit seasonal, annual, and long-term nutrient dynamics in response to management. The complexity of an entire watershed may obscure management effects on land-water interactions.
5. Distinct import and export pulses affect both the nature and rate of internal nutrient-cycling processes on agricultural watersheds.

The original objectives of the riparian ecosystem research on LRW were relatively straightforward. Hypotheses reflected the integration of approaches through a hierarchical framework (Figure 5.4). Objectives at the levels of large watershed, subwatershed, source or sink area, processes, and functional groups were identified and integrated.

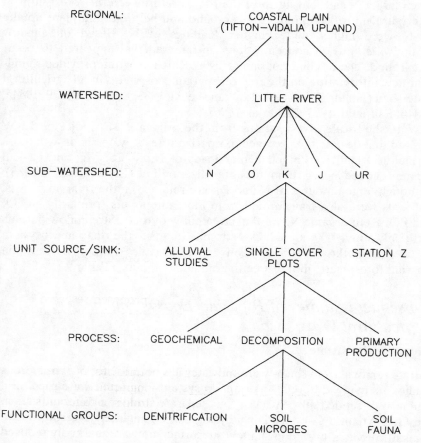

FIGURE 5.4. Hierarchical plan for research on Little River Watershed.

TABLE 5.1. Difference in nitrogen forms entering the riparian ecosystem in shallow groundwater and leaving in streamflow.[1]

	Groundwater Input	Streamflow Output	Difference (Input-Output)
NO_3^--N	10,153	1,000	9,153
NH_4^+-N	990	112	878
Organic N	2,325	4,470	−2,145

[1] All units are kg per year for the entire riparian zone (from Lowrance et al. 1983).

Nutrient budgets for the entire riparian zone showed that there was significant net retention of N and transformation of N from inorganic to organic forms. Denitrification and incorporation of N into plant woody biomass were approximately equal N sinks. Total N uptake by plants was a much larger N sink but much of the N was returned to the forest floor in litterfall. Denitrification and aboveground and belowground plant uptake of N were higher under higher loadings of N. Based on shallow groundwater nitrate concentrations, nitrate removal appeared to occur year-round. The original riparian forest research program provided significant new information on the role of riparian ecosystems in an agricultural watershed (Hendrickson 1982; Lowrance et al. 1983; Lowrance et al. 1984a, 1984b; Fail et al. 1985; Fail et al. 1986).

Watershed-scale investigations from the original NSF studies on LRW showed that despite the presence of riparian forests, watersheds with more agricultural land discharged higher loads of N, K, Ca, Mg, and Cl—all elements added in large amounts as fertilizer or lime (Lowrance et al. 1985). Although large amounts of N and P were retained in the riparian forests, nutrients were also transformed from inorganic inputs (primarily NO_3-N and PO_4-P) to organic N and P in streamflow outputs. Streamflow outputs were only about 1,000 kg of NO_3-N despite inputs to the riparian ecosystem of over 10,000 kg of NO_3-N. Conversely, organic N outputs from the riparian forest were about twice the organic N inputs. (Table 5.1).

Subsequent Studies of Riparian Ecosystems in Agricultural Watersheds

Groundwater Nitrate

Nitrate removal from shallow groundwater has been the focus of numerous studies in many different riparian ecosystems impacted by agricultural drainage or septic tanks. At least four separate studies at different sites in the Gulf-Atlantic Coastal Plain Physiographic Province have shown that concentrations of nitrate in shallow subsurface flow are markedly reduced after passage through portions of natural riparian forest (Lowrance et al. 1983, 1984a; Peterjohn and Correll, 1984; Jacobs and Gilliam 1985; Jordan et al. 1993). Several of these studies showed that most reduction in nitrate concentration takes place within the first 10 to 15 m of forest (Lowrance et al. 1984a; Peterjohn and Correll 1984; Jacobs and Gilliam 1985) and that the necessary forest buffer width for shallow groundwater nitrate removal could be relatively short. Nitrate removal from shallow groundwater in riparian forests was demonstrated in areas of Rhode Island receiving drainage from household septic tanks (Groffman et al. 1992; Simmons et al. 1992).

Understanding nitrate removal in riparian ecosystems requires understanding of hydrologic connections between source areas and streams (Hill 1996). The highest levels of nitrate removal were found in areas with high water tables that caused shallow groundwater to flow through or near the root zone to streams. The mechanisms for removal of nitrate in these study areas were thought to be a combination of denitrification and plant uptake. Linkages between plant uptake and denitrification in surface soils were postulated as a means for maintaining high denitrification rates in riparian ecosystems (Groffman et al. 1992; Lowrance 1992). In contrast, riparian systems without substantial contact between the biologically active soil layers and groundwater or with very rapid groundwater movement appear to allow passage of nitrate with only minor reductions in concentration and load. Correll et al. (1994) reported both high nitrate concentrations reaching streams and high nitrate removal rates. The seemingly contradictory findings occurred beneath a riparian forest where very high nitrate flux and rapid groundwater movement through sandy aquifer material limited nitrate removal efficiency. Staver and Brinsfield (1996) showed that groundwater flow beneath the biologically active zone of a narrow riparian buffer along a tidal embayment in Maryland resulted in little removal of nitrate. Phillips et al. (1993) indicated that groundwater nitrate might bypass narrow areas of riparian forest wetland and discharge into stream channels relatively unaltered when the forest is underlain by an oxygenated aquifer. This pattern of groundwater flow was supported by modelling of a small Coastal Plain watershed in Maryland (Reilly et al. 1994). Isotopic analysis of groundwater and surface water in this watershed suggested that denitrification was not affecting the nitrate concentrations of discharging groundwater. A riparian wetland had little effect on nitrate movement to the stream (Bohlke and Denver 1995). In these cases in which nitrate-enriched water surfaces in the stream channel, a wide riparian forest would have little effect on nitrate. Deeply rooted vegetation near the stream, however, might have some effect. Another means for high nitrate water to bypass the riparian zone is to enter as springs or streamlets. Hill (1990) estimated that 90% of the annual groundwater nitrate input to a small swamp in Ontario was contributed by seepage at the upslope riparian edge. The riparian ecosystem did not remove this nitrate because the streamlets flowed through the riparian zone in aerobic surface sediments that had low denitrification rates. Residence times in the riparian zone were also decreased drastically for groundwater that emerged and moved across the surface (Warwick and Hill 1988).

Nitrate removal is affected by soil heterogeneity. Studies in New Zealand showed that the majority of nitrate removal in a pasture watershed took place in organic riparian soils that received large amounts of nitrate-laden groundwater (Cooper 1990). The location of the high organic matter soils at the base of hollows caused a large proportion of groundwater (37 to 81%) to flow through the organic soils although they occupied only 12% of the riparian zone. A related study in New Zealand (Schipper et al. 1993) found very high nitrate removal in the organic riparian soils but streamflow was still enriched with nitrate. The authors speculated that water movement through mineral soils was responsible for most of the nitrate transport into streams. In riparian systems with intermingled organic and mineral soils we need to better understand where groundwater is moving and what types of soils it will contact, especially in seepage areas.

Although several studies have found plant uptake to be an important nutrient-removal mechanism in riparian forest ecosystems (Peterjohn and Correll 1984; Fail et al. 1986; Correll and Weller 1989; Groffman et al. 1992), several factors may reduce the importance of plants as nutrient sinks. Pollutants in groundwater flowing into the riparian buffer will only be accessible to plants if the water table is high in the soil profile (Ehrenfeld 1987) or if evapotranspiration moves water and solutes into the root zone. Coastal Plain riparian forests control localized downslope water transport by creating moisture gradients that move water in unsaturated flow from both the adjacent stream and the upland field (Bosch et al. 1996). Nutrients in surface runoff and in water percolating rapidly through soil macropores as "gravitational water" may not be available to plants. Large rainfall events, which may transport a high percentage of pollutants (Jaworski et al. 1992), often produce concentrated surface flow and macropore-dominated percolation.

Sequestering of nutrients by plants is also limited by seasonal factors. In temperate deciduous forest ecosystems, plant uptake will decline or stop during the dormant winter season. A high percentage of surface and groundwater flow occurs in the winter in most humid climates. There is also concern that nutrients in plant tissues can be released back into the soil solution following litterfall and decomposition. However, nutrients released from decomposing plant litter may be subject to microbial, physical, or chemical attenuation mechanisms in the root zone of forest soils. Storage of nutrients in woody tissue is a relatively long-term attenuation, but still does not result in removal of nutrients from the ecosystem unless biomass is removed. A final concern about plant uptake as a nutrient-removal mechanism arises from the possibility that the ability of trees in a buffer zone to sequester nutrients in woody biomass becomes less as trees mature

(Vitousek and Reiners 1975). If the average tree age in a riparian forest buffer is less than fifty to 100 years, as in many agricultural areas, the forest should be accumulating nutrients in woody biomass. Although net vegetation accumulation of nutrients may reach zero, net ecosystem accumulation may continue as nutrients are stored in soil organic matter.

In many cases, riparian zone retention of groundwater-borne nitrate may depend on a complex interaction of hydrology, plant, soil, and microbial factors. The potential importance of these interactions is hypothesized based on studies in which significant rates of nitrate removal from groundwater were measured, but the potential for denitrification in the subsurface was low. Groffman et al. (1992) and Hanson et al. (1994a) suggested that surface soil denitrification of groundwater-derived nitrate is an important route of N removal in riparian forests. Nitrogen removal depends on plant uptake of nitrate from groundwater, decomposition, N release from plant litter, and nitrification and denitrification of this N in surface soil (Figure 5.5). These interactions vary within and between riparian forests and should be strongly influenced by soil drainage class, vegetation and soil type, climate, and groundwater quality. Although soil denitrification should be sustainable indefinitely under proper conditions with a supply of nitrate and available C, Hanson et al. (1994b) found that long-term groundwater nitrate loading led to symptoms of N saturation in the surface soils of a riparian forest buffer.

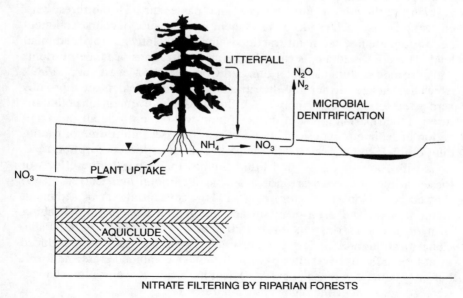

NITRATE FILTERING BY RIPARIAN FORESTS

FIGURE 5.5. Pathway for nitrate removal by plant uptake followed by eventual denitrification.

Pollutants in Surface Runoff

Fewer studies have been published on nonpoint-source pollution removal from surface runoff in riparian forests. The primary functions of the riparian forest buffer relative to surface runoff are to remove sediment and sediment-borne pollutants and to allow infiltration of runoff. Daniels and Gilliam (1996) found that mature riparian forests were effective for sediment-load reduction with removal of 50 to 80% of inputs from upland fields. Sediment trapping in riparian forest buffer zones is facilitated by physical interception of surface runoff that causes flow to slow and sediment particles to be deposited. Effective sediment trapping requires that runoff be primarily sheet flow. Channelized flow is not conducive to sediment deposition and can actually cause erosion of the riparian forest.

Two studies on long-term sediment deposition in riparian forests (Lowrance et al. 1986; Cooper et al. 1987; Lowrance et al. 1988) indicated that long-term deposition is substantial. In both these studies, two main actions occurred: (1) the forest edge fostered large amounts of coarse sediment deposition within a few meters of the field-forest boundary; (2) finer sediments were deposited further into the forest and near the stream. Both Cooper et al. (1987) and Lowrance et al. (1986) found much higher depths of sediment deposition at the forest edge than near the stream. A second peak of sediment depth was often found near the stream, possibly from upstream sediment sources deposited in overbank flows (Lowrance et al. 1986). Although the surface runoff that passes through the forest edge environment is much reduced in sediment load because of coarse sediment deposition, the fine sediment fraction is enriched relative to total sediment load. These fine sediments carry higher concentrations of labile nutrients and adsorbed pollutants (Peterjohn and Correll 1984; Magette et al. 1989), which are carried further into the riparian forest and are deposited broadly across the forest floor or may reach the stream. Movement of pollutants through the riparian forest in surface runoff will be controlled by a combination of sediment deposition and erosion processes, infiltration of runoff, dilution by incoming rainfall/throughfall, and adsorption/desorption reactions with forest-floor soil and litter. Studies that separate the effects of these various processes are not available. Peterjohn and Correll (1984) showed reductions by factors of 3 or 4 in concentrations of sediment, ammonium-N, and ortho-P in surface runoff that passed through about 50 m of a mature riparian forest buffer in the Maryland Coastal Plain. Concentrations and loads of two herbicides in surface runoff were reduced by at least a factor of 10 after passage through a combined grass and forest buffer system (Lowrance et al. 1997). Net herbicide concentration changes were greatest under the highest loading rates.

All studies of surface runoff through riparian forests agreed on the importance of minimizing channelized flow through the riparian forest. Flow in channels will bypass many of the nutrient removal processes that occur in surface runoff. Most studies have recommended spreading flow with either herbaceous filter strips or low berms before it reaches the riparian forest buffer.

Other Effects on Aquatic Ecosystems

As interest in the nonpoint pollution control value of riparian ecosystems increased, recognition of their importance to the physical and trophic status of streams also developed. Karr and Schlosser (1978) quantified the effects of riparian vegetation on sunlight penetration and temperature of streams. Research in the 1980s confirmed the importance of large woody debris and leaf litter inputs to the habitat and trophic status of most small streams (Meyer and O'Hop 1983; Benke et al. 1985; Harmon et al. 1986). Woody debris derived from riparian forests plays an important role in controlling channel morphology, the storage and routing of organic matter and sediment, and the amount and quality of fish habitat (Bisson et al. 1987). Although a thorough review of these studies is beyond the scope of this chapter, an understanding of riparian ecosystem effects on nonchemical water quality was also used to guide the development of management guidelines for riparian forest buffers. Gregory et al. (1991) provides an excellent review of interactions between riparian ecosystems and aquatic habitat.

Development of a Riparian Forest Buffer Specification

By the early 1990s, it was clear that a significant body of knowledge on riparian ecosystems and their effects on adjacent aquatic ecosystems had accumulated. At the same time, there was increasing interest in evaluating water quality on a watershed scale and in defining water quality in a way that clearly went beyond chemical constituents. These changes in how regulatory and advisory agencies viewed water quality were well-matched with the increasing knowledge and interest in riparian ecosystems. It was time to synthesize the existing knowledge into recommendations for the establishment, maintenance, and management of riparian ecosystems for a broad range of water quality functions. In 1991, USDAFS, with assistance from the USDA Agricultural Research Service, the USDA Soil Conservation Service, Stroud Water Research Center in Stroud, Pennsylvania, the Pennsylvania Department of Environmental Resources, Maryland Department of Natural Resources, and the USDI Fish and Wildlife Service described draft guidelines in a booklet entitled "Riparian Forest Buffers— Function and Design for Protection and Enhancement of Water Resources" (Welsch 1991). The report specified a riparian forest buffer system (RFBS) consisting of three zones (Figure 5.6), and gave management guidelines for each zone.

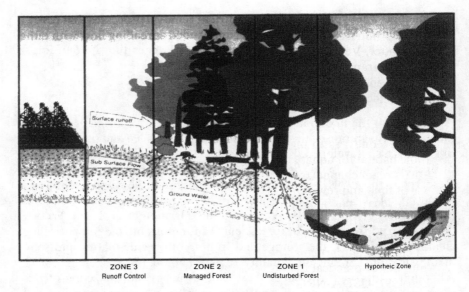

ZONE 3　　　　ZONE 2　　　　ZONE 1　　　　Hyporheic Zone
Runoff Control　　Managed Forest　　Undisturbed Forest

FIGURE 5.6. Schematic of the three-zone riparian forest buffer system.

Zone 1 is permanent woody vegetation immediately adjacent to the stream bank. Zone 2 is managed forest occupying a strip upslope from Zone 1. Zone 3 is an herbaceous filter strip upslope from Zone 2. The specification applies to areas where cropland, grasslands, or pasture are adjacent to riparian areas on permanent or intermittent streams, margins of lakes and ponds, margins of wetlands, or margins of such groundwater recharge areas as sinkholes. The original draft specification has been refined further and has been adopted as a Model State Standard, Riparian Forest Buffer (No. 391) by the USDA (USDA NRCS 1995). A similar three-zone riparian forest buffer model that incorporates shrub vegetation in Zone 2 has also been developed (Schultz et al. 1994; Dosskey et al. 1997).

The primary purposes of Zone 3 of the RFBS are to remove sediment from surface runoff and to convert channelized flow to sheet flow. These functions are especially important when reestablishing a riparian forest buffer using small trees planted on cleared land such as a field. To control channelized flow into a riparian forest, an herbaceous strip in Zone 3 could be much more easily reshaped and revegetated than a forest. Herbaceous buffers, especially grass filters, are effective at removing coarse suspended sediments and some sediment-borne pollutants but may require frequent maintenance and are not very effective at nutrient removal from shallow groundwater (Dillaha et al. 1989; Magette et al. 1989).

Long-term sequestering and removal of nutrients and other contaminants in the RFBS is the main purpose of Zone 2. This can occur by (1) accumulating sediment and adsorbed contaminants; (2) microbial transformations (for N) and biochemical degradation (for pesticides); and (3) incorporation of nutrients and other chemicals into woody biomass and soil organic matter. Vegetation and litter in Zone 2 forms a mechanical barrier to sediment transport. Plant roots take up chemicals that then become sequestered in growing biomass. Vegetation also produces organic matter that fosters chemical and biological processes that immobilize or transform pollutants.

Although most Zone 2 functions also occur in Zone 1, the primary purpose of Zone 1 is to maintain the integrity of the stream bank and a favorable habitat for aquatic organisms. After vegetation has been removed from the stream channel, recovery through plant succession may take long periods of time and revegetation may be dominated by undesirable species (Sweeney 1993). The need for permanent control of the stream's physical and trophic structure requires directed succession toward desirable permanent vegetation in those portions of the RFBS that directly influence the stream channel, in particular Zone 1.

A number of practical issues were considered in the RFBS specification (Welsch 1991; USDA-NRCS 1995; USDA 1997). Most of the RFBS should be available for management to provide an economic return without sacrificing water quality functions. Characteristics of soils, hydrology, and native riparian vegetation should guide design and planning of effective RFBS. The RFBS should be used in conjunction with sound upland-management practices including nutrient management and erosion control. Instream woody debris removal should be limited, but woody debris with potential to form dams that cause inundation of adjacent land uses or upstream areas may be removed. The dimensions of the RFBS should depend on the existing and potential nonpoint-source pollutant loads and the minimum size for sustained support of the aquatic environment.

Application of Riparian Forest Buffers as Ecosystem Management

Management recommendations for RFBS have followed a common pattern for development and dissemination of scientifically based ecosystem management. These are:

1. Preliminary studies of limited scope.
2. Initial synthesis and recognized need for more research.
3. Detailed studies.
4. Further synthesis and generation of new hypotheses.
5. Relation of research to policy objectives.
6. Distillation of research into scientifically based management approaches (achieved by consensus if possible).
7. Refinement of policy objectives based on available science.

8. Testing of approaches in controlled experiments.
9. Testing of approaches in real-world settings.
10. Further refinement of policy goals and management objectives based on new science and monitoring of real-world applications.

At each step in this dissemination, the process will lag unless individuals take the initiative to pursue creative new approaches or ideas that will further the process. A bureaucracy can sanction the new ideas and approaches, but it typically requires the commitment of a person or a small group of persons to take the necessary steps "outside the box." In the 1990's, a small group of scientists and technologists interested in the use of riparian ecosystems as buffers began to develop the types of information needed to transfer the concept of riparian buffers to decision-makers, land owners, and the general public.

When the condition of the nation's waters are assessed in another decade, the mid-1990s may be seen as the beginning of a movement toward restoration of the nation's riparian ecosystems. Numerous federal, state, local, and private initiatives have riparian buffers and riparian ecosystem restoration as major components of their plans. In the following section, I will describe two major federal initiatives that have riparian ecosystem-management components derived from the riparian ecosystem research of the past twenty years.

Conservation Reserve Program

The Conservation Reserve Program (CRP) was passed as part of the 1985 Farm Bill (i.e. Food Security Act) and was focused on controlling erosion on highly erodible cropland. Participation in CRP is voluntary and owners "bid" their eligible land for enrollment in CRP in return for an annual payment from the USDA. Upon reevaluation of the program for the 1996 Farm Bill, it was clear that the CRP had saved soil and had certainly idled large amounts of land, but had probably had little overall impact on environmental quality in general and water quality in particular. In the 1996 Farm Bill, Congress changed CRP from simply protecting highly erodible land to protecting the environment in agricultural landscapes. As described by the USDA "the new, environmentally focused CRP will provide immediate benefits through reduced soil erosion, improved water quality, and expanded wildlife habitat" (USDA 1997). The final rules for the CRP provided for evaluation of lands using a "comprehensive scoring system called the Environmental Benefits Index (EBI). The index explicitly evaluates the environmental benefits of acres offered for enrollment. This allows USDA to rank and select those acres that provide the most benefits to wildlife habitat, erosion control, water quality, and air quality at the least possible cost" (USDA 1997). The EBI provides for ranking of benefits based on wildlife, water quality, on-farm benefits, long-term benefits, and priority geographic areas. Clearly, most of these EBI factors are relevant to riparian forest buffer systems. The new rules for CRP also allow continuous enrollment of such "highly valuable environmental acreage" as filter strips and riparian buffers. Historically, much riparian forest has been lost resulting from the use of streamside areas as pastures. Pastures were not eligible for CRP until the new rules were announced, which provides for "making marginal pasture land eligible if it is suitable for use as a riparian buffer planted to trees."

For the first time, management of riparian ecosystems for multiple purposes will be integral to a USDA conservation initiative. As a reflection of this change, Secretary of Agriculture Dan Glickman commended the National Corn Growers for their part in promoting buffer systems for agriculture. Secretary Glickman went on to note that the focus of CRP was now such buffers as filter strips, shelter belts of trees, field windbreaks of trees, and riparian buffers. To carry out this change in CRP, the USDA Natural Resources Conservation Service (NRCS) launched a Conservation Buffer Initiative to encourage signing up for field and stream buffer systems. The field buffer systems depend primarily on such technology as vegetated filter strips and grass waterways that are used primarily to control sediment from erosion and to spread concentrated overland flow during storm events. The stream buffers will be primarily riparian forests.

Coastal Zone Management Act

The Coastal Zone Management Act Reauthorization Amendments of 1990 (CZARA) requires that state coastal programs, as well as state nonpoint-source programs, address nonpoint-source pollution affecting coastal water quality. Section 6217 of CZARA requires the EPA to provide guidance to states and territories "for specifying management measures for sources of nonpoint pollution in coastal waters" (US EPA, 1993). The management-measures programs are defined as "economically achievable measures for the control of pollutants from existing and new categories . . . which reflect the greatest degree of pollutant reduction possible" (US EPA, 1993). State coastal nonpoint pollution control programs must provide for the implementation of management measures that conform with the management-measures guidance.

Among the objectives of the management measures are protection of wetlands and riparian areas and restoration of wetland and riparian areas. The measures are designed to protect and restore the nonpoint-source pollution control functions of riparian areas and wetlands. Selection of these management measures was based on: (1) the opportunity to gain multiple benefits while reducing nonpoint-source pollution; (2) the effectiveness of riparian systems for reducing loadings of nonpoint-source pollutants; and (3) the localized increase in nonpoint-source pollutants associated with degradation of wetlands and riparian areas.

Case Study: Riparian Forest Buffers in the Chesapeake Bay Watershed

Because of serious and long-term water quality problems, the Chesapeake Bay has been the focus of considerable efforts to improve water quality by reducing nutrient loadings to the bay and by restoring aquatic habitats. In 1994, the Chesapeake Executive Council, consisting of the Governors of Maryland, Virginia, Pennsylvania, the Mayor of Washington, DC., and the Administrator of the EPA directed the Chesapeake Bay Program "to develop a policy which would enhance riparian stewardship and efforts to conserve and restore riparian forest buffers" (Chesapeake Executive Council, Directive, 1994). The Executive Council appointed a thirty-one-member Riparian Forest Buffer Panel to "develop goals based on sound science and to focus on voluntary incentive based programs" for conservation of riparian areas.

In 1994, the USDAFS, along with other federal, state, local, and private agencies began working with the Riparian Forest Buffer Panel and the Chesapeake Bay Program Forestry Work Group to increase outreach and information about riparian forests and their importance in controlling nonpoint-source pollution, instream water quality, and aquatic habitats. The Forestry Working Group provided technical support to the panel in reaching their recommendations for the Executive Council. USDAFS and the Forestry Work Group were responsible for producing technical reports and educational material that were integral to the riparian panel's deliberations. Estimates were made of the extent of riparian forest in the Chesapeake Bay watershed. As much as 50% of the streamside forest on the 110,000 miles of streams in the watershed has been removed or severely impaired (Alliance for the Chesapeake Bay, 1996). An assessment was made of the potential water quality functions of the three-zone RFBS as specified by USDAFS and USDA NRCS. The expected level of nonpoint-source pollution control for nitrate, sediment, sediment-borne pollutants, and phosphorus was estimated by physiographic province of the bay watershed (Lowrance et al. 1995).

In October of 1996, the Chesapeake Executive Council adopted the report of the Riparian Forest Buffer Panel. In their adoption statement, the Chesapeake Executive Council (1996) explained the rationale for adoption of the report:

In past commitments, we have agreed to reduce nutrients, to restore habitat, to improve access to thousands of miles of habitat for migratory fish, and to enhance watershed management by developing and implementing tributary-specific pollution reduction strategies. All of these are part of our effort to achieve our goals for improved water quality and living resources in the Chesapeake Bay. Building on these past commitments, we now highlight the role that conservation, restoration, and stewardship of our riparian areas, and in particular riparian forests, play in reaching our long term goal for restoration of the Chesapeake Bay.

Specifically, the Executive Council agreed to adopt the following additional Chesapeake Bay Program goals for states and federal agencies:

1. To assure, to the extent feasible, that all streams and shorelines will be protected by a forested or other riparian buffer.
2. To conserve existing forests along all streams and shorelines.
3. To increase the use of all riparian buffers and restore riparian forests on 2,010 miles of stream and shoreline by 2010, targeting efforts where they will be of greatest value to water quality and living resources.

The Executive Council also adopted five policy recommendations recommended by the Panel: (1) enhance program coordination; (2) promote private sector involvement; (3) enhance incentives for landowners; (4) promote education and information; and (5) support research, monitoring, and technology transfer (Chesapeake Executive Council, 1996).

Among the programs requiring coordination in the Bay States are such federal incentive programs as CRP and the Agricultural Conservation Program, as well as state incentive and regulatory programs. Maryland and Virginia both offer voluntary programs to make one-time payments to farmers and other landowners to establish minimum 50-foot forested buffers along streams and shorelines to reduce nonpoint-source pollution. Pennsylvania conducts a voluntary stream bank fencing program that is oriented toward stream bank stabilization by fencing off narrow riparian strips. The voluntary programs are hampered by concerns about paperwork requirements and by underfunding. The Maryland Buffer Incentive Program has averaged thirty sites and about 175 acres of buffer installed per year. The Virginia program has averaged only three sites and 15 acres per year (Forestry Workgroup, 1996). Maryland and Virginia also have regulatory programs and local zoning ordinances which address protecting existing riparian forests and establishing new forested buffers (Forestry Workgroup, 1996). Both the Chesapeake Bay Critical Area Act (Maryland) and the Chesapeake Bay Preservation Act (Virginia) mandate 100-foot riparian buffers for all tidal waters, tidal wetlands, and tributary streams in defined "critical areas." Ordinances in both states concentrate on developed land, with exemptions for agricultural and silvicultural lands.

The interaction of state and federal voluntary programs and state regulatory programs is likely to be altered by the emphasis on buffer systems under the new rules for CRP. The major USDA efforts to restore riparian ecosystems are funded through the continuous sign-up of the CRP and the Conservation Reserve Enhancement Program/State Enhancement Program (CREP/SEP) authorized in the 1996 Farm Bill. The continuous sign-up allows landowners to offer lands with high Environmental Benefits for the CRP at any time. Efforts spearheaded by Vice-President Gore to address the nation's lack of progress toward Clean Water Act goals have focussed on use of the CREP/SEP mechanism to address critical agricultural water quality problems. The first project funded for CREP/SEP is an effort to have up to 100,000 acres of riparian land along Maryland's streams and rivers set aside and maintained to protect water quality. The Maryland CREP/SEP will spend about $200 million over 15 years to restore riparian ecosystems. Clearly, USDA is planning to make large investments of funds in restoring streamside (riparian) ecosystems to control nonpoint source pollution, increase attainment of designated uses of streams, to provide wildlife habitat, and to restore aquatic ecosystems. The CRP will allow states such as MD and VA which are attempting to promote riparian buffers with limited state funds to concentrate state programs on non-agricultural areas needing buffers. One limitation common to all cost-share programs for practices on relatively small acreages, such as riparian forest buffers, is that the level of payments available to landowners are relatively small if based strictly on a set amount per acre. Although not reflected in any of the incentive programs for riparian forest buffers or other buffers, incorporating the idea of higher payments for higher environmental benefits is likely to be necessary.

Successes and Frontiers for Riparian Ecosystem Science

Successes

Scientifically, there are multiple lessons from the work on riparian ecosystems in agricultural watersheds. At the watershed scale, the integration of watershed hydrology, ecosystem ecology, and hillslope porous media-transport approaches lead to significantly more understanding than studies based in one or two disciplines. Integration of these approaches was a key factor in developing the idea of riparian ecosystems as control points for movement of materials at a watershed scale. A riparian ecosystem is probably the best example of a generalized landscape feature that serves as a controller of landscape or watershed response.

There are aspects of both choice and chance that influence the outcome and the scientific success of the original Coastal Plain studies of riparian ecosystems (Lowrance et al. 1984a; Peterjohn and Correll 1984; Jacobs and Gilliam 1985). The choice of relatively simple hydrologic systems with shallow confining layers provided an optimum opportunity to produce clear-cut tests of hypotheses and quantification of riparian functions. The availability of watersheds with relatively intact riparian forest buffers (see Figure 5.3) provided a further simplification of both the Little River and Rhode River studies and allowed the calculation of total nutrient budgets for the entire riparian area of these two watersheds. Later studies have attempted to elucidate the riparian functions in more complex hydrogeologic and land-use settings with limited success. The understanding of complex, hydrologic flow-paths in riparian ecosystems, the quantification of two-way feedback between vegetation and subsurface and stream hydrology, and the recognition of complex, hyporheic zone interactions adds to uncertainty concerning riparian ecosystem control of nonpoint-source pollution. Although as more complex systems are examined and nonpoint-source pollution control becomes less certain, the basic scientific approach undertaken in the earlier studies of simpler systems remains relevant. The main difference is that of scale in which larger-scale hydrologic interactions must be considered in order to explain riparian functions.

At the scale of the riparian forest ecosystem studies, there has been considerable success in determining both short-term and annual rates of biological processes that control N-retention. Hypotheses that explain nitrate removal through complex interaction of vegetation uptake, leaf and root litter incorporation in soil, and high rates of denitrification in relatively shallow soil horizons have been stated and tested, although not wholly accepted or rejected. The hydrologic, soil, and vegetation diversity and complexity associated with the studies attempting to test this hypothesis make generalization of results difficult. At this point in time, there is evidence that direct denitrification, direct vegetation uptake, and more complex N cycling are all responsible for a portion of nitrate removal in most riparian forest ecosystems. The focus on N as the element of interest is driven at least as much by the inherent and long-standing ecological interest in N-cycling processes instead of any prioritization of pollutant importance. Although sedimentation remains a much larger problem in surface waters than increased N-loading (obviously the two are linked through sediment transport of adsorbed N) there is only a relatively minor effort to study sediment deposition in riparian forest buffers.

The ecosystem management success of riparian forest buffer systems is ongoing and may not be fully realized for years. The management success is dependent on actual "on the ground" adoption of new procedures for riparian ecosystem management on public and private lands. The immediate measure of the ecosystem-management success will be relatively simple—the number of stream miles with restored, enhanced, or preserved riparian forest buffer. The ultimate measure of the ecosystem-management success will involve a much more complex evaluation of long-term function of riparian buffer systems, especially narrow buffers in intensive agricultural landscapes. Riparian forest buffer restoration is not widespread on private lands and it remains to be seen how acceptable it will be as a voluntary practice for private landowners.

The policy success is more immediate and although also ongoing, has been more fully realized than the actual ecosystem-management success. This is a typical situation, as actual management may often lag behind changes in policy, especially on private property. Such policy successes as the inclusion of riparian forest buffers in the CRP and the riparian initiative of the Chesapeake Bay Council show the general acceptance of riparian forests as sound environmental policy.

There are numerous reasons for the policy acceptance of riparian forest buffers. One strength of most riparian policies has been to integrate the findings from stream-ecosystem studies and nonpoint-source pollution control studies. The objectives are generally complementary and each benefits from association with the other. There will be situations in which nonpoint-source pollution control by riparian systems will not be substantial, nonetheless, there are substantial aquatic ecosystem benefits to be gained. Conversely, there will be situations in which aquatic-ecosystem restoration is of little concern but controlling nonpoint-source pollution is of primary importance. Use of an integrated practice that can provide both benefits creates more advocates for riparian forest-policy initiatives.

A second reason for policy successes is that flexibility built into scientifically based management recommendations has enhanced the acceptability of riparian forest buffer policy. Flexibility for management of riparian areas makes it more probable that policy initiatives can succeed despite resistance from some stakeholders. Additionally, incorporation of riparian forest buffers into broader buffer-system initiatives such as the USDA Conservation Buffer Initiative increases acceptance of riparian forest buffer policy. The general policy of building buffer systems into agricultural landscapes has support that goes beyond support for riparian forest buffers specifically.

Finally, riparian forest buffer policy has succeeded because the scientific community has provided the necessary interpretations of research in order to make recommendations on management practices and policy guidelines. The three-zone buffer system and regional guidelines for its application are based on incomplete but adequate knowledge bases. If the scientific community had not been willing to provide the best professional judgment decisions on management and policy, it is improbable that a riparian forest buffer policy would have moved forward. Complementing the scientific judgment is the obvious visual difference between watersheds with and without riparian forests.

Frontiers

Although riparian forest ecosystems will be increasingly important as a watershed ecosystem-management tool, there are still important ecosystem questions that relate to buffer effectiveness. There are also important buffer-system management questions that relate to ecosystem attributes and processes. These questions are posed in a hierarchy going from smaller to larger scales.

Two questions, related to ecosystem processes in the RFBS are:

1. How wide do buffers need to be for various purposes? Ecosystem attributes/processes of interest include nutrient-uptake masses and rates; nutrient-retention masses and rates; and resistance and resilience—essentially the ability to stay in place in spite of extreme events (i.e. floods, wind storms, ice storms).

2. What type of vegetation and vegetation management is best for buffers? Ecosystem attributes/processes of interest include: nutrient-uptake masses and rates; nutrient-retention masses and rates; canopy development and canopy cover for streams; ecosystem-feedback effects on soil—development of macropores, development of organic matter in surface soils, litter quality for chemical retention, and so forth; litter quality for aquatic secondary consumers; quantity and quality of coarse woody debris; and the ability to stabilize stream banks.

Two sets of questions, related to the buffer and the adjacent land uses (hillslope or landscape scale), are:

1. In a given setting, how should a buffer of fixed size be managed to remain effective under increased loadings? Can intensive management for nutrient uptake and removal allow increased loadings? What are the interactions between increased hydrologic loading and increased nutrient loading, especially for N?

2. What loadings or other perturbations from adjacent land use have significant negative impacts on buffer effectiveness? Do sediment, herbicide, or excessive nutrient loadings decrease buffer effectiveness? How can buffers be used effectively adjacent to more intensive land uses?

Two sets of questions, related to the upstream and downstream conditions controlling RFBS effectiveness, are:

1. In degraded systems, how do we determine when upstream conditions limit the effectiveness of downstream buffer systems? Does this differ among watersheds? How can other such watershed-management practices as impoundments, instream wetlands, drainage control, and so forth, be integrated with riparian ecosystem functions?

2. For a given watershed setting, what extent of buffer-system deployment is necessary for a "permanent" (long-term) change in water quality? Does this differ among pollutants (sediment vs. nutrient, for instance)? For a degraded watershed, how much of the aquatic systems must be buffered to assure the long-term stability of downstream buffers?

Scientific knowledge of such complex systems as riparian forests will always remain incomplete. The challenge for ecosystem science is to provide the best available information to those charged with developing ecosystem management approaches. This will often require extrapolation to situations that have not been studied in detail or at all. When this is necessary, basic principles that limit system function must be considered and limitations posed by lack of knowledge must be acknowledged and surmounted through new management-oriented ecosystem research.

References

Alliance for the Chesapeake Bay. 1996. *Riparian Forest Buffers—White Paper.* Alliance for the Chesapeake Bay. Annapolis, MD.

Benke, A.C., R.L. Henry III, D.M. Gillespie, and R.H. Hunter. 1985. Importance of snag habitat for animal production in southeastern streams. *Fisheries* 10:8–13.

Bisson, P.A., R.E. Bilby, M.D. Bryant, C.A. Dolloff, G.B. Grette, R.A. House et al. 1987. Large woody debris in forested streams in the Pacific Northwest: past, present, and future. Pages 143–190 in E.O. Salo and T.W. Cundy eds. *Streamside management forestry and fishery interactions.* Institute Of Forest Resources, University of Washington, Seattle.

Bohlke, J.K., and J.M. Denver. 1995. Combined use of groundwater dating, chemical, and isotopic analyses to resolve the history and fate of nitrate contamination in two agricultural watersheds. *Water Resources Research* 31:2319–2339.

Borman, F.H., and G.E. Likens. 1967. Nutrient cycling. *Science* 155:424–429.

Bosch, D.D., J.M. Sheridan, and R. Lowrance. 1996. Hydraulic gradients and flow rates of a shallow coastal plain aquifer in a forested riparian buffer. *Transactions of the American Society of Agricultural Engineers* 39:865–871.

Chesapeake Executive Council. 1994. Directive No. 94-1: Riparian Forest Buffers. Chesapeake Bay Program. Annapolis, MD.

Chesapeake Executive Council. 1996. Adoption Statement on Riparian Forest Buffers. Chesapeake Bay Program. Annapolis, MD.

Chesters, G., and L.J. Schierow. 1985. A primer on nonpoint pollution. *Journal of Soil & Water Conservation* 40:9–14.

Chichester, F.W. 1976. The impact of fertilizer use and crop management on nitrogen content of subsurface water draining from upland agricultural watersheds. *Journal of Environmental Quality* 5:413–416.

Cooper, A.B. 1990. Nitrate depletion in the riparian zone and stream channel of a small headwater catchment. *Hydrobiologia* 202:13–26.

Cooper, J.R., J.W. Gilliam, R.B. Daniels, and W.P. Robarge. 1987. Riparian areas as filters for agricultural sediment. *Soil Science Society of America Journal* 51:416–420.

Correll, D.L., and D.E. Weller. 1989. Factors limiting processes in freshwater wetlands: an agricultural primary stream riparian forest. Pages 9–23 in R. Sharitz and J. Gibbons, eds. *Freshwater wetlands and wildlife.* U.S. Department of Energy, Oak Ridge, TN.

Correll, D.L., T.E. Jordan, and D.E. Weller. 1994. Failure of agricultural riparian buffers to protect surface waters from groundwater contamination. In *Proceedings of the international conference on groundwater/surface water ecotones.* Cambridge University Press. London, England.

Daniels, R.B., and J.W. Gilliam. 1996. Sediment and chemical load reduction by grass and riparian filters. *Soil Science Society of America Journal* 60:246–251.

Dillaha, T.A., R.B. Reneau, S. Mostaghimi, and D. Lee. 1989. Vegetative filter strips for agricultural nonpoint-source pollution control. *Transactions American Society of Agricultural Engineers* 32:513–519.

Dosskey, M.G., R.C. Schultz, and T.M. Isenhart. 1997. *A riparian buffer design for cropland.* Agroforestry Notes No. 5. United States Department of Agriculture-Forest Service-National Agroforestry Center, Lincoln, NE.

Ehrenfeld, J.G. 1987. The role of woody vegetation in preventing ground water pollution by nitrogen from septic tank leachate. *Water Resources Research* 21:605–614.

Fail, J.L., Jr., B.L. Haines, and R.L. Todd. 1985. Riparian forest communities and their role in nutrient conservation in an agricultural watershed. *American Journal of Alternative Agriculture* 2:114–121

Fail, J.L., Jr., M.N. Hamzah, B.L. Haines, and R.L. Todd. 1986. Above and below ground biomass, production, and element accumulation in riparian forests of an agricultural watershed. Pages 193–224 in D.L. Correll, ed. *Watershed Research Perspectives*. Smithsonian Institution, Washington, DC.

Forestry Workgroup. 1996. Forestry best management practices in the Chesapeake Bay Watershed: Tracking accomplishments. Forestry Workgroup, Nutrient Subcommittee. Chesapeake Bay Program. Annapolis, MD.

Frissel, M.J. ed. 1977. Cycling of mineral nutrients in agricultural ecosystems. *Agro-Ecosystems* 4:1–354.

Gregory, S.V., F.J. Swanson, W.A. McKee, and K.W. Cummins. 1991. An ecosystem perspective of riparian zones. *BioScience* 41:540–551.

Groffman, P.M., A.J. Gold, and R.C. Simmons. 1992. Nitrate dynamics in riparian forests: microbial studies. *Journal of Environmental Quality* 21:666–671.

Hanson, G.C., P.M. Groffman, and A.J. Gold. 1994a. Denitrification in riparian wetlands receiving high and low groundwater nitrate inputs. *Journal of Environmental Quality* 23:917–922.

Hanson, G.C., P.M. Groffman, and A.J. Gold. 1994b. Symptoms of nitrogen saturation in a riparian wetland. *Ecological Applications* 4:750–756.

Harmon, M.E., J.F. Franklin, F.J. Swanson, P. Sollins, S.V. Gregory, J.D. Lattin et al. 1986. Ecology of coarse woody debris in temperate ecosystems. *Advances in Ecological Research* 15:133–302.

Hendrickson, O.Q., Jr. 1982. Flux of nitrogen and carbon gases in bottomland soils of an agricultural watershed. Ph.D. dissertation, University of Georgia, Athens.

Hill, A.R. 1990. Groundwater flow paths in relation to nitrogen chemistry in the near-stream zone. *Hydrobiologia* 206:39–52.

Hill, A.R. 1996. Nitrate removal in stream riparian zones. *Journal of Environmental Quality* 25:743–755.

Jacobs, T.C., and J.W. Gilliam. 1985. Riparian losses of nitrate from agricultural drainage waters. *Journal of Environmental Quality* 14:472–478.

Jaworski, N.A., P.M. Groffman, A. Keller, and A.C. Prager. 1992. A watershed-scale analysis of nitrogen loading: the Upper Potomac River. *Estuaries* 15:83–95.

Jordan, T.E., D.L. Correll, and D.E. Weller. 1993. Nutrient interception by a riparian forest receiving inputs from adjacent cropland. *Journal of Environmental Quality* 22:467–473.

Jackson, W.A., L.E. Asmussen, E.W. Hauser, and A.W. White. 1973. Nitrate in surface runoff and subsurface flow from a small agricultural watershed. *Journal of Environmental Quality* 2:480–482.

Karr, J.R., and I.J. Schlosser. 1978. Water resources and the land-water interface. *Science* 201:229–234.

Kohl, D.H., G.B. Shearer, and B. Commoner. 1971. Fertilizer nitrogen: contribution to surface water nitrate in a cornbelt watershed. *Science* 174:1331–1334.

Likens, G.E., F.H. Bormann, R.S. Pierce, J.S. Eaton, and N.M. Johnson. 1977. *Biogeochemistry of a forested ecosystem.* Springer-Verlag, New York.

Lowrance, R.R., R.L. Todd, and L.E. Asmussen. 1983. Waterborne nutrient budgets for the riparian zone of an agricultural watershed. *Agriculture Ecosystems and Environment* 10:371–384.

Lowrance, R., R.L. Todd, and L.E. Asmussen. 1984a. Nutrient cycling in an agricultural watershed: I. Phreatic movement. *Journal of Environmental Quality* 13:22–27.

Lowrance, R.R., R.L. Todd, J. Fail, Jr., O. Hendrickson, Jr., R. Leonard, and L. Asmussen. 1984b. Riparian forests as nutrient filters in agricultural watersheds. *Bioscience* 34:374–377.

Lowrance, R., R.A. Leonard, L.E. Asmussen, and R.L. Todd. 1985. Nutrient budgets for agricultural watersheds in the southeastern coastal plains. *Ecology* 66:287–296.

Lowrance, R., J.K. Sharpe, and J.M. Sheridan. 1986. Long-term sediment deposition in the riparian zone of a coastal plain watershed. *Journal of Soil and Water Conservation* 41:266–271.

Lowrance, R., S. McIntyre, and J.C. Lance. 1988. Erosion and deposition in a coastal plain watershed measured using CS-137. *Journal of Soil and Water Conservation* 43:195–198.

Lowrance, R. 1992. Groundwater nitrate and denitrification in a coastal plain riparian soil. *Journal of Environmental Quality* 21:401–405.

Lowrance, R., L.S. Altier, J.D. Newbold, R.R. Schnabel, P.M. Groffman, and J.M. Denver. 1995. *Water quality functions of riparian forest buffer systems in the Chesapeake Bay Watershed.* Report No. EPA 903-R-95-004; CBP/TRS 134/95. U.S. Em. Profection Hgencyl Chesapeake Bay Program Office Annapolis. MD.

Lowrance, R., G. Vellidis, R.D. Wauchope, P. Gay, and D.D. Bosch. 1997. Herbicide transport in a managed riparian forest buffer system. *Transactions American Society of Agricultural Engineers* 40:1047–1057.

Magette, W.L., R.B. Brinsfield, R.E. Palmer, and J.D. Wood. 1989. Nutrient and sediment removal by vegetated filter strips. *Transactions of American Society of Agricultural Engineers* 32:663–667.

Meyer, J.L., and J. O'Hop. 1983. The effects of watershed disturbance on dissolved organic dynamics of a stream. *American Midland Naturalist* 109:175–183.

Myers, C.F., J. Meek, S. Tuller, and A. Weinberg. 1985. Nonpoint-sources of water pollution. *Journal of Soil & Water Conservation* 40:14–19.

Peterjohn, W.T., and D.L. Correll. 1984. Nutrient dynamics in an agricultural watershed: observations on the role of a riparian forest. *Ecology* 65:1466–1475.

Phillips, P.J., J.M. Denver, R.J. Shedlock, and P.A. Hamilton. 1993. Effect of forested wetlands on nitrate concentrations in ground water and surface water on the Delmarva Peninsula. *Wetlands* 13:75–83.

Reilly, T.E., L.N. Plummer, P.J. Phillips, and E. Busenberg. 1994. The use of simulation and multiple environmental tracers to quantify ground water flow in a shallow aquifer. *Water Resources Research* 421–433.

Schipper, L.A., A.B. Cooper, C.G. Harfoot, and W.J. Dyck. 1993. Regulators of denitrification in an organic riparian soil. *Soil Biology and Biochemistry* 25:925–933.

Schultz, R.C., J. Colletti, W. Simpkins, M. Thompson, and C. Mize. 1994. Developing a multispecies riparian buffer strip agroforestry system. Pages 203–225 in *Riparian ecosystems in the humid U.S.: Functions, values and management.* National Association of Conservation Districts, Washington, DC.

Simmons, R.C., A.J. Gold, and P.M. Groffman. 1992. Nitrate dynamics in riparian forests: groundwater studies. *Journal of Environmental Quality* 21:659–665.

Staver, K.W., and R.B. Brinsfield. 1996. Seepage of groundwater nitrate from a riparian agroecosystem into the Wye River estuary. *Estuaries* 19:359–370.

Swank, W.T., and J.E. Douglass. 1977. Nutrient budgets for undisturbed and manipulated hardwood forest ecosystems in the mountains of North Carolina. Pages 343–362 in D.L. Correll ed. *Watershed research in eastern North America.* Smithsonian Institution, Edgewater, MD.

Sweeney, B.W. 1993. Effects of streamside vegetation on macroinvertebrate communities of White Clay Creek in eastern North America. *Proceedings Academy of Natural Sciences Philadelphia* 144:291–340.

Thomas, L.M. 1985. Management of nonpoint-source pollution: what priority? *Journal Soil & Water Conservaton* 40:8.

USDA. 1997. Clinton administration announces final conservation program rules. USDA Press Release No. 0046.97.

USDA NRCS. 1995. *Riparian Forest Buffer, 391. Model state standard and general specifications.* Watershed Science Institute, Seattle, WA.

US EPA. 1991. *Guidance for water quality based decisions: the TMDL process.* Report No. EPA 440/4-91-001. Washington, DC.

US EPA. 1993. *Guidance specifying management measures for sources of nonpoint pollution in coastal waters.* Report No. EPA-840-B-93-001c. Washington, D.C.

US EPA. 1994. *National Water Quality Inventory: 1994 Report to Congress.* Report No. EPA 503/9-94/006. Washington, DC.

Vitousek, P.M., and W.A. Reiners. 1975. Ecosystem succession and nutrient retention: a hypothesis. *BioScience* 25:376–381.

Warwick, J., and A.R. Hill. 1988. Nitrate depletion in the riparian zone of a small woodland stream. *Hydrobiologia* 157:231–240.

Welsch, D.J. 1991. *Riparian Forest Buffers.* United States Department of Agriculture—Forest Service Pub. No. NA-PR-07-91. United States Department of Agriculture—Forest Service, Radnor, PA.

Yates, P., and J.M. Sheridan. 1983. Estimating the effectiveness of vegetated floodplains/wetlands as nitrate-nitrite and orthophosphorus filters. *Agriculture, Ecosystems, and Environment* 9:303–314.

6
Ecological Research in Agricultural Ecosystems: Contributions to Ecosystem Science and to the Management of Agronomic Resources

G. Philip Robertson and Eldor A. Paul

Summary

Ecological research in agronomic systems has a long and rich history that has contributed much to our basic understanding of all terrestrial ecosystems. Of particular note are contributions to (1) soil organic matter concepts, (2) knowledge of specific nitrogen-cycling processes such as mineralization-immobilization, denitrification, nitrogen fixation, and biphasic nitrate leaching, and (3) understanding of soil resource heterogeneity. Recent concerns about the long-term sustainability of high productivity agriculture and the high environmental cost of many such systems provides managers opportunities for better application of ecosystem concepts. These opportunities include managing the temporal linkage between nutrient release from soil organic matter and the demand for nutrients by plant uptake, managing agronomic inputs such as fertilizers and pesticides with spatial precision, and using whole watershed input-output balances to assess and manage unwanted nutrient exports. The application of modern ecosystem concepts to emerging agronomic needs is an important frontier in ecosystem science with the potential to benefit both ecosystem theory and practical issues related to farm policy and management.

Introduction

Ecological research in agricultural systems dates from the 19th century—from our earliest efforts to understand the field biology of row crops such as barley and wheat. From the outset, this research has furthered our basic understanding of ecosystems in general. Early recognition by Leibig (1847) and others that plant resource needs can easily exceed available soil resources provided early insight into plant nutrient acquisition and the maintenance of soil nutrient pools, and furnished a basis for our current

142

understanding of nutrient interactions in ecosystems (e.g. Bolin and Cook 1983). These ideas continue to influence our approaches to ecosystem nutrient retention today. Early research by soil microbiologists working principally in agronomic soils laid the foundation for our understanding of the microbial basis for many ecosystem-level nutrient fluxes, especially those related to nitrogen, carbon, and sulfur. Dinitrogen fixation, mineralization-immobilization, nitrification, and denitrification are all critical nitrogen-cycling processes whose microbial ecology was first described or principally worked out in agronomic systems.

Prior to World War II, the agronomic and the ecological sciences were inextricably linked. Success in one often led to success in the other across a range of ecological hierarchies, from the physiological response of individual organisms to community and ecosystem-level dynamics. After 1945, however, with the advent of large-scale, mechanized, chemically based farming practices, ecology took a role subordinate to engineering in the design and implementation of food production systems. The availability of inexpensive fertilizer freed the agronomist from the need to understand the ecological basis for most microbial processes, and instead forced attention toward the fate and efficiency of alternative fertilizers. The availability of effective herbicides reduced the need to understand plant competition for limiting resources and the colonization and persistence of soil seed banks. The availability of pesticides obviated the need to understand the field and landscape-level population dynamics of most pest predators, instead directing attention toward the specificity of different chemicals and their application windows. Additionally, plant breeding changed harvest indices, root/shoot ratios, leaf areas and duration, and the secondary chemistry of leaf tissue, which concomitantly created an even greater need for more intensive crop management.

With the recent emergence of sustainable agriculture and environmental awareness as social and political issues in many countries, a balance is being restored, largely in response to social concerns that modern production-level agriculture carries an unnecessarily high environmental cost. Proponents of a more ecologically based agriculture cite problems associated with excess herbicide, pesticide, and fertilizer use: both the direct economic expense that makes many farms only marginally viable, as well as the indirect, externalized costs now paid by society as a whole. The long-term costs of erosion, excess water use, and salinization further subtract from the ledger.

For ecologists, reintegrating ecological concepts into high productivity farming systems presents a set of new challenges that can be considered an ecosystem frontier (sensu this volume) in its own right. As such, many of the contributions of agricultural ecosystem research to ecology are surely in the future. Nevertheless, even over the past three decades, when ecology in developed regions was as separate as possible from agronomic research, important contributions to ecosystem science have been made. Furthermore, ecosystem science has begun to significantly influence agronomic management, particularly in the areas of whole-system (both field-scale and watershed-scale) nutrient budgets and soil carbon management.

In this chapter we have attempted to outline three major contributions of research in agroecosystems to the general body of ecosystem science, as well as emerging ways in which ecosystem science is affecting the management of high productivity agricultural ecosystems.

Contributions of Agroecosystem Research to Ecosystem Science

Consider the effects of converting a prairie or native deciduous forest to row-crop agriculture, in which a mixed-species, slow-growing perennial plant community on a stable soil profile is replaced with a single fast-growing annual species on a soil profile that is inverted annually. This massive disturbance resets the successional clock on a yearly basis, introduces a major exotic plant (i.e. the crop) to the system, physically disrupts the soil food web and its structural support, and alters plant soil temporal relationships. This leads to a dramatic reorientation of many ecosystem processes and attributes. Among the most drastic are gradual but substantial changes in soil organic matter—its quantity, quality, and turnover rate—with concomitant changes in soil food web dynamics; rapid and major changes in soil structure and soil water and solute movement; major changes in certain nitrogen-cycling processes; and potentially major changes in the spatial distribution or heterogeneity of soil resources within the ecosystem. Consequently, it makes sense that advances in our general understanding of these ecosystem-level processes have been particularly accelerated in agronomic systems, and, therefore, that ecosystem research in row-crop systems has made particular contributions to ecosystem science in these areas.

Moreover, because annual crop ecosystems are highly pulsed, they are systems in which net fluxes of such nutrients as nitrogen are more readily interpreted, especially with the aid of carbon- and nitrogen-isotope technologies, largely developed in agronomic systems. Contrast this with the difficulty of understanding fluxes in perennial native ecosystems, which are dependent on the much less easily measured gross fluxes that occur without major changes in available nutrient pool sizes.

From the outset, we believe that the question "How has research in agricultural ecosystems benefited ecosystem science?" is a red herring. As many have pointed out in recent years, we should be moving beyond the applied vs. basic, disturbed vs. natural, managed vs. unmanaged paradigm of ecological research toward a perspective that is more gradient-oriented (Elliott and Cole 1989; Paul and Robertson 1989). All ecosystems today are affected by human intervention, whether intentional or not, and agricultural systems are simply closer to the intensely and intentionally disturbed end of the spectrum than are such remote ecosystems as arctic tundra, which, up to this time, may be subject mainly to the widespread but less intense, unintentional human disturbance of the chemical and physical climate.

In its purest sense, then, agricultural ecology is not targeted toward understanding ecological processes in row-crop ecosystems, but rather toward understanding ecological processes in extractive ecosystems that are actively maintained at an early stage of secondary succession. In this light, agricultural ecology becomes no different from forest ecology, restoration ecology, or the ecology of unmanaged ecosystems—the key is an understanding of the ecology of successional systems, for which, of course, there is a rich literature. The relevant question then becomes "How has agricultural ecosystem research contributed to our understanding of ecological processes in early successional, chronically disturbed ecosystems in general?" From a management standpoint, the converse question becomes, "What ecological knowledge about early successional ecosystems can we use to actively manage these systems to minimize the environmental and economic costs of a highly productive, extractive ecosystem?"

In the following three sections, we discuss in turn the particular contributions of agronomically related research to our present understanding of soil organic matter, specific nitrogen-cycling processes, and soil resource heterogeneity in early successional ecosystems in particular.

Soil Organic Matter

Models of soil organic matter dynamics now being applied in a wide variety of ecosystems (e.g. vanVeen and Paul 1981; Jenkinson et al. 1987; Parton et al. 1987) derive largely from an understanding of soil carbon dynamics developed in agricultural soils or in comparisons of grassland to agricultural soils. Results from long-term studies of carbon loss following the conversion of temperate-region ecosystems to agriculture typically show a rapid, several-decade decline in soil organic matter levels following the onset of cultivation (see Figure 6.1; Paul et al. 1997). Typically after two to three decades in warm climates, organic matter levels reach a new equilibrium level, often 50 to 80% of original levels in the whole profile and less in the topmost horizon. But both the rate at which carbon is lost from a soil and the level to which soil carbon pools decline depend on a large number of interacting factors: (1) climate; (2) soil texture; (3) the stability of soil aggregates; (4) the quality and quantity of new plant litter inputs; (5) the nature and timing of further soil disturbance; and (6) various other site-specific factors that affect feedbacks between soil processes and plant growth. These interactions make predictions of the rate of carbon loss and the potential rate of carbon recovery difficult in the absence of quantitative models that include these feedbacks.

Efforts to predict the rate and equilibrium level of soil carbon storage following disturbance have been successful only since modelers have also

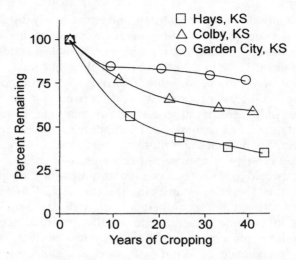

FIGURE 6.1. Soil organic matter loss following cultivation at three sites in the midwestern United States. Redrawn from Haas et al. 1957.

incorporated into their models the concept of multiple soil carbon pools, that is, pools of organic carbon that turn over at different rates. The most widely available models (e.g. Century, Rothamsted, Socrates) recognize at least three soil organic matter pools (see Figure 6.2); for example, Century (Parton et al. 1987) defines an active fraction that consists of microbial biomass and metabolites with a turnover rate on the order of months to years, a slow fraction of stabilized decomposition products with a turnover rate on the order of decades, and a passive fraction of recalcitrant, highly stabilized organic matter that turns over on the order of centuries to millennia.

Only recently have these fractions become analytically identifiable as well. A combination of acid hydrolysis and ^{14}C dating (Paul and vanVeen 1978; Paul et al. 1997) together with CO_2 release from long-term soil incubations (Figure 6.3; sensu Stanford and Smith 1972) now provide a reasonable means for quantifying passive, slow, and active soil carbon pools. The identification of these pools occurred principally through work in cultivated ecosystems or in native vs. cultivated comparisons (e.g. Shields and Paul 1973; Jenkinson 1977; Juma and Paul 1981). Similarly, the subsequent models (Hunt 1977; Jenkinson 1977; vanVeen and Frissel 1979; McGill et al. 1981; Molina et al. 1983; vanVeen et al. 1984; Parton et al. 1987; Grace et al. 1996) and thus far their most rigorous tests (e.g. Paustian et al. 1992; Figure 6.4) have been respectively developed and conducted primarily in agronomic systems. The application of these models in other systems and their utility for assessing atmosphere-soil carbon dioxide exchange is becoming especially important with respect to evaluating historical and future global change scenarios (e.g. Schlesinger 1984; Jenkinson et al. 1991; IPCC 1996).

Related to the conceptual development of soil carbon models is the associated development of soil food-web models. Although not developed initially in agronomic systems, experimentation in arable soils has provided substantial insight for identifying important members of soil food webs and their functional significance (Elliott et al. 1984; Hunt et al. 1987; Coleman and Crossley 1996). Of particular note has been the emerging recognition of the importance of protozoan and other faunal grazing for making pulses of nitrogen available to the plant community (Edwards 1983; Clarholm 1985),

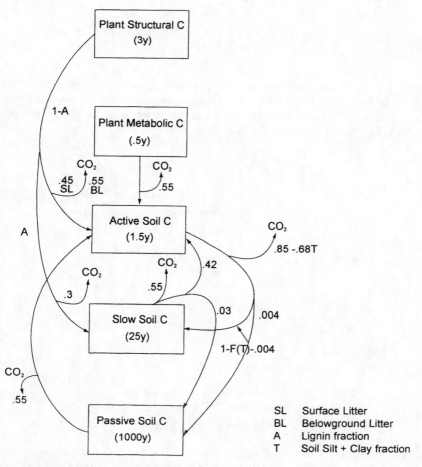

FIGURE 6.2. Structural components of the Century soil organic matter model. Values next to arrows are transfer coefficients for the carbon pools noted; values in parentheses are the approximate turnover times for the noted pool. From Paustian et al. 1992.

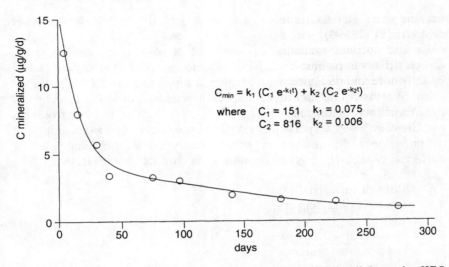

FIGURE 6.3. Long-term carbon mineralization in a cultivated soil from the KBS LTER site. The solid line is fit with the two-pool nonlinear model as noted in figure ($r^2 = 0.99$); units for the active fraction C1 and the slow fraction C2 are $g\,C\,g^{-1}$ soil; the passive fraction C3 (total soil C less C1 and C2) = $7,785\,g\,C\,g^{-1}$ soil.

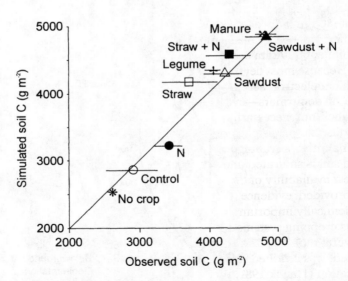

FIGURE 6.4. Actual vs. simulated soil carbon pools in a Swedish soil following thirty years of oat and barley cropping with the amendments noted. From Paustian et al. 1992.

and the potential importance of fungal- vs. bacterial-dominated food webs (Hendrix et al. 1986) with consequent potential effects on decomposition rates and nutrient availability (Beare et al. 1992). These dynamics are crucial drivers in trophic-based models of soil nitrogen cycling (Hunt et al. 1987; Moore and de Ruiter 1991). Soil biotic interactions also appear to be crucial regulators of soil aggregate structure as a consequence of their binding effects on soil mineral particles (Oades 1993); this phenomena was initially clarified in agricultural ecosystems.

Nitrogen Cycling: Process-Specific Advances

The history of terrestrial nitrogen-cycling research is no more specific to agricultural systems than it is to forest, grassland, wetland, or any other habitat. In fact, it can be argued that from a watershed standpoint, our understanding of whole-system nitrogen dynamics in agronomic systems has lacked a parallel understanding of nitrogen cycling in forest catchments, despite the early use of stable isotopes in agronomic systems (e.g. Allison 1955). For at least three—and possibly four—crucial nitrogen-cycling processes, however, the knowledge worked out in agricultural ecosystems represents a second major contribution of agroecological research to eco-system science. These processes include denitrification, dinitrogen fixation, and nitrate leaching via preferential flow, in addition to nitrogen-mineralization-immobilization. Denitrification, the biological reduction of soil nitrate to the dinitrogen gases nitrous oxide (N_2O) and nitrogen (N_2), is carried out by microorganisms that have been isolated from a wide variety of environmental habitats, including every terrestrial ecosystem examined thus far. Because most denitrifiers are facultative anaerobes that use nitrate as a terminal electron acceptor only when oxygen is limiting, however, the presence of denitrifiers—even if they can be shown to be metabolically active—does not necessarily imply active denitrification. Particularly in upland soils, the persistence of denitrifiers might be attributed to a short-lived competitive advantage present only during transient rainfall events when the soil is saturated and nitrate is freely available.

The new availability of ^{15}N in the 1950s and its application in agronomic studies provided evidence that denitrification in upland soils *might* be biogeochemically important. Only a portion of the ^{15}N applied as fertilizer in various cropping systems could be recovered in plant, soil, and leachate pools several months later; the remainder, sometimes totaling 50% of the ^{15}N applied, was ascribed to some combination of experimental error plus denitrification (Hauck 1981; Firestone 1982). Careful mass-balance studies without ^{15}N showed potential losses of a similar magnitude (e.g. Rolston et al. 1979) but with equal uncertainty. Not until a selective metabolic inhibi-tor of nitrous oxide reductase was discovered in the mid-1970s (Yoshinari et al. 1977) and tested largely in agronomic soils (Duxbury 1986) did it become possible to evaluate denitrification fluxes directly. These evalua-

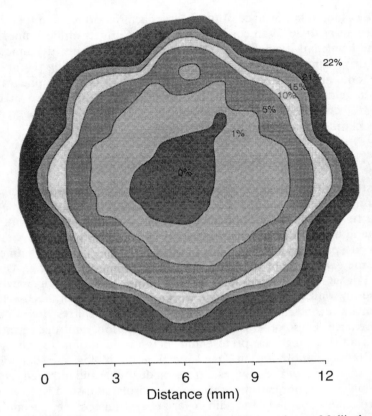

FIGURE 6.5. Oxygen profile in a 1.2 cm diameter aggregate from a Mollisol soil from Iowa; note the anaerobic center. Redrawn from Sexstone et al. 1985.

tions showed that denitrification can indeed account for 20 to 50% of the nitrogen lost from many cropping systems (Firestone 1982; Tiedje 1988).

Field-level denitrification research in row-crop ecosystems paved the way for similar research in other upland communities. Of particular importance has been research providing a conceptual basis for denitrifier habitat. As noted before, early recognition that denitrification is an anaerobic nitrate-requiring process kept many microbiologists from considering well-drained ecosystems as serious candidates for denitrification until the development of oxygen (O_2) models (e.g. Arah and Smith 1989) that showed the theoretical potential for persistent anaerobiosis in such soil microsites as aggregate centers and micropores. Confirmation of the anaerobic potential of soil aggregates came with the development of O_2 microelectrodes (Revsbech et al. 1981), applied to soil microsites in 1985 (Sexstone et al. 1985; Figure 6.5).

Why are aggregates anoxic? In part, anoxia is the result of limitations on gas diffusion across the water film that surrounds most aggregates; O_2 diffuses through water more slowly (more than an order of magnitude slower) than through air. Diffusion itself, however, is only a problem when it does not occur as quickly as O_2 consumption within the aggregate—which explains why most denitrification in soil may be associated with such small organic matter particles as relatively fresh plant litter (Parkin 1987) and invertebrate feces (Elliott et al. 1990).

With the conceptual groundwork for denitrification in well-aerated soils laid out by workers in agronomic systems, it has been a relatively short jump to apply these concepts to entire landscapes comprised of many different types of ecosystems. Tiedje (1988) presented a landscape-level conceptual framework for denitrification that identifies controls on denitrification at several different scales (Figure 6.6). In this context, it has not been surprising to find nitrogen gas fluxes from well-drained forested sites that appear to equal or exceed nitrogen lost from other pathways at these sites (e.g. Virginia et al. 1982; Mellilo et al. 1983; Robertson and Tiedje 1985, 1988; Robertson et al. 1987; Groffman and Tiedje 1989).

At a regional scale, denitrification in upland soils largely exceeds denitrification in other habitats and can play a large role in the overall nitrogen economy of the region. In their subcontinent nitrogen budget for

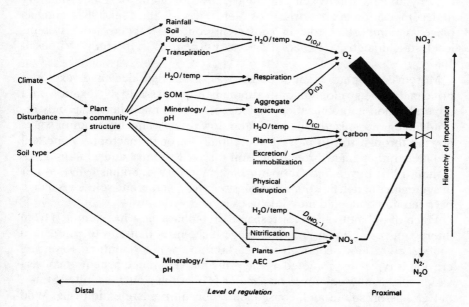

FIGURE 6.6. Proximal vs. distal controls on denitrification. From Robertson (1988) after Groffman et al. (1988).

West Africa, for example, Robertson and Rosswall (1986) estimated denitrification losses of nitrogen at $1.1\,TgNy^{-1}$, roughly equivalent to hydrologic exports of nitrogen from the region. Jordan and Weller (1996) estimated that only 5% $(0.7\,TgNy^{-1})$ of total anthropic nitrogen inputs to the United States $(15.3\,TgNy^{-1})$ are lost as river discharge—with the remaining $14\,TgN$ plus nonanthropogenic sources either stored in aggrading ecosystems, sediments, and groundwater, or (more probably) lost through denitrification and perhaps fire.

For at least three other important biogeochemical processes our ecological understanding has also been developed largely in agricultural ecosystems: biological nitrogen fixation, nitrogen mineralization-immobilization, and biphasic solute loss. The history of dinitrogen-fixation research dates from the 19th century (Havelka et al. 1982), with much of the terrestrial research focused on the ecology and biogeochemical impact of the legume-Rhizobia symbiosis, discovered in 1886 (Hellriegel 1886, in Havelka et al. 1982)—a dominant feature of many agronomic management practices especially prior to the widespread availability of nitrogen fertilizers (e.g. Pieters 1927). Similarly, much of what we know about mineralization-immobilization transformations in soils originated from work in agronomic systems, especially dating from isotope work beginning in the 1950s (Jansson and Persson 1982). For example, the capacity for litter of different carbon/nitrogen ratios to differentially immobilize and release nitrogen and thereby affect soil nitrogen availability now dominates our understanding of decomposition in terrestrial ecosystems (e.g. Melillo et al. 1982), and has played an important role in ecosystem concepts ranging from system-level stoichiometry (Reiners 1986) to nutrient-use efficiency (Vitousek 1982).

Moreover, biphasic solute flow was first identified and explored principally in agronomic soils, although in contrast to N_2 fixation and mineralization-immobilization, its biogeochemical importance has only recently been recognized. Biphasic water flow refers to the tendency of water flowing through soil to move preferentially through macropores and soil channels rather than through the soil matrix (Thomas and Phillips 1979; Bouma 1981; Beven and Germann 1982; White 1985; Sollins 1989). Water in aggregates and in fine pores remains relatively static, and solute exchange between immobile and mobile phases can be very slow.

Much of the motivation for research on biphasic flow has stemmed from interest in zero-till conservation tillage techniques that aim to conserve a soil structure more conducive to carbon and water retention. Reducing erosion is typically promoted as the principal agronomic benefit of conservation tillage, now employed on a majority of cropland in the United States, but effects on solute and, consequently, on nutrient retention (e.g. Wild 1972) and on loss of nitrogen to groundwater is also important. In nonagronomic systems the importance of biphasic flow and its sensitivity to disturbance is only now being enumerated (Sollins 1989); it is becoming apparent, however, that any disturbance that has the potential to affect aggregate structure, macroporosity, or the formation of soil channels also has the potential to affect solution nutrient losses via effects on soil water flow paths. Processes that affect aggregate structure and soil porosity include invertebrate (especially earthworm) activity, root turnover, fungal mycelia growth, and such larger-scale processes as root-throws, burrowing mammals, and harvest or site preparation activities.

Soil Resource Heterogeneity

A third area in which ecosystem-level research in agronomic systems has led research in other systems is that of soil resource heterogeneity. Ecologists have been in the forefront of efforts to understand plant resource variability since ecology first became a discipline. Recognizing plant distribution patterns and identifying their cause and consequence is a hallmark of plant community ecology; decades of work in successional systems has placed significant emphasis on the need to understand differences in soil resources among different communities and especially within seres. Historically, however, relatively little effort has been directed toward understanding within-community patterns of resource availability. Traditionally, such patterns have been ignored in favor of understanding mechanisms underlying between-community or between-ecosystem differences; within-system variability has been seen largely as a statistical hurdle to be overcome by adjusting a sample size upward just enough to statistically differentiate any between-system differences that may exist. Recently, however, led by advances developed in agronomic systems, new insights about in situ spatial variability are emerging for a broad variety of both terrestrial and marine ecosystems.

Although elucidating the variability of belowground resources has been the explicit subject of a number of studies dating from the 1950s (e.g. Downes and Beckwith 1951; Snaydon 1962; Pigott and Taylor 1964; Zedler and Zedler 1969; Allen and MacMahon 1985), only recently have the geostatistical tools that allow this variability to be closely examined become available (Robertson 1987; Rossi et al. 1992; Robertson and Gross 1994). Workers in agronomic ecosystems—as early adopters of these tools—led the way among life scientists. As a field, geostatistics was developed by mathematical geologists in the 1970s to describe mineral distributions for the mining industry (e.g. David 1977; Journel and Huijbregts 1978; Krige 1981). During the 1980s, geostatistical approaches were adopted by soil scientists seeking to understand soil morphological processes within agricultural landscapes, and in particular, patterns of such soil physical and chemical properties as stone content, infiltration capacity, and conductivity (e.g. Webster and Cuanalo 1975; Burgess and Webster 1980; Vieira et al. 1983; Trangmar et al. 1985; Webster 1985; Webster and Oliver 1990). Later in the decade, soil biologists began describing patterns of biological activity

across individual cropping systems (e.g. Folorunso and Rolston 1985;
Parkin et al. 1987; Trangmar et al. 1987; Aiken et al. 1991; Ambus and
Christensen 1995), and by the early 1990s, terrestrial ecologists had incorporated the technique and had begun applying geostatistics to questions of
resource heterogeneity in noncropped communities (e.g. Robertson et al.
1988, 1993; Jackson and Caldwell 1993; Gross et al. 1995; Fetcher et al. 1996;
Schlesinger et al. 1996).

Geostatistics offer substantial power for identifying the scale of environmental variation in landscapes and for describing patterns of variability at
different scales. The degree of structured variability that has been uncovered for important ecological processes has been surprising to many, and
in a variety of ways agronomic systems are well suited for addressing
questions about the ecological significance of the variability identified.
Robertson et al. (1997), for example, planted a single genotype of soybeans
across a 48-hectare (ha) row-crop ecosystem that—despite decades of
tillage and cropping—varied internally by up to an order of magnitude for
many important system-level attributes such as soil organic matter and
nitrogen availability (e.g. Figure 6.7). Although the variability for most
attributes examined was strongly structured, in some cases at multiple
scales (e.g. Figure 6.8), there was very little correspondence between specific groups of attributes and soybean productivity; at most less than 50% of

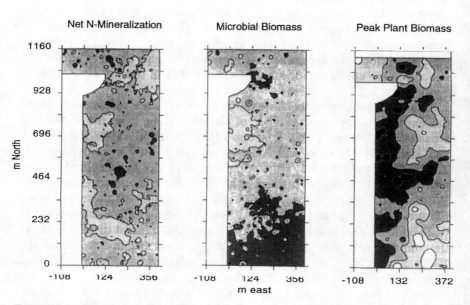

FIGURE 6.7. Spatial variability across a 48-hectare agricultural site in southwestern
Michigan; patterns correspond to even increments in levels for a given property: a)
net N mineralization ($0.30–0.70\,\mathrm{mg\,N\,g^{-1}\,d^{-1}}$); b) microbial biomass ($39–95\,\mathrm{mg\,C\,g^{-1}}$);
and c) peak plant (soybean) biomass ($170–347\,\mathrm{g\,m^2}$) (From Robertson et al. 1997).

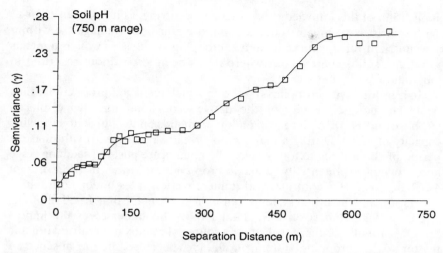

FIGURE 6.8. Nested spatial autocorrelation (semivariance) of soil pH within a 48-hectare row-crop ecosystem in southwestern Michigan. From Robertson et al. 1997.

the variation in plant productivity could be explained by variation in soil chemical and physical properties. Such findings challenge our understanding of controls on productivity at scales between the individual plant and the landscape, and underscore the value of agronomic systems for addressing questions of broad relevance to many ecosystems.

Applications of Ecosystem Science to Management Issues: Success in the Making

Inasmuch as several important advances in ecosystem science have been made in whole or in part in agronomic settings, it is not difficult to identify the application of these particular concepts as important contributions of ecosystem science to agronomic management. On the other hand, credit may be premature: in some cases, the effective application of ecosystem-level concepts predates their scientific understanding by centuries. Take, for example, organic matter management. The inclusion of legumes in long-term rotation strategies and the application of animal wastes to farm fields was part of the agronomic toolbox long before the development of active fraction concepts, in fact, long before nitrogen was even recognized as an essential nutrient (e.g. Oakley 1925; Francis and Clegg 1990). The effect of rotations on soil color, that is, on soil organic matter levels, was recognized and used to manage fallows even during the Roman Empire. It is therefore unfair to say that our new understanding of organic matter dynamics has contributed much to the design of modern cropping systems. In fact, the application of this knowledge is only now occurring, as it becomes clear to agronomists that a means is needed to more efficiently manage the within-year nutrient-release curve in annual cropping systems, as well as a means to rebuild soil organic matter where it has been lost without resorting to long fallow periods.

High productivity cropping systems in which managed productivity may be twice the productivity of the native ecosystem that was displaced (Robertson 1997) can place tremendous demands on the nutrient-supplying capacity of a soil. During a maize crop's rapid vegetative growth phase, a period of about three to four weeks when plants are past the seedling stage and growing exponentially, a single crop can withdraw on the order of $4\,kg\,N\,ha^{-1}\,d^{-1}$. This contrasts with mineralization rates under both native vegetation and disturbed systems that typically sum to some fraction of this (Keeney 1980; Robertson 1982; Paul 1989). This underscores the importance of providing either fertilizer or a source of readily mineralized organic matter to the cropping system at this growth stage. In the absence of abundant fertilizer, biologically derived nitrogen must be provided by aligning soil nitrogen-mineralization rates with plant nitrogen demand, and this is a major goal of efforts to reduce the present reliance on synthetic nitrogen in high productivity row-crop ecosystems (Figure 6.9; Hendrix et al. 1992; Robertson 1997). It is probable that emerging knowledge about the decomposition dynamics of crop residues (in particular, knowledge about controls on active fraction formation and turnover) will play a large role in allowing us to approach this nitrogen-management goal, and this topic is a major issue in soil fertility research today.

In contrast to efforts aimed at providing sufficient nitrogen for rapidly growing crop plants is the converse problem of restricting excess agricultural nutrients in surface and groundwater supplies and in the atmosphere. One of the greatest challenges facing the confined-animal industry today is the emerging environmental problem of excess nitrogen and phosphorus loading that stems from land application of animal waste. Environmental concerns in Europe and in some U.S. states have already led to specific restrictions on the amount and timing of manure applications to cropping systems, and these restrictions are apt to become even more common as poultry, swine, and dairy operations continue to become increasingly concentrated in large confined-animal operations. In Europe, several countries have adopted watershed input-output balances as the basis for assessment and regulation of agricultural production systems; further understanding of the basis for soil organic matter acretion and mineralization-immobilization relationships will aid efforts to optimize nutrient retention in these systems. It will be important to maintain a concomitant watershed-level focus in this effort, because many processes that can attenuate nutrient loss operate at different scales and along different portions of hydrologic flow paths (e.g. Lowrance et al. chapter 5; Peterjohn and Correll 1984; Robertson and Rosswall 1986; Barry et al. 1993).

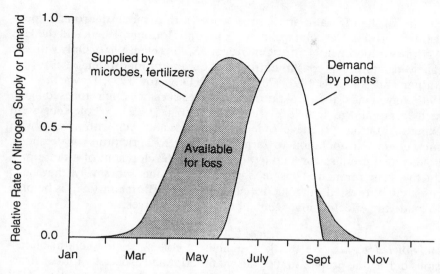

FIGURE 6.9. Temporal asynchrony between nutrient supply rate (from mineraliza-tion) and plant nutrient demand in a hypothetical row-crop ecosystem. From Robertson 1997.

Emerging knowledge about the relationship between plant productivity and the spatial distribution of soil resources will further aid efforts to increase ecosystem-level nutrient-use efficiency. New technologies that allow fertilizers to be applied differentially across individual fields—so-called site-specific farming practices (Robert et al. 1993)—have been touted for their potential ability to match resource inputs with resource needs. Without further information about the scale of soil biological activity in arable soils, however, and without matching application technology to this scale, it is improbable that the potential of the technology will ever extend beyond maps of such larger-scale soil properties as drainage class and topographic relief (Groffman 1997). Thus, knowledge of "biological precision" will be an important consideration in efforts to develop effective site-specific farming practices, and may well decide its theoretical effective-ness and, therefore, its economic viability.

The application of other ecosystem concepts should also provide the substantive insight needed to effectively manage the agricultural land base. For many concepts, the derivation of knowledge is coming full circle. Regional multi-watershed nutrient budgets were first developed for agricul-tural purposes (e.g. Lipman and Corybeare 1936; Ayers and Branson 1973; Miller and Smith 1976; Keeney 1979), and the whole-crop lysimeters first deployed in the Coshocton, Ohio watershed in the 1930s were arguably among the very first watershed-level biogeochemical experiments. Since then, however, most watershed-level experimentation has focused on forested regions, led by studies at the Coweeta Experimental Forest in

North Carolina (Swank and Crossley 1988), the Hubbard Brook Experimental Forest in New Hampshire (Likens and Bormann 1996), and the H.J. Andrews Experimental Forest in Oregon (Sollins et al. 1980). Only with the renewed interest in coastal eutrophication and groundwater quality (e.g. Turner and Rabalais 1991; Cole et al. 1993; Nixon 1995; Howarth et al. 1996) have landscape-level nutrient-cycling issues become refocused again at a farm-management level (e.g. Barry et al. 1993). Whole-ecosystem experimentation at the level of individual fields and even within fields, with subsequent extrapolation to farms and then to agricultural watersheds, holds much promise for helping to guide the development of a more productive, environmentally safe agriculture, in the same way that such research in forested watersheds (e.g. Likens and Bormann 1996) is helping to guide the effective management of forest resources.

Acknowledgments. We thank a number of colleagues for helpful discussions during the preparation of this manuscript and especially for constructive comments on earlier drafts. We are particularly indebted to A.R. Berkowitz, L.E. Drinkwater, P.M. Groffman, P.F. Hendrix, L.E. Jackson, C.G. Jones, J.F. Lussenhop, and K.H. Paustian. We acknowledge financial support from the NSF LTER Program at the Kellogg Biological Station and from the Michigan Agricultural Experiment Station.

References

Aiken, R.M., M.D. Jawson, K. Grahammer, and A.D. Polymenopoulos. 1991. Positional, spatially correlated, and random components of variability in carbon dioxide efflux. *Journal of Environmental Quality* 20:301–308.

Allen, M.F., and J.A. MacMahon. 1985. Impact of disturbance on cold desert fungi: comparative microscale dispersion patterns. *Pedebiologia* 28:215–224.

Allison, F.E. 1955. The enigma of soil nitrogen balance sheets. *Advances in Agronomy* 7:213–250.

Ambus, P., and S. Christensen. 1995. Spatial and seasonal nitrous oxide and methane fluxes in Danish forest-, grassland-, and agroecosystems. *Journal of Environmental Quality* 24:993–1001.

Arah, J.R.M., and K.A. Smith. 1989. Modeling denitrification in aggregated soils: relative importance of moisture tension, soil structure, and oxidizable organic matter. in J.A. Hansen, and K. Henriksen, eds. *Nitrogen in Organic Wastes Applied to Soils.* Academic Press, London, U.K.

Ayers, R.S., and R.L. Branson. 1973. *Nitrates in the Upper Santa Ana river basin in relation to groundwater pollution.* California Agricultural Experiment Station Bulletin Number 861. University of California, Riverside.

Barry, D.A., J.D. Goorahoo, and M.J. Goss. 1993. Estimation of nitrate concentrations in groundwater using a whole farm nitrogen budget. *Journal of Environmental Quality* 22:767–775.

Beare, M.H., R.W. Parmelee, P.F. Hendrix, D.C. Coleman, and D.A.J. Crossley. 1992. Microbial and faunal interactions and effects on litter nitrogen and decomposition in agroecosystems. *Ecological Monographs* 62:569–591.

Beven, K., and P.F. Germann. 1982. Macropores and water flow in soil. *Water Resources Research* 18:1311–1325.

Bolin, B., and R.B. Cook, eds. 1983. *The major biogeochemical cycles and their interactions.* John Wiley & Sons, Chichester, England.

Bouma, J. 1981. Soil morphology and preferential flow along macropores. *Agricultural Water Management* 3:235–250.

Burgess, T.M., and R. Webster. 1980. Optimal interpolation and isarithmic mapping of soil properties. II. Block kriging. *Journal of Soil Science* 31:315–331.

Clarholm, M. 1985. Possible roles for roots, bacteria, protozoa and fungi in supplying plants. Pages 355–365 in A.H. Fitter, D. Atkinson, D.J. Read, and M.B. Usher, eds. *Ecological Interactions in soil: plants, microbes, and animals.* Blackwell, Oxford, England.

Cole, J.J., B.L. Peoerls, N.F. Caraco, and M.L. Pace. 1993. Nitrogen loading of rivers as a human-driven process. Pages 141–157 in M.J. McDonnell and S.T.A. Pickett, eds. *Humans as components of ecosystems: the ecology of subtle human effects and populated areas.* Springer-Verlag, New York.

Coleman, D.C., and D.A.J. Crossley. 1996. *Fundamentals of Soil Ecology.* Academic Press, San Diego, CA.

David, M. 1977. *Geostatistical ore reserve estimation.* Elsevier Scientific Publishing Co., Amsterdam.

Downes, R.G., and R.S. Beckwith. 1951. Studies in the variation of soil reaction. I. Field variations at Barooga, NSW. *Australian Journal of Agricultural Research* 2:60–72.

Duxbury, J.M. 1986. Advantages of the acetylene method of measuring denitrification. Pages 73–92 in R.D. Hauck and R.W. Weaver, eds. *Field measurement of dinitrogen fixation and denitrification.* Soil Science Society of America, Madison, WI.

Edwards, C.A. 1983. Earthwork ecology in cultivated soils. Pages 123–137 in J.E. Satchell, ed. *Earthwork ecology.* Chapman & Hall, New York.

Elliott, E.T., and C.V. Cole. 1989. A perspective on agroecosystem science. *Ecology* 70:1597–1602.

Elliott, E.T., D.C. Coleman, R.E. Ingham, and J.A. Trofymow. 1984. Carbon and energy flow through microflora and microfauna in the soil subsystems of terrestrial ecosystems. Pages 425–434 in M.J. Klug and C.A. Reddy, eds. *Current perspectives in microbial ecology.* American Society Microbiology, Washington, DC.

Elliott, P.W., D. Knight, and J.M. Anderson. 1990. Denitrification in earthworm casts and soil from pastures under different fertilizer and drainage regimes. *Soil Biology and Biochemistry* 22:601–606.

Fetcher, N., B.L. Haines, R.A. Cordero, D.J. Lodge, L.R. Walker, D.S. Fernandez, and W.T. Lawrence. 1996. Response of tropical plants to nutrients and light on a lanslide in Puerto Rico. *Journal of Ecology* 84:331–341.

Firestone, M.K. 1982. Biological denitrification. Pages 289–326 in F.J. Stevenson, ed. *Nitrogen in agricultural soils.* American Society of Agronomy, Madison, WI.

Folorunso, O.A., and D.E. Rolston. 1985. Spatial variability of field-measured denitrification gas fluxes. *Soil Science Society of America Journal* 48:1214–1219.

Francis, C.A., and M.D. Clegg. 1990. Crop rotations in sustainable agricultural systems. Pages 107–122 in C.A. Edwards, R. Lal, P. Madden, R.H. Miller, and G. House, eds. *Sustainable agricultural systems.* Soil and Water Conservation Society, Ankeny, IA.

Grace, P.R., K. Bryceson, K. Hennessy, and M. Truscott. 1996. Climate change and soil carbon resources. Pages 19–23 in *Land use in a changing climate, Maurice Wyndham series*. University of New England Press, Armidale, Australia.

Groffman, P.M. 1997. Ecological constraints on the ability of precision agriculture to improve the environmental performance of agricultural production systems. Pages 52–64 in *Precision agriculture: spatial and temporal variability of environmental quality*. John Wiley and Sons, Chichester, U.K.

Groffman, P.M., and J.M. Tiedje. 1989. Denitrification in north temperate forest soils: spatial and temporal patterns at the landscape and seasonal scales. *Soil Biology and Biochemistry* 21:613–626.

Groffman, P.M., and J.M. Tiedje. 1989. Denitrification in north temperate forest soils: relationships between denitrification and environmental factors at the landscape scale. *Soil Biology and Biochemistry* 21:621–626.

Groffman, P.M., J.M. Tiedje, G.P. Robertson, and S. Christensen. 1988. Denitrification at different temporal and geographical scales: proximal and distal controls. Pages 174–192 in J.R. Wilson, ed. *Advances in nitrogen cycling in agricultural ecosystems*. CAB International, Wallingford, U.K.

Gross, K.L., K.S. Pregitzer, and A.J. Burton. 1995. Spatial variation in nitrogen availability in three successional plant communities. *Journal of Ecology* 83:357–367.

Haas, H.J., C.E. Evans, and E.F. Miles. 1957. Nitrogen and carbon changes in Great Plains soils as influenced by cropping and soil treatments. *USDA Technical Bulletin 1164*, Washington, DC.

Hauck, R.D. 1981. Nitrogen fertilizer effects in nitrogen cycle processes. Pages 551–562 in F.E. Clark and T. Rosswall, eds. *Terrestrial nitrogen cycles*. Swedish Natural Science Research Council, Stockholm.

Havelka, U.D., M.G. Boyle, and R.W.F. Hardy. 1982. Biological nitrogen fixation. Pages 365–422 in F.J. Stevenson, ed. *Nitrogen in agricultural soils*. American Society of Agronomy, Madison, WI.

Hendrix, P.F., D.C. Coleman, and D.A. Crossley, Jr. 1992. Using knowledge of soil nutrient cycling processes to design sustainable agriculture. *Journal of Sustainable Agriculture* 2:63–82.

Hendrix, P.F., R.W. Parmelee, D.A. Crossley, Jr., D.C. Coleman, E.P. Odum, and P.M. Groffman. 1986. Detritus food webs in conventional and no-tillage agroecosystems. *Bioscience* 36:374–380.

Howarth, R.W., G. Billen, D. Swaney, A. Townsend, N. Jaworski, K. Lajtha, J.A. Downing et al. 1996. Regional nitrogen budgets and riverine N & P fluxes for the drainages to the North Atlantic ocean. *Biogeochemistry* 35:75–139.

Hunt, H.W. 1977. A simulation model for decomposition in grasslands. *Ecology* 58:469–484.

Hunt, H.W., D.C. Coleman, E.R. Ingham, R.E. Ingham, E.T. Elliott, J.C. Moore, S.L. Rose, C.P.P. Reid, and C.R. Morley. 1987. The detrital food web in a short-grass prairie. *Biology and Fertility of Soils* 3:57–68.

IPCC. 1996. *Climate change: the intergovernmental panel on climate change (IPCC) scientific assessment*. University Press, Cambridge, UK.

Jackson, R., and M. Caldwell. 1993. The scale of nutrient heterogeneity around individual plants and its quantification with geostatistics. *Ecology* 74:612–614.

Jansson, S.L., and J. Persson. 1982. Mineralization and immobilization of soil nitrogen. Pages 229–252 in J.L. Smith and E.A. Paul, eds. *Process controls and nitrogen transformations in terrestrial ecosystems.* American Society of Agronomy, Madison, WI.

Jenkinson, D.S. 1977. Studies on the decomposition of plant material in soil. V. The effects of plant cover and soil type on the loss of carbon from ^{14}C labelled ryegrass decomposing under field conditions. *Journal of Soil Science* 28:424–434.

Jenkinson, D.S., D.E. Adams, and A. Wild. 1991. Model estimates of CO_2 emissions from soil in response to global warming. *Nature* 351:304–306.

Jenkinson, D.S., P.B.S. Hart, J.H. Rayner, and L.C. Parry. 1987. Modeling the turnover of organic matter in long-term experiments at Rothamsted. *INTECOL Bulletin* 15:1–8.

Jordan, T.E., and D.E. Weller. 1996. Human contributions to terrestrial nitrogen flux. *Bioscience* 46:655–664.

Journel, A.G., and C.J. Huijbregts. 1978. *Mining geostatistics.* Academic Press, New York.

Juma, N.G., and E.A. Paul. 1981. Use of tracers and computer simulation techniques to assess mineralization and immobilization of soil nitrogen. Pages 145–154 in M.J. Frissel and H.A. Van Veen, eds. *Simulation of nitrogen behavior in soil-plant* systems. PUDOC, The Netherlands.

Keeney, D.R. 1979. A mass balance of nitrogen in Wisconsin. *Wisconsin Academy of Sciences, Arts and Letters* 67:95–102.

Keeney, D.R. 1980. Prediction of soil nitrogen availability in forest ecosystems: a literature review. *Forest Science* 26:159–171.

Krige, D.G. 1981. Lognormal-de Wijsan geostatistics for ore evaluation. *South African Institute of Mining and Metallurgy Monograph Series Geostatistics* 1:1–51.

Liebig, J. 1847. *Chemistry in its application to agricultural and physiology,* 4th ed. Taylor and Walton, London.

Lipman, J.G., and A.B. Corybeare. 1936. Preliminary note on the inventory and balance sheet of plant nutrients in the United States. *New Jersey Agricultural Experiment Station Bulletin* 607, Rutgers, New Jersey.

Likens, G.E., and F.H. Bormann. 1996. *Biogeochemistry of a forested ecosystem. 2nd ed.* Springer-Verlag, New York.

McGill, W.B., H.B. Hunt, R.G. Woodmansee, and J.O. Reuss. 1981. PHOENIX, a model of the dynamics of carbon and nitrogen in grassland soils. *Ecological Bulletin* (Stockholm) 33:49–115.

Melillo, J.M., J.D. Aber, and J.M. Muratore. 1982. Nitrogen and lignin control of hardwood leaf litter decomposition dynamics. *Ecology* 63:621–626.

Melillo, J.M., J.D. Aber, P.A. Steudler, and J.P. Schimel. 1983. Denitrification potentials in a successional sequence of northern hardwood forest stands. *Ecological Bulletin* (Stockholm) 35:217–228.

Miller, R.J., and R.B. Smith. 1976. Nitrogen balance in the southern San Joaquin Valley. *Journal of Environment Quality* 5:274–278.

Molina, J.A.E., C.E. Clapp, M.J. Shaffer, F.W. Chichester, and W.E. Larson. 1983. NCSOIL, a model of nitrogen and carbon transformations in soil. Description, calibration and behavior. *Soil Science Society of America Journal* 47:85–91.

Moore, J.C., and P.C. deRuiter. 1991. Temporal and spatial heterogeneity of tro-phic interactions within below-ground food webs. *Agriculture, Ecosystems and Environment* 34:371–397.

Nixon, S.W. 1995. Coastal marine eutrophication: a definition, social causes, and future consequences. *Ophelia* 41:199–219.

Oades, J.M. 1993. The role of biology in the formation, stabilization and degradation of soil structure. *Geoderma* 56:377–400.

Oakley, R.A. 1925. The economics of increased legume production. *Journal of the American Society of Agronomy* 17:373–389.

Parkin, T.B. 1987. Soil microsites as a source of denitrification variability. *Soil Science Society of America Journal* 51:1194–1199.

Parkin, T.B., J.L. Starr, and J.J. Meisinger. 1987. Influence of sample size on measurement of soil denitrification. *Soil Science Society of America Journal* 51: 1492–1501.

Parton, W.J., D.S. Schimel, C.V. Cole, and D.S. Ojima. 1987. Analysis of factors controlling soil organic matter levels in Great Plains grasslands. *Soil Science Society of America Journal* 51:1173–1179.

Paul, E.A. 1989. Soils as components and controllers of ecosystem processes. Pages 353–374 in P.J. Grubb and J.B. Whittaker, eds. *Toward a more exact ecology, the 30th symposium of the British Ecological Society, London 1988.* Blackwell Scientific Publications, Boston, MA.

Paul, E.A., K.A. Paustian, E.T. Elliott, and C.V. Cole, eds. 1997. *Soil organic matter in agricultural ecosystems: long term experiments in North America.* CRC Publishers, Boca Raton, FL.

Paul, E.A., and G.P. Robertson. 1989. Ecology and the agricultural sciences: a false dichotomy. *Ecology* 70:1594–1597.

Paul, E.A., and J. vanVeen. 1978. The use of tracers to determine the dynamic nature of organic matter. *Transactions of the 11th International Congress of Soil Science* 3:61–102.

Paul, E.A., H.P. Collins, D. Harris, U. Schulthess, and G.P. Robertson. 1998. The influence of biological management inputs on carbon mineralization in ecosystems. *Applied Soil Ecology.* In press.

Paustian, K., W.J. Parton, and J. Persson. 1992. Modeling soil organic matter in organic-amended and nitrogen-fertilized long-term plots. *Soil Science Society of America Journal* 56:476–488.

Peterjohn, W.T., and D.L. Correll. 1984. Nutrient dynamics in an agricultural watershed: observations on the role of a riparian forest. *Ecology* 65:1466–1475.

Pieters, A.J. 1927. *Green manuring.* John Wiley & Sons, New York.

Pigott, C.D., and K. Taylor. 1964. The distribution of some woodland herbs in relation to the supply of nitrogen and phosphorus in the soil. *Journal of Ecology* 52:175–185.

Reiners, W.A. 1986. Complementary models for ecosystems. *American Naturalist* 127:59–73.

Revsbech, N.P., B.B. Jorgensen, and O. Brix. 1981. Primary production of microalgae in sediments measured by oxygen microprofile, $H^{14}CO_3^-$ carbon isotope labeled bicarbonate ion fixation, and oxygen exchange methods. *Limnology and Oceanography* 26:717–730.

Robert, P.C., R.H. Rust, and W.E. Larson, eds. 1993. *Soil specific crop management.* American Society of Agronomy, Madison, WI.

Robertson, G.P. 1982. Nitrification in forested ecosystems. *Philosophical Transactions of the Royal Society London* 296:445–457.

Robertson, G.P. 1987. Geostatistics in ecology: interpolating with known variance. *Ecology* 68:744–748.

Robertson, G.P. 1997. Nitrogen use efficiency in row-crop agriculture: crop nitrogen use and soil nitrogen loss. Pages 347–365 in L. Jackson, ed. *Agricultural ecology.* Academic Press, New York.

Robertson, G.P., J.R. Crum, and B.G. Ellis. 1993. The spatial variability of soil resources following long-term disturbance. *Oecologia* 96:451–456.

Robertson, G.P., and K.L. Gross. 1994. Assessing the heterogeneity of below-ground resources: quantifying pattern and scale. Pages 237–253 in M.M. Caldwell and R.W. Pearcy, eds. *Plant exploitation of environmental heterogeneity.* Academic Press, New York.

Robertson, G.P., M.A. Huston, F.C. Evans, and J.M. Tiedje. 1988. Spatial variability in a successional plant community: patterns of nitrogen availability. *Ecology* 69:1517–1524.

Robertson, G.P., K.M. Klingensmith, M.J. Klug, E.A. Paul, J.C. Crum, and B.G. Ellis. 1997. Soil resources, microbial activity, and primary production across an agricultural ecosystem. *Ecological Applications* 7:158–170.

Robertson, G.P., and T. Rosswall. 1986. Nitrogen in West Africa: the regional cycle. *Ecological Monographs* 56:43–72.

Robertson, G.P., and J.M. Tiedje. 1985. Denitrification and nitrous oxide production in successional and old growth Michigan forests. *Soil Science Society of America Journal* 48:383–389.

Robertson, G.P., and J.M. Tiedje. 1988. Denitrification in a humid tropical rainforest. *Nature* 336:756–759.

Robertson, G.P., P.M. Vitousek, P.A. Matson, and J.M. Tiedje. 1987. Denitrification in a clearcut loblolly pine plantation in the Southeastern U.S.: differences related to harvest intensity, site preparation, and cultivation practice. *Plant and Soil* 97:119–129.

Rolston, D.E., M. Fried, and D.A. Goldhamer. 1979. Denitrification measured directly from nitrogen and nitrous oxide gas fluxes. *Soil Science Society of America Journal* 40:259–266.

Rossi, R.E., D.J. Mulla, A.G. Journel, and E.H. Franz. 1992. Geostatistical tools for modeling and interpreting geological spatial dependence. *Ecological Monographs* 62:277–314.

Schlesinger, W.H. 1984. Soil organic matter: a source of atmospheric CO_2. Pages 111–127 in G.M. Woodwell, ed. *The role of terrestrial vegetation in the global carbon cycle.* John Wiley & Sons New York.

Schlesinger, W.H., J.A. Raikes, A.E. Hartley, and A.F. Cross. 1996. On the spatial pattern of soil nutrients in desert ecosystems. *Ecology* 77:364–374.

Sexstone, A.J., N.P. Revsbech, T.P. Parkin, and J.M. Tiedje. 1985. Direct measurement of oxygen profiles and denitrification rates in soil aggregates. *Soil Science Society of America Journal* 49:645–651.

Shields, J.A., and E.A. Paul. 1973. Decomposition of ^{14}C-labelled plant material under field conditions. *Journal of Soil Science* 53:297–306.

Snaydon, R.W. 1962. Micro-distribution of *Trifolium repens* L. and its relation to soil factors. *Journal of Ecology* 50:133–143.

Sollins, P. 1989. Factors affecting nutrient cycling in tropical soils. Pages 85–95 in J. Proctor, ed. *Mineral nutrients in tropical forest and savanna ecosystems.* Blackwell Scientific Publications, Oxford, England.

Sollins, P., C.C. Grier, F.M. McCorison, K. Cromack, R. Fogel, and R.L. Fredricksen. 1980. The internal element cycles of an old-growth Douglas-fir ecosystem in Western Oregon. *Ecological Monographs* 50:261–285.

Stanford, G., and S.J. Smith. 1972. Nitrogen mineralization potentials of soils. *Soil Science Society of America Journal* 36:465–472.

Swank, W.T., and D.A.J. Crossley. 1988. *Forest hydrology and ecology at Coweeta.* Springer-Verlag, New York.

Thomas, G.W., and R.E. Phillips. 1979. Consequences of water movement in macropores. *Journal of Environmental Quality* 8:149–152.

Tiedje, J.M. 1988. Ecology of denitrification and dissimilatory nitrate reduction to ammonium. Pages 179–244 in A.J.B. Zehnder, ed. *Biology of anaerobic microorganisms.* John Wiley & Sons, New York.

Trangmar, B., R. Yost, and G. Uehara. 1985. Application of geostatistics to spatial studies of soil properties. Pages 45–94 in N.C. Brady, ed. *Advances in agronomy.* Academic Press, New York.

Trangmar, B.B., R.S. Yost, M.K. Wade, G. Uehara, and M. Sudjadi. 1987. Spatial variation of soil properties and rice yield on recently cleared land. *Soil Science Society of America Journal* 51:668–674.

Turner, R.E., and N.N. Rabalais. 1991. Changes in Mississippi River water quality this century. *Bioscience* 41:140–147.

vanVeen, J.A., and M.J. Frissel. 1979. Mathematical modelling of nitrogen transformation in soil. Pages 133–157 in J.F.K. Gasser, ed. *Modelling nitrogen from farm wastes.* Applied Sciences Ltd., London.

vanVeen, J.A., J.N. Ladd, and M.J. Frissel. 1984. Modelling C and N turnover through the microbial biomass in soil. *Plant and Soil* 76:257–274.

vanVeen, J.A., and E.A. Paul. 1981. Organic carbon dynamics in grassland soils. I. Background information and computer simulation. *Canadian Journal of Soil Science* 61:185–201.

Vieira, S.R., J.I. Hatfield, D.R. Nielson, and J.W. Biggar. 1983. Geostatistical theory and application to variability of some agronomical properties. *Hilgardia* 51:1–75.

Virginia, R.A., W.M. Jarrell, and E. Franco-Vizcaino. 1982. Direct measurement of denitrification in a *Prosopis* (Mesquite) dominated Sonoran desert ecosystem. *Oecologia* (Berlin) 53:120–122.

Vitousek, P.M. 1982. Nutrient cycling and nitrogen use efficiency. *American Naturalist* 119:553–572.

Webster, R. 1985. Quantitative spatial analysis of soil in the field. Pages 1–70 in B.A. Stewart, ed. *Advances in soil science.* Springer-Verlag, New York.

Webster, R., and H.E. Cuanalo. 1975. Soil transect correlograms of north Oxfordshire and their interpretations. *Journal of Soil Science* 26:176–194.

Webster, R., and M. Oliver. 1990. *Statistical methods in soil and land resource survey.* Oxford University Press, Oxford, England.

White, R.E. 1985. The influence of macropores on the transport of dissolved and suspended matter through soil. *Advances in Soil Science* 3:95–120.

Wild, A. 1972. Nitrate leaching under bare fallow at a site in northern Nigeria. *Journal of Soil Science* 23:315–324.

Yoshinari, T., R. Hynes, and R. Knowles. 1977. Acetylene inhibition of nitrous oxide reduction and measurement of denitrification and nitrogen fixation in soil. *Soil Biology and Biochemistry* 9:177–183.

Zedler, J., and P. Zedler. 1969. Association of species and their relationship to microtopography within old fields. *Ecology* 50:432–442.

7
Progress in Understanding Biogeochemical Cycles at Regional to Global Scales

Ingrid C. Burke, William K. Lauenroth, and Carol A. Wessman

Summary

Global- to regional-scale studies have played an important role in the development of ecosystem ecology. Long before there was evidence of global-scale impacts by humans on biogeochemistry, ecologists recognized that there were strong, interactive forces at global scales that were responsible for the state of the earth. In addition to this knowledge, many early ecologists made observations of biogeochemical pools and processes at regional to continental scales; their interpretations of patterns and their causes made significant contributions to our understanding biogeochemistry. The work of these early scientists was characterized by creative induction and vision; many of the frontiers and questions identified long ago still remain the focus of our activities today.

In recent decades, immense progress has been made in understanding biogeochemical processes at regional to global scales. Considerable advances have been made in understanding global- and regional-scale budgets of carbon and of nitrogen, and the interactions of trace gas fluxes, biophysical processes, vegetation, and climate. These successes were partially the result of the development of new tools, new collaborations, and an imperative from the international public to solve important environmental issues. Although the linkage between our present-day scientific activities and regional- to global-scale environmental problems is strong and productive, there is a need for continued support for basic research that will identify new horizons.

Introduction

Nothing is more important than that which is neither probable nor demonstrable
Herman Hesse

The application of ecosystem science to large spatial scales, for example, from regional to global, is inherently problematical because many ecosystem processes are difficult—if not impossible—to measure at these scales. Nonetheless, biogeochemical processes are important determinants of regional to global earth-system dynamics and it is at these scales that their relevance to human activity is most important, because these are the socioeconomic and political scales at which economies, food, and other natural resource production, and atmosphere-biosphere interactions occur.

Our objective in this chapter is to review some of the significant successes in the area of global- and regional-scale biogeochemistry. We focus first on the early conceptual successes of ecosystem scientists working at these large scales, then summarize the advances that led to major new developments during the past two decades. Finally, we review several of the recent, key successes in global- and regional-scale biogeochemistry. Our intent is not to be exhaustive in our review, because the number of successes in this area has been enormous, but rather to review the conceptual breakthroughs, and assess the incentives for progress in the past and the present, to better understand our roots and perhaps to provoke discussion of the appropriate directions for the future.

Early Successes: The Dreamers and Thinkers

Global Biogeochemical Cycles

Very early in the history of ecosystem science, long before evidence of global-scale impacts by humans on biogeochemistry were evident, scientists recognized that there were strong, interactive forces at global scales that were responsible for the state of the earth. Individuals from a wide diversity of disciplines, including oceanography, evolutionary biology/microbiology, limnology, and atmospheric chemistry contributed to this vision of the earth as a system. The work of all these scientists was characterized not by detailed measurements and budgets constructed for large scales, but rather by creative induction and vision.

The early writings of G.E. Hutchinson demonstrate a remarkably early vision in global biogeochemistry. In 1944, he published an essay entitled "A Century of Atmospheric Biogeochemistry" (Hutchinson 1944a), in which he reviewed a published lecture by French chemists Dumas and Boussingault (1841). Certainly, once every fifty years or so, excerpts from such an elegantly written portrayal of the biogeochemical dynamics of the earth should be published:

As it is from the mouths of volcanoes, then, whose convulsions so often make the crust of our globe to tremble, that the principal food of plants, carbonic acid, is incessantly poured out; so is it from the atmosphere on fire with lightnings, from the bosom of the tempest, that the second and scarcely less indispensable aliment of plants, nitrate of ammonia is showered down for their behoof.

Might it not be said, that we have here a remembrance of that chaos mentioned in the Bible, of those periods of tumults and disorders which preceded the appearance of order and organization upon earth?

For, scarcely are carbonic acid and nitrate of ammonia formed, than a calmer, though not less energetic force begins to act upon them for new purposes: this force is LIGHT. By the agency of light, carbonic acid yields up its carbon, water its hydrogen, nitrate of ammonia its nitrogen. These elements combine, organic matters are formed, and the earth is clothed with verdure.

It is, in fact, from absorbing incessantly a true force, the light and heat of the sun, that vegetables perform their functions, and produce the vast quantities of organized or organic matter which are the destined food of the animal creation. . . . Then come animals, consumers of matter, and producers of heat and of force, true instruments of combustion. It is in them, unquestionably that organized matter acquires what may be called its highest expression. But it is not without detriment to itself that it becomes the instrument of sensation and of thought. In this new capacity organized matter is burnt; and in giving out the heat, or electricity, which constitutes and is a measure of our force, it is destroyed and returned to the atmosphere, from whence it had originally come.

The atmosphere, therefore, is the mysterious link that connects the animal with the vegetable, the vegetable with the animal kingdom.

One hundred years ago, the first scientists began to recognize the impact of humans on global biogeochemistry. Svante Arrhenius published the first work that addressed the influence of increasing global carbon dioxide (CO_2) concentrations on global temperature (Arrhenius 1896). Arrhenius's work was recently commemorated in both a special conference and an issue of *Ambio* (Arrhenius 1997; Crawford 1997; Rodhe et al. 1997). Although his initial interest in the greenhouse effect (which he called a "hothouse" (Rodhe et al. 1997)) was stimulated by work of others on geochemistry, and his interest in explaining the long-term variations in temperature during the glacial and interglacial periods (Crawford 1997), Arrhenius reached the conclusion via induction that fossil fuel burning could increase global temperatures. To test his induction, using very few data and thousands of calculations, Arrhenius proposed that doubling atmospheric carbon dioxide concentrations from coal burning would have the effect of increasing global temperatures by 3 to 4°C, a level that is not very different than our estimates today (IPCC 1996). He calculated that oceans would absorb five-sixths of the carbon dioxide emitted to the atmosphere. Crawford (1997) states "Arrhenius's final results are impressive both as an innovative exercise in model-building and as a first approximation of the influence of CO_2 on climate. This should not make one forget, however, that they hardly rested on solid empirical ground." Arrhenius's work, however, is currently viewed as a landmark in the geosciences. Interestingly, his work on anthropogenic influences was not driven by a concern for the global environment. He wrote "In any case it seems probably, however, that the carbonic acid in the atmosphere is at present gradually increasing. And we would then have some right to indulge in the pleasant belief that our descendants, albeit after many generations, might live under a milder sky and in less barren natural surroundings than is our lot at present." (quoted from Rodhe et al. 1997).

In 1926, Vernadsky, a Russian biogeochemist, wrote a paper on "The Biosphere" that first formalized the concept of the earth as a system, and proposed important interactions between life and the evolution of the atmosphere (Vernadsky 1944). He followed this work with an important examination of the matter and energy exchanges between living and inert parts of the biosphere (Vernadsky 1945). Vernadsky's ideas about global-scale processes provided important impetus for scientists to produce global estimates of carbon dioxide fluxes (Riley 1944), methane production (Hutchinson 1949), nitrogen fluxes (Hutchinson 1944b), and phosphorus fluxes (Hutchinson 1952). Global-flux estimates were produced by scaling up very scarce, small-scale measurements from two or three systems to the globe. As a specific example, Riley (1944) reported estimates of organic production of terrestrial environments using estimates from Shroeder (1919) and Noddack (1937). Global forest production was estimated to be 15.1 to 16.6 $\times$ 10^9 tons of organic carbon based upon data from Bavaria and the assumption that an average forest produces either two-thirds (Noddack 1937) or five-sixths (Shroeder 1919) as much ". . . as a good Bavarian forest" (Riley 1944). Hutchinson (1948) summarized the state of these calculations by saying, "It must be admitted that we are ignorant of many matters of importance here."

By 1944, this community of biogeochemists had detected an important human influence on the biosphere, and was writing expressively about it:

Chemically, the face of our planet, the biosphere, is being sharply changed by man, consciously, and even more so, unconsciously. The aerial envelope of the land as well as all its natural waters are changed both physically and chemically by man. . . . Man now must take more and more measures to preserve for future generations the wealth of the seas which so far have belonged to nobody. (Vernadsky 1945).

These early biogeochemists made numerous advances on the basis of induction. For example, Alfred Redfield (1958), observed ratios of elements in ocean water, and introduced two extremely important concepts for global biogeochemists. First, he proposed that the total amount of carbon and nitrogen in oceans was controlled by the amount of available phosphorus in ocean water and the stoichiometry of phytoplankton. Second, he suggested that atmospheric oxygen concentrations were dependent not only upon the long-term differences between the global integrals of photosynthesis and decomposition, but also upon the presence of sulfate in ocean waters as an oxidizing agent for organic matter. The idea that global-scale atmospheric dynamics were closely linked to the chemical composition of sea water and to the oceanic biota was a seminal one; Redfield (1958) continues to be cited at a rate of nearly thirty times per year, and his ideas are key in the recent Intergovernmental Panel on Climate Change (IPCC) review of marine biotic feedbacks to climate (Denman et al. 1996).

In 1974, James Lovelock and Lynn Margulis introduced the provocative Gaia hypothesis, the idea that the earth is an interactive system in which the biota play a major role in determining the physical and chemical climate (Lovelock and Margulis 1974; Margulis and Lovelock 1974; Lovelock 1988). Despite the fact that Vernadsky had published the same ideas four and a half decades earlier (Margulis and Hinkle 1991), the Gaia papers and books evoked significant controversy, partly because of ideas about homeostasis and implications about teleological forces in the evolution of the earth system, and partly because they identified the ideas as "hypotheses," which the scientific community perceived as not testable (Kirchner 1991). The concept of multiple, complex, global-scale interactions involving biotic and abiotic processes that drive the functioning of the biosphere was independently accepted by the scientific community; however, the Gaia work was important in popularizing the concept and initiating some interesting synthesis work. An example of this popularization is a recent Gaia conference and synthesis (Schneider and Boston 1991; Barlow and Volk 1992), twenty years after the original Gaia papers. An interesting question is whether Lovelock and Margulis's Gaia concept would have had the same impact had they not introduced nonscientific language into their hypothesis. An alternative explanation for its popularity is that the Gaia hypothesis was published at a time when public sensitivities to the environment were strong.

These are a few of the examples of the early work that led to the dramatic growth of an ecological discipline that examines the global-scale biogeochemistry. These early "dreamers and thinkers" used limited data and inductive reasoning to develop ideas about the global biogeochemical system. They recognized and popularized the ideas that a systems approach to such large scales as regions and the globe is important and appropriate, and that humans are impacting global biogeochemical cycles. This early work provided an important incentive and foundation for study at regional to global scales, but the tools of study were not available, and the uncertainties associated with scaling up were very high.

Regional Biogeochemistry

Regional-scale studies also have a long history in ecosystem science, and the early "dreamers and thinkers" provided important initiative to later studies. The early successes at the regional or continental scale involved the use of large-scale environmental gradients to study the controls over ecosystem structure and function. Much of this work was developed from a background provided by such plant geographers as Alexander von Humboldt and August Grisebach. With extremely limited opportunities to travel and very primitive tools, they produced maps of the vegetation of regions and the earth (Humboldt 1817; Grisebach 1872). Although this work was predominantly focused on the geographic distribution of vegetation types and their relationship to climate (e.g. Holdridge 1947), it has had an enormous impact on our present-day conceptual models of the distribution of ecosystems, net primary productivity, decomposition, and soil development. It does not take much imagination to see the connections between this early work and the multitude of recent analyses of the distribution of ecosystems (e.g. Emanuel et al. 1985; Prentice 1990; Prentice et al. 1992). Far more powerful than studying environmental variability in a single location through time or across limited spatial distances, these early scientists made observations as natural historians, or collected data from a large array of sites that expanded environmental gradients, and produced important relationships that have permanently altered the views of the world held by ecosystems scientists. Notably, this type of work still continues today, advancing our understanding of ecosystem structure and function.

The contributions of Dockuchaiev (1883), Shaw (1930) and Jenny (1941), made across large-scale regional gradients, greatly influenced the foundations of soil science and ecosystem science. These scientists observed changes in soil profiles and soil organic matter content across large-scale gradients in climate, parent material, geologic age, topography, and vegetation type, and developed the idea of the "5 state factors" that control soil formation. The state factor equation as conceptualized by Jenny (1941) has been widely used to guide and design research and has been extended by Jenny and others to apply directly to ecosystems (Major 1951; Perring 1958; Jenny 1961, 1980). These ideas have provided a very important conceptual framework for ecosystem scientists (Amundson et al. 1994), which is either explicitly or implicitly included in essentially every ecosystem-simulation model today.

In addition to Jenny's early work, many ecosystem scientists utilized spatial variation, which occurs at regional to subcontinental and even larger scales, to elaborate the controls over important aspects of ecosystem structure and function. For instance, Lieth and Whittaker (1975), Lieth (1978) and Rodin and Bazilevich (1967) collected data on net primary production across very large gradients, elucidating the large-scale relationships between net primary production and mean annual precipitation, mean annual temperature, and actual evapotranspiration. These relationships have been elaborated across regions within major biomes (Rosensweig 1968; Lauenroth 1979; Rutherford 1980; Webb et al. 1983; Sala et al. 1988). Meentemeyer (1984; Meentemeyer et al. 1982) studied litter decomposition rates across broad environmental gradients, and generated relationships among climate, litter quality, and litter decomposition. Recently, as will be discussed further in the next section, the availability of large spatial data sets has considerably furthered this type of study. These spatial relationships generated over regional, continental, and global scales have provided ecosystem science with important information about the controls over ecosystem structure and function, and have proven very useful for building ecosystem-simulation models.

More Successes: The Age of Global and Regional Biogeochemistry

There are several categories of major successes that have occurred over the past two decades that have led to very rapid progress in regional- and global-scale biogeochemistry. Numerous barriers (i.e. technological, logistical, and cultural) to such large-scale studies were removed, leading to major scientific developments in understanding the cycling of elements at regional to global scales. Additionally, some of our understanding has led to important changes in regional- to global-scale policies. In the following section, we summarize the successes of regional- to global-scale ecosystem science in each of these three major categories: (1) the removal of barriers, (2) major scientific developments, (3) and policy impacts.

The Removal of Barriers: Tools, Collaboration, and Funding

Systems analysis has provided many of the tools that have been critical to advancing global and regional biogeochemistry. The key tools of systems analysis were developed in concert with development of computers mainly by engineers during and following World War II. Their introduction to the atmospheric, oceanic, and ecological sciences occurred at approximately the same time. Energy exchange and biogeochemical problems were prominent in the early development of systems analysis in all of these fields although the spatial scales of concern were very different.

In the late 1950s, while atmospheric and oceanic scientists were building box and arrow diagrams and compartment models of carbon exchange between the ocean and the atmosphere (Revelle and Suess 1957), ecosystem scientists were using similar approaches to model the flow of energy through ecosystems (Odum 1955). For the next twenty-five years, oceanic and atmospheric scientists continued with problems and models of global scale, while, with a few exceptions, ecosystem scientists continued with ecosystem-scale (meters to hundreds of meters) problems and models. A symposium at Brookhaven National Laboratory in 1972 entitled "Carbon and the Biosphere" represented an early attempt by atmospheric, oceanic, and ecosystem scientists to collaboratively use systems analysis tools to analyze a global-scale biogeochemical issue (Woodwell and Pecan 1973). Box and arrow diagrams were both the dominant conceptual approach and the key technology.

Participation by the United States in the International Biological Programme (IBP) of the 1960s and 1970s provided a major influx of funding and talent to promote and develop systems analysis in ecosystem science (Van Dyne 1972). During this time, simulation modeling became the dominant ecosystem-analysis tool. Although most of this work was focused at small scales (1 to 100 m^2), the program was of great importance to the future of ecosystem modeling not only because of the advances in knowledge but also because of the large number of modelers that were trained during this time (Innis 1976; Van Dyne 1977). To a very large extent, the IBP experience provided the platform from which the present-day contributions of ecosystem modelers to regional- and global-scale biogeochemical issues developed. In some cases the models and in most cases our current ideas can be traced directly back to the models and the knowledge that were gained during IBP.

Coincident with the widespread recognition that human activities were having major regional and global impacts (i.e. during the 1980s) was the acceptance that simulation modeling represented one of the few tools available to investigate and analyze regional- and global-biogeochemical questions. This was also the time during which geographic information systems (GIS) became widely available. The result is that it is now possible to use GIS to manage spatial inputs and analyze spatial outputs from regional- and global-scale ecosystem simulations (e.g. Aber et al. 1993, Burke et al. 1991, 1997).

Early in the century, there were several significant barriers to global-scale studies. Perhaps the largest barrier was that the collection of sufficient data at large scales was not possible; rather, investigators either made natural history-type observations as they traveled across regions, collected sparse samples at several sites and extrapolated those to large scales, or collected data from several other researchers to generate meta-analyses. The development of remote sensing as a tool to assess ecosystem structure and function from regional to global scales launched an entirely new era beginning in the 1980s (reviewed by Matson et al. 1989; Matson and Ustin 1991; Roughgarden et al. 1991; Wessman 1992; Wessman and Asner chapter 14). Estimates at regional to global scales were now available for primary production (Tucker et al. 1983, 1985), phenological dynamics (Justice et al. 1985) vegetation type and landcover (Goward et al. 1985; Tucker et al. 1985), and canopy chemistry (Wessman et al. 1988). Multi-investigator efforts focused on coordinated ground and aircraft (and, in some cases, satellite) measurements have added significantly to atmosphere-biosphere research. Such experiments as the Amazon Boundary Layer Experiments (ABLE-2A and ABLE-2B, Harriss et al. 1988), the First International Satellite Land Surface Climatology Project (ISLSCP) Field Experiment (FIFE, Sellers et al. 1992; Hall and Sellers 1995), the Oregon Transect Ecosystem Research Project (OTTER, Peterson and Waring 1994), and the Boreal Ecosystem-Atmosphere Study (BOREAS, Sellers et al. 1995) have not only broadened the use of remote sensing by the scientific community but have increased the sophistication of its use in ecosystem science.

The availability of such data has contributed in two major ways to regional- and global-scale biogeochemistry. First, it is now possible to assess large-scale spatial and temporal patterns in the status and trends of ecosystems (e.g. Hall et al. 1991; Skole and Tucker 1993; Paruelo and Golluscio

1994; Nemani et al. 1996; Wessman et al. 1997). Second, the coupling of remotely sensed data both to ecological simulation models (Running 1986; Running and Coughlan 1988; Melillo et al. 1993; Potter et al. 1993; Running 1994; Running et al. 1994; Field et al. 1995; Prince and Goward 1995; Cohen et al. 1996) and global- or regional-circulation models (e.g. Pielke et al. 1997; Fennessy and Xue 1997) has allowed us to project our understanding of system function into the future as well as over large scales.

In addition to the availability of satellite imagery, in general, technology supporting geo-referenced data distribution and analysis has also contributed to major advances in regional- to global-scale ecosystem science. Geographic information systems permit not only the digital storage and overlay of geographic ecological data, but also can be used to organize data for multiple-variable input into ecological simulation models (Costanza et al. 1990; Burke et al. 1990, 1991; Coleman et al. 1994), and for assessment of atmosphere-biosphere interactions (Schimel and Burke 1992). The availability of large-scale geo-referenced data represents a major new technological advance from which many analyses are possible. For instance, large databases are currently available for land cover (e.g. Leemans and Cramer 1991; Steayart et al. 1997), precipitation (e.g. NEXRAD), and remotely sensed variables. This recently increased availability of regional and global data sets has fueled new modeling and high performance computational activities, heightening both the need and funding for more such data (Cramer and Fischer 1996). In addition, many advances have been made in information management and massive data storage.

The actual development of ecosystem-simulation models has had a crucial role in regional and global scale science. Process-level understanding generated from field experimentation is built into simulation models, tested against separate field data, and consequently, models are revised. The simulation models may then be applied to spatial databases (Figure 7.1). The geographic data, including remotely sensed data, give us the ability to extrapolate a present-day picture of our knowledge of ecosystems in space; simulation models allow us to add the temporal dimension (Burke et al. 1991). Even though simulation models have been used as tools for ecosystem analysis for over forty years, their application to large-scale spatial problems is quite recent. There is a need, however, for a comprehensive analysis of the role of modeling in ecosystem science that is beyond the scope of this paper (see Lanenroth et al. chapter 16).

Recent applications of models to regional- and global-scale biogeochemistry have provided our only available information about the potential sensitivity of ecosystems to effects of such large-scale perturbations as atmospheric deposition, land-use change, or global climate change (e.g. Burke et al. 1991; Leemans 1992; Aber et al. 1993; Melillo et al. 1993; Rosenzweig and Parry 1994; VEMAP Members 1995; Goudriaan 1996; Parry et al. 1996). These analyses are now forming the key links between global biogeochemistry and environmental policy. It is difficult to underestimate the importance of decisions regarding how to conduct such analyses.

Inputs   Outputs

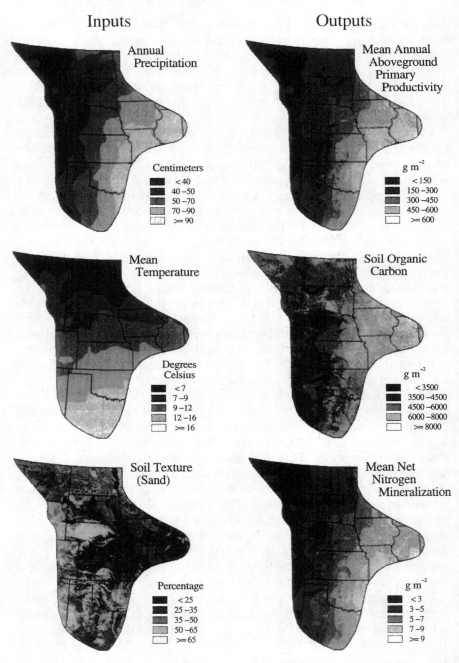

FIGURE 7.1. Georeferenced data were organized in a GIS and overlaid, then linked to the Century ecosystem model to simulate the pre-European condition of the central grasslands region (Burke et al. 1997). Input variables are shown on the left, and output variables on the right panel.

For example, is using the present output of general circulation models (GCMs) sufficient—or even desirable—as a complete set of scenarios for addressing the potential effects of climate change on ecological processes? How many such scenarios should we use? For that matter, how many simulation models should be used? How do we communicate the uncertainty regarding our models, the databases, and the input variables, to colleagues and to the public? As a recent example, VEMAP Members (1995) conducted a complete sensitivity analysis using several biogeography models, biogeochemistry models, and GCMs (Figure 7.2), providing the opportunity for readers to get a sense of the uncertainty associated with the analysis. These are among the largest challenges ahead for ecosystem scientists addressing regional- to global-scale questions.

Have the tools driven the science? There are a number of things that make an answer to this question complicated, not the least of which is the confounding of the appearance of a tool and the need for it. Some scientists reject out of hand the possibility that the appearance of tools could result in subsequent stimulation of new and interesting research. Others allow for the possibility, especially when the situation involves tools that have been created in other such disciplines as mass spectrometers or fast three-dimensional sonic anemometers. Our assessment is that many of the new tools that ecosystem scientists are currently using have altered the types of observations that are made and the kinds of questions that may be addressed. This has been especially true for the science conducted at regional to global scales. As an example, the appearance of GIS software that was developed for an engineering market of surveyors and city planners provided an opportunity for ecosystem scientists to wonder if their site-specific simulation models could be run for maps of input variables to produce maps of output variables (e.g. Burke et al. 1990, 1991). At the present time, remote sensing data, simulation models, regional and global data sets, and GIS software are the foundation of an ability to deal with regional- and global-scale questions. The large-scale data sets are only one of many components that were clearly developed to meet the need to answer regional to global questions. We submit that these tools have played and will continue to play a crucial role in eliciting new questions, and thus drive forward new regional- to global-scale science.

As our spatial scales of inquiry have increased, so have the needs for interfacing with other disciplines. It is not possible to understand biogeochemical cycles at large scales without explicit consideration of atmospheric dynamics, nor human interactions with ecological systems (Groffman and Likens 1994; Miller 1994; Riebsame et al. 1994; Schimel 1994). The integration of atmospheric science techniques into ecological studies of gaseous flux, through incorporation of eddy-flux measurements and aircraft measurements, has had a very significant impact on our understanding of large-scale cycling (Matson and Harriss 1988; Tans et al. 1990; Wofsy et al. 1993; Goulden et al. 1996). There is the potential to continue to

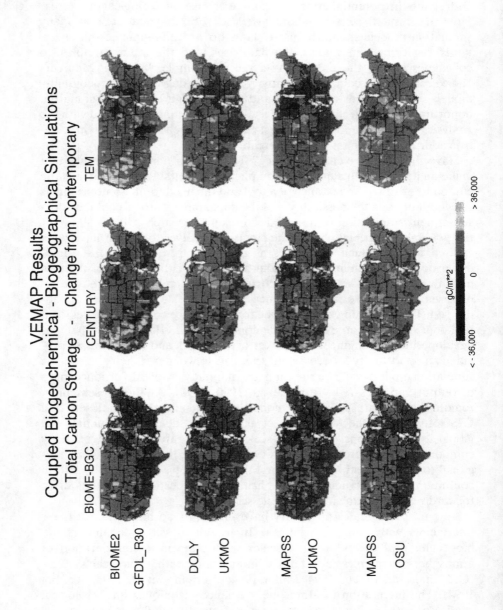

improve our knowledge with the implementation of a number of measurement sites with good geographic coverage (Tans et al. 1996). The advent of such large interdisciplinary projects as ABLE, FIFE, and BOREAS (Harriss et al. 1988; Sellers et al. 1992; Hall and Sellers 1995; Sellers et al. 1995) has led to substantial collaboration and information exchange among atmospheric scientists and ecosystem scientists. Increasingly, there are funding opportunities for interactions among ecosystem scientists, atmospheric scientists, and social scientists (e.g. the U.S. Global Change program); however, the majority of science funding is still focused on small-scale, short-term, single-investigator science (Kareiva and Anderson 1988; Tilman 1989).

Finally, ecological science has been enormously successful in the past decade in expanding the scale of international collaboration. Initiatives organized by the International Geosphere-Biosphere Program (IGBP 1988), the Global Change and Terrestrial Ecosystems project within IGBP (Walker and Steffen 1996), the International Council of Scientific Union's Scientific Committee on Problems of the Environment (e.g. Breymeyer et al. 1996), and the United Nations Environmental Program (e.g. Heywood 1995) have led to important scientific syntheses that have influenced funding priorities for individual governments. The reports of the Intergovernmental Panel on Climate Change (IPCC 1990; IPCC 1995; IPCC 1996) have represented an unprecedented level of interaction among ecosystem scientists, atmospheric scientists, and economists worldwide. Such international scientific discourse on the human impact on global-biogeochemical cycles has the highest probability of influencing both national and international policy.

Scientific Advances: What Are the Big Successes in Our Knowledge of Regional- and Global-Scale Biogeochemical Cycling?

Ecosystem scientists have made enormous progress in understanding regional- and global-scale biogeochemistry over the past two to three decades, and a simple review of the key advances would be a difficult if not

FIGURE 7.2. The results of a continental-scale sensitivity analysis evaluating the effects of global carbon dioxide increases and climate change on carbon storage of the United States (VEMAP Members 1995). The interdisciplinary team used three ecosystem simulation models (BIOME-BGC, CENTURY, and TEM), three biogeography models (BIOME2, DOLY, and MAPSS), and three GCMs (OSU, GFDL, and UKMO) to test the sensitivity of the region to long-term changes. This approach represents a powerful sensitivity analysis because it is not dependent upon the assumptions or relationships in any single model; commonalities that resulted from most combinations of models provide some confidence in model results.

overwhelming task. Rather than focusing on individual results, we will summarize what we consider to be the key conceptual advances that have fundamentally changed our understanding or altered the direction of our new pursuits. In the following sections, we summarize successes in three key areas: (1) global budgets, (2) atmosphere-biosphere interactions, and (3) regional impacts of land use on biogeochemical cycles.

Global Budgets

One of the important recent successes has been a substantial increase in our knowledge about global budgets. A major impetus for ecosystem scientists to work on global-biogeochemical budgets has been the need to understand the role of humans in influencing the concentrations of greenhouse gases in the atmosphere. Although global budgets have been an important area of basic study since the time of Arrhenius, concern about human-induced global climate change and of the need to develop predictions and assessments for use in international policy has been the key driving force for the present-day level of activity. The availability of funding to support this area of research has made rapid major advances possible.

Large uncertainties about the global carbon cycle have been major catalysts for research in the past two decades by both terrestrial and marine biogeochemists (reviewed in Schimel et al. 1994b; Heimann 1997). In addition, the realization that microbial reduction-oxidation reactions were important in determining the balance of other greenhouse gases, including nitrous oxide and methane, has led to research on the controls and spatial patterns of radiatively active trace-gas efflux over landscapes and regions (Andreae and Schimel 1989; Matson and Harriss 1995). More recently, large uncertainties in the global nitrogen cycle have stimulated a great deal of study on the global-scale interactions between carbon and nitrogen (Schindler and Bayley 1993; Vitousek et al. 1997).

Regional to global biogeochemistry is not dominated by scientists with a traditional ecological background; many, if not most, of the greatest contributions to our knowledge about regional to global budgets of greenhouse gases have been made by scientists in such disciplines as oceanography (e.g. Broecker and Peng 1991; Sarmiento 1993) and atmospheric science (e.g. Fung et al. 1987; Tans et al. 1990; Keeling and Shertz 1992). Contributions of ecosystem scientists have been primarily focused on understanding the terrestrial carbon budget and the magnitude of the terrestrial biotic sink for carbon (e.g. Melillo et al. 1993; Houghton 1995; Field et al. 1995).

Particularly over the past decade, biogeochemists have had numerous successes in increasing our understanding of the global carbon and nitrogen cycles, and the influence of humans on these cycles. The most important advances can be summarized in four statements, which have been reviewed in detail in recent publications (Schimel et al. 1994b; Melillo et al. 1996;

Heimann 1997). First, with respect to the global carbon budget, oceanic uptake may account for as much as 33 to 50% of the carbon dioxide added to the atmosphere by fossil fuel burning (between 1.1 and $3.6\,Gt\,Cy^{-1}$) (Keeling and Shertz 1992; Quay et al. 1992; Heimann and Maier-Reimer 1996; Keeling et al. 1996; Heimann 1997). These new estimates were possible from new techniques that assess the changes in atmospheric oxygen (oxygen/nitrogen ratios) relative to carbon dioxide (Keeling and Shertz 1992; Keeling et al. 1996), and from $^{12}C/^{13}C$ ratios in oceans and the atmosphere (e.g. Quay et al. 1992; Tans et al. 1993). These estimates are important advances because they suggest that geochemical absorption of carbon dioxide by oceans is a significant process in the global carbon budget, and the data provide boundary estimates of terrestrial carbon sinks, which may be the largest unknown of the global carbon cycle.

Second, human alterations of the global nitrogen cycle have important consequences for terrestrial and marine carbon storage. Because humans are increasing atmospheric deposition of nitrogen worldwide, many ecosystems may experience an increase in net primary production and net ecosystem production, leading to increased carbon storage (Schindler and Bayley 1993; Schimel et al. 1994a; McGuire et al. 1995; Melillo 1996; Asner et al. 1997; Vitousek et al. 1997). Changes in community structure relevant to biogeochemical cycling may also occur (Vitousek et al. 1997). Although our knowledge about nitrogen effects on ecosystems is substantial, there are still important research frontiers with respect to the effects of nitrogen on global carbon storage.

Third, land-use change is a very important influence on the global carbon and nitrogen cycles, with potentially 0.6 to 2.2 Gt C per year released into the atmosphere as a result of land conversions (Heimann 1997; Houghton 1997). Nitrogen trace-gas fluxes are dramatically increased with cultivation (Mosier et al. 1991) and with deforestation (Keller and Reiners 1994), and methane consumption is decreased with cultivation of grasslands and conversion of forests to pastures (Mosier et al. 1991; Keller and Reiners 1994). Recovery during succession over major regional and global scales may lead to significant carbon sinks (Smith et al. 1993; Houghton 1995), reduction of nitrous oxide fluxes, and increased methane consumption. The combination of these two important land-use forces— conversion to agricultural systems and successional recovery—may significantly influence the balance of many important gaseous constituents of the atmosphere.

Finally, ecosystem processes may have a strong response to increases in temperature and in carbon dioxide concentration, creating important biotic feedbacks to global climate change. Most modeling studies suggest that net ecosystem production will respond negatively in response to increases in temperature and/or carbon dioxide, although both net primary production and decomposition in some areas of the globe may increase (Townsend et al. 1992; Holland et al. 1995; Melillo et al. 1996).

Atmosphere-Biosphere Interactions: Changes in the Biota Alter Global Climate

An important success of large-scale biogeochemistry is represented by the current widespread recognition that biotic processes at the Earth's surface are important determinants of atmospheric processes and therefore climate (Koster and Suarez 1994; Henderson-Sellers et al. 1995; Pollard and Thompson 1995). Even though the concept of the effects of the biosphere on atmospheric processes has been recognized for a long time, it is only within the past decade that sufficient data have accumulated to make an irrefutable case. At the global scale, it has been largely through simulations with climate system models that land-surface properties have been shown to be important (Sato et al. 1989; Sellers 1987; Claussen 1994; Chase et al. 1996). At regional scales, although the effects have also been demonstrated in simulation model results, the processes involved have been much easier to couple to such particular surface characteristics as land-use changes (Pielke et al. 1991; Segal and Arritt 1992). Concern about understanding such effects by a combined atmospheric and ecosystem science community has resulted in large-scale field data collection efforts such as FIFE (Sellers 1987) in the tallgrass prairie, and BOREAS in the boreal forest (Sellers 1995).

At the present time, an effort that is being undertaken by ecosystem modelers largely through the efforts and encouragement of the Global Change and Terrestrial Ecosystems core project of IGBP is the development of a new class of ecosystem models that is being designed as components of climate system models. Currently identified as dynamic global vegetation models, they attempt to capture the behavior of ecosystems at the scale of a grid cell in a climate system model (Woodward 1996). The topic of coupling ecosystem responses to climate system models is an important frontier and challenge for ecosystem science.

Regional Impacts of Land Use

One of the most important successes of the past decade has been the regional-scale assessment of human land-use management impacts. Regions represent appropriate units for analysis because they can encompass major proportions of biomes that are managed similarly, and because they better approximate socioeconomic and political units than do globes, continents, or landscapes (Burke et al. 1991; Schimel 1994). Human influences may lead to important redistribution of elements within regions, and net losses or gains of elements. Many of the recent successes have addressed the influence of humans on regional storage of carbon, nitrogen, and phosphorus, and the implications for other important global pools. For instance, recent analyses on human land use has established that there are significant regional losses of carbon associated with deforestation (e.g. Houghton

1991; Skole and Tucker 1993), and with cultivation management (Burke et al. 1991, 1994). These regional changes in land use aggregate to form an important component of the global carbon cycle (Houghton 1995, in press). Further, ecosystem scientists have recently begun to assess the influence of human activities on large-scale watersheds, rivers, and marine bio-geochemical pools and processes. Human population is directly correlated with nitrate (Peierls et al. 1991; Cole et al. 1993; Howarth et al. 1996) and phosphate (Caraco 1995) export from terrestrial ecosystems to large rivers. River-nutrient export provides both an ideal integrated measure of regional terrestrial nutrient balance and an indication of aquatic eutrophication. Further, export from large rivers can be used to estimate total nutrient inputs to the ocean (Howarth et al. 1996). Much of the export is apparently the result of nitrogen and phosphorus fertilizer additions, sewage, and atmospheric deposition of nitrogen (Cole et al. 1993; Caraco 1995). Howarth et al. (1996) recently demonstrated that river fluxes of nitrogen are tightly related to agricultural inputs and inputs from combustion of fossil fuels (Figure 7.3). An important linkage that has yet to be made is the connection between carbon losses from cultivation as a result of stimulation of decomposition and nutrient exports to rivers.

In addition, great progress is being made on understanding the effects of long-term regional-scale atmospheric nitrogen and acidic deposition in lakes and terrestrial landscapes (Schwartz 1989; Aber et al. 1993). Progress in this area is reviewed in detail in Weathers and Lovett (chapter 8, this volume).

Policy Advances

Have ecosystem scientists changed the world with knowledge of regional- to global-scale biogeochemistry? One of the most important advances that could be made is to increase public awareness that humans alter the biogeochemical cycles, and that human welfare is in many ways linked to the alteration of these cycles. Clearly, general knowledge has increased regarding regional- and global-scale biogeochemistry: the newspapers frequently carry stories covering advances published in *Science, Nature,* or the IPCC reports, and a U.S. Senator (now Vice President Gore) produced a book on global change that was widely read. However, as might be expected, policy responses to ecological knowledge about human effects on regional and global element cycling have been greatest where human health is influenced. For instance, relatively clear evidence of lung disease fueled concern about air pollution standards. Ecological knowledge about acid deposition (see Weathers and Lovett chapter 8) and its effects on ecosystem structure and function positively influenced the renewal of the recent Clean Air Acts. Ecological understanding of the landscape and regional processes associated with eutrophication resulted in the Clean Water Act and increased availability of phosphate-free detergents. Again,

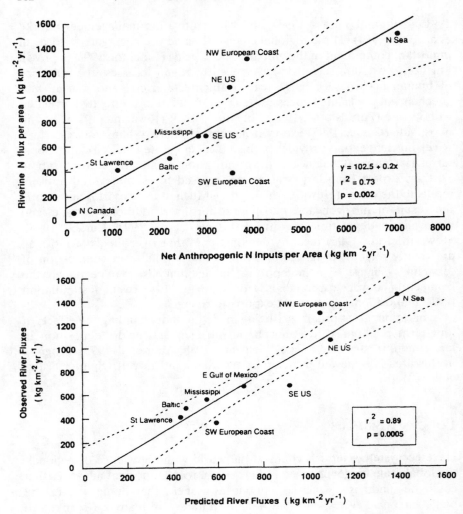

FIGURE 7.3. A. River-nitrogen export plotted against net anthropogenic nitrogen inputs to the temperate regions. B. Modeled vs. observed nitrogen river export (Howarth et al. 1996 with kind permission from Kluwer Academic Publishers). This graph demonstrates the large influence of humans on regional biogeochemistry, as well as the relatively high level of predictability of human impact.

the feedback to policy was probably related to such human values as health, esthetic value of streams and lakes, and availability of fish (see Smith chapter 2).

Policy changes that involve multinational collaboration are most difficult to affect, because levels of economic and technological development vary so widely among nations, even those that share borders. Perhaps the most important success to date has been the Montreal Protocol, an international agreement to limit the production of substances that deplete ozone. The Montreal Protocol was in direct response to the work of the scientific community (atmospheric chemists), demonstrating a strong probability that anthropogenic sources of halocarbons were depleting the ozone layer. As a result of this agreement, halocarbons in the atmosphere have apparently already peaked (Montzka et al. 1996), and impacts on the ozone layer are apt to have been significantly reduced (Anderson and Miller 1996; Prather et al. 1996). The work of biogeochemists with an ecosystem science background has not yet been globally effective, but the beginnings of an international agreement on carbon emissions were made at the 1992 Earth Summit, and the 1997 Kyoto conference. The presence of the Intergovernmental Panel on Climate Change (IPCC 1990, 1995, 1996) and its summaries for policy-makers provide a structure for the connection between international scientific consensus on the environment and international policy (Elzinga 1997). The challenge to the international scientific community is to provide a clear representation of our understanding, predictions, and uncertainties, to separate our political value judgments from the IPCC assessments, and maintain the integrity of the scientific assessment process (Bolin 1994; Elzinga 1997).

Surprises and Lessons: An Example

Have there been important surprises and lessons in our work to date on regional- to global-scale biogeochemistry? Here we present one example of a surprise that has important implications for the way in which we proceed with regional-scale biogeochemistry.

As we mentioned earlier, many of the regional spatial relationships between control factors and biogeochemical pools and processes have been important in the development of simulation models. In essence, an important part of our strategy has been to substitute spatial relationships for those that are temporal, which are incorporated into our models. Pickett (1989) and Lauenroth and Sala (1992) have suggested that regional-scale spatial relationships may not adequately represent the transient dynamics of ecosystems. Lauenroth and Sala (1992, Figure 7.4) compared the spatial relationship of average aboveground net primary production (ANPP) vs. average annual precipitation in the Great Plains of the United States with a temporal relationship of annual ANPP vs. annual precipitation at one site in the Great Plains. They found that the slope of the relationship is significantly lower for the temporal relationship, suggesting that there are constraints over ANPP that operate over short time scales. Clearly, the application of steady-state, spatial relationships to quantitative predictions of temporal dynamics is limited.

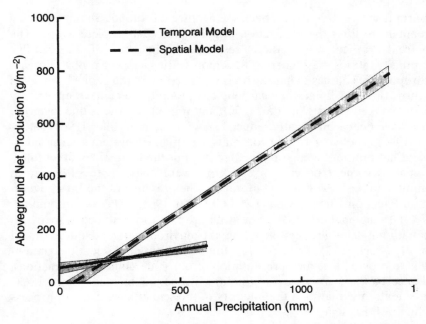

FIGURE 7.4. Two relationships of precipitation vs. aboveground net primary productivity for the central grasslands of the United States (Lauenroth and Sala 1992). The
"spatial" model was generated from a large set of points representing average
precipitation and average production across the central grasslands. The "temporal"
model was generated from a long time series of annual data (from individual years)
at a single location within the region. The large difference in the models suggests
that "steady-state" or average relationships generated from spatial datasets are not
necessarily directly applicable to predicting changes through time.

Moving Our Successes into the Future:
Back to the Dreamers and Thinkers

The field of regional and global biogeochemistry has made enormous
progress in the past several decades. Part of our thesis is that this success
was a direct result of the development of new tools, new collaborations, and
an imperative from the international public to solve important environmental issues that resulted in increased funding levels and an even larger increase in scientific activity. It is clear to all in the field that much research,
better funding, and stronger connections to social science and international
policy are desperately needed to address the state of the global environment today.

It is probably important to reflect on some of these forces, particularly
specific funding opportunities and the imperative from the public, and ask
if these are appropriately balanced by the kind of creative forces that evolve

from our obligation to science itself. Kuhn (1972) suggested that scientists who lead the field see far beyond the activities of the current scientific world and identify new horizons and frontiers, scientists such as Arrhenius, Vernadsky, and Hutchinson. The present-day forces of societal need and available funding may select for a character of activity that is qualitatively different from the type of activity that we might pursue in the absence of societal need.

It is possible to see our advances in the last several decades as the absolutely logical outgrowth of the early scientists who identified these frontiers; the removal of barriers has made it possible for us to follow through on many of their initial questions and ideas. Where within our own work is the dreaming and thinking today? Past experience suggests that we cannot predict all of the applied problems of the future, and a strong diversity of basic research unrelated to current problems is needed as a foundation for solving problems of the future. It will be critical to the future directions and advances in regional- to global-scale biogeochemistry that a segment of the community be engaged in the process of creative induction, identifying new ideas, and reaching beyond the present-day activities.

We have made remarkable progress as a discipline in the area of regional and global biogeochemistry over the past century and a half. In our enthusiasm to follow through on the needs of society, let us not forget our history and the demonstrated importance of identifying new horizons.

Acknowledgments. We gratefully acknowledge the helpful reviews of J. Gosz, P. Matson, B. Howarth, S. Running, S. Tartwoski, H. Epstein, P. Groffman, and M. Pace. Discussions with N. Caraco and J. Cole were particularly helpful during the early development of this chapter. Thanks to S. Collins for help with some of the literature search. The preparation of the manuscript was supported by NSF DEB 9350273 and the Institute of Ecosystem Studies.

References

Aber, J.D., C. Driscoll, C.A. Federer, R. Lathrop, G. Lovett, J.M. Melillo et al. 1993. A strategy for the regional analysis of the effects of physical and chemical climate change on biogeochemical cycles in northeastern (U.S.) forests. *Ecological Modelling* 67:37–47.

Amundson, R., J.W. Harden, and M.J. Singer. 1994. Factors of soil formation: a fiftieth anniversary perspective. *Soil Science Society of America Special Publication no. 33.* Madison, WI.

Anderson, S.O., and A. Miller. 1996. Ozone layer: the road not taken. Correspondence. *Nature* 382:390.

Andreae, M.O., and D.S. Schimel. 1989. *Exchange of trace gases between terrestrial ecosystems and the atmosphere.* John Wiley & Sons. Chichester, U.K.

Arrhenius, S. 1896. On the influence of carbonic acid in the air upon the temperature on the ground. *The Philosophical Magazine* 41:237–276.

Arrhenius, G. 1997. Carbon dioxide warming of the early earth. *Ambio* 26:12–16.

Asner, G.P., T.R. Seastedt, and A.R. Townsend. 1997. The decoupling of terrestrial carbon and nitrogen cycles. *BioScience* 47:226–234.

Barlow, C., and T. Volk. 1992. Gaia and evolutionary biology. *BioScience* 42:686–693.

Bolin, B. 1994. Science and policy making. *Ambio* 23:27.

Breymeyer, A.I., D.O. Hall, J.M. Melillo, and G.I. Agren. eds. 1996. *Global change: effects on coniferous forests and grasslands*. SCOPE 56. John Wiley & Sons. Chichester, U.K.

Broecker, W.S., and T.-H. Peng. 1991. Interhemispheric transport of carbon dioxide by ocean circulation. *Nature* 356:587–9.

Burke, I.C., T.G.F. Kittel, W.K. Lauenroth, P. Snook, C.M. Yonker, and W.J. Parton. 1991. Regional analysis of the central Great Plains, sensitivity to climate variability. *BioScience* 41:685–692.

Burke, I.C., W.K. Lauenroth, W.J. Parton, and C.V. Cole. 1994. Interactions of landuse and ecosystem structure and function: a case study in the central Great Plains. Pages 79–95 in P.M. Groffman and G.E. Likens, eds. *Integrated regional models*. Chapman & Hall, New York.

Burke, I.C., W.K. Lauenroth, and W.J. Parton. 1997. Regional and temporal variation in net primary production and nitrogen mineralization in grasslands. *Ecology* 78:1330–1340.

Burke, I.C., D.S. Schimel, C.M. Yonker, W.J. Parton, L.A. Joyce, and W.K. Lauenroth. 1990. Regional modeling of grassland biogeochemistry using GIS. *Landscape Ecology* 4:45–54.

Caraco, N.F. 1995. Influence of human populations on P transfers to aquatic systems: a regional scale study using large rivers. Pages 235–44 in H. Tiessen, ed. *Phosphorus in the global environment*. SCOPE. John Wiley & Sons, Ltd., New York.

Chase, T.N., R.A. Pielke, T.G.F. Kittel, R. Nemani, and S.W. Running. 1996. Sensitivity of a general circulation model to global changes in leaf area index. *Journal of Geophysical Research* 101:7393–7408.

Claussen, M. 1994. On coupling global biome models with climate models. *Climate Research* 4:203–221.

Cohen, W.B., M.E. Harmon, D.O. Wallin, and M. Fiorella. 1996. Two decades of carbon flux from forests of the Pacific Northwest. *BioScience* 46:836–844.

Cole, J.J., B.L. Peierls, N.F. Caraco, and M.L. Pace. 1993. Nitrogen loading of rivers as a human-driven process. Pages 163–74 in M.J. McDonnell and S.T.A. Pickett, eds. *Humans as components of ecosystems*. Springer-Verlag, New York.

Coleman, M.B., T.L. Bearly, I.C. Burke, and W.K. Lauenroth. 1994. Linking ecological simulation models to geographic information systems: an automated solution. Pages 397–412 in W. Michener and J. Brunt, eds. *Environmental information management and analysis: ecosystem to global scales*. E. Taylor and Francis, London, England.

Costanza, R., F.H. Sklar, and M.L. White. 1990. Modeling coastal landscape dynamics. *BioScience* 40:91–107.

Cramer, W., and A. Fischer. 1996. Data requirements for global terrestrial ecosystem modelling. Pages 529–565 in B. Walker and W. Steffen, eds. *Global change and terrestrial ecosystems*. Cambridge University Press, Cambridge, England.

Crawford, E. 1997. Arrhenius: 1896 model of the greenhouse effect in context. *Ambio* 26:6–11.

Denman, K., E. Hofman, and H. Marchant. 1996. Marine biotic responses to environmental change and feedbacks to climate. Pages 483–516 in J.T. Houghton, L.G. Meirra Filho, B.A. Callander, N. Harris, A. Kattenberg, and K. Maskell, eds. *Climate change 1995. The science of climate change.* Cambridge University Press, Cambridge, England.

Dockuchaiev, V.V. 1883, 1967. Russian chernozem. In *Collected writings, volume 3.* Israel Progress in Science Transactions, Jerusalem.

Dumas, J.B.A., and M.J.B. Boussingault. 1841. Lecon sur la statique chimique des estres organises. *Philosophical Magazine* 19:337–347, 456–469.

Elzinga, A. 1997. From Arrhenius to megascience: interplay between science and public decisionmaking. *Ambio* 26:72–80.

Emanuel, W.R., H.H. Shugart, and M.P. Stevenson. 1985. Climatic change and the broad-scale distribution of terrestrial ecosystem complexes. *Climatic Change* 7:29–43.

Fennessy, M.J., and Y. Xue. 1997. Impact of USGS vegetation map on GCM simulations over the U.S. *Ecological Applications* 7:22–33.

Field, C.B., J.T. Randerson, and C.M. Malmstrom. 1995. Global net primary production: combining ecology and remote sensing. *Remote Sensing of Environment* 51:74–88.

Fung, I.Y., C.J. Tucker, and K.C. Prentice. 1987. Application of advanced very high resolution radiometer vegetation index to study atmosphere-biosphere exchange of CO_2. *Journal of Geophysical Research* 923:2999–3015 .

Goudriaan, J. 1996. Predicting crop yields under climate change. Pages 260–274 in B. Walker and W. Steffen, eds. *Global change and terrestrial ecosystems.* Cambridge University Press, Cambridge, England.

Goulden, M.L., J.W. Munger, S.-M. Fan, B.C. Daube, and S.C. Wofsy. 1996. Exchange of carbon dioxide by a deciduous forest: response to interannual climate variability. *Science* 271:1576–1578.

Goward, S.N., C.J. Tucker, and D.G. Dye. 1985. North American vegetation patterns observed with the NOAA-7 advanced very resolution. *Vegetatio* 64:3–14.

Grisebach, A.R.H. 1872. *Die vegetation der Erde.* Engleman, Leipsig.

Groffman, P.M., and G.E. Likens. 1994. *Integrated regional models.* Chapman & Hall, New York.

Hall, F.G., D.B. Botkin, D.E. Strebel, K.D. Woods, and S.J. Goetz. 1991. Large-scale patterns of forest succession as determined by remote sensing. *Ecology* 72:628–640.

Hall, F.G., and P.J. Sellers. 1995. First International Satellite Land Surface Climatology Project (ISLSCP) Field Experiment (FIFE) in 1995. *Journal of Geophysical Research* 100:25, 383–25, 395.

Harriss, R.C., S.C. Wofsy, M. Garstang, L.C.B. Molion, R.S. McNeal, J.M. Hoell, R.J. Bendura et al. 1988. The Amazon boundary layer experiment. *Journal of Geophysical Research* 93:1351–1360.

Heimann, M. 1997. A review of the contemporary global carbon cycle and as seen a century ago by Arrhenius and Hogbom. *Ambio* 26:17–24.

Heimann, M., and E. Maier-Reimer. 1996. On the relations between the oceanic uptake of carbon dioxide and its carbon isotopes. *Global Biogeochemical Cycles* 10:89–110.

Henderson-Sellers, A., K. McGuthrie, and C. Gross. 1995. Sensitivity of global climate model simulations to increased stomatal resistance. *Journal of Climate* 8:1738–1756.

Heywood, V.H. ed. 1995. *Global biodiversity assessment.* United Nations Environment Programme. Cambridge University Press, Cambridge, England.

Holdridge, L.R. 1947. Determination of world plant formations from simple climatic data. *Science* 105:367–368.

Holland, E.A., A.R. Townsend, and P.M. Vitousek. 1995. Variability in temperature regulation of CO_2 fluxes and N mineralization from five Hawaiian soils: implications for a changing climate. *Global Change Biology* 1:115–123.

Houghton, R.A. 1991. Releases of carbon to the atmosphere from degradation of forests in tropical Asia. *Canadian Journal of Forest Research* 21:132–142.

Houghton, R.A. 1995. Land-use change and the carbon cycle. *Global Change Biology* 1:275–287.

Houghton, R.A. Emissions of carbon from land-use change. In T.M.L. Wigley and D. Schimel, eds. *The carbon cycle.* Cambridge University Press, Stanford, CA: in press.

Howarth, R.W., G. Billen, D. Swaney, A. Townsend, N. Jaworski, K. Lajtha, J.A. Downing et al. 1996. Regional nitrogen budgets and riverine N and P fluxes for the drainages to the North Atlantic ocean: natural and human influences. *Biogeochemistry* 35:75–139.

Humboldt, A. von. 1817. *De distributionae geographica plantarum.* Libraria Graeco-Latino-Germanica, Paris, France.

Hutchinson, G.E. 1944a. A century of atmospheric biogeochemistry. *American Scientist* 32:129–132.

Hutchinson, G.E. 1944b. Nitrogen in the biogeochemistry of the atmosphere. *American Scientist* 32:178–195.

Hutchinson, G.E. 1949. A note on two aspects of the geochemistry of carbon. *American Journal of Science* 247:27–32.

Hutchinson, G.E. 1948. On living in the biosphere. *The Scientific Monthly* LXVII:393–398.

Hutchinson, G.E. 1952. The biogeochemistry of phosphorus. Pages 1–35 in L.F. Wolterink, ed. *The biology of phosphorus.* Michigan State College Press, East Lansing, Michigan.

Innis, G.S. 1976. *Grassland simulation model.* Springer-Verlag, New York.

International Geosphere-Biosphere Programme. 1988. *A study of global change: a plan for action.* Special Committee for IGBP Report No. 4:200.

IPCC. 1990. *Climate change. The IPCC scientific assessment.* Cambridge University Press, Cambridge, England.

IPCC. 1995. *Climate change 1994. Radiative forcing of climate change and an evaluation of the IPCC IS92 Emission Scenarios.* Cambridge University Press, Cambridge, England.

IPCC. 1996. *Climate change 1995. The science of climate change.* Cambridge University Press, Cambridge, England.

Jenny, H. 1941. *Factors of soil formation: a system of quantitative pedology.* McGraw-Hill, New York.

Jenny, H. 1961. Derivation of state factor equations for soil and ecosystems. *Soil Science Society of America Proceedings* 25:385–388.

Jenny, H. 1980. *The soil resource: origin and behavior.* Springer-Verlag, New York.

Justice, C.O., J.R.G. Townshend, B.N. Holben, and C.J. Tucker. 1985. Analysis of the phenology of global vegetation using meteorological satellite data. *International Journal of Remote Sensing* 6:1271–1318.

Kareiva, P., and Anderson, M. 1988. Spatial aspects of species interactions: the wedding of models and experiments. Pages 35–50 in A. Hastings, ed. *Community ecology. Lecture Notes in Biomathematics 77.* Springer-Verlag, Berlin, Germany.

Keeling, R.F., S.C. Piper, and M. Heimann. 1996. Global and hemispheric CO^2 sinks deduced from changes in atmospheric O^2 concentration. *Nature* 381:218–221.

Keller, M., and W.A. Reiners. 1994. Soil-atmosphere exchange of nitrous oxide, nitric oxide, and methane under secondary succession of pasture to forest in the atlantic lowlands of Costa Rica. *Global Biogeochemical Cycles* 8(4):399–409.

Keeling, R.F., and S.R. Shertz. 1992. Seasonal and interannual variations in atmospheric oxygen and implications for the global carbon cycle. *Nature* 358:723–727.

Kirchner, J.W. 1991. The Gaia hypotheses: are they testable? are they useful? Pages 38–46 in S. Schneider and P. Boston, eds. *Scientists on Gaia.* MIT Press, Cambridge, MA.

Koster, R.D., and M.J. Suarez. 1994. The components of a "SVAT" scheme and their effects on a GCM's hydrological cycle. *Advances in Water Research* 17:61–78.

Kuhn, T.S. 1972. *The structure of scientific revolutions,* 2d ed. The University of Chicago Press, Chicago, IL.

Lauenroth, W.K. 1979. Grassland primary production: North American grasslands in perspective. Pages 3–24 in French NR, ed. *Perspectives in Grassland Ecology.* Springer-Verlag, New York.

Lauenroth, W.K., and O.E. Sala. 1992. Long-term forage production of North American shortgrass steppe. *Ecological Applications* 2:397–403.

Leemans, R. 1992. Modelling ecological and agricultural impacts of global change on a global scale. *Journal of Scientific and Industrial Research* 51:709–724.

Leemans, R., and W. Cramer. 1991. *The IIASA database for mean monthly values of temperature, precipitation, and cloudiness on a global terrestrial gird.* Research Report RR-91-18. International Institute of Applied Systems Analyses, Laxenburg, Austria.

Lieth, H. 1978. Primary productivity in ecosystems: comparative analysis of global patterns. Pages 300–321 in H.F.H. Lieth, ed. *Patterns of primary production in the biosphere.* Dowden, Hutchinson and Ross, Inc., Stroudsburg, PA.

Lieth, H., and R.H. Whittaker. 1975. *Primary production of the biosphere. Ecological Studies 14.* Springer-Verlag, New York.

Lovelock, J.E. 1988. *The ages of Gaia.* W.W. Norton Company, New York.

Lovelock, J.E., and L. Margulis. 1974. Atmospheric homeostasis by and for the biosphere: the Gaia hypothesis. *Tellus* 22:2–9.

Major, J. 1951. A functional factorial approach to plant ecology. *Ecology* 32:392–412.

Margulis, L., and G. Hinkle. 1991. The biota and gaia: 150 years of support for environmental sciences. Pages 11–18 in S. Schneider and P. Boston, eds. *Scientists on Gaia.* MIT Press, Cambridge, MA.

Margulis, L., and J.E. Lovelock. 1974. Biological modulation of the earth's atmosphere. *Icarus* 21:471–489.

Matson, P.A., and R.C. Harriss. 1995. *Biogenic trace gases: measuring emissions from soil and water.* Blackwell Science Ltd., Oxford, England.

Matson, P.A., and R.C. Harriss. 1988. Prospects for aircraft-based gas exchange measurements in ecosystem studies. *Ecology* 69:1318–1325.

Matson, P.A., and S.L. Ustin. 1991. Special Feature: the future of remote sensing in ecological studies. *Ecology* 76:1917.

Matson, P.A., P.M. Vitousek, and D.S. Schimel. 1989. Regional extrapolation of trace gas flux based on soils and ecosystems. Pages 97–108 in M.O. Andreae and D.S. Schimel, eds. *Exchange of trace gases between terrestrial ecosystems and the atmosphere.* John Wiley & Sons, Chichester, England.

McGuire, D.A., J.M. Melillo, and L.A. Joyce. 1995. The role of nitrogen in the response of forest net primary production to elevated atmospheric carbon dioxide. *Annual Review of Ecology and Systematics* 26:473–503.

Meentemeyer, V. 1984. The geography of organic decomposition rates. *Annals of the Association of American Geographers* 74:551–560.

Meentemeyer, V., E.O. Box, and R. Thompson. 1982. World patterns and amounts of terrestrial plant litter production. *BioScience* 32:125–128.

Melillo, J.M. 1996. Carbon and nitrogen interactions in the terrestrial biosphere: anthropogenic effects. Pages 431–50 in B. Walker and W. Steffen, eds. *Global change and terrestrial ecosystems.* Cambridge University Press, Cambridge, England.

Melillo, J.M., A.D. McGuire, D.W. Kicklighter, B. Vorosmarty III, C.J. Moore, and A.L. Schloss. 1993. Global climate change and terrestrial net primary production. *Nature* 363:234–240.

Melillo, J.M., I.C. Prentice, G.D. Farquhar, E.-D. Schulze, and O.E. Sala. 1996. Terrestrial biotic responses to environmental change and feedbacks to climate. Pages 449–81 in J.T. Houghton, L.G. Meira Filho, B.A. Callander, N. Harris, A. Kattenberg, and K. Maskell, eds. *Climate change 1994. The Science of Climate Change.* Cambridge University Press, Cambridge, England.

Miller, R.B. 1994. Interactions and collaboration in global change across the social and natural sciences. *Ambio* 23:19–24.

Montzka, S.A., J.H. Butler, R.C. Myers, T.M. Thompson, T.H. Swanson, A.D. Clarke et al. 1996. Decline in the tropospheric abundance of halogen from halocarbons: implications for stratospheric ozone depletion. *Science* 272:1318–1322.

Mosier, A.R., D. Schimel, D. Valentine, K. Bronson, and W. Parton. 1991. Methane and nitrous oxide fluxes in native, fertilized, and cultivated grasslands. *Nature* 350:330–332.

Nemani, R.R., S.W. Running, R.A. Pielke, and T.N. Chase. 1996. Global vegetation cover changes from coarse resolution satellite data. *Journal of Geophysical Research* 101:7157–7162.

Noddack, W. 1937. Der kohlenstoff im haushalt der natur. *Zeit. Ang. Chem.* 50:505–510.

Odum, H.T. 1955. Trophic structure and productivity of Silver Springs, Florida. *Ecological Monographs* 27:55–112.

Parry, M.L., J.E. Hossell, R. Bunce, P.J. Jones, R. Rehman, R.B. Tranter, J.S. Marsh et al. 1996. Global and regional land use responses to climate change. Pages 466–483 in B. Walker and W. Steffen, eds. *Global change and terrestrial ecosystems.* Cambridge University Press, Cambridge, England.

Paruelo, J.M., and R.A. Golluscio. 1994. Range assessment using remote sensing in northwest Patagonia (Argentina). *Journal of Range Management* 47:498–502.

Peierls, B.L., N.F. Caraco, M.L. Pace, and J.J. Cole. 1991. Human influence on river nitrogen. *Nature* 350:386–387.

Perring, F.H. 1958. A theoretical approach to a study of chalk grassland. *Journal of Ecology* 46:665–679.

Peterson, D.L., and R.H. Waring. 1994. Overview of the Oregon transect ecosystem research project. *Ecological Applications* 4(2):211–225.

Pickett, S.T.A. 1989. Space-for-time substitution as an alternative to long-term studies. Pages 110–135 in G.E. Likens, ed. *Long-term studies in ecology*. Springer-Verlag, New York.

Pielke, R.A., G. Dalu, J.S. Snook, T.J. Lee, and T.G.F. Kittel. 1991. Nonlinear influence of mesoscale land use on weather and climate. *Journal of Climate* 4:1053–1069.

Pielke, R.A., T.J. Lee, J.H. Copeland, J.L. Eastman, C.L. Zieglller, and C.A. Finley. 1997. Use of USGS-provided data to improve weather and climate simulations. *Ecological Applications* 7:3–21.

Pollard, D., and S.L. Thompson. 1995. The effect of doubling stomatal resistance in a global climate model. *Global Planet Change* 10:1–4.

Potter, C.S., J.T. Randerson, C.B. Field, P.A. Matson, P.M. Vitousek, H.A. Mooney, and S.A. Klooster. 1993. Terrestrial ecosystem production: a process model based on global satellite and surface data. *Global Biogeochemical Cycles* 7:811–841.

Prather, M., P. Midgley, F. Sherwood Rowland, and R. Stolarski. 1996. The ozone layer: the road not taken. *Nature* 381:551–554.

Prentice, I.C., W. Cramer, S.P. Harrison, R. Leemans, R.A. Monserud, and A.M. Solomon. 1992. A global biome model based on plant physiology and dominance, soil properties and climate. *Journal of Biogeography* 19:117–134.

Prentice, K.C. 1990. Bioclimatic distribution of vegetation for general circulation model studies. *Journal of Geophysical Research* 95:11811–11830.

Prince, S.D., and S.N. Goward. 1995. Global primary production: a remote sensing approach. *Journal of Biogeography* 22:815–835.

Quay, P.D., B. Tilbrook, and C.S. Wong. 1992. Oceanic uptake of fossil fuel CO^2: Carbon-13 evidence. *Science* 256:74–79.

Redfield, A.C. 1958. The biological control of chemical factors in the environment. *American Scientist* Autumn:205–221.

Revelle, R., and H.E. Suess. 1957. Carbon dioxide exchange between the atmosphere and ocean, and the question of an increase in atmospheric CO_2 during the past decades. *Tellus* 9:18–27.

Riebsame, W.E., K.A. Galvin, R. Young, W.J. Parton, I.C. Burke, L. Bohren, and E. Knop. 1994. An integrated model of causes of and responses to environmental change: land use/cover in the Central Great Plains. *BioScience* 44:350–356.

Riley, G.A. 1944. The carbon metabolism and photosynthetic efficiency of the earth as a whole. *American Scientist* 32:132–134.

Rodhe, H., R. Charlson, and E. Crawford. 1997. Svante Arrhenius and the greenhouse effect. *Ambio* 26:2–5.

Rodin, L.E., and N.I. Bazilevich. 1967. *Production and mineral cycling in terrestrial vegetation*. Oliver & Boyd. Edinburgh, London, England.

Rosenzweig, C.M., and L. Parry. 1994. Potential impact of climate change on world food supply. *Nature* 367:133–138.

Rosenzweig, M.L. 1968. Net primary productivity of terrestrial communities: prediction from climatological data. *American Naturalist* 102:67–74.

Roughgarden, J., S.W. Running, and P.A. Matson. 1991. What does remote sensing do for ecology? *Ecology* 72:1918–1922.

Running, S.W. 1986. Global primary production from terrestrial vegetation: estimates integrating satellite remote sensing and computer simulation technology. *Science of the Total Environment* 56:233–242.

Running, S.W. 1994. Testing forest-BGC ecosystem process simulations across a climatic gradient in Oregon. *Ecological Applications* 4:238–247.

Running, S.W., and J.C. Coughlan. 1988. A general model of forest ecosystem process for regional applications. 1. Hydrologic balance, canopy gas exchange and primary production processes. *Ecological Applications* 42:125–54.

Running, S.W., T. Loveland, and L.L. Pierce. 1994. A vegetation classification logic based on remote sensing for use in global biogeochemical models. *Ambio* 23:77–81.

Rutherford, M.C. 1980. Annual plant production precipitation relations in arid and semi arid regions. *South African Journal of Science* 76:53–56.

Sala, O.E., W.J. Parton, L.A. Joyce, and W.K. Lauenroth. 1988. Primary production of the central grassland region of the United States. *Ecology* 69:40–45.

Sarmiento, J.L. 1993. Carbon cycle: atmospheric CO_2 stalled. *Nature* 365:697–698.

Sato, N., P.J. Sellers, D.A. Randall, E.K. Schneider, J. Shukla, J.L. Kinter III et al. 1989. Effects of implementing the simple biosphere model in a general circulation model. *Journal of Atmospheric Sciences* 46:2757–2769.

Schimel, D.S. 1994. Introduction. Pages 3–10 in P.M. Groffman and G.E. Likens, eds. *Integrated regional models. Interactions between humans and their environment.* Chapman & Hall, New York.

Schimel, D.S., and I.C. Burke. 1992. Spatial interactive models of atmosphere-ecosystem coupling. Pages 284–289 in M.F. Goodchild, B.O. Parks, and L.T. Steyaert, eds. *Environmental Modeling with GIS.* Oxford University Press, New York.

Schimel, D.S., B.H. Braswell, E.A. Holland, R. McKeown, D.S. Ojima, T.H. Painter et al. 1994a. Climatic, edaphic, and biotic controls over storage and turnover of carbon in soils. *Global Biogeochemical Cycles* 8:279–293.

Schimel, D.S, I.G. Enting, M. Heimann, T.M.L. Wigley, D. Raynaud, D. Alves, et al. 1994b. CO^2 and the carbon cycle. Pages 35–72 in J.T. Houghton, L.G. Meira Filho, H. Lee, B.A. Callander, E. Haites, N. Harris, et al. eds. *Climate Change 1994. Intergovernmental Panel on Climate Change.* Cambridge University Press, Cambridge, England.

Schindler, D.W., and S.E. Bayley. 1993. The biosphere as an increasing sink for atmospheric carbon: estimates from increased nitrogen deposition. *Global Biogeochemical Cycles* 7:717–733.

Schneider, S., and P. Boston. 1991. *Scientists on Gaia.* MIT Press. Cambridge, MA.

Schwartz, S.E. 1989. Acid deposition: unraveling a regional phenomenon. *Science* 243:753–762.

Segal, M., and R.W. Arritt. 1992. Non-classical mesoscale circulations caused by surface sensible heat flux gradients. *Bulletin of the American Meteorological Society* 73:1593–1604.

Sellers, P. 1995. The Boreal Ecosystem-Atmosphere Study (BOREAS): an overview and early results from the 1994 field year. *Bulletin of the American Meteorological Society* 76:1549.

Sellers, P., F. Hall, H. Margolis, R. Kelly, D. Baldocchi, G. den Hartog, J. Cihlar et al. 1995. The Boreal Ecosystem-Atmosphere Study (BOREAS): an overview and early results from the 1994 field year. *Bulletin of the American Meteorological Society* 76:1549–1577.

Sellers, P.J. 1987. Modeling effects of vegetation on climate. Pages 133–62 in R.E. Dickson, ed. *The geophysiology of Amazonia*. John Wiley & Sons, New York.

Sellers, P.J., F.G. Hall, and G. Asrar. 1992. An overview of the First International Satellite Land Surface Climatology Project (ISLSCP) field Experiment (FIFE). *Journal of Geophysical Research* 97:18345–18371.

Shroeder, H. 1919. Die jährliche gesamptpruduktion der grünen pflanzendecke der Erde. *Naturwiss* 7:8–12.

Shaw, C.E. 1930. Potent factors in soil formation. *Ecology* 11:239–245.

Skole, D., and C. Tucker. 1993. Tropical deforestation and habitat fragmentation in the Amazon: satellite data from 1978 to 1988. *Science* 260:1905–1910.

Smith, T.M., H.H. Shugart, G.B. Bonan, and J.B. Smith. 1993. The transient response of terrestrial carbon storage to a perturbed climate. *Nature* 361:523–526.

Steayart, L.T., T.R. Loveland, and W.J. Parton. 1997. Land cover characterization and land surface parameterization research. *Ecological Applications* 7:1–2.

Tans, P.P., P.S. Bakwin, and D.W. Guenther. 1996. A feasible global carbon cycle observing system: a plan to decipher today's carbon cycle based on observations. *Global Change Biology* 2:309–318.

Tans, P.P., J.A. Berry, and R.F. Keeling. 1993. Oceanic 13C/12C observations: a new window on ocean CO^2 uptake. *Global Biogeochemical Cycles* 7:353–368.

Tans, P.P., I.Y. Fung, and T. Takahashi. 1990. Observational constraints on the global atmospheric CO^2 budget. *Science* 247:1431–1438.

Tilman, D. 1989. Ecological experimentation: strengths and conceptual problems. Pages 136–157 in G.E. Likens, ed. *Long-term studies in ecology*. Springer-Verlag, New York.

Townsend, A.R., P.M. Vitousek, and E.A. Holland. 1992. Tropical soils could dominate the short-term carbon cycle feedbacks to increased global temperatures. *Climatic Change* 22:293–303.

Tucker, C.J., J.R.G. Townshend, and T.E. Goff. 1985. African land cover classification using satellite data. *Science* 227:369–376.

Tucker, C.J., C. Vanpraet, E. Boerwinkel, and A. Gaston. 1983. Satellite remote sensing of dry matter production in the Sentalese Sahel. *Remote Sensing of Environment* 13:461–474.

Van Dyne, G.M. 1977. Content, evolution and educational impacts of a systems ecology course sequence. Pages 9–23 in G.S. Innis, ed. *New directions in the analysis of ecological systems. Part I.* The Society of Computer Simulation. La Jolla, California.

Van Dyne, G.M. 1972. Organization and management of an integrated ecological research program—with special emphasis on systems analysis, universities and scientific cooperation. Pages 111–72 in J.N. Jeffers, ed. *Mathematical models in ecology*. Blackwell Scientific Publishers, Oxford, England.

VEMAP Members. 1995. Vegetation/ecosystem modeling and analysis project: comparing biogeography and biogeochemistry models in a continental-scale study of terrestrial ecosystem responses to climate change and CO_2 doubling. *Global Biogeochemical Cycles* 9:407–437.

Vernadsky, W.I. 1944. Problems of biogeochemistry, II. *Transactions of the Connecticut Academy of Arts and Sciences* 35:483–517.

Vernadsky, W.I. 1945. The biosphere and the noosphere. *American Scientist* 33: 1–12.

Vitousek, P.M., J. Aber, R.W. Howarth, G.E. Likens, P.A. Matson, D.W. Schindler, W.H. Schlesinger, and D.G. Tilman 1997. Human alteration of the global nitrogen cycle: sources and consequences. *Ecological Applications* 7:737–750.

Walker, B., and Steffen, W. 1996. *Global change and terrestrial ecosystems.* Cambridge University Press, Cambridge, England.

Webb, W.L., W.K. Lauenroth, S.R. Szarek, and R.S. Kinerson. 1983. Primary production and abiotic controls in forests, grasslands, and desert ecosystems in the U.S. *Ecology* 64:134–151.

Wessman, C.A. 1992. Spatial scales and global change: bridging the gap from plots to GCM grid cells. *Annual Review of Ecological Systems* 23:175–200.

Wessman, C.A., J.D. Aber, D.L. Peterson, and J.M. Melillo. 1988. Remote sensing of canopy chemistry and nitrogen cycling in temperate forest ecosystems. *Nature* 6186:154–256.

Wessman, C.A., C.A. Bateson, and T.L. Benning. 1997. Detecting fire and grazing patterns in tallgrass prairie using spectral mixture analysis. *Ecological Applications* 7:493–511.

Wofsy, S.C., M.L. Goulden, J.W. Munber et al. 1993. Net exchange of CO_2 in a mid-latitude forest. *Science* 260:1314–1317.

Woodward, F.I. 1996. Developing the potential for describing the terrestrial biosphere's response to a changing climate. Pages 511–28 in B.H. Walker and W.L. Steffen, eds. *Global change and terrestrial ecosystems.* Cambridge University Press, Cambridge, England.

Woodwell, G.M., and E.V. Pecan. 1973. *Carbon and the biosphere.* United States Atomic Energy Commission, Springfield, VA.

8
Acid Deposition Research and Ecosystem Science: Synergistic Successes

KATHLEEN C. WEATHERS AND GARY M. LOVETT

Summary

In this chapter we have suggested that the relationship between acid deposition research and ecosystem science has been synergistic, and that both have benefited. Not only did ecosystem science help to initiate and contribute to acid deposition research, but significant advances have been made in the discipline of ecosystem science over the past few decades as a result of studying acid deposition and its effects.

It was through the study of ecosystems that the regional extent of air pollution and its potential ramifications were first brought to the attention of the scientific community as well as to the public. The national and international attention resulted in a high degree of public interest and substantial funding. The ecosystem research, synergism among disciplines and among scientists, and the research process itself yielded important benefits for the discipline of ecosystem science. These benefits fall into three categories of success: (1) scientific, including advances in our knowledge about the structure and function of ecosystems; (2) infrastructure, which concerns development of new physical, informational, and human resources for the field; and (3) experience on the public stage, which involves the discipline's first tentative steps toward achieving effective science about complex issues in the public spotlight.

Introduction

Humans arguably began polluting the atmosphere with the discovery of fire. In fact, the notion—and the evidence—that humans could negatively affect air quality is quite old:

As soon as I had gotten out of the heavy air of Rome and from the stink of smoky chimneys thereof, which, being stirred, poured forth whatever pestilential vapors and soot they had enclosed in them, I felt an alteration of my disposition . . . (Roman philosopher, Seneca, AD 61, cf. Stern et al. 1984)

195

Widespread regulatory action to curb air pollution, however, was not taken until strong links were made between human health and pollutants in the mid 1900s. Until the 1950s, air pollutants were emitted from relatively short smokestacks and had their most profound effects in areas immediately surrounding the source. In fact, point-source air pollution was a reasonably well-recognized problem (e.g. Smith 1872), and its effects on adjacent eco-systems and local rainfall chemistry (i.e., acid rain) were being examined by the mid-20th century (Schwartz 1989). In urban areas throughout the world, pollution events were both severe and frequent enough during this time period that they often led to human-health problems. One such event, the London smog of 1952, was purported to have caused 4,000 excess deaths (Brimblecombe 1987). This and other local pollution episodes attracted tremendous attention and considerable concern. In an effort to ameliorate the urban pollution situation, a relatively simple control measure was insti-tuted across Europe and North America between the 1950s and 1970s—smokestack height was increased so that pollutants would disperse more widely (e.g. Likens et al. 1979; Brimblecombe 1987). At roughly the same time, scientific interest in quantifying atmospheric inputs to ecosystems led to important insights about air pollution, both in Europe and the United States.

We now know that with increased stack height, much of the local air pollution became regional pollution, and nutrients and pollutants traveled longer distances before being deposited (e.g. Air Pollution 1971; Likens et al. 1979; Schwartz 1989). This information, as well as warnings about poten-tial detrimental ecological effects, was brought to the attention of the scientific community in the 1960s and 1970s through a series of papers written by scientists whose focus was on understanding ecosystem structure and function. They linked ecosystem inputs—pollution deposition—with emissions from distant sources (e.g. Odén 1968 cf. Likens 1989; Air Pollu-tion 1971; Likens et al. 1972; Rodhe 1972; Granat 1972 cf. Ottar 1976; Odén 1976; Likens and Bormann 1974; see also Likens et al. 1979). Further, they documented that in parts of Europe and North America there had been an increase in precipitation acidity over time (Cogbill and Likens 1974; Odén 1976). This groundbreaking work by scientists focused on eco-system function informed the scientific community and the public that air pollutants could travel some distance from their source, be deposited to downwind ecosystems, and possibly have deleterious effects on those ecosystems:

The increasing acidity of the atmosphere is in fact the most interdisciplinary envi-ronmental problem we have at the present. *Soils and surface waters are affected; plant growth is retarded; ecosystems are changed; the biota in lakes and rivers are changed; some organisms die, microorganisms, pathogens and the soil fauna change their activity and living patterns; deterioration of buildings takes place as well as corrosion in a wide sense; and human health is affected.* With very few exceptions, increasing acidity has proved to be detrimental in all these respects (Odén 1976 [italics Odén's]).

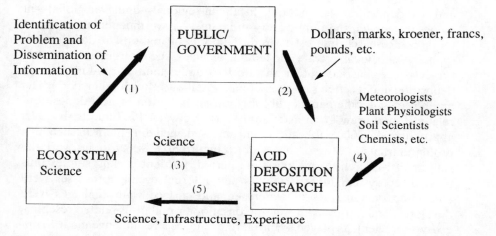

FIGURE 8.1. Synergism between ecosystem science, acid deposition research, and the public. (1) A focus on ecosystem science, in part, led to the discovery of acid deposition as a regional phenomenon, and ecosystem scientists helped to raise scientific and public awareness. (2) Eventually, public interest resulted in an infusion of funds to support acid deposition research. (3) Ecosystem science contributed significantly to acid deposition research, (4) as did science from other disciplines. (5) In turn, advances were made to the discipline of ecosystem science. Successes are identified in three broad categories: scientific (advances in understanding of ecosystem structure or function), infrastructure development (development of new physical, informational, and human resources for the field), and experience on the public stage (the discipline's first tenuous steps toward achieving effective science about complex issues in the public spotlight).

It was also the start of an era of extensive air pollution-related research that is still in progress today (see Likens et al. 1979; Likens 1989; Schwartz 1989; Erisman and Draaijers 1995 for more complete historical accounts).

In this chapter we suggest that the relationship between acid deposition research and ecosystem science has been synergistic, and that both have benefited (Figure 8.1). Not only did ecosystem science help to initiate and contribute to acid deposition research, but significant advances have been made in the discipline of ecosystem science over the past few decades as a result of studying acid deposition and its effects.

Acid Deposition Research: Benefits to Ecosystem Science

As noted before, ecosystem scientists identified the regional nature of acid deposition (e.g. Odén 1968; Air Pollution 1971; Likens et al. 1972; Likens and Bormann 1974). After scientific evidence mounted that air pollution and acid deposition were not only local, but regional—and even global—in scope, the media, the public and—ultimately—governments latched onto the idea. Research interest soared, hundreds of millions of dollars were spent by state, local, private and federal agencies, and thousands of research projects were carried out in an effort to understand the nature of acid deposition and its effects on ecosystems (Pitelka 1994; Schindler 1992). The number of scientific papers published on air pollution or acid deposition-related topics grew at an astounding rate. Between 1967 and early 1997, approximately 6,000 publications had amassed, including many in such high profile journals as *Science and Nature* (Figure 8.2). Many topics were the subject of research during this time period: nutrient cycling studies burgeoned, ecosystem inputs, outputs and internal cycling processes were elucidated, and models of watershed and surface water linkages were developed and improved. From the onset of air pollution-related research, ecosystem scientists played a central role, and, as a result, have made major contributions to the field (Figure 8.1).

Ecosystem science, as a discipline, has reaped many benefits from twenty years of acid deposition research. We suggest that these benefits fall into three categories of success: (1) scientific, which includes advances in our knowledge about the structure and function of ecosystems; (2) infrastructure, which includes development of new physical, informational, and human resources for the field; and (3) experience on the public stage, which involves the discipline's first tentative steps toward achieving effective science about complex issues in the public spotlight. In the next sections, we illustrate each of these categories of success with examples.

Scientific Advances in Ecosystem Science

Focused research on acid deposition led to marked improvements in our understanding of some key ecosystem processes. Our discussion of scientific successes in this section does not represent a thorough review of the literature, nor is it a review of the myriad ecosystem responses to air pollution. Rather, our purpose in this section is to outline very broadly the general knowledge gained about ecosystems through research on acid deposition.

Ecosystem Boundaries

An ecosystem may be simply defined as any unit of nature in which there is a functional and dependent interchange of energy and nutrients between living and nonliving components (Likens and Bormann 1972)

The knowledge that pollutants could be emitted and travel hundreds to thousands of kilometers before being deposited stretched our concept of ecosystem boundaries. That a molecule of nitrogen dioxide could be emitted in Ohio, travel in the atmosphere to New York, be deposited as nitrate in a rain droplet that acidifies streams in the Catskill Mountains, and ultimately contribute to the eutrophication of the Chesapeake Bay has enormous ramifications. Ecosystems, and their processes, are linked through airsheds, and these airsheds can be vast. Actions in one part of the country can have measurable effects on ecosystems thousands of kilometers away, and no amount of fencing or purchasing of land can keep the influence of humans at bay. Though not the topic of this chapter, the boundary

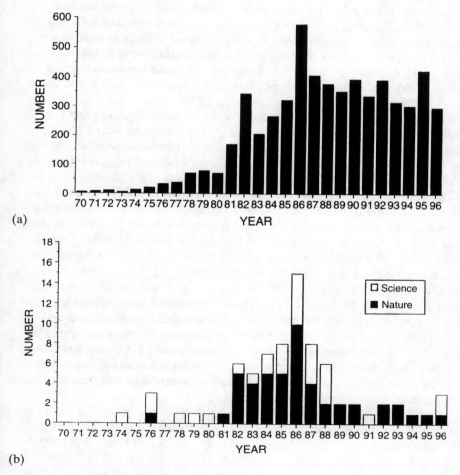

FIGURE 8.2. (a) Number of scientific publications on the topic of acid rain (key words: acid rain, acid precipitation, acid deposition, and atmospheric deposition) from 1970 through 1996. (b) Number of scientific publications in the journals *Science* and *Nature* on the topic of acid rain (key words: acid rain, acid precipitation, acid deposition, and atmospheric deposition) from 1970 through 1996.
Source: *BIOSIS ONLINE.*

issue has also had serious ramifications for the political and economic assessment of air pollution. Acid deposition research demonstrated that pollutants transcend scientific, geographic, and political boundaries. This fact has caused inter-regional and international tensions and considerable finger-pointing.

Inputs to Ecosystems

It had been understood for many decades—and in some cases for centuries—that precipitation (rain and snow), cloud droplet deposition, and dry deposition (direct deposition of dry particles and gases) could all deliver substances from the atmosphere to the surface of earth (e.g. Boussingault 1858). However, it was not until the 1980s that these avenues of deposition were widely recognized among ecologists, and quantitative measures of their relative contributions were made. The data that emerged showed that all three forms of deposition were critical components of the air pollution problem (e.g. Fowler 1980; Lovett et al. 1982; Mayer and Ulrich 1982; Lindberg et al. 1986; Weathers et al. 1986; Lovett 1994) and should be accounted for in decisions regarding such issues as critical loads of pollutants, air pollution abatement strategies, and development of transport models.

Significant ecological advances were also made in deposition research. To quantify pollutant and nutrient loads to ecosystems (i.e. the input terms), it was—and—is critical to understand the contributions of wet, dry, and cloudwater deposition. (Because it became clear in the 1980s that dry and cloudwater deposition were part of the air pollution problem, considerable research has been focused on quantifying these inputs, as well as their resultant interactions with organisms and ecosystems (e.g. Lovett and Kinsman 1990; Lovett 1994; Pitelka 1994; Winner 1994). From an ecosystem perspective, we now know that dry deposition is important in nearly all ecosystems and that cloudwater deposition can be the dominant chemical input mechanism for mountaintop forests (e.g. Lovett 1994). This finding has been crucial for watershed studies, especially for mass-balance calculations, where it is necessary to account for all inputs to an ecosystem.

Many of the advances, from both an air pollution and ecological perspective, were a result of cross-fertilization of ideas and techniques among plant physiologists, micrometeorologists, and ecosystem scientists (Lovett 1994). This multidisciplinary approach to understanding the mechanisms of atmospheric deposition was highly successful and represents important synergism between acid deposition research and ecosystem science (Figure 8.1). One embodiment of this progress can be seen in present-day atmospheric deposition models, which are descended from models of heat and water vapor transport developed by micrometeorologists decades ago (Lovett 1994). These models have been improved by better incorporation of plant physiological processes and adapted by ecosystem scientists to estimate dry deposition of critical nutrients and pollutants to ecosystems. To the extent that new techniques and models allow scientists to answer existing questions and ask new ones regarding air pollution, their development represents a significant advancement as well. Though there is still progress to be made, it is now possible to make much more accurate estimates of total deposition for ecosystems.

Soil Buffering Capacity and Weathering Processes

It has long been known that such natural processes as nitrification and cation uptake by plants, generate acidity in soils and that soils are generally well buffered against these acidifying processes (e.g. Reuss and Johnson 1986). In fact, in the early 1980s many soil scientists doubted that acid deposition could cause significant soil acidification because of the enormous buffering capacity of most soils. Similarly, many soil scientists doubted that acid deposition could reduce soil pools of base cations (e.g. K^+, Ca^{2+}, Mg^{2+}, and Na^+) because of the size of those pools and their continual replenishment by rock weathering. Research, however, has shown both soil acidification (e.g. van Breemen et al. 1982, 1983; Tamm and Hallbäcken 1986; Johnson et al. 1991) and cation depletion (e.g. Federer et al. 1989; Richter et al. 1992; Kirchner and Lydersen 1995; Likens et al. 1996) as a result of acid deposition. During this period of intensive research, sulfate adsorption in the mineral soil was identified as the major factor controlling sulfate retention and leaching in ecosystems receiving atmospheric deposition, and sulfate leaching was shown to be the major agent of base cation loss and aluminum transport in most systems (Overrein 1972; Reuss and Johnson 1986; Harrison and Johnson 1992). In the process of examining the effects of acid deposition on soils, the role of different buffering mechanisms has been elucidated (Ulrich et al. 1980; Ulrich 1983), weathering rates have been estimated for many more ecosystems (April and Newton 1985; Bailey 1996), and soil chemical processes have been more fully incorporated into ecosystem models (e.g. Reuss et al. 1986).

Though it is true that acidic deposition may not have a significant impact on all soils (Johnson et al. 1991), it has been clearly demonstrated that in some poorly buffered soils, in which both acidic deposition is prevalent and rainfall is greater than evapotranspiration, soil solutions are affected by strong mineral acidity and can show increased cation concentrations, as well as changes in the relative proportions of cations in solution (e.g. van Breemen et al. 1983; Last 1991). Further, recent evidence suggests that acidic deposition may have a "lagged" effect and that even after atmospheric deposition has been decreased, soil recovery from increased cation loss may take longer than had been assumed (Likens et al. 1996). These findings have raised concerns about the appropriate time scales over which environmental disturbance should be assessed.

Soil ecosystem research has benefited from the use of a combination of methods and tools including experimental acidification, both in the field and in the laboratory, and use of historical plots and long-term studies. Chemical equilibrium models were also extremely important tools for gaining a clearer understanding of important soil processes (e.g. Reuss and Johnson 1986). Finally, the call to focus on the measurement of both intensity factors (for example soil solution chemistry) and capacity factors (for example exchangeable base cations) has been a practical as well as conceptual advancement in the interpretation of acid deposition responses (Reuss and Johnson 1986).

Freshwater Response to Acidic Inputs

The case of freshwater ecosystem response to acidification may be one of the most compelling and straightforward in air pollution-related ecosystem research. More than twenty years of research has demonstrated that freshwater nutrient cycles are strongly affected by acidic inputs, especially in the northeastern United States, parts of Scandinavia (especially Sweden and Norway), and the United Kingdom (e.g. Wright et al. 1976; Schindler 1988). The discovery that lakes and streams in sensitive regions could be acidified through atmospheric deposition, and the elucidation of how biogeochemical cycles in these bodies of water could be altered by strong mineral acidity, have been important advances in ecosystem science (e.g. Drabløs and Tollan 1980; Rosenqvist et al. 1980; Dillon et al. 1987; Schindler 1988; Baker et al. 1991). That strong mineral acidity can cause (1) shifts in freshwater cation and anion budgets, (2) decreases in alkalinity, and (3) mobilization of trace metals is now well documented (e.g. Schindler 1988; Schindler et al. 1991). Aquatic nutrient cycles can be affected by mineral acidity either directly through deposition of sulfuric or nitric acids to freshwater ecosystems, or indirectly from the mobilization of aluminum and hydrogen ion from adjacent watersheds as a result of acid deposition (e.g. Wright et al. 1976; Drabløs and Tollan 1980; Seip 1980; Reuss and Johnson 1986; Schindler 1988). In regard to the latter, acid deposition research also has demonstrated that in-lake alkalinity generation through "biological buffering" can be more important than alkalinity generated from adjacent watersheds in acid-sensitive regions of North America and Scandinavia (e.g. Kelly et al. 1982; Schindler 1986).

Many significant advances to our understanding of the chemical and the biological effects of acidification were made as a result of experimental manipulations of lakes that began in the 1970s (e.g. Schindler 1988; Schindler et al. 1991). Additional evidence for and the demonstration of these effects has come from modeling efforts, evaluation of historical trends (Schofield 1976), watershed manipulations, laboratory studies, and combinations of all of these methods (e.g. Gherini et al. 1985; Asbury et al. 1989; Sullivan et al. 1990; Baker et al. 1991). Different techniques were used, and when these were compared they often corroborated each other. This represents a significant advance in ecosystem science and for its translation into public policy (e.g. Schindler et al. 1991).

Indirect Effects: Linkages Among Acid Deposition, Terrestrial, and Aquatic Ecosystems

As discussed in a prior section, it had long been understood that soil processes have important influences on surface- and groundwater chemistry (e.g. Reuss and Johnson 1985); it follows that some of the effects of acid deposition in terrestrial ecosystems would be manifested in surface- and groundwaters as well. Conceptual advances that were made in our understanding of the biogeochemical linkages between air pollution, terrestrial, and aquatic ecosystems are illustrated through the transport of monomeric aluminum from terrestrial to aquatic ecosystems (e.g. Cronan and Schofield 1979). That an extremely common, and usually biologically harmless, component of soils (i.e. aluminum) could be mobilized in a form and in quantities that would have deleterious effects on downstream ecosystems as a result of atmospheric inputs, is an important example of an indirect effect. The initial studies showing significant aluminum export in soil water in areas impacted by acidic deposition were part of ecosystem research aimed at understanding general nutrient cycles and budgets in high elevation ecosystems (e.g. Cronan et al. 1978). Much of the research on aluminum chemistry that followed was motivated by air pollution concerns, and it is now evident that in some acid deposition-impacted ecosystems, significant quantities of inorganic aluminum can be leached from soils of low base cation status, and that this aluminum can be transported to freshwater ecosystems in which aquatic biota are negatively affected and alkalinity generation is disrupted (e.g. Reuss and Johnson 1986; Schindler et al. 1991; Last 1991; Havas and Rossland 1995).

A further important development in our understanding of these linkages came with the demonstration that the effects of acidification in some freshwater ecosystems can be reversed when inputs of strong mineral acid are curtailed. Evidence now exists that in some systems, when inputs of sulfur via acid deposition have been reduced—either to the surrounding watershed or directly to surfacewater—alkalinity and pH increase, and concentrations of sulfate, monomeric aluminum, and trace metals decrease. Such factors as hydrologic recharge time are important in controlling the rate of this "reversibility." Thus, some freshwater systems have been shown to experience partial chemical reversal (Schindler 1988; Wright et al. 1988; Wright and Haughs 1991; Schindler et al. 1991). This has been demonstrated to be true whether the reduction in acidification was a result of manipulations (e.g. cessation of experimental acidification) or caused by a decrease in atmospheric inputs (e.g. Wright et al. 1988; Schindler et al. 1991).

Demonstration of ecosystem resilience is certainly a conceptual and practical advance, however, it is with some caution that we include this in our list, because it is not yet clear that *total ecosystem recovery* is ever possible. Whether complete biotic and chemical function can be recovered in most anthropogenically acidified lakes has not yet been determined.

Ecosystem Response to Air Pollution: Models

Many models of ecological systems and their responses to air pollution stresses were developed that addressed topics ranging from plant growth responses to ecosystem-level effects (e.g. Model of Acidification of Groundwater in Catchments (MAGIC), Nutrient Cycling Model (NuCM), Integrated Lake–Watershed Acidification Study (ILWAS); see Taylor et al. 1994). Models have myriad uses, but quite often they function as heuristic tools to understand sensitivities within ecosystems. In acid deposition research, they were often used to illuminate the relative importance of soil and surfacewater processes as well as to predict short- or long-term responses to changing deposition or exposure scenarios (e.g. Galloway et al. 1983; Gherini et al. 1985; Cosby et al. 1985; Liu et al. 1991). The use of and the results from these models have led to advances in our understanding of biogeochemical processes, especially in regard to freshwater and soil acidification, and have provided insights into potential long-term effects of acid deposition.

The longer-term implications of such processes as sulfate adsorption, the mobility of strong acid anions and aluminum in soils and surfacewaters, and rates of weathering in soils were illuminated through the use of acid deposition-related models (e.g. Wright and Henriksen 1983; Cosby et al. 1985; Reuss et al. 1986). Major insights arose from model predictions, for example the fact that acidification is, in large part, controlled both by sulfate adsorption in the soil as well as soil base cation status, and that water flow pathways are important (Gherini et al. 1985). In regard to the time frame over which acidification occurs, a common conclusion of ecologically based acid deposition research is that effects of atmospheric deposition may be delayed. Though it would be difficult to run an experiment long enough or make it big enough to measure these results, it has been possible to develop and test models on experimental systems and then run them under changing deposition conditions over time (Reuss et al. 1986). The model results have been important in elucidating ecosystem response and recovery times based on different deposition scenarios, and soil and catchment characteristics (Galloway et al. 1983; Cosby et al. 1985; Gherini et al. 1985; Reuss et al. 1986). The development of acid deposition models, especially those in which their predictions are compared to and coupled with existing databases, has contributed greatly to our knowledge of how different ecosystems function under present-day and future atmospheric pollution scenarios.

Infrastructure Benefits

An era of intensive acid deposition research caused, albeit not always by design, some very important additions and improvements to the "infrastructure" of ecosystem science. The following five examples are illustrative of the range of benefits ecosystem science reaped as a result of a period of focused research. We identify two types of infrastructure successes here. The first concerns such advances as the addition of new research sites, data sets, and personnel. These are illustrated in the first three examples. The second, though somewhat less tangible, is equally important and concerns interdisciplinary and international communication. Both of these are discussed in the final two examples. We suggest that improvements in these two areas resulted in—and in some cases will continue to cause—significant benefits to our discipline.

Long-Term Monitoring Sites and Programs Worldwide

Many questions were raised during the "acid deposition era" whose answers required that historical data be evaluated or ecosystem attributes be remeasured. Unfortunately, relatively few of the necessary data were available. There was a concerted effort at this time to institute long-term monitoring programs especially in Europe and North America, in a variety of such ecosystems as lakes, streams, and forests, and to quantify atmospheric inputs to these ecosystems (see Ford et al. 1993). Ecosystem scientists were often involved in selecting sites or setting up these networks, and the data from them were used to address ecosystem questions (e.g. Mitchell et al. 1996). The data from these programs will be crucially important in the studies of such slow ecological processes as forest growth and maturation and soil cation dynamics. For example, the network of permanent forest plots established during the 1980s is apt to be especially valuable to forest ecosystem research in the future.

Other examples of long-term monitoring programs are the precipitation, dry and cloudwater deposition, and air quality networks that now exist in locations around the world. Although many of them are currently supported and maintained by a variety of state and federal agencies, ecosystem scientists were involved in the early establishment and support of these networks because they recognized the need for quantifying atmospheric inputs to ecosystems over wide geographic regions and the importance of collecting continuous, long-term data. Some of these early networks include the Multi-state Atmospheric Power Production Study (MAP3S), National Atmospheric Deposition Program (NADP), and Global Precipitation Chemistry Program (GPCP). From an ecological perspective, the data from these networks have been critical for establishing spatial trends and gradients of deposition, for identifying key geographic areas where air pollution has its greatest impacts, and for estimating nutrient and pollutant inputs to ecosystems. To the extent that the networks are maintained over long time periods, they will provide temporal trend information as well. In many countries, data from monitoring networks will be used to measure the effectiveness of clean air legislation (e.g. Lynch et al. 1995).

The Nutrient Cycling Database

Much of the focus of acid deposition research was on response of nutrient cycles to increases in strong mineral acidity, and nutrient-cycling studies were an integral part of acid deposition research in most ecosystems. Not since the International Biological Program (IBP) had there been such an increase in nutrient-cycling studies in Europe and the United States. In the case of coordinated efforts, for example the Integrated Forest Study (e.g. Johnson and Lindberg 1992), during which the data were collected and analyzed in a consistent way, data can be compared and contrasted across sites. We suggest that these data are important resources for the future in much the same way as the IBP data sets are used extensively by ecologists today.

Graduate Students

One traditional, academic measure of success is the number of students trained in a field. The influx of funds for atmospheric deposition research resulted in a pulse of graduate training in American and European universities. In the United States, from the late 1970s through early 1997, at least 108 doctoral students published theses on acid rain, acidic or atmospheric deposition-related topics. From 1970 to 1979, only two dissertations were published, compared to seventy-one during the peak years (1980 to 1989). From 1990 to the present, thirty-five more have been recorded (Figure 8.3). Many of these students were probably involved in interdisciplinary research projects and therefore were introduced to collaborative, interactive research. Training through participation in cooperative, multidisciplinary research projects will serve the discipline well because it attends to the increasingly complex global environmental problems.

Communication Between Scientists and Government Agencies

The very nature of the research topic of acid deposition resulted in broader interactions and considerably more dialogue among scientists and other nonscientific groups. As an example, in Europe, teams composed of scientists and government officials worked diligently to define critical loads for ecosystems (e.g. Schulze et al. 1989; Last 1991). In the United States, ecosystem scientists were invited to serve on scientific advisory panels to the president, state governors, and to testify at congressional hearings about acid deposition. We suggest that setting the precedent and working out

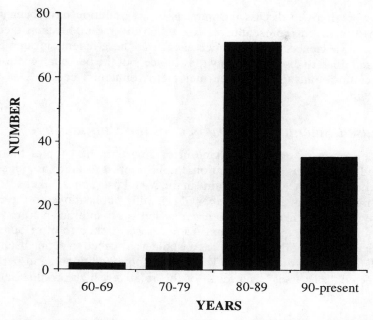

FIGURE 8.3. Number of Ph.D. dissertations on the topic of acid rain (key words: acid rain, acid precipitation, acid deposition, and atmospheric deposition) from 1970 to early 1997. Source: University of Michigan Dissertation Abstract Services.

effective strategies for establishing and maintaining direct communication were both successes.

Collaborations Across Disciplines, Interest Groups, and Countries

Multidisciplinary work is becoming the norm for ecosystem science. The era of acid deposition research did much to further interactions among scientists from different subdisciplines. For example, to understand ecosystem function in response to acid deposition, physiologists, geologists, meteorologists, chemists, and ecologists all contributed to and collaborated in research (e.g. Air Pollution 1971). Further, there was an upsurge in communication among ecosystem scientists worldwide (Brydges and Wilson 1991). Many North Americans were involved in the critical loads debate in Europe (Nilsson and Grennfelt 1988). Similarly, in the United States, Europeans participated in the National Acid Precipitation Assessment Program (NAPAP) research, evaluation, and discussion. In an era of global ecology and economy, establishing these relationships represents important advances for science and for society as well. Recently, the net has been cast wider: workshops to discuss the impacts of air pollution have been held that have included representatives from such diverse interests as electric utilities, law, science, and insurance agencies. These diverse groups have been assembled to discuss not only the science, but the personal, economic, and societal implications of such major environmental concerns as acid deposition.

Increased Scientific Experience on the Public Stage

Acid deposition is one of the handful of environmental topics that has received tremendous public attention. Public interest peaked (as judged by number of occurrences of "acid rain" in the *New York Times*, Likens 1997a) in the mid-1980s, as did the number of scientific publications on the topic (Figure 8.4). As such, it was necessary for scientists to interact in arenas that were otherwise foreign. Individual scientists were asked to offer opinions on policy issues, and the discipline as a whole was pushed to define itself and articulate and defend its worth. There were some additional, and perhaps unexpected, "successes" as a result of this relatively high profile scientific

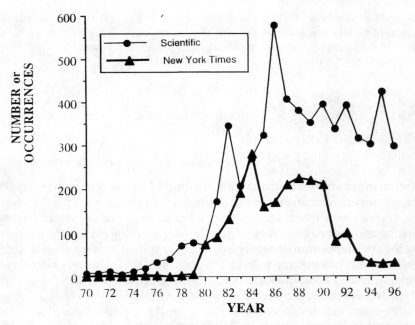

FIGURE 8.4. Number of scientific publications on the topic of acid rain (key words: acid rain, acid precipitation, acid deposition, and atmospheric deposition) from 1970 through 1996 and number of occurrences of the words acid rain and acid precipitation in articles appearing in The *New York Times* from 1970 to 1996. Sources: BIOSIS ONLINE; Likens 1997a.

pursuit. (In this case, we define success as experience gained.) Although this new-found experience has been very important, some of the lessons learned along the way may have been painful or unwelcome. The ecological research community, for example, learned how to do research in the face of hostile interests. It learned how to address—and in some cases how to quantify—uncertainty to a public with little understanding of it. The discipline also gained important experience in how to organize and present results from multinational, multidisciplinary research efforts. In addition, over the past few decades of air pollution-related research, the scientific community has been exposed to how the political process demands "timely" scientific results and either uses or ignores science. These experiences, no matter how difficult, painful, or tangential, have been important lessons for scientists and their interactions with society.

Scientists involved in acid deposition research were variously compelled, encouraged, or forced to discuss their research, as well as the general topic of acid deposition, in the public forum. At times, the impetus came from scientists, for example, through writing articles that were published in more "accessible" journals, such as *Scientific American, Natural History* and *Ambio*; more often it came from the news media, through interviews and articles, and periodically it was in angry response to "propaganda" from industry (e.g. Schindler 1992; Likens 1992). Regardless of the motivation, several "new" avenues of communication brought ecosystem science to the public, which we see as an important success.

Acid deposition research also represents one of ecology's first forays into "big science." With hundreds of millions of dollars being spent on research in the United States alone (although most of that was not spent on ecosystem science research), and billions of dollars in extra costs riding on the outcome, the scientific results were closely scrutinized by interest groups and the media. Ecosystem scientists were an important part of the overall research effort; many learned how to compete effectively for funding, how to present to policy-makers the issues that they considered important, and how to organize and communicate in a fast-moving and complex research environment. We suggest that these lessons are already being put to use as we grapple with other such complex environmental problems as climate change and loss of biodiversity.

The acid deposition debate also begged the question: Is scientific consensus necessary? Coming to scientific consensus is not, and has never been, an explicit goal of the scientific community. In fact, there are virtually no incentives for coming to consensus within the discipline! However, throughout the acid deposition debate, it became clear that presenting scientific consensus to the public was both useful and necessary (e.g. Driscoll et al. 1985). This lesson may have already been carried into the present (witness the Intergovernmental Panel on Climate Change reports).

Although learning by experience (a forced maturation process, in truth) is difficult to quantify, we contend that the training of this era will prove to have been good preparation for a new decade of environmental issues.

Contributions of Ecosystem Ecology to Policy Issues

A significant portion of the acid deposition-related research during 1980s and beyond was policy-driven, and one of the goals of the funding agencies was to obtain information on sensitivity of ecosystems to acidic deposition (Pitelka 1994; NAPAP 1993). It was hoped that by the 1990s the political and legislative processes regarding acid deposition would be informed by scientific results (e.g. NAPAP 1993). Although the connection between science and policy may not have been a resounding success (Schindler 1992; Likens 1992), in the broader perspective, ecosystem scientists identified the various scientific problems of acid deposition and brought them to the attention of the public. It was the public's concern about acid deposition that eventually led to changes in emission regulations, and this, in our view, constitutes a major policy success for ecosystem science. In addition to this general success, there have been several instances in which the work of ecosystem scientists directly influenced the debate about specific policy issues on air pollution topics. Particularly relevant to policy-makers were research results on the topics discussed in the following three sections.

The Link Between Emission Reduction and Deposition

An important aspect of documenting the effectiveness of emission reductions, such as those required by the Clean Air Act (CAA) and its 1990 amendments in the United States, and similar clean air legislation in Europe, has been the demonstration of commensurate changes in the delivery of those pollutants to downwind ecosystems. Sulfur dioxide has been a target pollutant for reductions worldwide. Because there are many steps from the emission of sulfur dioxide to the deposition of sulfuric acid downwind, questions were posed in the early 1980s about the effectiveness of sulfur dioxide emission reductions in decreasing sulfur deposition. In the United States and Europe the answer to this question came from comparing long-term databases of precipitation chemistry with sulfur dioxide emission trends (e.g. National Research Council 1983; Hedin et al. 1987; Butler and Likens 1991; Likens 1992; Erisman and Draaijers 1995). These data clearly illustrated a direct relationship between atmospheric deposition and the results of air pollution legislation, in this case a reduction in sulfur dioxide (National Research Council 1983). A similar relationship has been demonstrated for a decrease in lead deposition after leaded gasoline was banned in the United States (e.g. Likens 1992).

The Role of Nitrogen and Base Cations in Air Pollution

Initially, air pollution-related research focused primarily on the effects of sulfur. Although oxides of nitrogen were of concern in relation to their role in tropospheric ozone production, the acid deposition problem was assumed to be almost entirely the result of sulfuric acid in much of the United States and Europe, and thus abatement measures were designed primarily to reduce sulfur dioxide emissions (e.g. NAPAP 1993).

Since the mid-1980s, however, evidence has been mounting in North America and Europe that suggests other areas of atmospheric deposition are of concern: increasing emissions and subsequent deposition of nitrogen is an increasing part of the air pollution problem (e.g. van Breemen and van Dijk 1988; Vitousek et al. 1997) even though a decline in base cations has been shown to offset sulfate declines in precipitation (e.g. Hedin et al. 1994). Excess nitrogen is now considered one of the most serious pollution-related environmental problems (e.g. Galloway 1995; Vitousek et al. 1997), and reductions in the deposition of base cations may have serious ramifications for the buffering capacity of some sensitive ecosystems (Likens et al. 1996). The research efforts supporting these assertions have been manifold, including: (1) evidence that severe nitrogen pollution is leading to acidification of soils and increased concentrations of nitrate in ground- and surfacewaters (e.g. van Breemen et al. 1983); (2) long-term monitoring data that show that although sulfate deposition has decreased since clean air legislation has been instituted, in many regions there have *not* been commensurate decreases in surfacewater acidity, and that nitrate has partially taken the place of sulfate as a mobile anion (e.g. Driscoll et al. 1989); and (3) long-term monitoring data that also show base cation input in precipitation has declined in Europe and North America (Hedin et al. 1994).

The first two examples listed, combined with other evidence, have been used in support of legislation to control nitrogen oxides as well as sulfur oxides. As a result, more recent clean air legislation has specific recommendations to control nitrogen oxide emissions.

Critical Loads of Deposition on Ecosystems

Ecosystem scientists contributed substantially to the process of setting critical deposition loads in Europe (Nilsson and Grennfelt 1988), and these critical loads define air pollutant emission policies throughout western Europe. Similarly, in the United States, ecosystem scientists were major contributors to the definition of critical pollutant loads in wilderness areas, an effort that has served to define United States Forest Service policy for evaluating proposals for new emission sources near wilderness areas (Fox et al. 1989).

Our point with these three examples is that ecosystem scientists, through sound science and diligence, can and do affect policy.

What Led to These Successes?

An examination of the myriad successes surrounding the era of acid deposition research shows that there were some common contributing factors. As is often the case, a combination of basic and applied research (i.e. tackling interesting scientific problems), communication and collaboration among researchers inside and outside of the discipline (i.e. discussing ideas, hypotheses, and data with other researchers), adequate funding, and synergism among disciplines and individuals all have contributed to scientific advances. Further, the use of a wide variety of techniques and tools (i.e. experimentation, long-term monitoring, modeling, gradient studies) over a range of different scales, have been instrumental. In the case of acid deposition research, communicating with the public and the (eventual) response from the public, which resulted in research monies, have been important (Figure 8.1). Finally, there was an element of serendipity—as in many cases of scientific success—that played a central role as well.

An example of the interaction among many of these factors is found in the events leading to the identification of acid rain as a regional phenomenon. In Scandinavia, concern over the increasing acidity of precipitation and its effects on aquatic and terrestrial ecosystems was articulated in the mid 1960s (Odén 1976). Experts, including soil scientists, meteorologists, economists, and others joined forces to understand the effects of acid deposition on ecosystems (e.g. Air Pollution 1971). In the United States, the scientific questions being pursued were not air pollution oriented, at least not initially. It was scientific inquiry aimed at understanding ecosystem structure and function through quantifying and evaluating nutrient cycles and input-output budgets—essentially developing the mass-balance approach—that led to measuring atmospheric inputs (Likens et al. 1972). The necessary ingredients for putting the scientific pieces together seem to have been the availability of continuous, high quality data, a willingness to see unanticipated results when examining the data (the "aha" factor), and the discussion of those data with the broader scientific community. Communication between scientists from different regions of the world was also a critical factor: a conversation in which concerns and ideas about regional acid rain in Europe was influential in the discovery of acid rain in North America (Likens 1989). In regard to communication with the public, one of many important events was that Sweden brought their case, Air Pollution Across National Boundaries: The Impact on the Environment of Sulfur in Air and Precipitation, before the United Nations Conference on the Human Environment in 1972 (Air Pollution 1971). This report was crucial for raising public awareness about acid deposition and its effects. Essentially, Sweden's case study put the problems of air pollution and acid deposition on the world map. Personal crusades to make scientific information available, accessible, and ensure that is a part of the scientific and public agendas have had demonstrable impacts.

Conclusions

We suggest that it was through the study of ecosystems that the regional extent of air pollution and its potential ramifications were first brought to the attention of the scientific community as well as to the public. The national and international attention resulted in a high degree of public interest and substantial funding. The ecosystem research, synergism among disciplines and among scientists, and the research process itself yielded important benefits for the discipline of ecosystem science: through advances in science, infrastructure development, and hard-won lessons about science and society. This process—and the successes themselves—were not without mistakes, pitfalls, shortcomings, misunderstandings, and a measure of anxiety! These facts notwithstanding, we argue that the successes of more than twenty years of acid deposition research have been manifold.

But what, ultimately, are the important lessons from more than two decades of acid deposition research? Are these lessons that science and society might use in the future?

Lessons for Science

From a strictly scientific perspective, the study of air pollutants (acid deposition included) and ecosystems has provided ample evidence that complexity is both the norm and a formidable challenge with which to contend. The issue of forest response to air pollution is, perhaps, a case-in-point challenge for the future. Despite extensive effort, the connection between air pollution and forest decline is still tenuous at best. Although many air pollutants have been shown to affect tree performance, and exposure to combinations of air pollution with other environmental stresses are apt to have even greater effects, establishing cause and effect has remained elusive. Solving such complex environmental problems as these is probably the wave of the future (see Likens, 1997b). Given that ecological complexity is here to stay, scientists face an additional challenge of how to communicate scientific expertise, wisdom, and limitations—with credibility—to the public; the experience with acid deposition research suggests that our ability to meet this challenge will be extremely important in the future.

Lessons for Society

We think that the era of acid deposition research resulted in several lessons for society as well. First, the solution to pollution is not dilution! It should now be abundantly clear that sending air pollutants higher into the atmosphere did not solve the acid deposition problem, but simply changed its scope. As common sense would dictate, reducing emissions seems the simplest and most efficient means of solving the problem.

Further, we know that complexity is the norm in science, but it is also true with the media's presentation of science. A tremendous number of interest groups and disciplines weighed in on the acid deposition debate in the public forum, and many of them had vested interests in the outcome of research, and especially of legislation (Likens 1992). A general public that has some understanding of the strengths and limitations of science, as well as the ability to see and evaluate motivations behind particular representations of science or scientific results, would facilitate more informed decision-making.

Finally, we suggest that wise decisions can be made in the face of uncertainty. Public officials often waffled on proposing air pollution control measures partly because of a perceived lack of certainty that acid deposition could cause widespread ecological damage. The fact is that there are shockingly few things we know with 100% certainty: death and taxes are the usual candidates. However, humans make prudent decisions in the face of a high degree of uncertainty every day: we buy life insurance, opt for elective surgery, and invest in retirement plans based on incomplete evidence that these measures will ensure a healthy and happy future. The lesson for society is to put science in a practical framework. Complete certainty is unreasonable; making wise decisions based on sound scientific evidence is not only possible but probably a very sound investment in the future.

Acknowledgments. We thank J.J. Cole, A. Kinzig, L.O. Hedin, and N. van Breeman for helpful comments on an earlier draft of this manuscript. We are grateful to M. Spoerri and A. Frank for their assistance.

References

Air Pollution Across National Boundaries. 1971. *The impact on the environment of sulfur in air and precipitation. Sweden's case study for the United Nations conference on the human environment.* Kungl. Boktryc Keriek. P.A. Norstedt & Söner, Stockholm.

April, R., and R. Newton. 1985. Influence of geology on lake acidification in the ILWAS watersheds. *Water, Air and Soil Pollution* 26:373–386.

Asbury, C.E., F.A. Vertucci, M.D. Mattson, and G.E. Likens. 1989. Acidification of Adirondack lakes. *Environmental Science and Technology* 23:362–365.

Bailey, S.W., J.W. Hornbeck, C.T. Driscoll, and H.E. Gaudette. 1996. Calcium inputs and transport in a base-poor forest ecosystem as interpreted by Sr isotopes. *Water Resources Research* 32:707–719.

Baker, L.A., A.T. Herlihy, P.R. Kaufmann, and J.M. Eilers. 1991. Acidic lakes and streams in the United States—the role of acidic deposition. *Science* 252:1151–1154.

Boussingault, J.B. 1858. Recherches sur la quantité de l'acide nitrique contenue dans la pluie, le brouillard, la rosée. *Compretative Rend* 46:1123–1130, 1175–1183.

Brimblecombe, P. 1987. *The big smoke: a history of air pollution in London since medieval times*. Metheun, London and New York.

Brydges, T.G., and R.B. Wilson. 1991. Acid rain since 1985—times are changing. In F.T. Last and R. Watling, eds. *Acidic deposition: its nature and impacts*. Proceedings of the Royal Society of Edinburgh 97B:1–16.

Butler, T.J., and G.E. Likens. 1991. The impact of changing regional emissions on precipitation chemistry in the eastern United States. *Atmospheric Environment* 25A:305–315.

Cogbill, C.V., and G.E. Likens. 1974. Acid precipitation in the northeastern United States. *Water Resources Research* 10:1133–1139.

Cosby, B.J., G.M. Hornberger, J.N. Galloway, and R.F. Wright. 1985. Time scales of catchment acidification. *Environmental Science and Technology* 129:1144–1149.

Cronan, C.S., W.A. Reiners, R.C. Reynolds, Jr., and G.E. Lang. 1978. Forest floor leaching: contributions from mineral, organic, and carbonic acids in New Hampshire subalpine forests. *Science* 200:309–311.

Cronan, C.S., and C.L. Schofield. 1979. Aluminum leaching response to acid precipitation: effects on high-elevation watersheds in the northeast. *Science* 204:304–305.

Dillon, P.J., R.A. Reid, and E. DeGrosbois. 1987. The rate of acidification of aquatic ecosystems in Ontario, Canada. *Nature* 329:45–48.

Drabløs, D., and A. Tollan, eds. 1980. *Ecological impact of acid precipitation* [Proc. Internaf. Conf. Sandefjord, Norway 383 p.] SNSF Project, Oslo-Aŝ, Norway.

Driscoll, C.T., J.N. Galloway, J.F. Hornig, G.E. Likens, M. Oppenheimer, K.A. Rahn, and D.W. Schindler. 1985. *Is there scientific consensus on acid rain? Excerpts from six governmental reports. Ad hoc committee on acid rain: science and policy*. Institute of Ecosystem Studies, Millbrook, New York.

Driscoll, C.T., G.E. Likens, L.O. Hedin, J.S. Eaton, and F.H. Bormann. 1989. Changes in the chemistry of surface waters: 25-year results at the Hubbard Brook Experimental Forest, New Hampshire. *Environmental Science and Technology* 23:137–143.

Erisman, J.W., and G.P.J. Draaijers. 1995. *Atmospheric deposition in relation to acidification and eutrophication*. Elsevier, Amsterdam, The Netherland.

Federer, C.A., J.W. Harbeck, L.M. Tritton, C.W. Martin, and R.S. Pierce. 1989. Long-term depletion of calcium and other nutrients in eastern U.S. forests. *Environmental Management* 13:593–601.

Ford, J., J.L. Stoddard, and C.F. Powers. 1993. Perspectives on environmental monitoring. An introduction to the U.S. EPA long-term monitoring project. *Water, Air and Soil Pollution* 67:247–255.

Fowler, D. 1980. *Removal of sulphur and nitrogen compounds from the atmosphere in rain and dry deposition*. SNSF Project, Oslo-Aŝ, Norway. Ecological Impact & Acid Precipitation Proce Infernet Conf. Sande Effects Norway. Project, Oslo, Norway.

Fox, D.G., A.M. Bartuska, J.G. Byrne, E. Cowling, R. Fisher, G.E. Likens, S.E. Lindberg et al. 1989. *A screening procedure to evaluate air pollution effects on*

Class I wilderness areas. USDA Forest Service General Technical Report RM-168. US Dept & Agriculfore Rocky Mantain forst & Range Experiment station. Fort Collins, CO.

Galloway, J.N. 1995. Acid deposition: perspectives in time and space. *Water, Air and Soil Pollution* 85:15–24.

Galloway, J.N., S.A. Norton, and M.R. Church. 1983. Freshwater acidification for atmospheric deposition of sulfuric acid: a conceptual model. *Environmental Science and Technology* 17:541A–545A.

Gherini, S.A., L. Mok, R.J.M. Hudson, F.F. Pauls, C.W. Chen, and A. Goldstein. 1985. The ILWAS model: formulation and application. *Water, Air and Soil Pollution* 26:425–457.

Harrison, R.B., and D.W. Johnson. 1992. Inorganic sulfate dynamics. Pages 104–118 in D.W. Johnson and S.E. Lindberg, eds. *Atmospheric deposition and forest nutrient cycling: a synthesis of the Integrated Forest Study.* Springer-Verlag, New York.

Havas, M., and B.O. Rosseland. 1995. Response of zooplankton, benthos, and fish to acidification: an overview. *Water, Air and Soil Pollution* 85:51–62.

Hedin, L.O., L. Granat, G.E. Likens, T.A. Buishand, J.N. Galloway, T.J. Butler, and H. Rodhe. 1994. Steep declines in atmospheric base cations in regions of Europe and North America. *Nature* 367:351–354.

Hedin, L.O., G.E. Likens, and F.H. Bormann. 1987. Decrease in precipitation acidity resulting from decreased SO_4^{2-} concentration. *Nature* 325:244–246.

Johnson, D.W., M.S. Cresser, S.I. Nilsson, J. Turner, B. Ulrich, D. Binkley, and D.W. Cole. 1991. Soil changes in forest ecosystems: evidence for and probable causes. *Proceedings of the Royal Society of Edinburgh* 97B:81–116.

Johnson, D.W., and S.E. Lindberg, eds. 1992. *Atmospheric deposition and forest nutrient cycling: a synthesis of the Integrated Forest Study.* Ecological studies #91. Springer-Verlag, New York.

Kelly, C.-A., J.W.M. Rudd, R.B. Cook, and D.W. Schindler. 1982. The potential importance of bacterial processes in regulating rate of lake acidification. *Limnology and Oceanography* 27:868–882.

Kirchner, J.W. and E. Lydersen. 1995. Base cation depletion and potential long-term acidification of Norwegian catchments. *Environmental Science and Technology* 29:1953–1960.

Last, F.T. 1991. Critique. Pages 273–324 In F.T. Last, and R. Watling, eds. Acidic deposition: its nature and impacts. *Proceedings of the Royal Society of Edinburgh.*

Likens, G.E. 1989. Some aspects of air pollution effects on terrestrial ecosystems and prospectus for the future. *Ambio* 18:172–178.

Likens, G.E. 1992. *The ecosystem approach: its use and abuse.* Ecology Institute, Oldendorf/Luhe, Germany. 166p.

Likens, G.E. 1997a. (in press) Eugene Odum, the ecosystem approach, and the future. In G.W. Barrett, T.L. Barrett, and M.H. Smith, eds. *The Institute of Ecology: past, present and future.* University of Georgia Press.

Likens, G.E. 1997b. Limitations to intellectual progress in ecosystem science. In M.L. Pace, and P.M. Groffman, eds. *Successes, limitations and frontiers in ecosystem science.* Springer-Verlag, New York.

Likens, G.E., and F.H. Bormann. 1972. Nutrient cycling in ecosystems. Pages 25–67 in J. Wiens, ed. *Ecosystem structure and function.* Oregon State University Press, Corvallis.

Likens, G.E., and F.H. Bormann. 1974. Acid rain: a serious regional environmental problem. *Science* 184:1176–1179.

Likens, G.E., F.H. Bormann, and N.M. Johnson. 1972. Acid rain. *Environment* 14:33–40.

Likens, G.E., C.T. Driscoll, and D.C. Buso. 1996. Long-term effects of acid rain: response and recovery of a forest ecosystem. *Science* 272:244–246.

Likens, G.E., R.F. Wright, J.N. Galloway, and T.J. Butler. 1979. Acid rain. *Scientific American* 241:43–51.

Lindberg, S.E., G.M. Lovett, D.D. Richter, and D.W. Johnson. 1986. Atmospheric deposition and canopy interactions of major ions in a forest. *Science* 231:141–145.

Liu, S., R. Munson, D. Johnson, S. Gherini, K. Summers, R. Hudson, K. Wilkinson et al. 1991. Application of a nutrient cycling model (NuCM) to a northern mixed hardwood and a southern coniferous forest. *Tree Physiology* 9:173–184.

Lovett, G.M. 1994. Atmospheric deposition of nutrients and pollutants in North America: an ecological perspective. *Ecological Applications* 4:629–650.

Lovett, G.M., and J.D. Kinsman. 1990. Atmospheric pollutant deposition to high elevation ecosystems. *Atmospheric Environment* 244:2767–2786.

Lovett, G.M., W.A. Reiners, and R.K. Olson. 1982. Cloud droplet deposition in sub-alpine balsam fir forests: hydrological and chemical inputs. *Science* 218:1303–1304.

Lynch, J.A., V.C. Bowersox, and C. Simmons. 1995. Precipitation chemistry trends in the United States: 1980–1993, *National atmospheric deposition program summary report*. NADP/NTN, Fort Collins, CO.

Mayer, R. and B.L. Ulrich. 1982. Input of atmospheric sulfur by dry and wet deposition to two central European forest ecosystems. *Atmospheric Environment* 12:375–377.

Mitchell, M.J., C.T. Driscoll, J.S. Kahl, G.E. Likens, P.S. Murdoch, and L.H. Pardo. 1996. Climatic controls of nitrate loss from forested watersheds in the northeast United States. *Environmental Science and Technology* 30:2609–2612.

NAPAP. 1993. *1992 Report to Congress*. NAPAP, Washington, D.C.

National Research Council. 1983. *Acid deposition: atmospheric processes in eastern North America*. National Academy Press, Washington, DC.

Nilsson, J., and P. Grennfelt. 1988. *Critical loads for sulphur and nitrogen*. Report 1988:15. Nordic Council of Ministers.

Odén, S. 1968. The acidification of air and precipitation and its consequences on the natural environment. *Swedish National Science Research Council, Ecology Committee, Bulletin 1*.

Odén, S. 1976. The acidity problem—an outline of concepts. *Water, Air and Soil Pollution* 6:137–166.

Ottar, B. 1976. Organization of large range transport of air pullution monitoring in Europe. In *proceedings of the first international symposium on acid precipitation and the forest ecosystem*. Columbus, Ohro May 12–15, 1975, USDA Forest Service general technical report NE-23.

Overrein, L.N. 1972. Sulfur pollution patterns observed: leaching of calcium in a forest soil determined. *Ambio* 1:145–147.

Pitelka, L.F. 1994. Introduction to special section on Air Pollution. *Ecological Applications* 4(4):627–628.

Reuss, J.O., N. Christophersen, and M. Seip. 1986. A critique of models for freshwater and soil acidification. *Water, Air and Soil Pollution* 30:909–930.

Reuss, J.O., and D.W. Johnson. 1986. Effect of soil processes on the acidification of water by acid deposition. *Journal of Environmental Quality* 14:26–31.

Reuss, J.O., 1986. *Acid deposition and the acidification of soils and waters*. Ecological studies #59 Springer-Verlag, New York.

Richter, D.D., D.W. Johnson, and K.H. Rai. 1992. Cation exchange reactions in acid forested soils: effects of atmospheric pollutant deposition. Pages 341–357 in D.W. Johnson and S.E. Lindberg, eds. *Atmospheric deposition and forest nutrient cycling: a synthesis of the Integrated Forest Study*. Springer-Verlag, New York.

Rosenqvist, I. Th., P. Jørgensen, and H. Rueslátten. 1980. The importance of natural H+-production for acidity of soil and water. Pages 240–242 in D. Drabløs, and A. Tollan, eds. *Ecological impact of acid precipitation*. Proceedings Internat. Conf. Sanlefjord, Norway SNSF Project, Oslo-Äs.

Schindler, D.W. 1986. The significance of in-lake production of alkalinity. *Water, Air and Soil Pollution* 30:931–944.

Schindler, D.W. 1988. Effects of acid rain on freshwater ecosystems. *Science* 239:149–157.

Schindler, D.W. 1992. NAPAP from north of the border. *Ecological Applications* 2:124–130.

Schindler, D.W., T.M. Frost, K.H. Mills, P.S.S. Chang, I.J. Davies, L. Findlay, D.F. Malley, et al. 1991. Comparisons between experimentally- and atmospherically-acidified lakes during stress and recovery. In F.T. Last and R. Watling, eds. *Proceedings of the Royal Society of Edinburgh* 97B:193–226.

Schofield, C. 1976. Acid precipitation: effects on fish. *Ambio* 5:228–230.

Schulze, E.D., W. Devries, M. Haughs, K. Rosen, L. Rasmussen, C.O. Tamm, and J. Nilsson. 1989. Critical loads for nitrogen deposition of forest ecosystems. *Water, Air and Soil Pollution* 48:451–456.

Schwartz, S.E. 1989. Acid deposition: unraveling a regional phenomenon. *Science* 243:753–763.

Seip, H.M. 1980. Acidification of freshwaters: sources and mechanisms. Pages 358–366 in D. Drabløs and A.M. Tollan, eds. *Ecological impact of acid precipitation*. Proc. Interction & Conf., Sandefjord, Norway, SNSF Project, Oslo-Äs, Norway.

Smith, R.A. 1872. *Air and rain: the beginnings of chemical climatology*. Longmans, Green and Co., London.

Stern, A.C., R.W. Boubel, D.B. Turner, and D.L. Fox. 1984. *Fundamentals of air pollution (2nd ed)*. Academic Press, Inc., Orlando, FL.

Sullivan, T.J., D.F. Charles, J.P. Smol, B.F. Cumming, A.R. Selle, D.R. Thomas, J.A. Bernert, and S.S. Dixit. 1990. Quantification of changes in lakewater chemistry and response to acidic deposition. *Nature* 345:54–58.

Tamm, C.O., and L. Hallbäcken. 1986. Changes in soil acidity in two forest areas with different acid deposition: 1920s–1980s. *Ambio* 17:56–61.

Taylor, G.E., Jr., D.W. Johnson, and C.P. Andersen. 1994. Air pollution and forests. *Ecological Applications* 4:629–650.

Ulrich, B., R. Mayer, and P.K. Khenna. 1980. Chemical changes due to acid precipitation in a loess-derived soil in central Europe. *Soil Science* 130:193–199.

Ulrich, B. 1983. Soil acidity and its relations to acid deposition. Pages 127–146 in B. Ulrich, and J. Pankrath, eds. *Effects of accumulation of air pollutants in forest ecosystems*. D. Reidel, Boston, MA.

van Breemen, N., P.A. Burrough, E.J. Velthorst, H.F. van Dobben, Toke de Wit, T.B. Ridder, and H.F.R. Reijnders. 1982. Soil acidification from atmospheric ammonium sulphate in forest canopy throughfall. *Nature* 299:548–550.
van Breemen, N.J., J. Mulder, and C.T. Driscoll. 1983. Acidification and alkalinization of soils. *Plant and Soil* 75:283–308.
van Breemen, N.J., H.F. van Dijk. 1988. Ecosystem effects of atmospheric deposition of nitrogen in The Netherlands. *Environmental Pollution* 54:249–274.
Vitousek, P.M., J.D. Aber, R.W. Howarth, G.E. Likens, P.A. Matson, D.W. Schindler, W.H. Schlesinger et al. 1997. Human alteration of the global nitrogen cycle: sources and consequences. *Ecological Applications* 7:737–756.
Weathers, K.C., G.E. Likens, F.H. Bormann, J.S. Eaton, W.B. Bowden, J.L. Anderson, D.A. Cass et al. 1986. A regional acidic cloud/fog water event in the eastern United States. *Nature* 319:657–658.
Winner, W.E. 1994. Mechanistic analysis of plant responses to air pollution. *Ecological Applications* 4:651–661.
Wright, R.F., and A. Henriksen. 1983. Restoration of Norwegian lakes by reduction in sulfur deposition. *Nature* 305:422–424.
Wright, R.F., and M. Haughs. 1991. Reversibility of acidification, soils and surface waters. *Proceedings of the Royal Society of Edinburgh* 97B:169–191.
Wright, R.F., E. Lotse, and A. Semb. 1988. Reversibility of acidification shown by whole-catchment experiments. *Nature* 334:670–675.
Wright, R.F., T. Dale, E.T. Gjessing, G.R. Hendrey, A. Henriksen, M. Johannessen, and I.P. Muning. 1976. Impact of acid precipitation on freshwater ecosystems in Norway. In *proceedings of the first international symposium on acid precipitation and the forest ecosystem*. USDA Forest Service general technical report NE-23.

9
Empirical and Theoretical Ecology as a Basis for Restoration: An Ecological Success Story

JAMES A. MACMAHON

Summary

Within this chapter, some samples of the contributions of ecological science to the process of disturbed system restoration are discussed. At nearly every stage of the development of restoration science, ecological studies have made significant contributions to developing more effective restoration procedures. Increasingly, ecosystem science and its intellectual approaches are being adopted by restorationists. Whole-system perspectives are now commonly included in restoration plans, modelling as a knowledge-organizing activity appears in many projects, especially those of large areal extent, and attention to such ecosystem processes as cycles of chemicals and not merely the list of component species is evolving.

Just as restoration has benefited from ecological knowledge, so has ecology benefited from restoration activities. In this chapter, insight into many ecological processes and phenomena have been highlighted in the context of restoration. Ecological theory is easily tested in the milieu of the highly managed restoration process. Additionally, restoration projects are often of a size and intensity that could neither be duplicated nor funded to conduct normal ecological research. In a sense, restoration projects are some of the large experiments that ecologists cannot afford to conduct. We must avail ourselves of these unique opportunities.

Ecology-based restoration will probably become more important in the future for two reasons. First, there will be more disturbed areas to restore than at present, and second, ecology-based restoration is apt to provide efficient methods (in terms of both time and money) of restoration. Various examples of ecology-based restoration projects have been presented in this chapter, and some comments about the future are offered both to alert ecologists to the future, and to bring decision-makers, restoration practitioners, and ecologists into closer partnerships.

Introduction

Disturbance is more characteristic of ecosystems than stability. Natural disturbances are integral parts of, and are responsible for, some characteristic aggregations of species. Anthropogenic disturbance, planned or unplanned, is so pervasive that it also characterizes, to some degree, the ecosystems of the earth from the depths of the oceans to the tops of the highest mountain peaks. Hannah et al. (1994) determined that 90 million km^2 (52%) of undisturbed land remains on the Earth's surface. When they excluded uninhabitable areas from analyses, however, nearly 75% of the Earth was disturbed. Although similar data are not available for oceans, they are probably comparable if chemical disturbances are included.

Often, when humans view disturbed systems, regardless of the cause of disturbance, they decide to restore all or some of the characteristics of those systems to a desired state. The states chosen vary from an idealized view of a pristine assemblage of species and landscapes to simply restoring one or a very few desirable attributes of the landscape. All of these acts of restoration are based on human goals and are mediated by human activities.

In this chapter, I attempt to show how knowledge derived from ecology has aided in the restoration process. My coverage is not uniform, however. Only a few examples are given and these are derived mainly from North America despite extensive work elsewhere. I know terrestrial systems better than those that are aquatic, and mineland restoration receives more of my attention than other topics, even though prairie restoration may be one of the most successful examples of ecology-based restoration procedures (Packard and Mutel 1997).

The world, with its ever-increasing population, will be more ecologically disturbed in the future than at present. Restoration activities will become more important, and the science of ecology will probably provide many of the bases for restoration procedures and technologies (Dobson et al. 1997). More restoration will include ethical issues, as well as those that are technical, and "good" restoration will include historical, social, cultural, political, aesthetic, and moral aspects (Higgs 1997). Because of the probability of these future developments, I attempt to look ahead to some drastic, ecology-based restoration activities and also provide some insight into lessons learned thus far that will make restoration easier to implement in the future.

Some Historical Observations

The acts of humans attempting to repair damaged ecosystems (whether natural or managed) can be traced, with some imagination, as far back as when human activities were first recorded historically. A conscious effort to restore disturbed sites, however, became more prominent during the late 1800s and early 1900s, as people in Australia, Europe, and North America began to develop an interest in conservation activities. This interest was occasioned by the observation of dwindling populations of animals, disturbance of "pristine" habitats, and gross mismanagement of landscapes, including those that were managed for a variety of human-related uses. Many observers, especially those interested in conservation, took note of these events and called for study. Meanwhile, new government agencies were formed to manage, protect, or conserve resources and to restore degraded lands to a more useful condition. In the United States, the Ecological Society of America (ESA) established the Committee on the Preservation of Natural Conditions for Ecological Study chaired by Victor Shelford, shortly after its founding in 1917. Under Shelford's guidance, the work of this committee led to the production of the *Naturalist's Guide to the Americas* in 1925, creating an awareness of the degree of landscape alteration. With other efforts, this led to the application of ecological ideas to a variety of human activities. An unfortunate footnote to history is that Shelford's attempts at developing a science of applied ecology were not fully endorsed by the ESA, probably for both economic and political reasons. During 1946, Shelford founded the Ecologist's Union, a separate group that became the Nature Conservancy in 1950. Although the goal of this organization was specifically to preserve natural areas, the restoration of natural systems and their components were concerns of its founder from the very beginning (Croker 1991). Shelford fought for restoration of predator species following government control programs, and worked to restore the American grasslands. It is interesting how we have come full circle and now have active programs to restore wolf (*Canis lupus*) and black-footed ferret (*Mustela nigripes*) populations.

Early applied ecology efforts did not fit into a formal fabric of restoration ecology, although reintroduction of species was clearly restoration, as were a variety of other activities. In the United States, one of the largest ecosystem restoration efforts involved members of the Civilian Conservation Corps. In 1935, they attempted to replant prairie on degraded farmlands of the Midwest as well as other restoration projects. One of these projects was fostered by the prominent conservationist, Aldo Leopold, who wanted an ecological restoration project that would return part of the University of Wisconsin–Madison–Arboretum to its former tallgrass prairie status (Jordan et al. 1987). Even today, this project is used as an example of the earliest successful efforts at restoration of ecological systems.

During the last forty years, scientists, governments, and nongovernmental organizations have turned their attention specifically to restoration of disturbed ecosystems around the world, but especially in Europe (Bradshaw and Chadwick 1980; National Research Council 1986), North America (Box 1996), and Australia (Majer 1989). Restoration ecology has matured from simply being either a point of view or an approach to some management problems to gaining the status of a scientific discipline that is a subset of ecology (Bradshaw 1993).

The rapid development of restoration ecology as a discipline is attested to by the development of many new journals. In 1961, the British Ecological Society began to discuss the need for a conservation journal. This ultimately was broadened to include all of applied ecology, and in 1964, the *Journal of Applied Ecology* was launched. By 1972, the *Journal of Environmental Management* was initiated, followed by *Environmental Management*, indicating a level of interest sufficient to attract commercial publishers. *Reclamation and Revegetation Research* was initiated in 1982 by Elsevier Publishers, only to be discontinued in 1986. Meanwhile, the University of Wisconsin Arboretum started its valuable journal, *Restoration and Management Notes*. As recently as 1991, the Ecological Society of America first published its journal, *Ecological Applications*, and, in 1993, the Society for Ecological Restoration initiated its journal, *Restoration Ecology*. Interestingly, the Applied Ecology section of the ESA has been the fastest-growing portion of that organization. Other forms of publications about ecology-based restoration appeared also, which include hundreds of books, monographs, and federal documents. Today, restoration projects are often widely publicized in the press and are even the topic of popularized science books (Stevens 1995).

What Is Restoration?

Before proceeding, we need to establish what is meant by restoration ecology. There are a variety of methods to manage disturbed landscapes and their ecosystems and bring them back to a state that is more or less like that which is considered to be the predisturbed condition, or at least, a state that has a suite of characteristics that are valuable to humans. In some cases, the objective is to bring them back to some "pristine condition"; in other cases, the goals are much more modest and may only involve restoration of a single species (Bowles and Whelan 1994; Mladenoff et al. 1997). The differences in these efforts are usually pigeonholed into the following five categories of restoration ecology.

Restoration, per se, generally implies the activities necessary to bring a disturbed site into its former, or original, state (Cairns 1993, 1995). This approach is the basis for the formal definition endorsed by the National Academy of Science Committee that describes restoration as "the return of an ecosystem to a close approximation of its condition prior to disturbance" (National Research Council 1992). In contrast, the term *rehabilitation* refers to attempts to restore some elements of structure or function to an ecological system. The term *reclamation* is reserved for work carried out on severely degraded sites; often, measures are taken to prevent further degeneration and to implement a first step to restoring a more or less natural ecosystem. Historically, reclamation (often an engineering activity) and restoration (an activity more related to biological systems) have developed quite independently as fields of endeavor, a circumstance that has not served the best interests of the two fields. *Re-creation* is an attempt to reconstruct an ecosystem on a site so severely disturbed that there is virtually nothing left to restore. This is an extremely difficult task that is usually not financially possible. Finally, *ecological recovery* is simply an approach to restoration whereby nothing is done to the system, the assumption being that given sufficient time the system itself will return to its natural state. (MacMahon 1997). For the purposes of this chapter, I will use the word *restoration* generally to cover all of these possibilities and will only differentiate them when necessary.

The Process of Restoration and the Science of Ecology

Restoration activities have been implemented for a long time. The degree of ecological input to restoration activities has grown over time, but in retrospect seems to have increased its influence more rapidly since ecologists adopted an ecosystem perspective, that is, since they have addressed problems relating to the flow of matter and energy and not just species composition or structure of an area. This is not surprising because practitioners of ecosystem science often address applied problems.

Overall, the contributions of ecology to the restoration process have varied from a simple increase in our understanding of the requirements of various species to an appreciation of the important and varying interactions of organisms and the abiotic environment. In some cases, the contribution has been in the form of a sophisticated understanding of biogeochemical processes and an ability to express these in the mathematical form of simulation models. The need for restoration is often recognized by the use of some metric of the biological integration of a system. This has been especially true for aquatic systems, for instance, the development of the highly successful and widely used Index of Biological Integrity (Karr 1991) in the United States and other such indices in South Africa, the United Kingdom, France, and Belgium (Petts and Calow 1996).

The ecosystems of greatest interest have often been those that represent "natural" systems. Increasingly, however, scientists recognize that there are few, if any, pristine environments. Humans dominate earth processes and landscapes, and examples of many ecosystems in need of restoration exist.

There are a variety of ways that ecological science, especially from the ecosystems perspective, has positively influenced the process of restoration. It is also important to note that some ecological input to restoration has not produced positive results, and sometimes, what appears to be an ecologically sound practice fails. Ecologists sometimes fail to state the certainty of their statements and their utterances are accepted, uncritically, by practitioners, who then obtain undesirable results. Conversely, some practitioners have based their restoration approaches on unsound science or weak empiricism.

The process of developing models of a restoration project, not just the implementation of models, is an obvious way that the ecosystem approach is aiding a variety of restoration projects. The initiation of the inclusion of models in the planning of restoration projects (Reith 1986), a phenomenon that may have been fostered, in part, by the widespread influence of the International Biological Program (IBP), is an outgrowth of ecosystem science. Modelling continues to grow as a component of ecosystem restoration planning, however, today it is used with more insight than in the past. For example, it is understood that models do not necessarily provide accurate forecasts of the exact time frame or the precise outcome of a restoration project (Loucks 1985), but they do help us to understand the operant mechanisms in the ecosystems intended for restoration, as well as providing a blueprint that tests our understanding of the dynamics of those systems. Our ability to conceive of realistic model structures is, in part, what Bradshaw (1987) called the "acid test for ecology."

Modelling has been especially important in highlighting considerations that ecologists might not have included in their thinking thirty years ago. For example, systems exhibit thresholds (or sudden changes in states that occur because of minor perturbations), which are not intuitive (Friedel 1991), and the levels of biological organization—individuals, communities, and ecosystems—are not merely a biological ladder, but are interacting in much more detailed and complex manners. To manage just one of these levels, we often must look at the level below it to find the mechanism by which the level of interest operates, and then at the level above to find the context within which the level exists (O'Neill 1986).

The realization that systems are ever-changing (Luken 1990; Sprugel 1991; Hobbs and Norton 1996) underlines the need to understand that many restoration efforts, with all good intentions, may be designed to achieve an impossible goal of mimicking some reference site, which is not following the same trajectory of change as the altered site that is the object of concern (Pickett and Parker 1994). Figure 9.1 shows some of these possibilities.

The increased knowledge of the variation of ecosystems in time and space has highlighted the need for constant assessment of the degree to which restoration goals have been met. Many examples exist in which "successful" completion of a restoration project has been assumed, only to find later that the system is an imperfect mimic of the model system (Zedler 1996c). These assessments often require new mathematical analyses (Ames 1993), different approaches to thinking about community-level processes and using these as indicators of ecosystem development (Allen 1992), or measurements of such novel indicators of success as soil microbes (Bentham et al. 1992); or invertebrates in rainforests (Jansen 1997). Simply measuring certain system characteristics, especially species number, diversity, and other system morphological or pattern characteristics does not

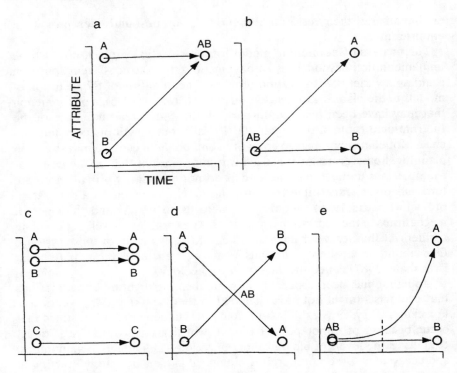

FIGURE 9.1. Potential successional trajectories of reclaimed and reference areas. The y-axis represents any attribute measured to indicate the progress of succession or the success of reclamation. The x-axis is the time interval required to obtain climax. Different scenarios include: (a) the reclaimed area converges with the reference area; (b) the reclaimed area diverges from the reference area; (c) the reclaimed area parallels the reference area without divergence or convergence; (d) the reclaimed and reference areas may change in different directions (e.g., as the result of new or different types of management); and, (e) changes may exhibit various functions. The point in time when the attribute is measured affects perceived reclamation success (from Chambers et al. 1992).

guarantee that the system will have returned to a functioning state (Naveh 1994), that is, that the restored system is structurally and functionally identical to some reference system. Zedler (1995, 1996a and b) has shown that in California coastal restoration projects, systems were brought to a state of being superficially homomorphic (i.e. similar in structure) to their reference areas and yet such integrated system characteristics as nutrient pools, had not been returned to predisturbance levels. Soil nitrogen, foliar nitrogen and biomass, soil organic carbon and nitrogen fixation were lower in these newly constructed marshes compared to the natural systems, and there was no indication that these would change over short time periods (Langis et al. 1991).

Similarly, a system being restored to an approximation of its former status, may lack important ecological components. In the case of Zedler's work in coastal wetlands, a constructed marsh and a natural marsh both contained thirty-seven epibenthic invertebrates but only shared 46% of the species lists and at least one exotic was more numerous in the constructed marsh (Scatolini and Zedler 1996). Similarly, although Holl (1996) found that most butterfly species were restored during a mine restoration project, the rare species had not returned and thus, while the system in a general sense appeared to have recovered, this important component had not. For all of these cases, success is measured against a specific goal and often these goals do not include restoration to the status of an intact, functioning natural system, but merely to a system that looks natural.

An interesting comparison of the effects of three disturbances on forest plots (Foster et al. 1997) shows how tenuous the appearance of normality can be. A site blown down by a hurricane looked totally devastated, but because the internal processes were not significantly altered, it is on the path to recovery. In contrast, sites subjected to chronic nitrogen and soil warming looked intact, but expressed significant imbalances in fundamental processes, which will probably lead to serious problems in the future (Foster et al. 1997).

Ecosystem restoration tends to be directed to reestablishment of plant communities, ignoring animals and a variety of belowground organisms and processes. It must be kept in mind that restoring plant diversity does not always guarantee that the animal component of an ecosystem will be restored. Also, to restore plant diversity, it may be important to give consideration to managing animals, which can play a significant role in the restoration process; restoration of these animals may require more than just a particular variety of plants (Majer 1989). Recent examples of the important role of animals indicate that such simple manipulation of ecosystem components as providing perches for birds may enhance the recovery of plant species (McClanahan and Wolfe 1993; Robinson and Handel 1993). In both of these studies, areas beneath structures where birds perched had higher concentrations of certain seedlings than other areas. Plants whose seeds had been carried in bird feces were established; without perches, these plants might have required a longer time to occupy the site. In contrast, some animals may be pests and can slow the restoration process. Finally, it is possible that animals may be major controllers of ecosystem processes. Carpenter and Kitchell (1988) demonstrated that productivity of some lakes may be controlled in direct and indirect ways by fish. Clearly, lake restoration must take such relationships into account.

In a very real sense, restoration is, in some cases, the application of what is known about the process of succession to real-world problems. In many instances, especially in terrestrial systems, the goal of restoration is to shorten the process of succession. Several restoration workers have alluded to this parallel (Bradshaw and Chadwick 1980; MacMahon 1981; Majer

228   J.A. MacMahon

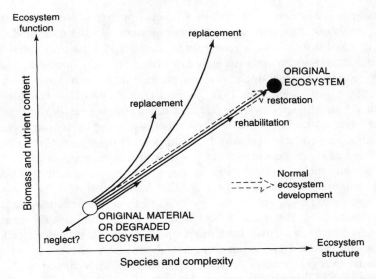

FIGURE 9.2. Relationships between ecosystem structure and function during natural succession and various types of reclamation (Reprinted from Bradshaw 1984, with kind permission from Elsevier Science-NL, Sara Burgerhartstrant 25, 1055 KV Amsterdam, The Netherlands.). Note the alternate end-points (replacement) and the highly directed trajectory of restoration to original system status.

1989; Luken 1990). Bradshaw (1984) depicted some differences between succession and restoration (Figure 9.2). His figure emphasized alternate end-points (replacement) and a very direct restoration trajectory. Essentially, for many applications of ecology to real-world problems we are either trying to shorten the time course of succession, or we are trying to maintain a particular community at some specific stage of succession that we deem desirable. The use of an understanding of succession as a guide to appropriate restoration protocols does not always apply. For many systems, especially aquatic systems, morphometry, flow rates, hydroperiod, and so forth may be more important to the restoration process (Hammer 1997) than succession. In some cases, application of ecological assembly rules may offer effective guides to restoration (Lockwood 1997). Nonetheless, many restoration successes have been based on a successional paradigm and even common guidelines for terrestrial restoration include this approach (Munshower 1994). A brief introduction to succession will set the stage for some of our success stories.

Succession

Clements (1916) elucidated the processes of succession. Even though I will not use his unusual terminology, I will summarize his observations. He referred to a disturbance phase that was characterized by the attributes of the disturbing agents and the amount of damage done to the existing system. This stage is best defined by the residual organisms or entities that remain after the disturbance because these are apt to be components of the early successional community. Next, he envisioned a rain of migrants onto the site. Both the migrants and the residuals must be able to establish (i.e. grow and reproduce) in the newly altered site; this process of establishment, though not obvious, is different from merely occupying a site and is extremely important. Organisms that develop on a particular site begin to change the site characteristics. For example, an area opened by a disturbance, which favors plants that have high sunlight requirements, may actually become so shaded by these plants that the site exhibits a new set of conditions that favor another set of residuals that may have been dormant, or may favor yet a different set of migrants. During all of these changes, organisms develop a variety of interactions ranging from neutralisms to predator–prey relationships. Any of these may alter the trajectory of succession or at least its time course. Finally, until recently it was assumed that succession came to some conclusion that was termed a climax. It is now known that communities and ecosystems are seldom in a stable state, but rather are even-changing, often dramatically, over long periods of time. Other changes, although perhaps imperceptibly, occur over very short time periods.

Restoration as a practice may involve the direct manipulation of any of these processes of succession to meet a particular goal. For example, the restorationist often acts as the agent of migration, bringing desirable plants and animals to a site to initiate the restoration process. The decision of whether or not to store topsoil in a mining situation is predicated upon the probability that stored topsoil that has been removed from a mine site is apt to become anaerobic and will probably kill many of the residual seeds and microorganisms that might otherwise benefit a site (Dvorak 1984). Thus, the process of direct application of topsoil (i.e., immediately moving soil from a site about to be mined to another site that is being reclaimed) has gained popularity in the mining industry (Munshower 1994).

The establishment phase is also often extremely important. As will be noted later, in extreme environments, the restorationist may have to develop strategies that will ensure the establishment of plants, not just the availability of seed. After being established, many plants are resilient to later environmental perturbations. Given how generalizable the process of succession is to many situations, it is no wonder that resource managers often use management of succession as the basis for their protocols and yet they do not realize it. Some managers, and scientists for that matter, use a knowledge of succession to direct many of their management activities. A few management problems that have been addressed through consideration of succession appear in Table 9.1.

TABLE 9.1. Some resource management problems in which succession can be manipulated to achieve management goals (from Luken 1990. Reprinted by permission of Chapman & Hall).

Conserving rare or endangered species.
Conserving and restoring communities.
Manipulating the diversity of plant and animal communities.
Creating relatively stable plant communities on rights-of-way.
Revegetating drastically disturbed lands.
Minimizing the impact and spread of introduced species.
Maximizing wood production from forests.
Minimizing adverse environmental effects of forestry.
Predicting fuel buildup and fire hazards.
Increasing animal populations for recreation and aesthetics.
Developing multiple-use plans for parks and nature reserves.
Determining the minimum size of nature reserves.
Minimizing the impact of roads, parking areas, trails, and campsites.
Preserving scenic vistas in parks.
Minimizing the cost of grounds maintenance on public lands.
Controlling water pollution.
Minimizing erosion.
Maintaining high quality forage production.
Minimizing the cost of crop production in agricultural communities.
Developing vegetation in wetlands or on the edges of reservoirs.

Restoration at Three Scales

Restoration occurs at different intensities depending on the particular goals of the restoration effort and the nature of the prerestoration disturbances. In extreme cases, the process might start from nearly ground zero to the reestablishment of a functioning ecosystem. In other cases, the restoration is merely an effort to return a species or a few species that have disappeared. Similarly, the restoration prescription can vary widely. The process can be as simple and small as preparing the soil surface of a site that will receive seeds of a target species, or as complex and large as the case of the Grand Canyon in which water was released at 1,270 cubic meters per second to mimic a natural flood regime of the Colorado river. This was done to change the river's geomorphological characteristics to a condition that favors some native fish and a more "natural" riparian vegetation (Collier et al. 1997). Additionally, projects falling everywhere between these extremes have also been attempted. In the next sections, three different restoration attempts are discussed, all of which are based on, to varying degrees, an ecologically explicit basis for planning and implementation. The three examples include a small-scale example of the reintroduction of a few species, a middle-scale experiment to reclaim mine spoils, and finally, restoration of an entire river system and its associated wetlands that dominate the southern third of Florida.

A Small-Scale Example

Richard Primack (1996) describes some lessons that ecological theory offered concerning the reintroduction of plant species in the northeastern United States. His overview emphasizes ecological insights concerning dispersal, establishment, and population structure. Many areas of the Northeast have lost plant species since the turn of the century. Middlesex Fell, a 400-hectare conservation area near Medford, Massachusetts, had 338 species of flowering plants in 1894 but only 227 native species in 1992. Many attempts to reintroduce species failed. At Hammond Woods, Massachusetts, Primack found that no plants were established for thirty-two of thirty-six species sown by broadcast seeding and 167 of 176 sites showed no establishment success.

In essence, Primack found that mimicking the natural processes of dispersal and colonization was a more successful technique for reintroducing a variety of species than traditional seeding methods. In this approach, he chose ecologically appropriate planting techniques and effectively executed these in ways that paralleled the natural dispersal and establishment processes for the plants of interest rather than merely broadcasting seeds. Throughout his discussion, Primack goes back to the general framework of succession that was discussed previously. One of the most important components of his approach was to effectively choose sites where plants were to be reintroduced. Such sites had to be characterized as "safe sites"; that is, the site had to exhibit conditions that were conducive to germination, establishment, and the fostering of the competitive ability of the plants that he was attempting to reintroduce. Because successful establishment is not a constant process from year to year, and often depends on the vagaries of weather, he recommends successive attempts at establishment, a view that is consistent with data on the dynamics of natural ecosystems where the year-to-year variability is significant.

Based on his findings, Primack suggests an ecologically based approach, which emphasizes that reintroductions are more apt to be successful for annuals than perennials. Because of the extremely high initial mortality observed in natural systems, the introduction of large numbers of transplants or seeds is also necessary. He alludes to the probable increase in the rate of establishment by the use of transplanted seedlings or adult plants rather than attempting reintroduction from seeds. In effect, he wants to guarantee the success of the establishment phase of succession. These observations are consistent with general knowledge that is derived from succession theory and our knowledge about seed life-history strategies (Chambers and MacMahon 1994).

Interestingly, animal reintroductions have been implemented for a considerable period of time. Griffith et al. (1989) reviewed reintroduction attempts of 198 bird and mammal species that occurred between 1973 and 1986. In general, their observations suggested that they had greater success for the following. (1) game species as opposed to threatened species; (2) introductions within the normal geographic range of a species rather than outside of the range; (3) wild-caught rather than captive-raised animals; (4) herbivores rather than carnivores; and (5) larger rather than smaller numbers of animals being reintroduced. The results of these comparative studies are consistent with what would be predicted from a knowledge of ecological data. Reintroducing animals into their normal range probably means that their tolerance-response curves and acclimation abilities are genetically within the range of variation of the environments into which they are being reintroduced. The use of wild-caught rather than captive-raised animals may imply a greater range of genetic variation in the wild animals than in the captive animals because captive breeding alone may act as a force of selection that decreases the genetically based range of variation. Herbivores being more successfully reintroduced rather than carnivores may relate to the greater accessibility of food and the generally smaller home ranges of herbivores. The success of reintroduction of species by use of greater numbers of animals is simply a "hedging your bet" strategy, which is similar to way that Primack attempted several successive reintroductions. In discussions, we often record and marvel at the establishment of a species from a single-chance propagule. In nature, I suggest that most species that successfully establish do this as a result of numerous failed "attempts," which, when conditions are exactly correct, culminate in establishment.

A Middle-Scale Example

Throughout the world, surface-mined lands cover large areas (Bradshaw and Chadwick 1980). In the United States alone, some 1.5 million hectares of surface have been altered by mining. The restoration of mine sites in the western United States and in other arid areas of the world has been considered to be an especially difficult task. This was highlighted in 1974, when a panel from the National Academy of Sciences suggested that the rehabilitation of western coal lands required new restoration approaches. This awareness rested, in large measure, on observations that attempts to reclaim mine lands in arid areas were often unsuccessful both here and abroad (Bradshaw and Chadwick 1980). A cursory analysis of these unsuccessful projects indicated that the procedures used were based on traditional agricultural practices, for example, row-crop seeding. The projects also were, for the most part, devoid of techniques based on ecological observations and theory, which were tailored to specifically address the unique problems of arid lands.

It should be of little surprise that because arid lands represent an environmental extreme, the use of traditional agricultural practices, generally developed for mesic environments, might not be as successful as they would be for areas with more suitable climate patterns and weather. The high year-to-year variation of rainfall in arid lands seems to preclude the estab-

lishment of plants from seeds except in the infrequent case in which sufficient rainfall for germination and establishment occur in successive years
(MacMahon 1981). To again use the succession analogy, this suggests that
more attention should be given to the establishment phase of an arid-land
restoration project.

Additionally, observations of arid-land plants indicate that, contrary to
conventional wisdom, community dispersion patterns are often clumped
rather than regular (Phillips and MacMahon 1978). Traditional seeding
would not normally produce clumped dispersion patterns. However, clumping could convey a variety of advantages to plants. Among these advantages
are that in open environments plants growing in clumps act as epicenters for
the deposition of finely divided wind blown materials. These materials
include organic matter, seeds, invertebrates, and the spores of mycorrhizal
fungi, bacteria, and protozoans. In such cases, the deposition of these materials within the clump actually enriches the clumps, and provides (1) more
potential propagules for reestablishment, (2) access by the plants to a
mutualist organism that confers an advantage to most vascular plants (i.e.
mycorrhizal fungi) (Allen 1991), and (3) a concentrated nutrient source in
the form of trapped organic matter.

During the 1980s a group of scientists from Utah State University attempted to implement several cultural practices that derived from empirical
and theoretical studies of desert plants and animals, and their interactions.
These studies and pertinent references are reviewed in more detail in
MacMahon (1997). Several hypotheses were tested. First, because the establishment phase of succession was so critical in arid areas (MacMahon
1981, 1987, 1997) and because conditions for establishment did not occur
yearly, any effort to enhance the probability of establishment might reap
dividends. Additionally, the observation that plants occurred in clumps in
arid areas and that these clumps might confer a variety of chemical and
biological advantages suggested that mimicking this clumping pattern could
show more success than using traditional, regular-spacing patterns for
planting.

Briefly, in this study, plants were arrayed in three dispersion patterns, and
each of these patterns was planted at four different densities. These included densities that were equivalent to the natural density of the surrounding vegetation, one density that was significantly higher, and two that were
lower than the natural densities. The study was carefully implemented to
exactly position seedlings that had been started in plastic soil tubes so that
fates of each individual plant could be followed. The results of this study
(Figure 9.3) indicated that clumped planting showed better survivorship
and better recruitment of new individuals than any other dispersion pattern. In examining the clumps, the rates of formation of mycorrhizal associations were higher than in any other dispersion pattern, soil organic
matter was higher, and thus, soil water-holding capacity was increased
(Allen and MacMahon 1985; Allen et al. 1989). Comparing the establishment rate of young transplants to seedlings clearly indicated the superiority

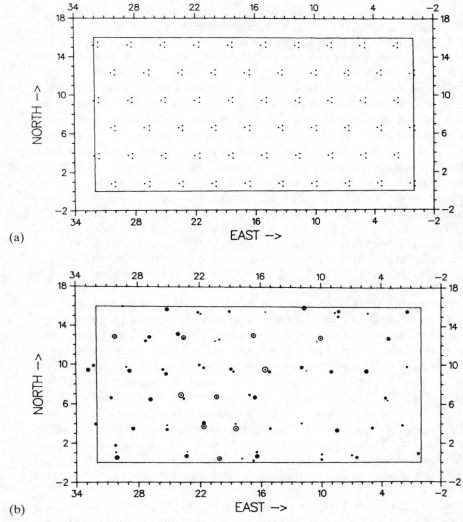

FIGURE 9.3. Results of a study of two plant dispersion pattern effects on the success of establishment and recruitment of plants on a reclaimed coal mine site at Kemmerer, Wyoming: (a) Planting array of triads of three plant species in a low density, regular dispersion plot. Each plot is 16×32 m; (b) status of plants in a after three years. Positions and sizes of circles indicate actual positions and sizes of plants on the site; (c) planting array of triads in a high density, clumped dispersion plot; (d) Status of plants in c after three years. Note the degree of clumping, the larger sizes of some plants, and more recruitment of new plants from seeds compared with the low density, regular plantings (a and b) (from MacMahon 1997. Reprinted by permission of Sinaeur Associates, Inc.).

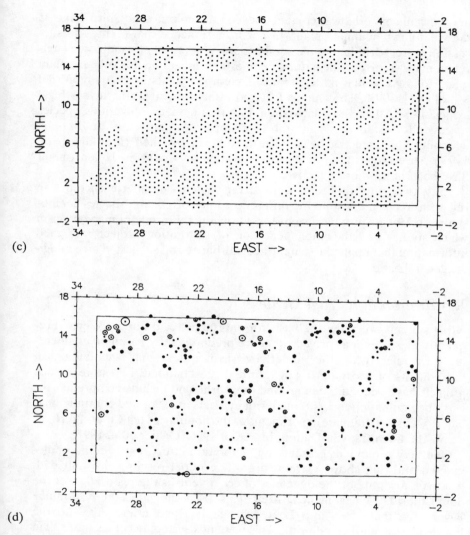

(c)

(d)

FIGURE 9.3. *Continued*

of the transplants and thus avoided, to a great extent, the heavy losses incurred by newly germinated seedlings. All in all, this design, based both on the empirical observations and on the theoretical arguments concerned with succession, provided a planting that was more successful than the use of the traditional reclamation techniques.

It should be noted that restoration studies often lead to interesting ecological questions about natural systems. In 1980, when the Mount St. Helens volcano erupted in the state of Washington, a barren, windswept plain of new volcanic material was created between the volcano and Spirit Lake. If the windblown, clumped-plant, mycorrhizae practice that seemed to successfully direct arid-land mine spoil reclamation (another windblown environment) was correct, the natural plant establishment on this volcanic plain might follow a similar pattern. Indeed, clumped plants were common (del Moral et al. 1995). The clumps fostered establishment of the mycorrhizal association (Allen and MacMahon 1988), and seeds of various species successfully established in these clumps. Attempts at traditional, agriculture-like seeding to stabilize land surfaces failed (Franklin et al. 1995). An ecological approach to restoration at Mount St. Helens might have been much more effective.

Thus, two quite different, wind-driven systems parallel each other in regard to recovery despite the differences in environment, species composition, and the disturbance agent. Additionally, this is one of many cases in which studies and the implementation of restoration projects provided information that applied to a natural disturbance and suggested an ecological generalization.

Restoration at a Huge Scale

River systems are often the basis for enormous restoration projects. Frequently these begin as projects that are predominantly based on engineering principles with the underlying assumption that the biological components of ecosystems will return if the "habitat" is restored. Such projects have been initiated around the world, and include comprehensive work beginning in the 1970s in Germany (Larsen 1996) and continue unabated with such dramatic experiments as the flooding of the Colorado River below the Glen Canyon Dam in March of 1996 (Collier et al. 1997).

One very prominent present-day restoration effort that is being implemented and that shows extremely positive initial results consistent with what is known about the dynamics of ecosystems is the restoration of the South Florida Management District. This district includes the Kissimmee River area that feeds into Lake Okeechobee and ultimately, into the Everglades in southern Florida. This wetlands system occupies more than 3.6 million hectares (Davis and Ogden 1994a,b). The Kissimmee River project, already initiated, covers 70 kilometers of river channel, 11,000 hectares of wetland, and will take two decades to fully implement. The background and some of the ecological considerations for this study are the basis for an entire issue of the journal *Restoration Ecology* (1995, Vol. 3, No. 3). A summary of the ecosystem aspects of the project is available in Dahm et al. (1995) and their simple conceptual model of the ecosystem is presented as Figure 9.4. The details of developing a restoration plan for such a huge, diverse area are documented by Harwell (1997).

Dramatic changes in this ecosystem occurred as a result of river engineering between 1962 and 1971 that channelized the river, introduced control

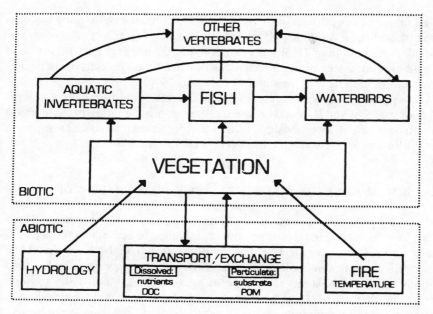

FIGURE 9.4. A general conceptual model of ecosystem structure and trophic interactions for the Kissimmee River and floodplain. The proposed dominant direction of interactions is shown, but multiple interactions are commonplace (from Dahm et al. 1995. Reprinted by permission of Blackwell Science, Inc.).

structures, and regulated water levels, all of which caused alterations of the hydrology in ways that changed the biota. Starting in the 1970s, eutrophication of Lake Okeechobee became obvious and public concern was raised. Massive planning occurred, resulting in the present program of restoration. Similar negative effects were also obvious in the Everglades and even in the near-shore ocean habitats of Florida Bay. The underlying ecological assumptions raised are that the abiotic variables of river hydrology are driving the entire system. The assumed causal relationships are: (1) a reinstitution of normal hydrologic regimes will cause vegetation to reestablish developing a structure that, in turn, will control the establishment of aquatic invertebrates; and (2) these will act as a food source for the redevelopment of historically important fisheries. Also, other vertebrate species dependent on fish will, of course, respond (Toth 1995). This massive undertaking is in its infancy, but direct observations by a variety of restoration workers (personal communication) at the Kissimmee River Project indicate that the highly vagile marsh birds are returning (although the migration phase is easy), fish populations have increased, and a "normal" plant community is rapidly reappearing. Each of these observations is an indirect indicator that the causal pathway through the ecosystem to each component is being enhanced as planned. Subsequently, the success of the

Kissimmee River portion of the project will be linked to a detailed restoration plan of the Everglades, which is based, in large measure, on ecosystem-simulation modelling (Holling et al. 1994). An integral part of this is a detailed set of water-policy alternatives that are based on ecosystem principles and modelling (Walters and Gunderson 1994).

This project is especially noteworthy because it has required the cooperation of national and state government agencies, private industry, academics, and many others. Relationships had to be forged that, in some cases, were as important as the scientific facts (Harwell 1997).

Restoration in the Extreme: Designer Ecosystems

Up to this point, the discussion has addressed successful restoration efforts that have as their underpinnings some semblance of ecological theory or at least an ecological perspective. Using information from both empirical and theoretical studies in ecology, it is conceivable that enough is understood about certain ecosystems that some more ambitious, albeit risky projects might be attempted. Starting from scratch, on extremely disturbed sites, attempts could be made to develop what I term as "designer ecosystems," which could serve a specific purpose for humans. Take the case of a particular site, which has been extensively altered, which occurs near a population center, and which will probably need to be used to provide some services for humans. One possibility is to design an ecosystem structure that mimics some natural system, but which is composed of a variety of species that do not all co-occur in nature. Each species would be handpicked to have characteristics that help achieve the goal for this designer ecosystem. For example, the structure of a forest might be re-created, layer by layer, using a series of species that can survive on a soil with specific chemical problems. Although native species would be best, it is possible that not all layers could be constructed using them and non-native species would have to be used. In the ideal case, the structure of the forest and even such integrated ecosystem characteristics as net annual primary production and nutrient cycling can be re-created. The physical structure might allow the introduction of a variety of animals that respond to this particular structure, given the caveat that there is food available. The mix of species could have some economic benefits because they were chosen to provide for sustainable harvesting of a desired commodity, or it may be used merely as an aesthetic enhancement for visitors to such an area. A situation is conceivable in which a rainforest might actually be "built", the sole function of which would be to allow visitors to experience the aesthetics of a rainforest while not causing the damage that occurs by human visitation to pristine rainforests elsewhere. The various plant species would be chosen to provide the structure that would support animal communities, and would be durable in the sense of being resilient to human traffic. On a minor scale, the naturalistic animal-habitat displays in zoological gardens and the displays of various plants by vegetation type in botanical gardens are of this kind of designer ecosystem. This could be taken to an even greater extreme in that commodity production might be accomplished by designing ecosystems whose parts are productive for humans, yet the overall site may harbor a variety of threatened species. Coffee plantations of the traditional shade-type, in which layers of vegetation are fostered, are a primitive example of this. These sites are often rich in native species that respond to the artificially layered structure that mimics natural forests (Greenberg et al. 1997). As coffee plantations have been changed from the shade- or layered-type to the more industrial, nonshade-type, biodiversity has been lost (Perfecto et al. 1996).

There are precedents for this radical approach. Miyawaki and Golley (1993) report on 285 forest sites that were "ecologically engineered" to hide industrial complexes, control pollution, and stabilize soils. Similar projects in aquatic systems occur in the form of created or constructed wetlands. These wetlands have been developed on sites previously supporting terrestrial vegetation. Many examples that are at least partially successful have been completed and "how to" manuals exist (Hammer 1997). These sites are often developed in response to mitigation needs. In some cases, small degraded sites are exchanged for larger, high value systems.

There is a risk of "playing God" in creating ecosystems. For example, introduced species could escape and these aliens could harm or even displace native species, causing unknown evolutionary consequences for a community (Abrams 1996). All of the knowledge of the ecology and life-history characteristics of species would have to be used to avoid this (Ruesink et al. 1995). Recently, some progress has been made in attempting to evaluate the potential ability to invade by various plant species based on their life-history characteristics (Reichard and Hamilton 1997). It is a challenge to our understanding of ecosystems to attempt this type of invasion resistant designer ecosystem.

Biosphere II is a small-scale, limited example of one approach to this type of management scheme. It contrasts with the approach I envision because it is containerized and attempts to use species that co-occur. In many ways, trying to recreate a natural mix of species is more difficult than my less ambitious designer ecosystem because the designer approach selects candidate species apt to be successful. Many efforts at re-creation using natural species mixes are currently underway, however. In addition to sites being developed in response to wetland mitigation, there is large-scale re-creation of forests (Ferris-Kaan 1995), prairies (Packard and Mutel 1997), and riparian systems (Briggs 1996) among others.

In a world that is so rapidly being altered, extreme measures must be taken to simultaneously accommodate human needs and to conserve its natural systems and species. The radical triage of developing designer ecosystems may prove to be one small step in halting a worrisome trend of habitat and species loss, however imperfect the process seems now.

Some Lessons for Science and Society

Reflecting on the development of restoration ecology brings into focus some of our mistakes and some challenges that we must address if ecology is going to provide the bases for managing the future of this planet. I divide my comments into three areas that represent three groups of people who should be concerned about restoration as a science.

1. Ecologists
   - Ecologists must understand that academic and applied ecology are not mutually exclusive fields. We need to be sure that we do not build barriers between practitioners and theorists that mimic the problems that Victor Shelford faced in the first half of this century.
   - Every investigator should ask the question "Does my research have some applied possibilities?" This question can certainly be asked after research is done, not as a criterion for whether the research should be done, because it is possible that even the most esoteric study may have some significance as we try to understand ecosystems well enough to restore them.
   - Ecologists must be proactive at what I will term "technology transfer." That is, we must make the results of our research available in a palatable form to a variety of constituencies that need this information (Figure 9.5). Too often we sit back and assume that somebody will find the information that they need in our journals (Pickett and Parker 1994); however, we know a lot more about the functioning of ecosystems from our research than is being used by managers and legislators.
   - We must understand restoration ecology in the broader context of being a major component of the conservation of biodiversity (Jordan et al. 1988), of extending nature reserves, and ultimately as the basis for something as dramatic as designing *de novo* ecosystems.
   - We must consider a variety of alternative outcomes to our restoration efforts. These must, of necessity, include consideration of interaction between science and policy, and may also include ethical considerations (Wyant et al. 1995).
   - Ecologists must be certain to provide scientifically sound information and distinguish between what is known and what are working concepts. Sometimes such concepts as species diversity, ecosystem management, biodiversity, and so forth, become buzzwords and the basis for fuzzy management plans (West 1993) if we do not clearly state the level of certainty of our knowledge.
2. Managers and Legislators
   - It must be conceived as possible that ecology-based procedures and prescriptions for managing natural and damaged systems may be more time- and cost-efficient than some of the present-day contrived prescriptions, if we take into account the long-term costs of restoration.

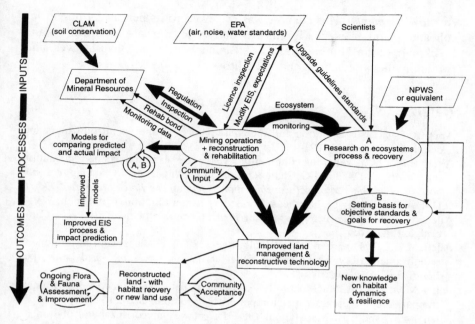

FIGURE 9.5. Recommendations for relationships and information flow between the mining industry and other agencies or sectors with an indication of improved outcomes that would result. The need for stronger links (dialogue) are indicated by bold lines, dotted paragon boxes indicate inputs, ovals indicate processes, and single-lined squares indicate outcomes. Several improved information feedback loops are shown by circular arrows (from Woodside and O'Neill 1995. Reprinted by permission of Surrcy Beatly & Sons Pty. Limited).

• Restoration is a broad field that demands that managers, legislators, and ecologists work together on issues of policy, science, and the implementation of management procedures.

• Managers and legislators must define specific, desired end-points for envisioned projects, and provide long-term assessment of how well the objectives of those projects are met. These objectives should be consistent with what is known about the functioning of natural ecosystems.

3. Society as a Whole

• Because of the past and future gains that will accrue to society because of restoration activities, society cannot assume that the earth's surface can continue to be atered and that scientists will always be able to undo or repair damage. This single lesson is perhaps the most difficult to learn.

Acknowledgments. This chapter was reviewed to its benefit by numerous colleagues and friends. My sincere thanks to Barbara Bedford, Diana Freckman, Sarah Hobbie, Jim Karr, Elaine Matthews, Dave Strayer, and two anonymous reviewers and the editors of this book.

References

Abrams, P.A. 1996. Evolution and the consequences of species introductions and deletions. *Ecology* 77:1321–1328.

Allen, E.B. 1992. Evaluating community level processes to determine reclamation success. Pages 47–58 in J.C. Chambers, and G.L. Wade, eds. *Evaluating reclamation success: the ecological consideration. General Technical Report NE-164.* USDA Forest Service, Northeast Forest Experiment Station, Charleston, WV.

Allen, M.F. 1991. *The ecology of mycorrhizae.* Cambridge University Press, Cambridge, England.

Allen, M.F., L.E. Hipps, and G.L. Wooldridge. 1989. Wind dispersal and subsequent establishment of VA mycorrhizal fungi across a successional arid landscape. *Landscape Ecology* 2:165–167.

Allen, M.F., and J.A. MacMahon. 1985. Impact of disturbance on cold desert fungi: comparative microscale dispersion patterns. *Pedobiologia* 28:215–224.

Allen, M.F., and J.A. MacMahon. 1988. Direct VA mycorrhizal inoculation of colonizing plants by pocket gophers (*Thomomys talpoides*) on Mount St. Helens. *Mycologia* 80:754–756.

Ames, M. 1993. Sequential sampling of surface-mined land to assess reclamation. *Journal of Range Management* 46:498–500.

Bentham, H., J.A. Harris, P. Birch, and K.C. Short. 1992. Habitat classification and soil restoration assessment using analysis of soil microbiological and physicochemical characteristics. *Journal of Applied Ecology* 29:711–718.

Bowles, M.L., and C.J. Whelan. 1994. *Restoration of endangered species. Conceptual issues, planning, and implementation.* Cambridge University Press, Cambridge, England.

Box, J. 1996. Setting objectives and defining outputs for ecological restoration and habitat creation. *Restoration Ecology* 4:427–432.

Bradshaw, A.D. 1984. Ecological principles and land reclamation practice. *Landscape Planning* 11:35–48.

Bradshaw, A.D. 1987. Restoration: an acid test for ecology. Pages 23–29 in W.R. Jordan III, M.E. Gilpin, and J.D. Aber, eds. *Restoration ecology. A synthetic approach to ecological research.* Cambridge University Press, Cambridge, England.

Bradshaw, A.D. 1993. Restoration ecology as a science. *Restoration Ecology* 1:71–73.

Bradshaw, A.D., and M.J. Chadwick. 1980. *The restoration of land.* Blackwell Scientific Publications, Oxford, England.

Briggs, M.K. 1996. *Riparian ecosystem recovery in arid lands.* University of Arizona Press, Tucson, AZ.

Cairns Jr., J. 1993. Is restoration ecology practical? *Restoration Ecology* 1:3–7.

Cairns Jr., J. 1995. Restoration ecology: protecting our national and global life support systems. Pages 1–12 in J. Cairns Jr., ed. *Rehabilitating damaged ecosystems, 2nd edition.* Lewis Publishers, Boca Raton, FL.

Carpenter, S.R., and J.F. Kitchell. 1988. Large-scale experimental manipulations reveal complex interactions among lake organisms. *BioScience* 38:764–769.

Chambers, J.C., J.A. MacMahon, and G.L. Wade. 1992. Differences in successional processes among biomes: importance in obtaining and evaluating reclamation success. Pages 59–72 in J.C. Chambers and G.L. Wade, eds. *Evaluating reclamation success: the ecological consideration—Proceedings of a symposium*. American Society for Surface Mining. USDA Forest Service, Northeastern Forest Experiment Station, Radnor, PA.

Chambers, J.C., and J.A. MacMahon. 1994. A day in the life of a seed: movements and fates of seeds and their implications for natural and managed systems. *Annual Review of Ecology and Systematics* 25:263–292.

Clements, F.E. 1916. Plant succession: an analysis of the development of vegetation. *Carnegie Institution of Washington*, Publication Number 242, 1–512.

Collier, M.P., R.H. Webb, and E.D. Andrews. 1997. Experimental flooding in Grand Canyon. *Scientific American* 276:82–89.

Croker, R.A. 1991. *Pioneer ecologist: the life and work of Victor Ernest Shelford 1877–1968*. Smithsonian Institution Press, Washington, DC.

Dahm, C.N., K.W. Cummins, H.M. Valett, and R.L. Coleman. 1995. An ecosystem view of the restoration of the Kissimmee River. *Restoration Ecology* 3:225–238.

Davis, S.M., and J.C. Ogden, eds. 1994a. *Everglades: the ecosystem and its restoration*. St. Lucie Press, Delray Beach, FL.

Davis, S.M., and J.C. Ogden. 1994b. Toward ecosystem restoration. Pages 769–796 in S.M. Davis and J.C. Ogden, eds. *Everglades: the ecosystem and its restoration*. St. Lucie Press, Delray Beach, FL.

del Moral, R., J.H. Titus, and A.M. Cook. 1995. Early primary succession on Mount St. Helens, Washington, USA. *Journal of Vegetation Science* 6:107–120.

Dobson, A.P., A.D. Bradshaw, and A.J.M. Baker. 1997. Hopes for the future: restoration ecology and conservation biology. *Science* 277:515–522.

Dvorak, A.J. 1984. *Ecological studies of disturbed landscapes: a compendium of the results of five years of research aimed at the restoration of disturbed ecosystems*. National Technical Information Service, Springfield, VA.

Ferris-Kaan, R. 1995. *The ecology of woodland creation*. John Wiley & Sons, Chichester, England.

Foster, D.R., J.D. Aber, J.M. Melillo, R.D. Bowden, and F.A. Bazzaz. 1997. Forest response to disturbance and anthropogenic stress. *BioScience* 47:437–445.

Franklin, J.F., P.M. Frenzen, and F.J. Swanson. 1995. Re-creation of ecosystems at Mount St. Helens: contrasts in artificial and natural approaches. Pages 287–333 in J. Cairns Jr., ed. *Rehabilitating damaged ecosystems, 2nd edition*. Lewis Publishers, Boca Raton, FL.

Friedel, M.H. 1991. Range condition assessment and the concept of thresholds: a viewpoint. *Journal of Range Management* 44:422–426.

Greenberg, R., P. Bichier, A.C. Angon, and R. Reitsma. 1997. Bird populations in shade and sun coffee plantations in central Guatemala. *Conservation Biology* 11:448–459.

Griffith, B., J.M. Scott, J.W. Carpenter, and C. Reed. 1989. Translocation as a species conservation tool: status and strategy. *Science* 245:477–480.

Hammer, D.A. 1997. *Creating freshwater wetlands, 2nd edition*. Lewis Publishers, Boca Raton, FL.

Hannah, L., D. Lohse, C. Hutchinson, J.L. Carr, and A. Lankerani. 1994. A preliminary inventory of human disturbance of world ecosystems. *Ambio* 23:246–250.

Harwell, M.A. 1997. Ecosystem management of south Florida. *BioScience* 47: 499–512.

Higgs, E.S. 1997. What is good ecological restoration? *Conservation Biology* 11: 338–348.

Hobbs, R.J., and D.A. Norton. 1996. Towards a conceptual framework for restoration ecology. *Restoration Ecology* 4:93–110.

Holl, K.D. 1996. The effect of coal surface mine reclamation on diurnal lepidopteran conservation. *Journal of Applied Ecology* 33:225–236.

Holling, C.S., L.H. Gunderson, and C.J. Walters. 1994. The structure and dynamics of the everglades system: guidelines for ecosystem restoration. Pages 741–756 in S.M. Davis, and J.C. Ogden, eds. *Everglades: the ecosystem and its restoration.* St. Lucie Press, Delray Beach, FL.

Jansen, A. 1997. Terrestrial invertebrate community structure as an indicator of the success of a tropical rainforest restoration project. *Restoration Ecology* 5:115–124.

Jordan III, W.R., M.E. Gilpin, and J.D. Aber. 1987. *Restoration ecology. A synthetic approach to ecological research.* Cambridge University Press, Cambridge, England.

Jordan III, W.R., R.L. Peters II, and E.B. Allen. 1988. Ecological restoration as a strategy for conserving biological diversity. *Environmental Management* 12:55–72.

Karr, J.R. 1991. Biological integrity: a long-neglected aspect of water resource management. *Ecological Applications* 1:66–84.

Langis, R., M. Zalejko, and J.B. Zedler. 1991. Nitrogen assessments in a constructed and a natural salt marsh of San Diego Bay. *Ecological Applications* 1:40–51.

Larsen, P. 1996. Restoration of river corridors: German experiences. Pages 124–143 in G. Petts and P. Calow, eds. *River restoration.* Blackwell Science Ltd., Oxford, England.

Lockwood, J.L. 1997. An alternative to succession. *Restoration & Management Notes* 15:45–51.

Loucks, O.L. 1985. Looking for surprise in managing stressed ecosystems. *BioScience* 35:428–432.

Luken, J.O. 1990. *Directing ecological succession.* Chapman & Hall, London, England.

MacMahon, J.A. 1981. Successional processes: comparisons among biomes with special reference to probable roles of and influences on animals. Pages 277–304 in D.C. West, H.H. Shugart, and D.B. Botkin, eds. *Forest succession: concept and application.* Springer-Verlag, New York.

MacMahon, J.A. 1987. Disturbed lands and ecological theory: an essay about a mutualistic association. Pages 221–237 in W.R. Jordan III, M.E. Gilpin and J.D. Aber, eds. *Restoration ecology.* Cambridge University Press, Cambridge, England.

MacMahon, J.A. 1997. Ecological restoration. Pages 479–511 in G.K. Meffe, and C.R. Carroll, eds. *Principles of conservation biology, 2nd edition.* Sinauer Associates, Inc., Sunderland, MA.

Majer, J.D. 1989. *Animals in primary succession. The role of fauna in reclaimed lands.* Cambridge University Press, Cambridge, England.

McClanahan, T.R., and R.W. Wolfe. 1993. Accelerating forest succession in a fragmented landscape: the role of birds and perches. *Conservation Biology* 7:279–288.

Miyawaki, A., and F.B. Golley. 1993. Forest reconstruction as ecological engineering. *Ecological Engineering* 2:333–345.

Mladenoff, D.J., R.G. Haight, T.A. Sickley, and A.P. Wydeven. 1997. Causes and implications of species restoration in altered ecosystems. *BioScience* 47:21–31.

Munshower, F.F. 1994. *Practical handbook of disturbed land revegetation*. Lewis Publishers, Boca Raton, FL.

National Research Council. 1986. Ecological knowledge and environmental problem-solving. National Academy Press, Washington, DC.

National Research Council. 1992. *Restoration of aquatic ecosystems: science, technology, and public policy*. National Academy Press, Washington, DC.

Naveh, Z. 1994. From biodiversity to ecodiversity: a landscape-ecology approach to conservation and restoration. *Restoration Ecology* 2:180–189.

O'Neill, R.V., D.L. De Angelis, J.B. Waide, and T.F.H. Allen. 1986. A hierarchical concept of ecosystems. Princeton University Press, Princeton, NJ.

Packard, S., and C.F. Mutel, eds. 1997. *The tallgrass restoration handbook*. Island Press, Washington, DC.

Perfecto, I., R.A. Rice, R. Greenberg, and M.E. Van der Voort. 1996. Shade coffee: a disappearing refuge for biodiversity. *BioScience* 46:598–608.

Petts, G., and P. Calow, eds. 1996. *River restoration*. Blackwell Science, Oxford, England.

Phillips, D.L., and J.A. MacMahon. 1978. Gradient analysis of a Sonoran desert bajada. *Southwestern Naturalist* 23:669–680.

Pickett, S.T.A., and V.T. Parker. 1994. Avoiding the old pitfalls: opportunities in a new discipline. *Restoration Ecology* 2:75–79.

Primack, R.B. 1996. Lessons from ecological theory: dispersal, establishment, and population structure. Pages 209–233 in D.A. Falk, C.I. Millar, and M. Olwell, eds. *Restoring diversity. Strategies for reintroduction of endangered plants*. Island Press, Washington, DC.

Reichard, S.H., and C.W. Hamilton. 1997. Predicting invasions of woody plants introduced into North America. *Conservation Biology* 11:193–203.

Reith, C.C. 1986. Understanding reclamation with models. Pages 85–107 in C.C. Reith, and L.D. Potter, eds. *Principles & methods of reclamation science*. University of New Mexico Press, Albuquerque, NM.

Robinson, G.R., and S.N. Handel. 1993. Forest restoration on a closed landfill: rapid addition of new species by bird dispersal. *Conservation Biology* 7:271–278.

Ruesink, J.L., I.M. Parker, M.J. Groom, and P.M. Kareiva. 1995. Reducing the risks of nonindigenous species introductions. *BioScience* 45:465–477.

Scatolini, S.R., and J.B. Zedler. 1996. Epibenthic invertebrates of natural and constructed marshes of San Diego Bay. *Wetlands* 16:24–37.

Sprugel, D.G. 1991. Disturbance, equilibrium, and environmental variability: what is "natural" vegetation in a changing environment? *Biological Conservation* 58:1–18.

Stevens, W.K. 1995. *Miracle under the oaks. The revival of nature in America*. Pocket Books, New York.

Toth, L.A. 1995. Principles and guidelines for restoration of river/floodplain ecosystems—Kissimmee River, Florida. Pages 49–73 in J. Cairns Jr., ed. *Rehabilitating damaged ecosystems, 2nd edition*. Lewis Publishers, Boca Raton, FL.

Walters, C.J., and L.H. Gunderson. 1994. A screening of water policy alternatives for ecological restoration in the Everglades. Pages 757–767 in S.M. Davis, and J.C. Ogden, eds. *Everglades: the ecosystem and its restoration*. St. Lucie Press, Delray Beach, FL.

West, N.E. 1993. Biodiversity of rangelands. *Journal of Range Management* 46: 2–13.

Woodside, D., and D. O'Neill. 1995. Mining in relation to habitat loss and reconstruction. Pages 76–93 in R.A. Bradstock, T.D. Auld, D.A. Keith, R.T. Kingsford, D. Lunney, and D.P. Sivertsen, eds. *Conserving biodiversity: threats and solutions.* Surrey Beatty & Sons Pty. Limited, Chipping Norton, NSW, Australia.

Wyant, J.G., R.A. Meganck, and S.H. Ham. 1995. A planning and decision-making framework for ecological restoration. *Environmental Management* 19:789–796.

Zedler, J.B. 1995. Salt marsh restoration: lessons from California. Pages 75–95 in J. Cairns Jr., ed. *Rehabilitating damaged ecosystems, 2nd edition.* Lewis Publishers, Boca Raton, FL.

Zedler, J.B. 1996a. Ecological function and sustainability in created wetlands. Pages 331–342 in D.A. Falk, C.I. Millar, and M. Olwell, eds. *Restoring diversity. Strategies for reintroduction of endangered plants.* Island Press, Washington, DC.

Zedler, J.B. 1996b. *Tidal wetland restoration: a scientific perspective and southern California focus.* California Seat Grant College System, University of California, La Jolla.

Zedler, J.B. 1996c. Ecological issues in wetland mitigation: an introduction to the forum. *Ecological Applications* 6:33–37.

10
Limitations to Intellectual Progress in Ecosystem Science

GENE E. LIKENS

Summary

Are ecosystem scientists poised to make great leaps forward in developing understanding, and in applying it to real-world problems? The challenge is to deal simultaneously with the multidimensional complexity of time and space. An intellectual limitation is the false assumption that there will be simple, all-inclusive answers to complex, ecological questions, whereas the elegance of creative, innovative thought and approach is to discover ways to simplify questions or systems being studied to obtain fundamental, inclusive answers. Ecology must be an integrative science and will flourish from creative integration and synthesis. How are vision and creativity enhanced? Is the rate of breakthroughs in ecosystem science appropriate for a healthy, robust scientific field? Teams and team-building are critical components of successful ecosystem science, but efforts should be made to promote inter-disciplinary teams. There is a continuing need to attract the brightest and best into ecosystem science by utilizing the popularity of the ecosystem concept, the challenge of solving complex problems, and the awareness of how such solutions are of value to humanity. More undergraduate courses, textbooks, and scientific journals focused on ecosystem science are needed, however time limitations have become serious problems for scientists. Being busy is good, but if being busy leads to fragmentation of effort, loss of focus, superficial scholarship, and inability to meet commitments, then being busy represents a major intellectual limitation. The weak connection between ecosystem science and policy in the United States is frustrating, primarily because of fragmentation of approach and implementation. However, many opportunities exist for intellectual progress in ecosystem science.

Introduction

Two major goals of ecology, and especially ecosystem ecology, are to develop an integrated understanding of nature and to provide approaches and guide solutions to environmental problems. It is important to ask the questions: How well is ecosystem science fulfilling these goals? Are we poised in ecosystem science to make great leaps forward in developing understanding, and to apply that understanding to real-world problems? Will we just plod along collecting more information and focusing on the details? Can we synthesize and integrate the massive amounts of ecological information available into significant breakthroughs of understanding within the near future? If not, what are the intellectual barriers? Tackling such questions obviously is daunting, but I agreed to try in this chapter.

Ecosystem science has come a very long way since the early formulations of the concept (Möbius 1886; Forbes 1887; Thienemann and Kieffer 1916; Friederichs 1927; Tansley 1935; see Jax 1998). The developments in the field have ranged from early descriptions, primarily of the structure, and to a lesser extent function, of a diverse variety of ecological systems, to more sophisticated analyses of functions (e.g. food-web analyses, flux and cycling of chemicals, regulation of productivity and other processes, microbial loops), of interactions among structures and functions and among interlinked ecosystems, and of change with time. With such various modern tools and approaches as radioactive and stable isotopes, remote sensing, experimental manipulation of entire ecosystems, comparative analyses and simulation modeling, the ecosystem's "black box" is being opened more and more, and we are learning about the roles of ecosystems in global processes.

The term "ecosystem" grew from formulation (Tansley 1935) to being a household word in only about fifty years. Conceptually, the ecosystem approach has great appeal because the "units of nature" (Tansley 1935), for example, a lake, an island, or a watershed, can be considered holistically, and because results based upon the analysis of an entire ecological system often have obvious, direct, and relevant management applications (e.g. Likens 1992). "Ecosystem" is a convenient word, capturing the linkages and complexity of abiotic and biotic components of these "units of nature." It is also convenient for capturing the common (i.e. the public's) understanding of the concept, as well as the professional components (i.e. the science) of the concept, and the management aspects of landscapes and regions.

As is frequently the case in the development of a field, new ideas were common, new approaches were tried, much information was accumulated, hypotheses refined, and modeling attempted (Golley 1993). The question now remains, however, at this somewhat giddy stage in the development of the field (at least in terms of public attention and expectation): What are the limitations to the intellectual advancement of this area of ecological science, and what are the principal intellectual challenges for the future?

By limitations, I mean barriers or problems that reduce or constrain the intellectual advancement toward developing either understanding or management solutions at the ecosystem level.

At the outset, this question could be turned around to ask: Is the advancement of the field limited by intellectual ability? That is, are we "smart enough" to pose and then tackle the extremely complex and multifaceted questions that are generated at the ecosystem level of organization? How many ecosystem ecologists does it take to screw in a lightbulb (or measure a flux)? However legitimate the question, it will be assumed that we are smart enough, and that question will not be considered in this chapter.

Multiple-Factor Complexity

A major difficulty for ecosystem scientists is the need to analyze simultaneously the multidimensional issues of time and space. Normally, only one or a few components are considered explicitly at any one time by humans, whereas ecosystem structure, function, and change with time result from the interactions of many interconnected factors, including many simultaneous stresses. A major challenge for ecosystem science therefore is to evaluate and understand the integrated effect of these multiple factors through time.

Simulation modeling is an important tool for coping with the complexity of ecosystem-level questions, because in modeling, a scientist is forced to deal with the interactions between components and functions, if not entire disciplines. But evaluation of modeling attempts must be honest. If modeled predictions and reality differ by two- to ten-fold, we are not doing very well.

In my opinion, there are no easy answers when studying ecosystems. I believe we need to acknowledge that fact up front—then get on with embracing the complexity to our advantage (Polis and Strong 1996), and do what Kohm and Franklin (1997b) suggest: "... [look] for new ways of thinking about ... problems" in dealing with or thinking about ecosystems. Kohm and Franklin (1997b) go on to state: "... 21st century forestry will be defined by understanding and managing complexity, providing a wide range of ecological goods and services, and managing across large landscapes." Similarly, Uhl et al. (1997) state that "... environmental problems are many-faceted clusters of problems, and technical information represents only a part of the total information needed ...". Perhaps a major, if not the overriding intellectual limitation, is the false assumption or belief that there will be simple, all-inclusive answers to complex ecological problems. The elegance of creative and innovative thought and approach is to discover ways to simplify questions or the system being studied to obtain fundamental answers. This is the essence of developing breakthroughs. The major breakthroughs of the future will depend on stating and addressing questions in new ways, spending the time (and hard work) to think in new ways, all approached with humility, if not awe, about the complexitity that we face. But the challenge to understand this complexity is what drives us forward in ecosystem science.

Complex Linkages—An Example

As an example of how complex linkages in ecosystem structure and function can become clarified with long-term research and monitoring, how this increased complexity can confuse the public and decision-makers (particularly if vested interests are looking for obfuscations), and how improved public and professional education could help in understanding the importance of these linkages and interconnections in complex multidisciplinary issues, I want to examine briefly the development of the acid rain issue. In my experience, this issue is one of the most complex environmental problems faced by humans because of the various scientific, political, economic, legal, and social issues involved.

In the late 1960s and early 1970s when acid rain was first identified in Europe and North America as a regional problem (Odèn 1968; Likens et al. 1972; Likens and Bormann 1974), the problem was characterized naïvely, primarily as emissions of sulfur dioxide from the industrial combustion of

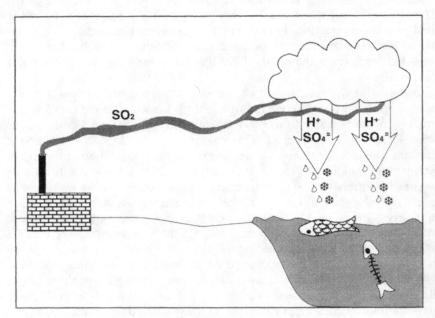

FIGURE 10.1. The "simple" view of ecosystem response to acid rain in late 1960s to early 1970s.

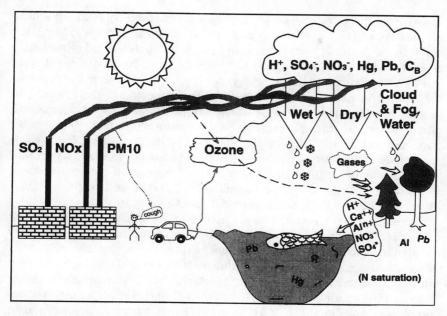

FIGURE 10.2. Some of the structural, functional, and temporal complexity of ecosystem response to acid rain in the 1990s.

coal and oil, which then were oxidized and hydrolyzed in the atmosphere to form sulfuric acid. When this acid fell to the Earth's surface in precipitation (i.e. rain or snow), lakes were acidified, and fish died (Figure 10.1).

With time and further research, it became clear that the acid deposition issue was much more complex (Figure 10.2). Not only was sulfuric acid involved, but acid rain contains a mixture of nitric acid and other chemicals such as calcium and magnesium. In addition to precipitation, inputs of acids via dry deposition as well as cloud- and fogwater also are important, often equalling or exceeding those in precipitation (e.g. Lovett and Lindberg 1984; Lindberg and Garten 1988; Likens et al. 1990; Weathers et al. 1986; Weathers and Likens 1997). Additionally, toxic metals (e.g. Lead and Mercury) and other pollutants, as well as nutrients (e.g. nitrogen and phosphorus), are contained in atmospheric deposition (see Likens 1992).

The effects of these chemicals reverberate throughout the receiving aquatic and terrestrial ecosystems, because biogeochemical cycles are interlinked. For example, there is evidence that methylation of Mercury to its more toxic form in lakes is enhanced through acidification (e.g. Lindberg et al. 1987; Grieb et al. 1990; Bloom et al. 1991; Watras et al. 1991). Toxic aluminum is mobilized as soils are acidified (e.g. Baker and Schofield 1982; Driscoll et al. 1985; Shortle and Smith 1988; Bredemeier and Ulrich 1992), and effects on the nitrogen cycle (e.g. nitrogen saturation in terrestrial ecosystems and nitrate in groundwater) are enhanced by polluted atmospheric deposition (e.g. Vitousek et al. 1997). Base cations (Ca^{2+}, Mg^{2+}, K^+, Na^+) can be depleted from forest soils in large amounts by acid rain, making ecosystems less well-buffered to further atmospheric inputs of acids (Likens et al. 1996). Coniferous tree species generally are more effective at scavenging acidifying gases and particles from the atmosphere than deciduous species (e.g. Harriman et al. 1994). Deleterious interactions between atmospheric ozone and atmospheric acids on vegetation can be large, but variable (e.g. Reich et al. 1988; Lovett and Hubbell 1991). Not surprisingly, some species are much more sensitive than others to various atmospheric pollutants or to chemicals mobilized in the soil, and in general, species responses to these pollutants differ considerably (e.g. Baker and Schofield 1982; Havas and Likens 1985; Schindler 1988; Løkke et al. 1996).

Eutrophication of aquatic ecosystems can be affected negatively by direct effects of atmospheric deposition of acids and toxic metals, or positively through interactions with other biogeochemical cycles (e.g. sulfate control of phosphorus release from lake sediments, Caraco et al. 1991a, b; elevated nitrate in estuaries, Fisher and Oppenheimer 1991). Human health effects from gaseous, acidic aerosols and other small particles are significant, but complicated and variable (e.g. Hanson 1997). Indeed, the complexity shown (Figure 10.2) and referred to in this chapter is only a small part of the overall complexity involved within ecosystems impacted by acid rain; disease, exotic pests, drought, and many other environmental factors must be considered and evaluated as well.

We typically tackle only those pieces of a puzzle with which we can cope, about which we can think productively, and which we hope to solve. This approach is understandable, but the goal for the future in ecosystem science should be to address these problems in a much more holistic and integrated way, again recognizing the complexity involved. Nature (real ecosystems) is not a simple machine (see Merchant 1989), in which simple tinkering will "fix" the machine. Piecemeal "solutions" to such a complex problem as acid rain have not been satisfactory from a regulatory point of view (Likens 1992; Likens et al. 1996; Allen 1996), and do not characterize a considered ecosystem approach or view of the overall problem. Because of the complexity of ecosystems, as well as the complexity of the acid rain issue, politicians and the public who are looking for simple answers have been confused by the diverse response of ecosystems to this widespread problem. Such diversity of response is not unexpected scientifically, however (Likens 1992). Moreover, it was widely assumed that the acid rain problem had been solved as soon as legislation was passed in 1990. Unfortunately, this assumption was not correct. Acid rain continues and ecosystems continue to be damaged by it. We can do a much better job of educating the public about this complexity and variability of response in natural ecosystems.

When the approach to an environmental problem is fragmented (by not using the ecosystem approach), management and policy are also frag-

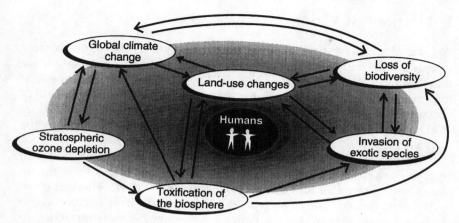

FIGURE 10.3. Interactions among the major components of human-accelerated environmental change (from Likens 1994).

mented (e.g. Wargo 1997). The examples of fragmented regulatory programs are common: acid rain, global climate change, separate and isolated pollution control programs with examples such as the Safe Drinking Water Act, the Clean Air Act, the Toxic Substances Control Act, and so forth. To be more efficient and effective, regulation and management solutions to such problems need to be integrated, based on the ecosystem concept (Train et al. 1993).

Given the human-accelerated environmental changes (Likens 1991) occurring now at all spatial scales throughout the globe and especially their complicated interactions (Figure 10.3), it becomes ever more important to develop a more integrated and complete understanding of ecosystems to guide and to find solutions to these pressing environmental problems (Likens 1994; Vitousek 1994). For example, Schneider (1997) has suggested that one of the most serious global concerns is "the combined or synergistic effects of habitat fragmentation and climate change." Clearly, this complex interaction is important, but it is only part of what is happening globally (Figure 10.3). Individual environmental problems lose prominence as multiple problems interact and as proposed solutions conflict.

If, in 1970, we had started to manage and to educate concerning the acid rain problem more holistically and based upon the ecosystem approach, I believe we would have been much more efficient and effective in developing viable solutions. It is clear to me, however, that there were intellectual, economic, and political limitations that prevented such an approach (Likens 1992). Are we training students with a view that is broad enough to study, manage, and politically handle such problems in the future, yet also provides the expertise for them to do quality, in-depth science?

Limitations of Vision

Are we asking innovative, creative, tough questions to drive our ecosystem research or are we only filling in details? (I want to make it abundantly clear that I am not belittling "filling in details"; I believe that obtaining new information is an extremely important component for advancing the field of ecosystem science). Odum's Second Edition of *Fundamentals of Ecology* (1953) spawned a revolution in ecology by promoting the value of the ecosystem concept. But what will be the next revolution?

In my opinion, an example of both an extremely important and difficult issue in ecosystem science relates to scale and scaling. Is there an optimum size or configuration of ecosystems for ecosystem study? How do we appropriately and realistically scale up or scale down results from one ecosystem study to a larger area or to a smaller one, for example, from plot studies to a region (e.g. Johnson et al. in prep; Levin et al. 1997)? Are results generated for one spatial scale misleading to management when applied to another spatial scale? How can information on processes from a plot be scaled quantitatively to a watershed, to a region, or to the globe?

We may understand process-level interactions for a leaf or a 2×2 m plot very well, but which processes are critical for a watershed-based model? How do we manage the increased heterogeneity and complexity of larger scales? Or conversely, how do we apply large-scale information to a small plot? The world, clearly, is extremely heterogeneous. How do we relate quantitatively the heterogeneity of a leaf (see examples in Hartley and Jones 1997) to the heterogeneity of a landscape? The key is understanding heterogeneity. What are the common themes? How do we identify them? In my opinion, these questions represent a major intellectual challenge. This issue becomes critical as most detailed studies are done at a small scale and often for relatively short periods (e.g. 1 to 5 years), whereas many major environmental problems, for instance, the pollution of air and groundwater, and climate change are regional or larger in area. Therefore, a clear challenge is to interrelate and synthesize results quantitatively among individuals, plot scales, watershed scales and remotely sensed regions.

There also are problems in interpreting different temporal scales. For example, long-term biogeochemical data from the Hubbard Brook Experimental Forest show clearly that short-term data (1 to 5 years) can be quite misleading relative to actual long-term trends (e.g. Likens 1992; Likens and Bormann 1995) that are required for decision-makers and resource managers. Management policies that are based on short-term records can be quite inappropriate or faulty, thus leading to both costly mistakes in regulation and an erosion of confidence in the underlying science. Clearly, when different temporal scales are combined with the spatial scaling problem, understanding and management become exceedingly complex. On the other hand, long-term studies and the long-term data they generate, provide one valuable opportunity to address complexity in eco-systems. Fisher (1994) has suggested that the quantitative integration of scale with pattern and process is especially difficult for ecology, but has also proposed that "Creativity can be learned, fostered, nurtured, and devel-oped" (Fisher 1997). If correct, such efforts should be focused on this problem and pursued vigorously in ecosystem ecology.

What have been the major breakthroughs in ecology, and particularly ecosystem ecology, during the past decade? Do we have an adequate or high rate of breakthroughs in ecosystem science relative to other fields? Is the rate of breakthroughs in ecosystem science appropriate for a healthy, robust scientific field? These are very difficult questions to answer, but one indirect "answer" was provided recently in an analysis done by *Science* magazine during 1996 in which they list no breakthroughs for either ecology or ecosystem ecology for the year (Bloom 1996; News and Editorial Staffs 1996). (This analysis, however, also may represent in part *Science*'s recent bias against publishing ecological papers!).

The question concerning breakthroughs is important because it relates at least indirectly to intellectual limitations for the science. If we are merely filling in details, but not providing intellectual breakthroughs, it could be argued that the field is not advancing rapidly and is becoming stagnant. However, it is extremely difficult, and often subjective, to define what exactly a breakthrough is. It should be pointed out that ecology and ecosystem science usually consider interactions over long time periods (e.g. seasons, years, decades), and as a result it may take many years to identify a breakthrough. In other fields, the identification of a new particle or new gene sequence may be considered a breakthrough. Would becoming more goal-oriented produce more breakthroughs in ecosystem science?

So, what would be an approach for enhancing our ability to generate breakthroughs? One clear possibility is to have better hypothesis formula-tion. Interdisciplinary teams, discussed in a later section, must be important components of modern ecosystem analysis and should be able to generate important, testable hypotheses on the basis of observations, common sense, collective expertise and experience, management considerations, concep-tual models, and simulation predictions. Too often, however, hypotheses are generated by "vote" within teams or dictated by the funding agency, which make no sense scientifically.

We need better approaches within multidisciplinary teams, for consensus-building in hypothesis formulation. These approaches should be based on vision, sound science, and team strength. In this regard, models serve both as an important reality check and as a direct opportunity to work at the interfaces and linkages between disciplines involved in the ecosystem or the problem being considered. Conceptual models and simulation mod-els can be especially helpful in crystallizing what is known and what is not known, and in guiding the formulation of testable hypotheses.

Generation of trendy hypotheses probably will not be useful in advancing the field, whereas considered, thoughtful hypotheses and idea generation will (sensu Fisher 1997). At the same time, there must be an allowance (i.e. funding) for the pursuit of the innovative, offbeat, "risky" idea.

This issue of hypothesis formulation and breakthroughs obviously is very complicated and may be undermined by a variety of factors, including a recent, worrisome trend for shorter, more media-type publications, rather than more comprehensive, monographic considerations. Many present-day journals have decided to accept only shorter papers, primarily for financial and other reasons. Will the Internet exacerbate this problem?

This trend toward shorter papers also relates to the issue of "need to know" vs. "want-to-know" in ecology—and probably in science, in general. Our need-to-know answers to important environmental problems can lead to breakthroughs in understanding and management, but a more cynical view would suggest that the "need-to-know" approach may be driven by other needs, such as a "need" to be promoted, a "need" to have salary increases, a "need" to be famous. This approach contrasts with the more Darwinian or Hutchinsonian "want-to-know" approach, which is based firmly in thoughtful, considered natural history observation, and hypothesis testing, with a longer time frame for maturation of ideas, hypothesis generation, integration, and conclusions.

What signals are we giving to young people in the field in this regard? Are we suggesting that papers should be split into their component parts for publication to add to the number of publications produced per year, or are we suggesting that the more complete, integrated analysis resulting in scholarly monographs is a better way of synthesizing and integrating information at the ecosystem level of analysis (see Wetzel 1992)? I believe that the pendulum has swung too far toward the short paper; that is, there is too much emphasis on how many papers a thesis can be split into for publication, or how many papers can be obtained from a particular study. Are these so-called "least publishable units" even compatible with high standards of scholarship and ethical conduct in science (see Gunsalus 1997)? It seems that there also is a soft correlation between the increase in the number of shorter papers and being too busy (Figure 10.4). Would anyone have wanted 137 separate papers by Darwin rather than one *On the Origin of Species*?

It is my view that ecology must be an integrative science, and that it only flourishes on the basis of the integration and synthesis that is accomplished. So, finding ways to enhance this type of integration, synthesis, and analysis, which lie at the intersection of disciplines, is our overriding goal in analyzing ecological systems, and is the basis for breakthroughs in the science. Hopefully, the recent formation of the National Center for Ecological Analysis and Synthesis (NCEAS) will help to foster a new wave of synthesis in ecology. Overall, breakthroughs in the formulation of clarifying theories will be based on successful integration and synthesis.

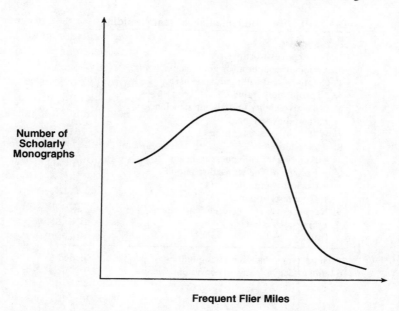

FIGURE 10.4. Proposed relation between the production of scholarly monographs and amount of travel for "busy" ecosystem ecologists.

Team Building

Gorham (1996) has reminded us that few scientists have expertise in more than one ecosystem, that is, "... generalists in ecosystem studies are rare ..." It is uncommon for an ecosystem ecologist to study and have expertise simultaneously in stream ecosystems and lake ecosystems, or even more unusual to have expertise in lake ecosystems and terrestrial ecosystems. As a result, usually there is the need in ecosystem "projects" to put together multidisciplinary teams of scientists to achieve the necessary scope of expertise. The trend toward multidisciplinary, team approaches is common now (if not required) in such large, National Science Foundation-funded projects as Long-Term Ecological Research, Land-Margin Ecosystems Research, and the various efforts focused on problems or sites within the world's oceans.

Focusing such teams of scientists on an ecosystem project or landscape unit is useful, but requires that the team members have special attributes if the team is to be successful. The formation of a research team is not necessarily straightforward. How then should a team be formed? How should it be led? What is the optimum size? Is there a relation between scientific productivity and team size? Little formal attention has been given to the requirements for establishing the most efficient and effective team structure for ecosystem science.

TABLE 10.1 Ten fundamentals of team-building.[1]

1. Brightness.
2. Trusting/trustworthy (trust).
3. Abundant common (or good) sense.
4. Creativity and willingness to share.
5. Appropriately trained.
6. Collective ability to make up deficiencies.
    • Shared experiences.
7. Willing to give team time.
8. Personality.
    • Ability and willingness to listen.
    • Enjoy working with other people.
    • Curiosity and interested.
    • Openness of mind.
9. Keeping eyes open (serendipity reigns).
10. Liking each other.
    [Luck helps!]

1. These types/characteristics/qualities of people are needed to build effective teams in ecosystem science.

Toward this end, I have proposed here (Table 10.1), some fundamental characteristics that are needed for successful team-building, particularly for long-term studies. These characteristics are based primarily on my experience of more than three decades as an ecosystem ecologist of a successful (e.g. Golley 1993; Frost and Blood 1996) ecosystem project, rather than on an analysis of the literature related to group dynamics or social behavior. It is my belief that successful team-building represents one of the most critical, yet difficult challenges for productive ecosystem ecology in the future.

That team members must be "bright" goes without saying, so it is placed in the number one position on the list, but can be set aside in this discussion as was done at the outset of this chapter. But the next two qualities (Table 10.1), trust and common (or "good") sense, are extremely important in my opinion, particularly trust. I believe that trust is at the heart of all successful teams. Trust, combined with a commitment of making adequate time to focus on science, undergird all of the other attributes on the list. How can trust be fostered and enhanced within a group of busy, diverse, often egocentric individuals comprising a team? It is a difficult challenge, but in my experience, open and frequent communication is the most important ingredient for promoting trust. The remaining qualities on the list are all essential for successful team-building because teams are composed of individuals. The goal is that the sum (i.e. the team) is greater than the individual parts (i.e. individual team members).

Teams provide an opportunity to share the excitement of ecosystem research, and to learn and gain understanding from this sharing. Sharing requires openness, responsibility, and especially trust. A collegial decision-making process is critical to successful team function. Bright, aggressive scientists must believe that they have ownership in the decision-making process and will receive fair recognition for their ideas and efforts.

Are we providing appropriate education and training for researchers, particularly students and postdoctoral associates, to improve team-building? What specific measures could be taken to enhance team-building in ecosystem ecology? Some obvious steps include (1) training of team leaders, (2) better mentoring by experienced, senior team members, (3) increased face-to-face communication about team and individual expectations, (4) more opportunities (i.e. time) for significant face-to-face interactions (time management of individuals is a major problem to be faced and solved by teams), (5) more discussions about responsibilities, setting priorities, and openness and trust, (6) listing titles of potential publications and authors, (7) establishing the order of authors in advance of drafting papers (this is a good practice at the beginning of each year), and (8) honing and appreciating common sense, based on experience and commitment, which is a major ingredient for successful serendipity.

The problem of multidisciplinary vs. interdisciplinary approach should also be mentioned. Developing interdisciplinary understanding is our primary goal in ecosystem science, and is the necessary ingredient for integrated and comprehensive management. However, such understanding is rarely achieved when addressing complex ecological or environmental problems. Attempting to think, work, and understand at the true interfaces among disciplines is exceedingly rare, even in the best modeling efforts. At best, we are currently multidisciplinary in approach. How can we improve or change our education and training efforts to foster work at these disciplinary interfaces? Is there an intellectual barrier for some mix of disciplines to achieve this interdisciplinary understanding? For example, multiauthored papers are often necessary and certainly acceptable for ecosystem research, but often frowned upon for population-level efforts. Do such limitations reduce our ability to work at the interfaces?

Recently Pimm (1994), a population ecologist, stated that "Ecosystem ecology for too long has operated in a dream world with few hypotheses and even fewer data." The reasons for such thoughtless attacks on ecosystem science are unclear, except that ideological considerations and competition for financial support must be involved. The schism between population ecologists (Pimm) and ecosystem ecologists has been growing for some time (Likens 1992; Jones and Lawton 1995), although it makes no sense in terms of ecological integration and synthesis. Developing interdisciplinary programs and reducing fragmentation of effort certainly are major challenges for ecosystem science in the future.

Similar characteristics are required for individuals in teams facing complex, natural-resource management problems as summarized in the following statement: "Guiding principles for holistic management perspective include: cooperation, balance, fairness, integration, trust, responsibility, communication and adaptability" (Naiman et al. 1997).

Finally, there is a need to revise and enhance the present-day system of rewards. It is okay to work in teams! There are suggested guidelines for accepting responsibility, rewarding, and giving proper credit to multiple authors of multidisciplinary papers (e.g. White 1997; Knight 1997; Gilson 1997; Gunsalus 1997). There is also a need to reward outstanding teams instead of or in addition to individuals only, while also rewarding strong individual input to teams.

Limitations of Education

It is obvious that we want to attract the "brightest and best" into ecosystem science. But how can this be done, and how do we determine who are the brightest and best? This is another very difficult question that has only fuzzy answers. I conducted a brief analysis of the graduate students applying to Cornell University's graduate field of Ecology and Evolutionary Biology during the last four years. Of the 135 to 170 students that applied each year, some 13% of them indicated that they wanted to pursue graduate work in ecosystem ecology (several, however, had not yet become brainwashed and indicated both population and ecosystem ecology!). The following questions arise from these data: (1) Is this a high percentage? (2) Does this 13% include the very best that have applied to Cornell? (3) Of the graduate students applying to the field of Ecology and Evolutionary Biology, are they the best and brightest, or are the brightest and best going into computer studies or investment banking or some other field in which the potential for jobs and financial gain is greater? (I would assume these questions would apply to all graduate programs in the United States).

Given my stated views about the enormous complexity and the need to work in even deeper aspects of this complexity in the future within ecosystem science, I believe that it is imperative that we attract the brightest and best, that is, the most creative, innovative, and scholarly, to ecosystem science. How do we do this? Certainly we should not rely solely on such current, widely used, standardized tests as the Graduate Record Examination.

Because ecosystem ecology (at least the word, ecosystem) is presently popular, have we taken advantage of this popularity to attract the brightest and best? I would propose that an effort be generated to capitalize on this current popularity and to develop an aggressive program to attract the brightest and best who indeed should be and would be challenged by the complexity of the discipline. This effort could be done by publicizing the challenge of solving complex questions faced in ecosystem science (e.g. in textbooks, highly visible scientific and popular articles) and how solutions to these complex issues would be of value to humanity.

Major limitations in education for ecosystem science are a lack of ecosystem courses at the undergraduate level in United States colleges and universities, lack of comprehensive, up-to-date textbooks in ecosystem sciences, and lack of professional journals. The key to understanding environmental problems still is an in-depth understanding of natural ecosystems. This premise should guide our educational systems in ecosystem ecology. Ehrlich (1992) has called for an end to the training of graduate students who do nothing but basic research. I agree with a need for a broader approach, but maybe then there is too much for graduate students to do. They also must have in-depth expertise to compete in today's world. Are there ways to change the educational system to improve and broaden graduate training in ecosystem science? The answer is clearly "Yes!", but this could be the topic for another Cary Conference. The development of the new, rigorous journal *Ecosystems* (edited by S.R. Carpenter and M. Turner, Springer-Verlag) has the opportunity to rise to this challenge.

Limitations of Funding, Particularly for Large-Scale, Multidisciplinary Projects

There are several approaches to the study of ecosystem ecology, including an empirical or natural-history approach, a balance or budgetary approach, an experimental approach, a comparative approach, and a modeling or computer simulation approach (Likens 1992). Because of the scale and complexity of ecosystems, all are relatively expensive to do. At the present time, funding from various sources, including federal agencies, is extremely competitive. Limitation of funds increases competition, which may be intellectually stimulating to a point, but severe constraints in funding will limit availability of intellectual power and attention, especially for large-team efforts. Additionally, too much time may be spent pursuing dollars instead of ideas. Obviously, excessive time spent in securing funds will also limit the focused intellectual activity on ideas related to research and education.

Experimental manipulation can be a very powerful tool in ecosystem ecology (see Schindler 1973; Likens 1985; Schindler et al. 1985; Lawton 1995; Carpenter et al. 1995; Carpenter, Chapter 12), and particularly when the experimentation is done at the whole-ecosystem level, such as for entire lakes or watersheds. A major limitation for entire-system manipulation, however, relates directly to the large cost in time and money involved in doing such experiments, and also the availability of large ecosystems for such manipulations. As far as I know, there is only one Experimental Lakes Area (ELA), such as that found in Ontario, Canada, where a large variety of lakes is available for experimental manipulation, and adjacent, similar lakes are available for reference to those manipulations. The University of

Notre Dame's Ecological Research Center (UNDERC) in the upper
midwestern portion of the United States provides some opportunities for
this type of ecosystem research. Unfortunately and shortsightedly, funding
in support of the ELA facility was reduced severely last year.

There are a few more examples of watershed-ecosystems available for
experimental manipulations, such as the Hubbard Brook Experimental
Forest in New Hampshire, the Coweeta Hydrologic Laboratory in North
Carolina, and the Andrews Experimental Forest in Oregon, but these systems also are limited in number. For example, at Hubbard Brook there are
only nine gauged watersheds available, of which three already have been
experimentally manipulated. Six of those nine are located on south-facing
slopes, and these include the three that have been manipulated. A planned
manipulation is scheduled for an additional south-facing watershed, leaving
only two for long-term reference. So, at Hubbard Brook, we have essentially exhausted our options, at least in the short term, for experimental
studies on entire watersheds. The cost in time and money to establish and
calibrate a watershed-ecosystem prior to experimental manipulation is very
large.

The availability of such experimental units doesn't represent a direct
intellectual limitation, but certainly does limit the intellectual studies that
can be accomplished and the questions that can be asked, and as a result,
restricts intellectual activity. I strongly would suggest that a concerted effort
be made to find ecosystem units, including urban areas, on public and
private lands throughout the world that could be made available for experimental manipulation. There are possibilities. For example, there are lakes
within the Adirondack Mountain Region of NewYork State, other such
lake districts as in southern Chile, watersheds on lands held by The Nature
Conservancy or paper-producing companies, and areas within national
parks around the world. (There is a great opportunity to educate the public
about the functions of ecosystems through such studies as well). These
areas will become increasingly valuable for these important studies in the
future.

We need to identify, set aside, and make available whole ecosystems that
can be manipulated experimentally to gain a more integrated understanding about how these systems function overall. This represents a kind of
scientific conservation and provides a different point of view from the
current Long Term Ecological Research (LTER) effort of site selection.
This approach could be incorporated as a part of the present-day analysis
being undertaken by the Federal government to find and establish an environmental monitoring system across the United States. Low-impact manipulations would be more palatable for private lands than high-impact
studies, but ecosystem scientists need to be innovative and clever in designing such experiments to be compatible with the scientific need, the resource
available, and the owner's sensitivities.

Valuable experimental manipulation clearly is not limited to large experimental units such as watersheds or lakes. Small-scale, intensively stud-

ied ecosystems and computer simulations of ecosystems have been extremely valuable for developing understanding about ecosystems. Results from larger ecosystems are somewhat easier to relate directly to realistic management units, however.

Limitations of Time and of Public Understanding

Time

Obvious problems related to intellectual limitation (although not necessarily unique to ecosystem ecology) are based on the well-known fact that scientists have become too busy, are traveling too much, are attending too many meetings, and are "needing" to do too many reviews. My own experience (Figure 10.5), with busyness, was to have an increase in scientific productivity with increasing busyness, but then at some very high level of busyness, my scholarship probably declined as I became more fragmented and didn't have time to focus on sustained scholarly efforts. Being busy is good, but if being busy leads to fragmentation of effort, loss of focus, superficial scholarship, and inability to meet commitments, then being busy represents a major intellectual limitation and a real worry for the future of the field of ecosystem science. It is my opinion that we are seeing more and more examples of this bad side of being "too busy". Being too busy has become a commonplace complaint—far more, in my opinion, than forty years ago, and particularly serious for young scientists. Consider the waffle ad on TV, in which the farmer doesn't have time to eat waffles for breakfast

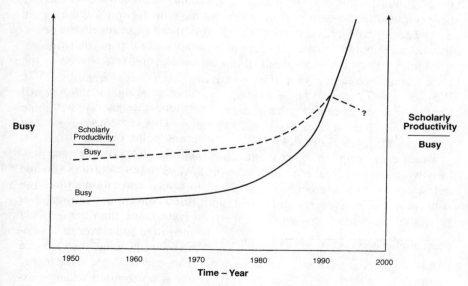

FIGURE 10.5. The Likens's experience with busyness.

so the fast-food waffle is proposed as the solution to being too busy. The ad is intended to be humorous, but frankly I think it is a sad, if not serious, commentary on the present-day situation. Indeed, this "time-bind" problem seems to be common at many levels of our society (Hochschild 1997).

Unfortunately, it appears that much of this busyness in ecosystem science is driven by increasing competition and "need" for funding. Such problems of too little time, too much travel, and too many meetings are clear and will not be discussed further here, except to note again that these problems seem to have been increasing at some exponential rate during the last ten years or so (e.g. see Rigler and Peters 1995), and may be the basis of many intellectual limitations related to focus.

The proliferation of shorter papers, the increase in number of journals and the development of unwieldy conflict-of-interest webs, largely as a result of more and larger teams and other multidisciplinary efforts, all lead to increased demands on time for peer review. Is this escalation a wise use of precious time? Is the quality of peer review decreasing as a result?

Public Understanding of Complex Ecosystem Issues

Problems somewhat unique to ecosystem science and ecology, in general, are related to the popularization of these very words. On the one hand, ecology and ecosystems are concerned with many common human experiences under the general heading, "environment," for instance, weather, habitat adjustment (e.g. too hot, too humid, too cold), pollution, and so forth. Laypersons are quite familiar with these features, yet these issues are the scientific domain of ecologists. It seems to me that there is a serious gap between what ecosystem ecologists do and have professional expertise in, and what the public thinks about ecosystems, about what ecosystem scientists do, and about the value of ecosystem management. This public misunderstanding may be widespread about science in general. By now, the public usually equates ecologist with activist (Likens 1992; Westoby 1997). Westoby (1997) goes so far as to state, ". . . ecology is one of those words that has escaped from its academic cage of definitions, and we are not going to be able to herd it back in again." We need to retain, if not regain, the professional status of our field.

Take ecosystem management, currently a very popular concept, particularly with some Federal bureaucrats in the United States. There are strong detractors to this approach, often generating misinformation about the concept for whatever reason(s). Fitzsimmons (1996), for example, has suggested that the ecosystem concept is:

. . . "unsuitable as operational guides for spatially-based public policies because:

1. location of ecosystems on the landscape is arbitrary, imprecise, and variable over time;

2. ecosystems can be of any size or shape;
3. the location of ecosystem boundaries are usually little more than geographic best guesses and generally do not really exist on the landscape;
4. ecosystem boundaries reflect a very small amount of ecological (and no socioeconomic) information about the areas they enclose;
5. there are an unlimited number of ecosystems and ecosystem patterns associated with the United States or any of its regions;
6. there are no protocols to select a best set of ecosystems to use as the basis for planning or management decisions; and
7. a particular tract is simultaneously in an unlimited number of different ecosystems and criss-crossed by an unlimited number of ecosystem boundaries."

Unfortunately, Fitzsimmons points are incorrect (3,4,5,6), specious (1,2,5,7) or irrelevant (2,5,7). For example, point number 5 is naïve and not correct because a fundamental task for ecosystem science is to find patterns and understand processes that apply to many ecosystems. Fitzsimmons's statements provide a good example (now all too common) of developing a position first (i.e. against ecosystem management) and then developing post hoc arguments in support of this position. Fitzsimmons (1996) even cites me (Likens 1992) as a source when discussing the problems associated with selecting ecosystem boundaries by stating that "Researchers select ecosystem boundaries chiefly for their own convenience since they are not constrained by theoretical or methodological considerations . . .". What I did say (Likens 1992) was, "Ecosystem boundaries are usually determined for the convenience of the investigator rather than on the basis of some known functional discontinuity with an adjacent ecosystem." However, the theoretical and methodological constraints are given in Likens (1992) and elsewhere (e.g. Bormann and Likens 1967; Likens 1975; Bormann and Likens 1979; Likens and Bormann 1985; Wiens et al. 1985). Ecosystem scientists must delineate the boundaries of an ecosystem to make quantitative measurements of inputs and outputs. Indeed, this convenient delineation normally represents a powerful advantage for making quantitative ecosystem analyses, such as mass-balances, and we need to make this point clear.

Regardless of the misrepresentation and hype currently associated with the concept of ecosystem management—the holistic consideration of entire ecosystems and the protection of entire ecosystems are useful goals. For example, evaluation of ecosystem inputs and outputs are extremely useful measures for both defining and reducing the effects of pollution, for the conservation of habitat and species, and for quantitatively evaluating the effects of disturbances such as forestry or human development on watershed-ecosystems, and for intelligently managing landscapes or regions.

At the present time, there is a weak connection between ecosystem science and policy-making and implementation of policy in the United States. It seems to me that this linkage is much better developed in some other countries, for example in Australia. Lack of confidence in the concept of ecosystem management in large segments of the U.S. public, in Congress, and in industry makes this problem even worse. I believe that the ultimate goal of ecosystem ecology is to develop an integrated understanding of ecological systems. There is an urgent need, however, to apply this understanding to real-world, management issues. "Legislation which ignores the biospheric perspective or the complexity of the landscape mosaic is ultimately naïve" (Likens and Bormann 1974). Thus, ecosystem management, stripped of all the hype, can be a powerful approach (e.g. Christensen et al. 1996; Kohm and Franklin 1997a; Yaffee et al. 1996). We need to build effective, adaptive partnerships between scientists and resource managers if we are to be successful (Likens 1989). Ecosystem "services" are real and vital contributors to maintaining a high quality of human life. We need clear descriptions and valuations of these services (Westman 1977; Daily 1997; Costanza et al. 1997).

Challenges for the Future

Overall, the field of ecosystem science has done very well since 1935. But, the challenges for the future of the field are clear (at least to me). The following list sets forth some of these challenges and concerns:

- Find ways to solve complexity through integration and synthesis;
- Identify and focus on central/fundamental problems;
- Set and maintain priorities;
- Develop understanding of linkages, connections, and interfaces;
- Don't run before we walk; theory will come if integration and synthesis are comprehensive and scholarly; systems thinking is a crucial part of the foundation of ecosystem science;
- Learn how to work successfully in interdisciplinary teams;
- Remember to take risks with ideas and approaches;
- Train for the future, not the past;
- Don't overlook urban ecosystems;
- Use complexity to attract the brightest and best to the field;
- Reclaim our science;
- Provide intellectual leadership;
- Exude the highest passion toward finding the integrated answer.

Acknowledgments. Financial support was provided by the Institute of Ecosystem Studies and The Andrew W. Mellon Foundation. I am especially appreciative of ideas, suggestions and reactions provided by I.C. Burke, J.F. Franklin, P.M. Groffman, R.O. Hall, Jr., C.G. Jones, W.K. Laurenroth, M.L. Pace, R. Pulliam, S.L. Tartowski and K.C. Weathers during the preparation of this manuscript. Thoughtful comments on the manuscript were made by F.B. Golley, R.O. Hall, Jr., J.E. Hobbie, M. Levin, S.L. Tartowski, S.F. Tjossem and J.S. Warner.

References

Allen, S. 1996. Acid rain: the problem that won't go away. *The Boston Globe*, December 23, 1996, pp. C1, C4.

Baker, J.L., and C.L. Schofield. 1982. Aluminum toxicity to fish in acidic waters. *Water, Air, and Soil Pollution*. 18:289–309.

Bloom, F.E. 1996. Breakthroughs of the year, 1996. *Science* 274:1987.

Bloom, N.S., C.J. Watras, and J.P. Hurley. 1991. Impact of acidification on the methylmercury cycling of remote seepage lakes. *Water, Air, and Soil Pollution* 56:477–492.

Bormann, F.H., and G.E. Likens. 1967. Nutrient cycling. *Science* 155:424–429.

Bormann, F.H., and G.E. Likens. 1979. *Pattern and process in a forested ecosystem*. Springer-Verlag, New York.

Bredemeier, M., and B. Ulrich. 1992. Input/output-analysis of ions in forest ecosystems. Pages 229–271 in A. Teller, P. Mathy, and J.N.R. Jeffers, eds. *Responses of forest ecosystems to environmental changes*. Commission of the European Communities, Elsevier Applied Science, Essex, UK.

Caraco, N., J.J. Cole, and G.E. Likens. 1991a. Phosphorus release from anoxic sediments: lakes that break the rules. *Verhandlungen Journal Internationale Vereinigung für Theoretische und Angewandte Limnologie* 24(5):2985–2988.

Caraco, N., J.J. Cole, and G.E. Likens. 1991b. A cross-system study of phosphorus release from lake sediments. Pages 241–258. in J.J. Cole, G.M. Lovett, and S.E.G. Findlay eds. *Comparative analyses of ecosystems: patterns, mechanisms, and theories*. Springer-Verlag, New York.

Carpenter, S.R., S.W. Chisholm, C.J. Krebs, D.W. Schindler, and R.F. Wright. 1995. Ecosystem experiments. *Science* 269:324–327.

Christensen, N.L., A.M. Bartuska, J.H. Brown, S. Carpenter, C. D'Antonio, R. Francis, J.F. Franklin et al. 1996. The report of the Ecological Society of America Committee on the scientific basis for ecosystem management. *Ecological Applications* 6(3):665–691.

Costanza, R., R. d'Arge, R. deGroot, S. Farber, M. Grasso, B. Hannon, K. Limburg et al. 1997. The value of the world's ecosystem services and natural capital. *Nature* 387:258–260.

Daily, G.C. ed. 1997. *Nature's services*. Island Press, Washington, DC.

Driscoll, C.T., N. van Breemen, and J. Mulder. 1985. Aluminum chemistry in a forested spodosal. *Soil Science Association American Journal* 49:437–444.

Ehrlich, P.R. 1992. One ecologist's view on the so-called Stanford scandals and social responsibility. *BioScience* 42:702–703.

Fisher, D.C., and M. Oppenheimer. 1991. Atmospheric nitrogen deposition and the Chesapeake Bay Estuary. *Ambio* 20(3):102–108.

Fisher, S.G. 1994. Pattern, process and scale in freshwater systems: some unifying thoughts. Pages 575–591 in P.S. Giller et al. eds. *Aquatic ecology: scale, pattern and process*. Blackwell Scientific Publ., London.

Fisher, S.G. 1997. Creativity, idea generation, and the functional morphology of streams. *Journal of the North American Benthological Society* 16:365–318.

Fitzsimmons, A.K. 1996. Sound policy or smoke and mirrors: does ecosystem management make sense? *Water Resources Bulletin* 32(2):217–227.

Forbes, S.A. 1887. The lake as a microcosm. *Bulletin Peoria (Illinois) Science Association.* Reprinted in *Bulletin of the Illinois Natural History Survey* 15:537–550 (1925).

Friederichs, K. 1927. Grundsätzliches über die Lebenseinheiten höherer Ordnung und den ökogischen Einheitsfaktor. *Naturwissenschaften* 8:153–157, 182–186.

Frost, T.M., and E.R. Blood. 1996. The role of major research centers in the study of inland aquatic ecosystems. Pages 279–288 in *Freshwater ecosystems*. National Academy Press, Washington, DC.

Gilson, M.K. 1997. Responsibility of co-authors. *Science* 275:14.

Golley, F.B. 1993. *A history of the ecosystem concept in ecology*. Yale University Press, New Haven, CT.

Gorham, E. 1996. Linkages among diverse aquatic ecosystems: a neglected field of study. Pages 203–217 in *Freshwater ecosystems*. National Academy Press, Washington, DC.

Grieb, T.M., C.T. Driscoll, S.P. Gloss, C.L. Schofield, G.L. Bowie, and D.B. Porcella. 1990. Factors affecting mercury accumulation in fish in the upper Michigan peninsula. *Environmental Toxicology and Chemistry* 9:919–930.

Gunsalus, C.K. 1997. Ethics: sending out the message. *Science* 276:335.

Hanson, D. 1997. EPA reports continued air quality improvement. *Chemical and Engineering News* 75:22.

Harriman, R., G.E. Likens, H. Hultberg, and C. Neal. 1994. Influence of management practices in catchments on freshwater acidification: afforestation in the United Kingdom and North America. Pages 83–101 in C.E.W. Steinberg, and R.F. Wright eds. *Acidification of Freshwater ecosystems: implications for the future.* Dahlem Konferenzen, Berlin, Germany. John Wiley & Sons Ltd, New York.

Hartley, S.E., and C.G. Jones. 1997. Plant chemistry and herbivory, or why the world is green. Pages 284–324 in M.J. Crawley ed. *Plant ecology, 2nd edition.* Blackwell Science Ltd., Oxford, UK.

Havas, M., and G.E. Likens. 1985. Toxicity of aluminum and hydrogen ions to *Daphnia catawba, Holopedium gibberum, Chaoborus punctipennis* and *Chironomus anthracinus* from Mirror Lake, New Hampshire. *Canadian Journal of Zoology* 63:1114–1119.

Hochschild, A.R. 1997. There's no place like work. *The New York Times Magazine*, April 20, 1997: 51–55, 81, 84.

Jax, K. 1998. Holocoen and ecosystem. On the origin and historical consequences of two concepts. *Journal of the History of Biology* 31:113–142.

Johnson, C.E., C.T. Driscoll, T.G. Siccama, and G.E. Likens. Element fluxes and landscape position in a northern hardwood forest watershed-ecosystem. (In Preparation)

Jones, C.G. and J.H. Lawton eds. 1995. *Linking Species & Ecosystems.* Chapman & Hall, New York.

Kohm, K.A., and J. Franklin eds. 1997a. *Creating a forestry for the 21st century.* Island Press, Washington, DC.

Kohm, K.A., and J.F. Franklin. 1997b. Introduction. Pages 1–5 in K.A. Kohm, and J.F. Franklin eds. *Creating a forestry for the 21st century.* Island Press, Washington, DC.

Knight, J. 1997. Multiple authorship. *Science* 275:13.

Lawton, J.H. 1995. Ecological experiments with model systems. *Science* 269:328–331.

Levin, S.A., B. Grenfell, A. Hastings, and A.S. Perelson. 1997. Mathematical and computational challenges in population biology and ecosystems science. *Science* 275:334–343.

Likens, G.E. 1975. Nutrient flux and cycling in freshwater ecosystems. Pages 314–348 in F.G. Howell, J.B. Gentry, and M.H. Smith eds. *Mineral cycling in southeastern ecosystems.* ERDA Symp. Series CONF-740513. May 1974. Augusta, GA.

Likens, G.E. ed. 1985. *An ecosystem approach to aquatic ecology: Mirror Lake and its environment.* Springer-Verlag, New York.

Likens, G.E. ed. 1989. *Long-term studies in ecology. Approaches and alternatives.* Springer-Verlag, New York.

Likens, G.E. 1991. Human-accelerated environmental change. *BioScience* 41(3): 130.

Likens, G.E. 1992. *The ecosystem approach: its use and abuse. Excellence in ecology, book 3.* Ecology Institute, Oldendorf/Luhe, Germany.

Likens, G.E. 1994. *Human-accelerated environmental change: an ecologist's view.* 1994 Australia Prize Winner Presentation. Murdoch University, Perth, Australia.

Likens, G.E., and F.H. Bormann. 1974. Linkages between terrestrial and aquatic ecosystems. *BioScience* 24(8):447–456.

Likens, G.E., and F.H. Bormann. 1985. An ecosystem approach. Pages 1–8. in G.E. Likens ed. *An ecosystem approach to aquatic ecology: Mirror Lake and its environment.* Springer-Verlag, New York.

Likens, G.E., and F.H. Bormann. 1995. *Biogeochemistry of a forested ecosystem, 2nd edition.* Springer-Verlag, New York.

Likens, G.E., F.H. Bormann, L.O. Hedin, C.T. Driscoll, and J.S. Eaton. 1990. Dry deposition of sulfur: a 23-yr record for the Hubbard Brook Forest Ecosystem. *Tellus* 42B:319–329.

Likens, G.E., F.H. Bormann, and N.M. Johnson. 1972. Acid rain. *Environment* 14(2):33–40.

Likens, G.E., C.T. Driscoll, and D.C. Buso. 1996. Long-term effects of acid rain: response and recovery of a forest ecosystem. *Science* 272:244–246.

Lindberg, S.E., and C.T. Garten, Jr. 1988. Sources of sulphur in forest canopy throughfall. *Nature* 336:148–151.

Lindberg, S., P.M. Stokes, and E. Goldberg. 1987. Group Report: Mercury. Chapter 2. Pages 17–33 in T.C. Hutchinson, and K.M. Meema eds. *Lead, mercury, cadmium and arsenic in the environment.* SCOPE, John Wiley & Sons, Chichester, U.K.

Løkke, H., J. Bak, U. Falkengren-Grerup, R.D. Finlay, H. Ilvisniemi, P. Holm Nygaard, and M. Starr. 1996. Critical loads of acidic deposition for forest soils: is the current approach adequate? *Ambio* 25(8):510–516.

Lovett, G.M., and J.G. Hubbell. 1991. Effects of ozone and acid mist on foliar leaching from eastern white pine and sugar maple. *Canadian Journal of Forestry* 21:794–802.

Lovett, G.M., and S.E. Lindberg. 1984. Dry deposition and canopy exchange in a mixed oak forest as determined by analysis of throughfall. *Journal of Applied Ecology* 21:1013–1027.

Merchant, C. 1989. *Ecological revolutions. Nature, gender and science in New England.* University of North Carolina Press, Chapel Hill.

270   G.E. Likens

Möbius, K.A. 1886. Die bildung, geltung und bezeichnung der artbegriffe und ihr verhältnis zur abstammungslehre. *Zoologische Jahrbücher. Abteilung für Systematik, Okologie und Geographie der Tiere* 1:247.

Naiman, R.J., P.A. Bisson, R.G. Lee, and M.G. Turner. 1997. Approaches to management at the watershed scale. Pages 239–253 in K.A. Kohm, and J. Franklin eds. *Creating a forestry for the 21st century.* Island Press, Washington, DC.

News and Editorial Staffs. 1996. Breakthrough of the year. *Science* 274:1988–1991.

Odèn, S. 1968. *The acidification of air and precipitation and its consequences on the natural environment.* Bulletin 1, Swedish National Science Research Council, Ecology Committee.

Odum, E.P. 1953. *Fundamentals of ecology. 2nd Edition.* W.B. Saunders, Philadelphia, PA.

Pimm, S. 1994. An American tale. *Nature* 370:188–189.

Polis, G.A., and D.R. Strong. 1996. Food web complexity and community dynamics. *American Naturalist* 147(5):813–846.

Reich, P.B., A.W. Schoettle, H.F. Stroo, and R.G. Amundson. 1988. Effects of ozone and acid rain on white pine (*Pinus strobus*) seedlings grown in fine soils. II. Nutrient relations. *Canadian Journal of Botany* 66:1517–1531.

Rigler, F.H., and R.H. Peters. 1995. *Science and limnology. Excellence in Ecology, Book 6.* Ecology Institute, Oldendorf/Luhe, Germany.

Schindler, D.W. 1973. Experimental approaches to limnology—an overview. *Journal of the Fisheries Research Board of Canada* 30:1409–1413.

Schindler, D.W. 1988. Effects of acid rain on freshwater ecosystems. *Science* 239:149–157.

Schindler, D.W., K.H. Mills, D.F. Malley, D.D. Findlay, J.A. Shearer, I.J. Davies, M.A. Turner et al. 1985. Long-term ecosystem stress: the effects of years of experimental acidification on a small lake. *Science* 228:1395–1401.

Schneider, S.H. 1997. *Laboratory Earth—The planetary gamble we can't afford to lose.* Basic Books, New York.

Shortle, W.C., and K.C. Smith. 1988. Aluminum-induced calcium defining syndrome in declining red spruce. *Science* 240:1017–1018.

Tansley, A.G. 1935. The use and abuse of vegetational concepts and terms. *Ecology* 16:284–307.

Thienemann, A. and J.J. Kieffer. 1916. Schwedische chironomiden. *Archiv für Hydrobiologie und Planktonkunde.* 2:483–554.

Train, R.E., P.A.A. Berle, J.E. Bryson, A.W. Clausen, D.M. Costle, M.M. Kunin, G.E. Likens et al. 1993. *Choosing a sustainable future. The report of the National Commission on the Environment.* Island Press, Washington, DC.

Uhl, C., P. Barreto, A. Veríssimo, E. Vidal, P. Amaral, A.C. Barros, C. Souza, Jr. et al. 1997. Natural resource management in the Brazilian Amazon. *BioScience* 47(3):160–168.

Vitousek, P.M. 1994. Beyond global warming: ecology and global change. *Ecology* 75(7):1861–1876.

Vitousek, P.M., J.D. Aber, R.W. Howarth, G.E. Likens, P.A. Matson, D.W. Schindler et al. 1997. Human alteration of the global nitrogen cycle: sources and consequences. *Ecological Applications* 7:737–750.

Wargo, J. 1997. *Our children's toxic legacy: how science and law fail to protect us from pesticides.* Yale University Press, New Haven, CT.

Watras, C.J., N.S. Bloom, W.F. Fitzgerald, J.P. Hurley, D.P. Krabbenhoft, R.G. Rada, and J.G. Wiener. 1991. Mercury in temperate lakes: a mechanistic field study. *International Association of Theoretical and Applied Limnology* 24(4): 2199.

Weathers, K.C., and G.E. Likens. 1997. Clouds in southern Chile: an important source of nitrogen to nitrogen-limited ecosystems? *Environmental Science and Technology* 31(1):210–213.

Weathers, K.C., G.E. Likens, F.H. Bormann, J.S. Eaton, W.B. Bowden, J. Andersen, D.A. Cass et al. 1986. A regional acidic cloud/fog water event in the eastern United States. *Nature* 319:657–659.

Westman, W.E. 1977. How much are nature's services worth? *Science* 197:960–964.

Westoby, M. 1997. What does "ecology" mean? *Trends in Ecology and Evolution* 12(4):166.

Wetzel, R.G. 1992. Professional altruism: remarks by R.G. Wetzel. *Limnology and Oceanography* 37(8):1835–1836.

White, B. 1997. Multiple authorship. *Science* 275:13.

Wiens, J.A., C.S. Crawford, and J.R. Gosz. 1985. Boundary dynamics: a conceptual framework for studying landscape ecosystems. *Oikos* 45:421–427.

Yaffee, S.L., A.F. Phillips, I.C. Frentz, P.W. Hardy, S.M. Malecki, and B.E. Thorpe. 1996. *Ecosystem management in the United States. An assessment of current experience.* Island Press, Washington, DC.

11
Improving Links Between Ecosystem Scientists and Managers

CARL J. WALTERS

Summary

Misunderstandings between ecosystem scientists and managers arise from several sources, ranging from confusion of science with ethics to misplaced faith that process research can help resolve key uncertainties about policy impacts. Both scientists and managers routinely confuse questions about what will happen with questions about what should happen. Managers expect too much of scientists, and scientists often promise far more than they can deliver. There are also serious institutional barriers to cooperation, arising from reward systems for basic science and from the organization and reward systems of management institutions. Scientists are generally not rewarded quickly enough to sustain their interest in the large spatial scales and long time scales in which important management policy impacts unfold, and management institutions are also not structured to deal with these scales effectively. The following four tactics are suggested for helping to overcome these misunderstandings: (1) development of improved communication through shared development of policy models and research programs; (2) active advocacy for legislative change needed to reduce institutional barriers; (3) development of partnerships for knowledge that link science and management and extend to broader stakeholder groups as well; and (4) development of case studies to determine the effectiveness of ecosystem approaches to management.

Introduction

Most textbooks in applied ecology have at least some admonition about the need to maintain an ecosystem perspective in the design of management policies. Fisheries texts urge consideration of food-chain relationships and "environmental variation" as key determinants of productivity. Wildlife texts emphasize that managed populations require ecosystem support services (appropriate habitat structure), and that maintenance of ecosystem habitat structure is the first and primary requisite for successful management. Watershed-management texts emphasize the role of vegetation communities and human disturbances of these communities as key determinants of hydrologic processes ranging from interception to stream-channel development. Ecosystem management is therefore not a new idea, and many practicing natural-resource managers would be offended to think that they are doing anything else, or are somehow promoting policies that are inimical to ecosystem function and sustainability. But we repeatedly see management policies that are obviously risky from an ecosystem perspective, for example, myopic and species-specific fisheries harvest plans and forest-logging practices that cause major changes in hydrologic regimes. We cannot explain such policies as being driven purely by economic pressures against the advice of professional managers and scientists; there are government and academic defenders (or at least apologists) for all of them. It appears that there is a communication problem: those who see a need for better integration of policy at the ecosystem level, or at least a change in attitudes about how policies should be screened in terms of impact on ecosystem structure and function, are somehow not getting their message across to practicing managers.

This chapter reviews some of the barriers to communication between ecosystem scientists and managers. It suggests that the barriers arise in part from misunderstandings at the level of individual people, and in part from the structure of the larger institutions within which we work. It further suggests a series of tactics that have been helpful in breaking through the barriers so as to produce both better ecosystem management and better ecosystem science.

Sources of Misunderstanding Between Scientists and Managers

There are now many case studies in which scientists have tried to work with managers to develop ecosystem-scale policies. Generally these cooperative efforts have been driven either by frustration with the efficacy of such existing policies as the failure of water management tactics to restore bird populations in the Florida Everglades (Davis and Ogden 1994), or by deliberate attempts to use such techniques as Adaptive Environmental Assessment Modeling (AEAM) workshops to promote shared visions of management by having scientists and managers work together to develop management models (Holling 1978; Walters 1986). Generally these case studies have not been particularly successful, and it is my experience there are at least five reasons why substantive collaboration has been difficult to achieve. These five reasons are illustrated in the next sections.

Confusion of Science and Ethics

The essence of management is making choices among policy alternatives; a manager who simply observes without intruding, or mindlessly defends a status quo regime, is an observer rather than a manager. The logic of making choices necessarily involves two components: predictions about the relative impacts of alternative choices (i.e., a manager cannot avoid prediction), and a weighting of these impacts in terms of value or utility. This logic defines a sensible and objective role for science in the management-decision process: to help find better models and experimental designs for use in the prediction part of policy comparison. Scientists may also provide useful experience and creative ideas in the identification of policy alternatives, and in defining appropriate utility measures for policy comparison. It also defines a clear role for natural resource managers: to orchestrate the identification and comparison of policy alternatives, generally in which the final choice will involve a political process of debate and negotiation among stakeholders (e.g. resource users, public interest groups, and so forth).

Unfortunately, scientists and natural-resource managers rarely enter management policy analysis and debate with anything like a dispassionate and objective attitude. We are usually strong personal advocates of conservation, maintenance of natural community structure and diversity, and avoidance of unnecessary ecological risk. When we enter policy analysis with these attitudes, we make exactly the same mistakes as a scientist who sets out to prove a hypothesis: we distort evidence, fail to consider alternative hypotheses and value measures, and seek to avoid even the discussion of alternatives that are inimical to our values. In approaching policy analysis in these ways, we become no different from any other policy or value advocate: neither our scientific advice or our ethical judgment can be trusted nor should it be given particular weight relative to the views of nonscientific stakeholders in management decision processes.

There is nothing wrong with being conservation-minded or risk-averse. But trouble arises when we allow these attitudes to color our intuitive judgments about those policy alternatives that should be considered, or about the consequences of these alternatives. This has been a particular problem in debates about such relatively risky and ecologically disruptive policies as clear-cut logging and intense fisheries exploitation, in which human activities are apt to substantially simplify ecosystem structure and function. In these situations, it is common to hear both scientists and managers make pronouncements about ecosystem "collapse" and the importance of maintaining "natural community structure and function." We cross the line between science and ethical advocacy, and lose our credibility as experts about ecology, when we speak of these risks as facts rather than highly uncertain possibilities, and when we openly advocate a risk-averse policy without admission that acceptance of high risk is a perfectly valid ethical position for at least some resource stakeholders to adopt.

Extreme risk-aversion has, in fact, worked against development of the large-scale field experiments that are needed in many settings to provide tests of predictions from ecosystem science about the efficacy of integrated policy options. When we perhaps should be deliberately "sacrificing" parts of the ecological landscape as vivid demonstrations and warnings of the need for better management systems, we instead are watching much larger landscapes deteriorate slowly toward the same end-points as we argue without convincing evidence to slow the overall change.

Mistaken Belief That Mechanistic Studies Can Be Successfully Integrated for Overall Understanding

It is a frightening and humbling experience to try to construct an explicit model that integrates "mechanistic" biophysical understanding into a framework for making predictions about the consequences of specific policy alternatives. Such exercises in synthesis of mechanistic understanding have revealed gross gaps in information, particularly about processes (e.g. dispersal) whose impacts are only clearly evident on large space-time scales. There are some very well-trodden scientific paths (we know a great deal about such processes as bioenergetics of fish growth, nesting behavior of birds, and water circulation in coastal lagoons), which generally involve mechanisms that are easy for scientists to study. Therefore, in coastal marine systems we can now do a good job at modeling basic hydrodynamics and primary production processes. But when we step beyond these easy paths, particularly into modeling dynamics at higher trophic levels and over large landscapes, the mechanistic research fails us entirely. Scientists have not, and probably never will, study many of the difficult processes that are key to management prediction but are too expensive or time-consuming for attack by traditional scientific methods of investigation.

Unfortunately, most scientists and managers have not been forced into situations that require systematic integration of understanding (i.e., ecosystem modeling), and so have no personal experience with just how gross the gaps in "mechanistic understanding" really are and how difficult it would be to fill these gaps through field studies. So the silly belief persists that good science will ultimately provide the information needed for effective management, and that managers can purchase this information just by contracting scientists to gather more data. This is great for scientists (who get more research funding) and for managers (who get thick reports filled with charts and tables to impress the politicians), but the end result is very much like fiddling while Rome burns: the gaps are not being filled, the models are not being better constructed, and the really valuable information is coming not from mechanistic science but rather from hard experience when policies are actually implemented and fail miserably.

Mistaken Belief That Process Understanding Will Untangle Confounded Effects of Multiple Changes in System Inputs Over Time

In many ecosystem-management settings, bitter debate has arisen over the relative importance of several natural and anthropogenic changes that have occurred simultaneously. Scientists working on the Alaskan oil spills of the 1980s argue about whether salmon and marine mammal populations would have changed enormously anyway resulting from the influences of fishing and climate change. Practically every fisheries collapse has been blamed on environmental changes as well as overfishing (Walters and Collie 1988). In the Laurentian Great Lakes, changes in fishing practices coincided with increases in pollution inputs and with invasion of exotic species like the sea lamprey (Milliman et al. 1987). In the Florida Everglades, water diversions have been accompanied by changes in water quality (nutrient loading), in natural rainfall patterns, and in shrinkage of overall ecosystem size as a result of agricultural and urban development (Walters et al. 1992). Such confounding of effects has made it practically impossible to learn much from historical management experience, and has provided excuse for policy failure (and hence, the continued use of the same policies).

The obvious scientific answer to confounding of effects is to set up planned experimental comparisons in which one or more inputs are deliberately manipulated to provide as much contrast in input circumstances as is practical (given that ecosystems are open to various uncontrollable changes). But instead of admitting that this is ultimately the only scientific way to resolve debates about causality, often scientists and managers pretend that they can substitute mechanistic (i.e. process) studies of particular inputs for real experimental manipulation. Fisheries oceanographers claim to be able to understand how changes in marine environmental regimes have and will influence fish populations. Avian ecologists claim to be able to study how water-regime changes in the Florida Everglades have influenced nesting success of wading birds, and hence to separate water-regime effects from effects of such factors as eutrophication from agricultural runoff.

Process studies can be valuable in helping to narrow the possible explanations for historical changes and policy failures. But it is a serious and wasteful mistake to assume that such studies are *sufficient* to sort out causality in ecosystems that have been subject to multiple changes in inputs. Unfortunately, there are plenty of scientists who are unscrupulous enough to make such claims as a way to secure research funding, and plenty of managers who are naïve or unscrupulous enough to think that spending money on basic research is a way to avoid the difficult choices and risks associated with sound experimental management.

Faith in the Existence of "Magic Bullets": Single Limiting Factors and Technological Solutions

It is extremely unfortunate that terminology and attitudes from medical research have crept into the analysis of ecosystems. We use terms such as ecosystem "health," and we often seek answers to ecosystem management problems in such simplistic goals as "restoration of ecosystem function" as though a disturbed ecosystem were an unhealthy organism. Forest managers use such words as "prescription" as though they were medical doctors tending to routine patient needs. These terms and attitudes invite the hope that we will have the same success as medical researchers at finding "magic bullets" to restore or repair particular physiological functions that are limiting organismal health, and further invite a dependence on technology for finding ways around particular limiting factors. Consequently, giant fish hatcheries in the Pacific Northwest continue to spew hundreds of millions of juvenile chinook and coho salmon into the coastal ecosystem, in spite of very clear evidence that adding these fish has done nothing to increase total abundance available for harvesting (Emlen et al. 1990; Walters 1994). Water managers in the Florida Everglades continue to expect that manipulation of water releases through control structures will provide the "natural" flooding patterns needed for successful nesting by wading birds. Forest entomologists continue to seek better but more environmentally "friendly" pesticides and biological control agents (Clark et al. 1979).

The search for magic bullets and technological solutions plays directly into the hands of scientists who want to study particular processes and relationships, and provides managers with politically visible and easily understood opportunities for investment of management funds and humanpower. Over time, the providers of magic bullets can become a dominant component of management institutions, so that much of a management institution's resources may be directed to simply maintaining the bureaucratic apparatus and people needed to keep applying the bullet. People employed in or emotionally committed to such institutional functions have little or no incentive to be honest about the efficacy of what they are doing. To such people, ecosystem management is not only unnecessary, it is a direct threat to their well being.

Arrogance About Proper "Scientific" Standards for Measurement

As we have started to see the need for large-scale experimentation in ecosystem management, we have also been forced to admit the need for massive increases in information gathering, routine monitoring, and local enforcement of policies and regulations. For example, we recently completed a design study for an experiment to measure the effects of fishing on the Great Barrier Reef, Australia (Walters and Sainsbury 1991). This ecosystem contains some 2,000 reef structures that can be used as experimental and management units, and we proposed a factorial experimental design of reef openings and closures to fishing that would involve monitoring fish community responses on up to sixty-four reefs each year for at least a decade. Were this experiment to be conducted entirely by Australian government agencies and scientists, the annual cost for even simple fish-community surveys (i.e. visual counts, tagging experiments, and so on) would easily exceed the total present budget of the Great Barrier Reef Marine Park Authority (the central management agency for that ecosystem). Clearly, we are only going to be able to conduct management experiments of such scale if we are willing to reach far beyond traditional information-gathering methods and people. Our recommendation in this case was to enlist the assistance of people who are already using the system regularly (i.e. recreational fishing and diving clubs, marine tour operators, and commercial fishermen) to do as much of the local survey work as possible, under supervision from trained government scientists and consultants.

Although the idea of reaching out to ecosystem stakeholders for help in conducting experiments and gathering information has much intuitive appeal, it is treated with scorn by many "professional" scientists and managers. These people seem to think that it is necessary to have a university degree to take an oxygen sample, correctly identify fish along a transect, map vegetation types, or record behavioral observations in a clear and systematic way. This is the worst kind of arrogance and deception. It is clear that the bulk of observations that ecologists make in the field could be done equally well by high school students with a bit of training and supervision. It is in the design and overall conduct of sound sampling and observation systems (and most professional biologists are not very good at this in the first place) that professional guidance and ongoing supervision are needed.

Institutional Barriers to Cooperation

There has been much discussion about institutional barriers to more effective ecosystem management (Gunderson et al. 1995). These barriers are many and complex, ranging from incentive systems for scientists and managers to bureaucratic "inertia" (i.e. standard operating procedures, Allison 1971) to lack of appropriate legal mandates. The following sections summarize a few of these barriers that are particularly important for communication and cooperation between scientists and managers.

Reward Systems for Ecosystem Scientists

Important ecosystem responses to management often only unfold on time scales of decades, particularly for terrestrial vegetation communities, long-lived animal species, and nutrient-cycling processes. Sometimes we can anticipate these changes by conducting comparative studies on systems that were disturbed in "similar" ways in the past, but such comparative studies are of limited utility; often the disturbance options available today have no immediate historical counterpart in terms of scale and extent of change, and almost always the historical record has confounded changes in multiple factors. In short, ecosystems often do not respond on scales that are convenient for scientific study. Scientists can avoid this problem by choosing systems and questions that are amenable to short-term study, but managers do not have this luxury.

As noted before, there are great and persistent gaps in ecological process research, centered on processes that are too large in spatial scale or too slow for scientists to study profitably in terms of the usual academic reward systems. There are a few brave scientists who have accepted reduced publication rates and other academic hazards in order to devote significant parts of their careers to long-term studies (e.g., people who have developed the Long Term Ecological Research [LTER] studies). But most often, only established scientists have been able to afford to take such risks, and it has been difficult or impossible for these people to establish and defend institutional arrangements for long-term study that will extend beyond their own careers. The youngest, brightest scientists are understandably opting for such research areas as evolutionary population biology in which intellectual and academic rewards are quickest to come. This lament is of course not new to ecologists; it has been a major concern in our discipline for decades.

Reward Systems for Managers

Myopic research priorities are not unique to scientists. Few of our management institutions have the budget or personnel structure to reward long-term, patient work, or to coordinate large-scale studies involving cross-institutional cooperation. The government or international agency resource manager who tries to develop a long-term program is apt to be viewed as a slacker who is not paying attention to immediate pressures on the agency (i.e. is not being a good fire fighter), or as a pseudoacademic visionary who should seek a university position somewhere.

Management institutions reward people who are good administrative organizers ("team players") and who can make the organization look good to political funding sources and review systems. The net effect of this reward system is that most management agencies have an "inverse pyramid of competence" from a scientific and resource-management perspective: those people who are most honest about what is really happening in the field, and who work the hardest at actually doing effective resource management (e.g. gathering data, formulating hard policy choices, and so forth), end up with the slowest professional advancement. The inverse pyramid is self-reinforcing, because the honest and conscientious people are not only pesky, but are also a direct threat (by the example they set) to people who have moved up through the system rapidly by employing other talents. As with problems with academic reward systems, these institutional pathologies are something about which we are all acutely embarrassed but have been largely powerless to change.

Legal Mandates for Integration Versus the Defense of Management Practice

In a number of recent case situations, for instance, the Florida Everglades, the Columbia River Basin, and forests in the Pacific Northwest, legal mandates have been developed to try and force cross-institutional cooperation and ecosystem approaches to management. Such arrangements have been developed in the hope that fragmentary management programs have been the result of fragmentary agency mandates and goals, which, in principle, can be corrected by such devices as creation of new management authorities or bureaus with appropriately broad mandates.

By and large, it appears that integration on paper and integration in practice are very different entities indeed. Integrative efforts have been relatively successful (at least at getting integrated management plans on paper) in situations in which there is a clear win-win opportunity for all government and academic cooperators (i.e. more money, more humanpower, more recognition for everyone). But not all management situations are like this. In particular, it has often resulted that upon analysis, one or more of the original agency mandates or practices is in fact ineffective or actively causing damage (e.g. some fish hatcheries for pacific salmon, and some protective policies for endangered species in the Florida Everglades); in these cases, a wise ecosystem-management plan would involve a win-lose combination of bureaucratic changes, with a fair number of people either losing their jobs or having to develop new expertise and experience. In these cases, it is not surprising that effective integrative policies have been virtually impossible to implement. People who are threatened by proposed changes find many ways to thwart the changes, through such tactics as misinformation and direct appeal for public or political support. Misinformation campaigns and direct appeals for public support have been particularly common tactics for people who have built careers upon various technological "fixes" to management problems. Probably the best example of such a mess is the Columbia River Basin (Lee 1993).

Tactics to Improve Ecosystem Science for Management

There are at least four steps that can be taken to break down some of the barriers to more effective use of ecosystem science in management. These involve improved communication, advocacy for legislation and regulatory change, development of partnerships for knowledge, and documentation of case studies demonstrating successful changes in management as the result of taking ecosystem perspectives. The next sections review these steps.

Employ Tactics to Improve Understanding Between Scientists and Managers

Two types of activities have established at least modest track records for improving communication and mutual understanding. First, formal policy design and analysis exercises involving computer modelling, for example, the AEAM modeling workshop process (Holling 1978; Walters 1986) have helped to clarify and redefine management problems and options. Second, "role reversal" situations, in which scientists have taken management positions and vice versa, with all the attendant responsibilities, accountability, and institutional baggage that such role changes imply, have produced a fairly large community of people who understand both perspectives and can help with problems of communication and differences in priorities and objectives.

The most critical step in the processes for improving understanding is to discard the naïve scientific presumption that "good science" should be the starting point for "good management". This is a deadly and wasteful intellectual trap into which scientists repeatedly fall. Operationally, this trap results in: (1) development of models based on variables and factors that scientists find interesting and can say something about, rather than models based on deliberate and focused representation of policy options and value measures identified by managers; (2) expensive field research programs that are only vaguely related to policy questions, and hence have gross gaps when it comes time to use the research results for policy analysis; and (3) design of policies that have fatal flaws when applied in the field because of insufficient attention to the details and dynamics associated specifically with the field implementation. For example, fisheries-assessment scientists often develop policy models to predict overall catches that should be taken, but without careful attention to whether these catches can actually be achieved through such available regulatory tactics as individual quotas or closed seasons; understanding how these regulations will work or fail involves understanding the dynamics of how fishermen respond to regulation. These three mistakes are really just symptoms of a general point that systems analysts have been making for many years: natural systems are complex with many response scales and variables, and therefore the variety of predictions and models that can be developed is practically endless. If you do not start analysis (and communication) with very specific objectives in terms of which "outputs" (i.e. value measures) that are to be predicted from which "inputs" (i.e. policy options), the result will almost certainly be a model that is not only incomplete, but does not even address the correct questions.

When communication between scientists and managers begins with analysis of policy options and value measures rather than scientific questions, resulting policy models and field research programs are apt to be drastically different from what scientists would produce if left alone to look at "interesting" relationships. In particular, the result is generally a much greater focus on questions related to precisely how policies are implemented, and how stakeholders who cause ecological impact will respond to the implementation (i.e., on how to effectively regulate people as well as ecological function). These questions are not those that have been interesting to ecologists (they are often not biology at all, or they concern details of local biology and animal behavior that are of little general scientific interest). Some scientifically interesting new questions for applied ecological research have been uncovered through cooperation between scientists and managers, but any scientist who seriously tries to work closely with managers must be prepared to spend considerable time working on "tactical" questions and relationships that are not going to result in papers for traditional ecological journals.

Advocate Legislative and Regulatory Changes to Remove Barriers to Effective Management

There is no lack of interest in better management through science and broadening of the set of management instruments or tools used in the field. But very often, legislative restrictions or regulatory conventions simply do not provide a mandate or authority for managers to use new tools, to regulate human activities at scales (in both space and time) appropriate to the ecological dynamics being regulated, and to define management objectives or priorities more broadly. At times, such restrictions are so bizarre and unreasonable that we have difficulty even believing that they exist as real barriers to change. A good example is in the Florida Everglades, where AEAM and other collaborative processes over the past decade have led to broad consensus that ecosystem-restoration goals can only be met by very substantial "de-plumbing" of the whole wetlands system in South Florida (Davis and Ogden 1994; Gunderson et al. 1995) by removing various water control and storage structures from the heart of the system. Yet in the physical heart of this system sits a particular "water conservation area" that has been used in recent years for shallow storage of water for urban uses, functioning as a reservoir of sorts. This near-permanent water pool has developed a population of snails, and has attracted an endangered species, the snail kite. Endangered species legislation and regulatory practice now dictate that this reservoir area be protected in its present-day state (proven productive for the snail kite). If protection involves maintaining current operating practice for the levees and canals that form the "reservoir," the whole concept of large-scale ecosystem restoration by de-plumbing will have to be discarded; the plumbing that protects the snail kite is right in the center of the larger system.

Practically all AEAM modeling exercises have revealed similar regulatory constraints that were put into place with good intentions but have turned out to be ineffective, inappropriate, or inimical to new and broadened management goals. Science can help to make it clear that these constraints are now part of the problem rather than part of the solution, but scientists and managers are going to have to step out in public and actively become advocates of change if anything is to be done about the situation. With every regulatory system and approach, there are people who will champion the status quo for personal reasons (e.g. jobs depend on continuing the same policy, involvement in that particular policy-making and hence take personal pride in it, emotional commitment to it for various other reasons, as so on). There is no use pretending that such people can be made to change their views or actions through "rational" argument; change must generally be forced on them by debate and decision at other institutional levels.

Partnerships for Knowledge

As noted in a prior section, effective ecosystem approaches to management will involve radical increases in information gathering, and sometimes drastic change in the spatial scales in which action is taken (much more microscopic manipulation and regulation as well as broader, larger-scale landscape-management initiatives). The public is not going to provide scientists and managers with the financial resources to make these changes. We can only achieve the changes that are needed and learn from this experience if we can develop really effective partnerships between scientists, managers, and other stakeholders. Such partnerships must begin with a really basic change in attitudes—especially by scientists—about the ability of people in general to make systematic, scientific observations and to engage productively in the process of designing local management policies. Bluntly, good partnerships are going to require that scientists be much less arrogant, self-serving, and protective of our activities. We must stop hiding behind (and trying to impress others with) both our methods of observation, and our ways of communicating results. This recommendation is particularly important for partnerships between scientists and the large community of people that Franklin (1995) has called "resource specialists," that is, those resource agency personnel who have often spent lifetimes accumulating detailed data about local management situations and changes.

Effective partnerships for long-term monitoring and management will not be built upon a basis of simple cooperation and volunteer involvement. We need to develop effective, sustainable economic incentive systems that make cooperation worthwhile for stakeholders. Such systems might be as simple as per-observation payments for measurements, funded by government, or as complex as carrot-stick license requirements for resource users to both gather information and fund research as conditions for holding a license. In situations in which particular observational outcomes could be seen as dangerous by stakeholders (e.g. if logging companies in the Pacific Northwest were involved with monitoring spotted owl distributions, and if discovery of an owl nesting site might be used as grounds to stop logging), we need further to develop systems to cross-check and validate observations, and these systems could be quite expensive.

One key scientific challenge we face today is to make use of the explosive development in remote sensing and information movement, both to gather more data cheaply and to provide geo-referenced validation checking on information gathered through partnership arrangements (e.g., use of Global Positioning Systems [GPS] and automatic movement track recording to check on fishing vessels that are engaged to conduct marine ecosystem surveys, and to automatically record such information as temperature, salinity, and chlorophyll concentration as the vessel moves about). All sorts of information gathering will become radically cheaper if we can even automate just the most tedious and boring components of geo-referencing and recording data.

Case Studies to Demonstrate Effect of Ecosystem Perspective

Natural-resource managers really cannot be blamed for saying "what is new," or "if it isn't broke, don't fix it" in response to academic appeals for new management perspectives. We frankly have not been able to make many serious and constructive recommendations about precisely what it means to do "ecosystem management," and much of what we have said falls dangerously closer to arm-waving about ecosystem ethics rather than about how *in practice* to sustain and improve the functioning of managed ecosystems. Anyone who wishes to act as an advocate for ecosystem management must be prepared to describe specific case examples, and to list precisely what differences in management practice would be entailed by the "new" approach.

There are relatively few case examples today in which we can so precisely list just those differences that an ecosystem approach would make. In the

Florida Everglades, an ecosystem-modeling approach developed using AEAM has led to substantial changes in viewpoints about the spatial scales and kinds of management changes that are needed for ecosystem restoration. On the Great Barrier Reef in Australia, a variety of use-zoning changes have been driven by ecosystem-scale analysis of such factors as dispersal of larval fish and invertebrates among reef structures (i.e. "seed reef" protection concept), and the various risks of human impact through fishing (Mapstone et al. 1996). In northern Canada, attempts are underway to design wood-harvesting policies for the boreal forest zone that will mimic natural disturbance regimes by fire and insects, though in this case it is not yet clear whether it will be economical for the forest industry to harvest in similar patterns to the historical fire pattern (Cumming et al. 1994).

Conclusion

Barriers to communication and cooperation between ecosystem scientists and managers are not particularly formidable, and can be overcome through relatively simple tactics for encouraging clarity in problem analysis and policy design. For many important and unique ecosystems, the barriers have already been substantially broken down through cooperative efforts to develop policy models and design-management experiments. I personally believe that most of the remaining misunderstandings are centered on the failure of both scientists and managers to distinguish clearly between scientific and ethical issues, particularly in relation to the question of acceptable risk in those settings in which more intense ecosystem use may lead to increased risk of major ecosystem change. Today, both scientists and managers should be seeking understanding of the use-risk relationship, through analysis and experimentation; such challenges as how to develop partnerships for more effective information gathering are far more important today than debates about ecosystem values.

References

Allison, G.T. 1971. *The essence of decision: explaining the Cuban missile crisis.* Little, Brown & Co., Boston, MA.

Clark, W.C., D.D. Jones, and C.S. Holling. 1979. Lessons for ecological policy design: a case study of ecosystem management. *Ecological Modelling* 7:1–53.

Cumming, S.G., P.J. Burton, S. Prahacs, and M.R. Garland. 1994. Potential conflicts between timber supply and habitat protection in the boreal mixedwood of Alberta, Canada: a simulation study. *Forest Ecology and Management* 68:281–302.

Davis, S.M., and J.C. Ogden, eds. 1994. *Everglades: the ecosystem and its restoration.* St. Lucie Press, Del Ray, FL.

Emlen, J.M., R.R. Reisenbichler, A.M. McGie, and T.E. Nickelson. 1990. Density-dependence at sea for coho salmon (*Oncorhynchus kisutch*). *Canadian Journal of Fisheries and Aquatic Sciences.* 47:1765–1772.

Franklin, J.F. 1995. Scientists in wonderland: experiences in development of forest policy. *Biosciences Supplement* S74–S78.

Gunderson, L.H., C.S. Holling, and S.S. Light. 1995. *Barriers and bridges to the renewal of ecosystems and institutions.* Columbia University Press, New York.

Holling, C.S. ed. 1978. *Adaptive environmental assessment and management.* John Wiley & Sons, New York.

Lee, K.N. 1993. *Compass and gyroscope: integrating science and politics for the environment.* Island Press, Washington, DC.

Mapstone, B.D., R.A. Campbell, and A.D.M. Smith. 1996. *Design of experimental investigations of the effects of line and spearfishing on the Great Barrier Reef.* CRC Reef Research Centre, James Cook University, Townsville, Technical Report. No. 7.

Milliman, S.R., A.P. Grima, and C. Walters. 1987. Policy making within an adaptive management framework, with an application to lake trout (*Salvelinus namaycush*) management. *Canadian Journal of Fisheries and Aquatic Sciences* 44:425–430.

Walters, C.J. 1986. *Adaptive management of renewable resources.* McMillan Public Company New York.

Walters, C.J. 1994. Use of gaming procedures in evaluation of management experiments. *Canadian Journal of Fisheries and Aquatic Sciences* 51:2705–2714.

Walters, C.J. and J.S. Collie. 1988. Is research on environmental factors useful to fisheries management? *Canadian Journal of Fisheries and Aquatic Sciences.* 45:1848–1854.

Walters, C.J., L. Gunderson, and C.S. Holling. 1992. Experimental policies for water management in the Everglades. Ecological Applications 2:189–202.

Walters, C.J., and K.J. Sainsbury. 1991. *Design of a large scale experiment for evaluating effects of fishing on the Great Barrier Reef.* Great Barrier Reef Marine Park Authority, Townsville, Australia.

12
The Need for Large-Scale Experiments to Assess and Predict the Response of Ecosystems to Perturbation

Stephen R. Carpenter

Summary

Ecosystem experiments are field experiments in which the experimental unit is large enough to include the relevant physical, chemical, and biotic context of the processes being studied. Whole-ecosystem experiments have yielded insights about many processes in a diversity of habitats. Successful design and interpretation of ecosystem experiments depend on connections to theoretical, long-term, and comparative studies. Ecologists have overcome many problems of inference for ecosystem experiments. However, the issue of replication is far less important than the need to compare alternative explanations for the results, which may involve reference ecosystems, premanipulation data, and additional measurements or experiments designed to compare possible explanations. Potential limitations of ecosystem experimentation include the variability and slow dynamics of ecosystems, certain aspects of academic and management culture, and institutional shortcomings. Progress in ecosystem experiments can be accelerated through dedicated sites and funding, and keystones that foster collaborations between management and science for adaptive ecosystem management.

Introduction

Scientific learning takes place through cycles of inspiration, trial, and evaluation. In ecosystem ecology, the trials can take place in the computer and the mind of a theorist, through observations of natural fluctuation in long-term studies, or through comparisons of contrasting ecosystems. Deliberate experimental manipulation of whole ecosystems is a particularly powerful aid to learning. The experimenter causes something interesting to happen, and therefore does not have to wait for an informative event. Moreover, the experimenter knows what was manipulated; this knowledge greatly simplifies the interpretation of events. It is not surprising that ecosystem experimentation has been an insightful and influential component of ecology. It is surprising, however, that ecosystem experimentation is not used more widely to advance the discipline and improve management policies.

Ecosystem experiments are field experiments in which the experimental unit is (1) defined by a useful natural boundary (such as a shoreline or hydrologic divide); (2) large enough to include the relevant physical, chemical, and biotic context of the processes being studied; and (3) deliberately manipulated. Ecosystem experimenters assign high importance to matching the scale of the study to natural or management processes.

This chapter explains the place of ecosystem experimentation in comparison with the other dominant modes of learning about ecosystems, which include theory, long-term observation, and comparison. The diverse accomplishments of ecosystem experiments are also briefly reviewed, and subsequently, the limitations to ecosystem experimentation are discussed: barriers that have been overcome, and barriers that remain. Finally, some steps that could be taken to accelerate progress in ecosystem experimentation are suggested.

Learning About Ecosystems Requires Multiple Approaches

Ecosystem ecology is a table borne by four strong legs, each essential for the intellectual support of the whole (Figure 12.1). The legs are the four major approaches that scientists use to learn about ecosystems: theory, long-term study, comparisons, and ecosystem experiments. Each approach is applied to many types of ecosystems, involves diverse and multidisciplinary practitioners, and creates "invisible colleges" of scientists who read and cite each other's work. Each approach garners substantial intellectual support from the scientific community, and significant financial support from funders of science. Each approach has also been the subject of several influential books in the past decade, and is well represented in this book. Although individual publications of ecosystem ecology have diverse goals, these studies are often motivated by, and find their greatest intellectual significance in, one or more of the four fundamental legs of the science.

The four approaches to ecosystem science have complementary strengths and weaknesses. Because of this complementarity, most ecosystem researchers employ two or more approaches in their own work (Table 12.1). Each approach tends to be applied at certain spatial and temporal scales, and the results tend to be used for certain kinds of inference (Figure 12.2). The four approaches are related to two distinct philosophical perspectives in ecology (Pickett et al. 1994; Holling 1995). One perspective

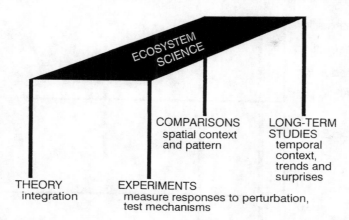

FIGURE 12.1. Ecosystem ecology is similar to a table with four legs essential to the integrity of the whole: theory, comparison, long-term study, and experiment.

is integrative, synthetic, and creative of new explanations and is marked by fertile development of multiple-alternative hypotheses, and comparisons of the degree of corroboration of alternatives. Integration and description represent this philosophy in Figure 12.2. The second philosophical perspective is deductive, experimental, and focused narrowly on the elimination of potential explanations. Deduction and hypothesis testing represent this

TABLE 12.1. Selected strengths and limitations of the four approaches to learning about ecosystems.[1]

Approach	Some strengths	Some limitations
Theory	Flexibility of scale Integration Deduction of testable ideas	Cannot develop without continuous linkage to observation, experiment
Long-term observation	Temporal context Detection of trends and surprises Test hypotheses about temporal variation	Potentially site-specific Difficult to determine cause
Comparison	Spatial or inter-ecosystem context Detection of spatial pattern Test hypotheses about spatial variation	Difficult to predict temporal change or response to perturbation
Ecosystem experiment	Measure ecosystem response to perturbation Test hypotheses about controls and management of ecosystem processes	Potentially site-specific Potentially difficult to rule out some explanations Institutional structures limit scope

1. The limitations arise when one approach is used alone. Usually they can be overcome by combining two or more approaches.

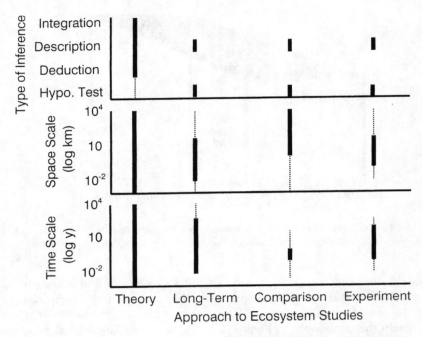

FIGURE 12.2. Comparison of approaches to ecosystem science with respect to type of inference (Pickett et al. 1994), spatial scale, and temporal scale. Solid bars denote primary applications of an approach; dashed lines denote potential or more limited applications of an approach.

philosophy in Figure 12.2. Although both perspectives are essential for progress, ecology has been dominated by deduction and hypothesis testing in recent years.

Theory is essential for integration and deduction, and can be extremely flexible with respect to spatial and temporal scales (Figure 12.2). However, theory must be coupled with observational or experimental research to develop (Table 12.1). Long-term studies have been extremely successful in describing system dynamics and establishing the temporal context for other types of ecosystem studies (Likens 1989). The principal limitation of long-term studies is that the causes of temporal variability may not be apparent. In some cases, mechanisms can be inferred through other approaches, including ecosystem experimentation. Comparative studies have a long and successful record of describing broad patterns in ecosystems and establishing the spatial context for ecosystem studies (Cole et al. 1991). The principal limitation of comparative studies is that they are often based on regressions fit to data from many ecosystems and may be unable to predict how a given ecosystem will respond to a change. In some cases, the capacity for site-specific predictions can be improved by using theory, ecosystem experiments, or local long-term studies. Ecosystem experiments have successfully measured effects of manipulations and tested hypotheses at relatively large spatial scales, including scales relevant to environmental management (Likens 1985; Mooney et al. 1991; Carpenter et al. 1995). The principal limitations and challenges of ecosystem experiments are elaborated in the next sections. Design of successful ecosystem experiments requires theory (to provide intellectual context and suggest those questions that are worthy of experimentation) and the spatial and temporal context provided by comparative and long-term studies.

Large-Scale Experiments Have a Unique and Essential Role in Ecology

Experimentation at the spatial and temporal scales of natural ecosystem processes has unique and important advantages. This type of experiment is the only research approach that tests responses to known perturbations directly at the ecosystem scale. Ecosystem experiments can match the scale of management, and can provide results directly applicable to environmental policy and management.

Ecological systems do not have a single characteristic scale, and ecological concepts, measurements, and experiments are usually scale-dependent (Levin 1992). Insightful ecological experiments have been performed at a wide range of scales, and it is often valuable to nest experiments at several different scales (Frost et al. 1988). This chapter does not promote any particular scale as being optimum for ecological experiments; however, it does argue that relatively large-scale experiments provide vital information that cannot be obtained by other approaches. Ecosystem experiments are the only approach that measure effects of changing key variables in the context of other major controls of ecosystem processes. Comparative and long-term studies assess context without direct manipulation, and small-scale experiments study mechanisms out of context.

Small-scale experiments cannot substitute for ecosystem experiments because crucial ecosystem processes are distorted or excluded by the small size and short duration of microcosm experiments. Microcosm experiments cannot include such wide-ranging or slow processes as large-scale turbulence, wide-ranging predators, or large, slow-growing trees. Some small-scale processes quickly reach unrealistic rates in microcosms ("bottle effects"); these include microbial production and biomass, and phytoplankton production. Some examples of misleading results from experiments that were too brief or too small are given by Tilman (1989) and Carpenter (1996). These points, however, do not negate the value of small-scale experiments for testing and measuring mechanisms. They only emphasize the difficulty of extrapolating measurements from small-scale experiments to larger scales in which the key controls of ecosystem processes may be different.

The fundamental problem with learning from small-scale experiments is that results must be translated across scales to draw conclusions about ecosystems. Projection across scales is a very difficult process, and the models are complicated, context-specific, and subject to considerable debate and research (Allen and Hoekstra 1992; Levin 1992). Conversely, inference from ecosystem experiments requires fewer assumptions because appropriate scaling can be taken into account by the experimental design. Thus, ecosystem experiments are convincing tests of basic ecological ideas. The scale of ecosystem experimentation is especially useful in environmental management because it is very difficult to convince managers or other stakeholders to change policies using complex arguments and extrapolations (Lee 1993). Direct results from appropriately scaled experiments are simpler, obviously relevant, and, therefore, more apt to influence policy.

Large-Scale Experiments Have Succeeded in Diverse Circumstances

Insightful ecosystem experiments have succeeded in diverse circumstances (Table 12.2). Indeed, many different types of ecosystems have been manipulated. Independent variates have included physical, chemical, and biotic factors, and in some cases, tracers have been added to large systems to measure process rates (e.g. Bower et al. 1987; Coale et al. 1996). Dependent variates have included such ecosystem processes as production or nutrient cycling, as well as community and population processes. Additionally, large-scale experiments have considerable value in community ecology as well as ecosystem ecology (Lodge et al. 1997).

Inadvertent disturbances, caused by either natural or human events, have also yielded valuable insights. Inadvertent disturbances are especially useful when predisturbance data exist, and when undisturbed reference areas have been studied (i.e. when they resemble deliberate manipulative experiments). Consequently, the most useful information from inadvertent disturbances probably occur at sites of ongoing, long-term studies. For the purposes of this chapter, inadvertent disturbances are classified as a component of long-term studies, and the focus is instead on deliberate manipulations.

Measuring the magnitude of response to manipulation and understanding our capacity to manage are the two most common reasons for ecosystem experimentation. Some ecosystem experiments, however, are performed purely to test ecological ideas. One example of this type of experiment is the research on trophic cascades, which was designed primarily to test ideas about the effects of aquatic community structure on ecosystem processes (Carpenter and Kitchell 1993). Other experiments are primarily environmental demonstrations. For example, the Risdalsheia acidification experi-

TABLE 12.2. Selected examples of ecosystem experiments.[1]

Ecosystem Type	Independent Variates	Dependent Variates	Reference
Lakes	Lime addition	Fish production	Hasler et al. 1951
Lakes	Nutrient addition	Primary production	Schindler 1977
Lakes	Acid addition	Biogeochemical cycles, food webs	Schindler et al. 1991
Lakes	Fish-community manipulations	Primary production, nutrient cycles, food webs	Carpenter and Kitchell 1993; Gulati et al. 1990
Open ocean	Iron addition	Primary production	Coale et al. 1996
Great Barrier Reef	Fishing	Fish production, community composition	Walters 1993
Streams	Insect removal	Carbon export	Wallace et al. 1996
Streams	Acid addition	Chemical and biological exports	Hall and Likens 1981
Catchment and stream (boreal forest)	Acid addition, exclusion	Chemical exports	Wright et al. 1988
Boreal forest	Acid addition, liming	Soil chemistry, biology, forest production	Abrahamson et al. 1994
Boreal forest	Nutrients, forage, predator exclusion	Snowshoe hare population	Krebs et al. 1978
Deciduous forest	Forest harvest, herbicides	Hydrology, nutrient export	Likens et al. 1978
Tallgrass prairie	Fire frequency	Nitrogen and carbon cycle, species composition	Ojima et al. 1994; Gibson et al. 1993

1. This table is intended to illustrate the diversity of ecosystem experiments, and is far short of a comprehensive list.

ments showed, in a clear and dramatic way, that watersheds could recover rapidly if the acid was removed from precipitation (Wright et al. 1988). Some experiments are conducted in a management context, in which the primary goal is to improve management of a resource and learn in the process. Fisheries experiments, for example, have been used to explore effects of contrasting harvest policies on fish production (McAllister and Peterman 1992; Walters 1993). In fact, many ecosystem experiments have multiple goals. A good example of a multiple-goal experiment is the Hubbard Brook deforestation studies (Likens et al. 1978). Initial studies were motivated by very basic questions, but later experiments were intended to compare forest-harvest practices to determine how forests could be harvested in a sustainable way.

Some Significant Barriers Have Been Overcome

Ecologists have identified and overcome some important barriers to ecosystem experimentation. Practical barriers include access to sites, sustained financial support, and the capacity to manipulate and study large systems. The examples reviewed here (Table 12.2) and many others show that sometimes these barriers have been overcome, often by ingenious means but occasionally by persistence and good luck. Nevertheless, access to sites, sustained financial support, and mechanisms for collaboration of scientists and managers often pose barriers to ecosystem experimentation.

Ecosystem experimenters often must choose between the need to apply strong, sustained manipulations that cause a clear, interpretable response and the need to understand effects of more modest or gradual perturbations. For example, an experiment may double the atmospheric carbon dioxide concentration immediately, but the global impacts of doubled carbon dioxide concentrations will actually develop gradually over decades. In this particular case, models and long-term trend studies provide additional information that is crucial for interpreting and applying the experimental results to predict ecosystem change. Studies of the effects of nutrients and food-web structure on lake productivity demonstrate a combination of experiments with comparisons and theory (see Smith, chapter 2). Comparisons show that whole-lake manipulations of nutrients and food webs are substantial, yet within the ranges known for many lakes (Carpenter et al. 1991). In general, extensions of experimental results will require some integration with theory, and long-term and comparative studies (Figure 12.1).

Inference poses significant intellectual challenges that have been met in many ecosystem experiments. Inference is the process of deciding upon an explanation for the experimental results. When a scientist argues for a particular explanation, there is an obligation to show that the favored explanation is more probable than all plausible alternatives. This process involves results of the experiment itself and external information (e.g. observations and theory from other studies). Hilborn and Mangel (1997) discuss the process of inference using multiple alternative hypotheses.

Reference ecosystems are unmanipulated ecosystems that are expected to experience the same regional trends (e.g. climate) as the manipulated ecosystems. A.D. Hasler introduced the term "reference ecosystem" (instead of "control") to acknowledge the fact that no two ecosystems are identical, therefore pretreatment conditions in experimental and reference ecosystems cannot be controlled as thoroughly as in a laboratory experiment (Likens 1985). Reference ecosystems are used to check the possibility that responses of the manipulated ecosystem were caused by regional trends and not by the manipulation; they are also used to measure the variability of an undisturbed ecosystem, to help determine whether the response of the manipulated ecosystem can be explained by random variations (Carpenter 1993). Since the pioneering work of Hasler et al. (1951), most ecosystem experimental designs have included reference ecosystems.

The possibility that random variations can explain any apparent responses to the manipulation can be checked using statistical methods. All such methods depend on having multiple observations of the experimental ecosystem both before and after the manipulation. "Before and after" data (i.e. premanipulation and postmanipulation) are necessary even if there is a reference ecosystem (Stewart-Oaten et al. 1986) because ecosystems are never identical, and it is necessary to have a measure of the differences between reference and experimental ecosystems prior to manipulation. In general, premanipulation and postmanipulation data are compared using a statistic that accounts for the variability in the data, which is measured using repeated observations through time. The analysis can include information from both reference and manipulated ecosystems (Stewart-Oaten et al. 1986; Carpenter et al. 1989). Serial dependency in the time series of obser- vations can strongly affect the results but time-series methods can correct for any biases caused by serial dependency (Carpenter 1993; Rasmussen et al. 1993). The most commonly used time-series methods assume that observations are evenly spaced and that there are no missing data. Bayesian time-series methods relax these assumptions and appear to have great promise for analysis of ecosystem experiments (Pole et al. 1994; Cottingham and Carpenter 1998).

The issue of pseudoreplication (Hurlbert 1984) has created considerable confusion in ecosystem ecology. The definition of a replicate is a matter of scale and if the research question applies to a singular ecosystem, then no genuine replicate is possible. Sampling should measure the relevant spatial and temporal variability of the ecosystem before and after manipulation, and methods cited previously can be used to check the possibility that the responses are merely random (Stewart-Oaten et al. 1986). If the research question applies to a larger universe (e.g. "all dimictic lakes"), then repli- cates could be drawn randomly (i.e. from the world's dimictic lakes), assigned randomly to treatments, and compared statistically (Hurlbert 1984). In practice, of course, this is impossible. Ecosystem experiments are usually unreplicated, and may be viewed as case studies or system- specific investigations. In a few situations, however, particularly interesting ecosystem experiments have been repeated in many parts of the world. Comparative analyses of such sets of experiments can be valuable (e.g. Schindler et al. 1991) and statistical methods do exist for pooling data to evaluate hypotheses across multiple experiments (Gurevitch and Hedges 1993; Gelman et al. 1995).

Measuring the magnitude of ecosystem response is generally far more important than testing the null hypothesis. Manipulations are designed to produce the effect so that it can be studied (Carpenter 1989); the null hypothesis is obviously false. By the time we are convinced to invest in an ecosystem experiment, there is usually considerable evidence that the manipulation will have some effect. The important questions have to do with the nature and magnitude of the effect. The appropriate statistics are often descriptive, for example, means or confidence intervals (Hurlbert 1984) and Bayesian analyses provide far more useful information than conventional statistics (Box 12.1).

Nevertheless, replication can substantially increase the value of ecosystem experiments. In the most obvious case, a replicate is appropriate if

---

Box 12.1

Ecosystem experiments are often used to measure the response to a specified manipulation. Because ecosystem responses are inherently variable, it is natural to describe them using probability distributions. Bayesian analysis is the branch of statistics used to calculate probability distributions for experimental responses (Gelman et al. 1995).

An example of a probability distribution for an ecosystem response is depicted in Figure 12.3. In this experiment, dense beds of an invading aquatic plant (Eurasian water milfoil, *Myriophyllum spicatum* L.) were harvested in channels designed to increase habitat for fishes (bluegill, *Lepomis macrochirus* L.) (Carpenter et al. 1997). Annual bluegill growth was measured in four experimental lakes and six reference lakes before and after manipulation. The probability distribution for the improvement in growth in the manipulated lakes relative to the reference lakes was calculated (Gelman et al. 1995).

The distribution presents a substantial amount of useful ecological information. Most of the area of the distribution lies above zero (about 98%), indicating a 98% probability that bluegill growth increased and only a 2% probability that bluegill growth decreased. The peak of the distribution, indicating the most probable magnitude of the increase, lies at about 10 mm. The probability of a growth increase of 10 mm is about 0.11, while the probability of a growth increase of 0 is about 0.01, therefore we are eleven times more apt to see a growth increase of 10 mm than we are to see no growth increase. On the other hand, a growth increase as large as 20 mm is much less probable than a growth increase of about 10 mm. An additional 10 mm of growth is a substantial increase for bluegill, which typically grow about 25 mm in their third growing season (Carpenter et al. 1995).

In contrast, the information from a typical statistical test (in this case, a simple student t-test comparing manipulated and reference

Box 12.1 *Continued*

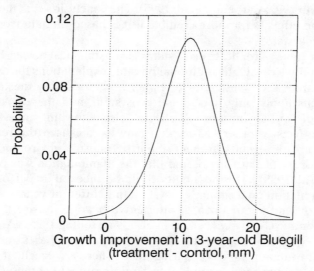

FIGURE 12.3. Probability vs. growth improvement in three-year-old bluegill caused by removal of Eurasian water milfoil.

ecosystems) contains little useful ecological information. Unlike the Bayesian analysis, the conventional analysis does not determine the probability that we most want to know (i.e. the probability that the manipulation caused the effect). Rather, all probabilities are risks of error associated with particular decisions toward the null hypothesis that the manipulation had no effect. If we reject the null hypothesis of no growth improvement in manipulated lakes, the probability of making an error is 0.0243. If we accept the null hypothesis, the probability of making an error depends on the actual size of the improvement in bluegill growth. If the actual growth improvement is 10 mm, then the probability of error if we accept the null hypothesis is about 0.9 (Carpenter et al. 1995). The t-test calculations reveal that the most probable growth improvement is 10.4 mm, but provide no estimate of the odds of this growth improvement relative to other such values as 0 mm or 20 mm. The information provided by the conventional statistical analysis is considerably more complex than the information provided by the Bayesian analysis, and is far less relevant to the ecological question that motivated the experiment.

For introductions to Bayesian inference in ecology, see Ellison (1996) and Hilborn and Mangel (1997). Gelman et al. (1995) present methods and computational approaches; Howson and Urbach (1989) discuss the use of Bayesian evidence in scientific reasoning.

there is a good chance that the manipulation will have no effect (McAllister and Peterman 1992; Carpenter et al. 1995). The conclusion that the manipulation has no effect is far more credible if it is based on data from several different ecosystems.

The great value of replicate ecosystems is not statistical power, but rather the opportunity to evaluate additional possible explanations for the results. Replicates allow the investigator to explore the possibility that these particular results apply only to a single ecosystem, and thereby assess the generality of the conclusions. Replicates also allow the investigator to assess the effects of other factors that may have caused the results. For example, were the results brought about by unusual environmental conditions in the particular time period that the manipulation was performed (Walters et al. 1988)? This possibility could be checked by performing the same manipulation on a similar ecosystem in a different year.

Generation and comparison of alternative hypotheses are the cornerstone of scientific creativity (Pickett et al. 1994; Holling 1995). An effective ecosystem experiment anticipates the alternative hypotheses and includes methods for evaluating and comparing them. There are excellent statistical methods for comparing the credibility of alternative hypotheses (Jefferys and Berger 1992; Hilborn and Mangel 1997), but unfortunately these methods are underused by ecologists. However, the key arguments depend not on statistics, but on effect sizes (and their variability), careful checking of alternative possibilities, and ecological reasoning.

Some Significant Barriers Remain

Ecologists have clearly demonstrated that insightful experiments can be performed on large, complex systems. There are, however, a number of limitations or barriers to ecosystem experimentation.

Some barriers cannot or should not be overcome and not every ecology question requires an ecosystem experiment. Ecosystem experiments will always be relatively costly, and we need to choose carefully those experiments that have priority. There are also ethical and environmental limitations; perhaps there are some ecosystem experiments that simply should not be done.

Substantial barriers may exist even for experiments that have high priority for science and society. The next section considers limitations created by the dynamics of ecosystems, academic culture, the culture of management, or the institutions that conduct ecosystem experiments.

Limitations Posed by Ecosystems

Variability, slow dynamics, and multiple-interacting controls of ecosystems pose significant challenges to experimentation. Because of these properties, some questions cannot be answered by ecosystem experimentation and must be addressed by other approaches. There are perhaps questions that are not susceptible to any approach presently available, but that issue is beyond the scope of this chapter.

Ecosystems can be highly variable in time and space. Long-term and spatially intensive studies are an effective (but potentially costly) way of coping with variability. Alternatively, ecologists can choose to focus on variates that have tractable variability. In lake phytoplankton, for example, ecosystem variates (total biomass or production) and allometric variates (aspects of community-size structure) are far less variable and more predictable than population sizes or the taxonomic composition of communities (Cottingham and Carpenter 1998). Consequently, clear inferences are possible in lake ecosystem experiments focused on biomass, production, and size structure of plankton. The responses of populations of species are more variable and hard to interpret, however. In terrestrial ecosystems, there may be comparable patterns that would suggest those variates that are most appropriate for experimentation.

Ecosystem dynamics can be very slow, and therefore it is difficult to study them for a long enough period of time to perceive the response. Ecologists have coped with this problem through experimental designs that accelerate change (Carpenter and Kitchell 1993), comparative studies that substitute spatial patterns for temporal ones (Pickett 1989), and direct long-term observation (Likens 1989). When ecological dynamics take longer than the careers of investigators, it is essential to create institutions capable of sustaining the research. The U.S. Long-Term Ecological Research (LTER) network is an effort to create such institutions.

Controls of ecosystem processes also change over time (Levin 1992; Holling 1995). Transitions in key controlling processes pose special problems for long-term experiments, because it is very difficult to account for them in the experimental design. Walters (1986) has studied this problem for fisheries where the goal is to manipulate fishing intensity to measure effects on fish production rate. But fish production rate can also be controlled by other "hidden variables," which are not set by the experimenter and may not even be measured (e.g. climate, forage fish production, and so forth). If the experiment can be conducted in a time span that is long enough to account for several natural fluctuations in the hidden variables, or short enough that the hidden variables do not change, then the experiment is apt to succeed. But if the time span of the experiment is near the cycle of fluctuation in the hidden variables, the experiment could be confounded. A good description of the system through long-term studies, or repeated manipulations over several cycles of the hidden variables, may help to resolve this problem.

Barriers from Academic Culture

Academic culture poses several barriers that need to be considered and overcome by ecosystem experimenters. Academic reward systems often favor research that is sharply focused and of a narrow discipline, instead of broadly relevant and interdisciplinary. In ecological terms, intense competition for limited research resources drives academics into narrower and more rigorously defended niches. This type of environment favors research on systems that are well controlled, carefully defined, and simple. It is not an environment that selects experiments about complex, variable, and unpredictable ecosystems, which require long-term studies and cross-disciplinary collaborations in areas in which knowledge is incomplete and evolving.

Academic standards for hiring, promotion, and tenure create pressure for fast publication (relative to slow ecosystem responses) and aversion to interdisciplinary multiauthored publications. In some departments, applied research is viewed as less valuable than basic research. There has been progress in breaking down these barriers, but young scientists are sometimes averse to multiauthored publication. The community of ecologists can accelerate progress by recognizing and rewarding research that is appropriately scaled, collaborative, cross-disciplinary, and useful.

Ecosystem experimentation can also be an excellent context for graduate training. Even though the experiment itself may take too much time for a thesis project, insightful student projects can often be designed around pretreatment studies, modeling, retrospective analyses or other aspects of the experimental program. The multidisciplinary nature of ecosystem experiments and interactions with management agencies can create extraordinary opportunities for graduate students. Graduate programs, however, may not be well-positioned to obtain the maximum benefit from these opportunities.

Traditionally, graduate training has been viewed as a simple student–mentor process, and academic rewards and resources have been allocated accordingly. However, in modern interdisciplinary research it "takes a whole village to train a graduate student." We need to improve our mechanisms for training students in team research. Academic mechanisms and funding should be developed for multi-mentored, cross-disciplinary doctoral projects and students should be rewarded and recognized for their contributions to teams.

Barriers from the Culture of Management

The culture of management also poses significant barriers to ecosystem experimentation. Gunderson et al. (1995) observed that management systems follow an adaptive cycle through time. Peaks of adaptability and receptivity to learning are followed by troughs of bureaucratic stasis when learning is impossible. During periods when policies appear to be working, management systems become more streamlined and narrow. Monitoring and research programs may be abandoned in order to focus resources more efficiently on policy objectives. Even as the management system is becoming more rigid, the ecosystem and society's expectations are changing in ways that can make the policies myopic and irrelevant. Eventually it can become clear that the policies are failing. A crisis, brought on by resource collapse or social conflict, often brings about widespread acknowledgement of management failure. Recognition of fundamental problems can lead to a period of experimentation, learning, and innovation. The resulting insights set the stage for a new round of policy development and the next adaptive cycle.

The adaptive cycle describes some factors that can impede experimentation by management agencies. Learning, by ecosystem experimentation or any other means, may be viewed as a threat to dogma or entrenched policy during the rigidly bureaucratic phase of the adaptive cycle. Symptoms of this phase include "command-and-control" management styles, diminished research and monitoring budgets, and the belief that science is mainly useful for enforcement and litigation (Holling 1995). Creative science will not be favored under bureaucratic management regimes. On the other hand, science can make crucial contributions during the reorganization phase of the adaptive cycle.

Even when agencies are receptive to learning and open to the prospect of new management policies, progress may be halting and inefficient (Table 12.3). Hilborn (1992) documents several factors that affect the capacity of management agencies to learn from experience. Informative experimentation may require a radical change in management for a period of time and conservative managers may view these changes as too risky. An urgent need for results could prompt manipulation of the ecosystem before adequate pretreatment data are collected. Without premanipulation data, the experiment cannot be interpreted and therefore no learning takes place. Lack of shared vision may lead to complex manipulations of many factors, which yield results that are hard to interpret. Useful ecosystem experiments change one aspect of an ecosystem in a strong sustained way, for a long enough time period to learn the consequences. The institution may be unable to maintain the experiment for the required time to learn from it (Hilborn 1992). Turnover of personnel, changes in political pressures, or unforeseen budget problems may cause an experiment to be abandoned or compromised before the objectives are achieved. Some of these limitations could be overcome by improvements in training that prepare agency personnel to take greater advantage of opportunities to learn by doing.

Institutional Limitations

Ecosystem experimentation is also limited by the kinds of institutional arrangements available to operate experiments. Institutional challenges include coordination of management personnel and scientists in ways that achieve overall objectives in addition to fostering individual creativity and

TABLE 12.3. Some barriers that can prevent effective management experiments.[1]

Barrier	Explanation	Solutions
Excessive bureaucracy	Learning perceived as threat to entrenched policy	Wait until the need for change is obvious
Excessive caution	Manipulations that would be informative are viewed as risky	Subdivide the resource and experiment with part of it; wait until the potential value of better management outweighs the risks
Manipulation is done too early	Premanipulation data are not sufficient to interpret the experimental results	Collect adequate premanipulation data; describe the system adequately before experimenting
Dithering	The manipulation is overly complex; too many variables are changed at one time	Change one aspect in a strong sustained way, for a long enough time period to learn the consequences
Learning disabilities	The institutional attention span is too short; the experiment is compromised before the ecosystem can respond	Sustain the experimental design a time period long enough to detect ecosystem responses

1. See Hilborn 1992; Gunderson et al. 1995.

productivity, the logistics of the experiment itself, and sustaining the program for the time period required to determine the ecosystem response and interpret the results.

One indicator of institutional problems may be the relatively narrow range of staff sizes associated with successful ecosystem experiments. If the research team is too small, the necessary disciplines cannot be included and the experiment will fail to meet scientific goals. If the research team is too large, the level of bureaucracy necessary to conduct the experiment can become excessive and obstruct creativity or even the integrity of the experiment itself. Schindler (1992) describes some pathologies that can arise in extremely large environmental research programs. Adaptive environmental management, which explicitly views management activities as experiments, may work best with small, flexible institutions (Lee 1992). Evidence from well-known, successful ecosystem experiments suggests that the optimum team size is approximately ten to thirty people (i.e. senior scientists and managers, plus essential technical staff, graduate students, and postdoctoral researchers). If this is true, it suggests that there may be a limit to the scope of successful ecosystem experimentation. Alternatively, we may need to devise innovative new management approaches that facilitate success of larger ecosystem experiments.

Progress from Ecosystem Experiments Can Be Accelerated

Ecosystem experiments are a nexus of growth for ecosystem science and management. Human pressures on the environment will grow for the foreseeable future and experimentation is an efficient way to learn the effects of human action and compare the consequences of alternative management policies. At the same time, large-scale management experiments can stimulate vital, relevant science (Slobodkin 1988).

It is surprising that so few management actions or developments are viewed as large-scale experiments. In truth, management plans are hypotheses masquerading as answers. Orians (1986) points out that every environmental impact statement is a hypothesis for which the necessary manipulation will be carried out. With postmanipulation monitoring and a reference ecosystem, every environmental impact study could be an ecosystem experiment. The same is true for planned timber harvests, fish stocking, toxin remediations, and so forth. Paine et al. (1996) suggest that coastal oil spills are opportunities for large-scale experimentation which, however tragic, should not be wasted. Of course, more proactive ecosystem experiments are also needed on a great number of questions. In the next sections, some actions are suggested that could accelerate progress through ecosystem experimentation.

Leadership

The best-known and most productive ecosystem experiments have been led by a handful of close colleagues (Likens chapter 10) and productive collaborations seem to owe as much to luck as to design. Because of this, ecosystem experimentation may be limited by the numbers of scientists willing and able to form the necessary teams. Additionally, graduate training programs should be modified to develop research styles suited to ecosystem experimentation. Training in the history, opportunities, and challenges of ecosystem experimentation is needed for both academics and agency personnel.

Ecosystem experiments have generally been shorter in duration than the careers of the lead investigators, however, there are many questions that will require longer-term experiments. For example, what are the effects of chronic inputs of contaminants on ecosystems? Planning for longerterm experiments will require that issues of succession in leadership and continuity of vision be addressed. Ecologists have rarely considered these challenges.

Sites

Dedicated sites for ecosystem experiments are rare. Not all ecosystem experiments need to be performed in experimental reserves; management experiments, for example, may be performed in areas accessible to the public. Experimental reserves should be used to perform experiments that are risky to the resources, involve hazardous substances that should not be released in areas open to the public, or investigate crucial basic questions of ecosystem science. We are running out of time to create and sustain experimental ecosystem facilities like Coweeta, Hubbard Brook, Konza Prairie, the Experimental Lakes Area, Risdalsheia, Rothamsted, and the University of Notre Dame Environmental Research Center. Even these existing sites will not necessarily be maintained in the future. Canada's Experimental Lakes Area, for example, has experienced severe cutbacks in recent years.

Funding

Funding, or rather the lack of it, limits ecosystem experimentation. It can be difficult to persuade granting agencies to fund the extensive pretreatment studies needed for successful experimentation. It can be difficult to sustain grants for the length of time needed to see slow ecosystem responses. At the planning stage, the time horizon necessary to answer the questions should be clearly stated. The funders should be committed to sustaining the experiment for the necessary length of time, and the scientists should be committed to ending the experiment when it is over.

Management agencies should greatly increase their participation in critical ecosystem experiments. In some cases, management agencies may fund ecosystem experiments in experimental reserves about topics in which basic and applied questions intersect. However, the greatest growth opportunity lies in management experiments to compare and ultimately improve policies. To succeed, such experiments require carefully constructed partnerships between management agencies and academic scientists.

Keystones for Adaptive Ecosystem Management

Ecosystem experiments are critically needed to improve environmental management practices and policies. Adaptive ecosystem management is the process of adapting policies to changing ecosystems and changing societal expectations (Holling 1978; Gunderson et al. 1995) and tracking the changes in ecosystems that result from alternative policies (in other words, ecosystem experimentation) is central to adaptive ecosystem management (Walters 1986; Lee 1993; Walters chapter 11). Research relevant to adaptive ecosystem management must be flexible enough to address emerging questions, yet stable enough to sustain the long-term observations required to learn about ecosystems.

There are tremendous opportunities in combining the capabilities of management agencies and academic institutions for adaptive ecosystem management (Kitchell 1992). Agencies offer management expertise and authority or responsibility for the sites to be managed, and can dedicate staff and funding for careful, consistent measurements over the relatively long-time horizons needed for some ecosystem experiments. Academics bring scientific expertise and the talents of students and postdoctoral researchers. Student and postdoctoral staffs can be mobilized quickly around emerging scientific questions, yet the individuals (unlike permanent agency staff) move on when the project is over. Recognizing the potential power of agency–academic collaborations, attendees of the Second Cary Conference in 1987 called for a "new partnership between scientists and resource managers" (Likens 1989). Ten years later, at the Seventh Cary Conference, this partnership still remains largely unrealized.

In a common model for the interaction of agencies and academics, the relationship is asymmetric and can easily become adversarial (Figure 12.4a). Society has very different expectations of management agencies (which are funded to maintain ecosystem services) and academic institutions (which are funded to educate people and create new knowledge). In theory, academics can become informed of management needs, propose research to an agency, and eventually perform research and provide findings to the agency. In practice, however, this interaction is often fragile.

Interactions between management agencies and academic institutions can break down in many ways, much as a nonlinear, unstable interaction between two populations in an ecosystem. Academics and managers have different professional goals that are brought about by the different goals of their respective institutions. Departure of a key individual, political interventions, budget shortfalls, discoveries inconvenient to entrenched policy, or unprecedented ecological events can break the linkage between agencies and academics. In ecosystems, potentially unstable interactions can be stabilized by keystone species, which hold in check some component that would otherwise destabilize the system (Paine 1969). What are the keystones that might stabilize the interaction of scientists and managers?

A system that could sustain agency–academic partnerships should include a keystone responsible for adaptive ecosystem management (Figure 12.4b). The keystone could be a well-placed individual, a unit of an agency, or a freestanding institution. Society would fund the keystone to assess the status of ecosystem services, anticipate surprise, compare the performance of alternative policies, and adapt research and management to the evolving state of the ecosystems and societal expectations. Management agencies and academic institutions must collaborate to propose adaptive management projects, and conduct the necessary management actions and research. Westley (1995) notes that power dispersal is essential to successful collaboration. The keystone model distributes power in a potentially stabilizing way.

The keystone also encourages academics and managers to develop convergent goals. They must work together to propose projects, satisfy review

a

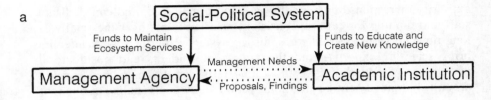

b

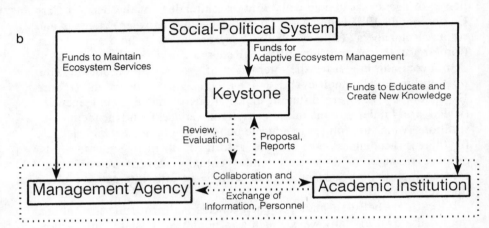

FIGURE 12.4. (a). A common model for the interactions of management agencies and academic institutions. (b). Proposed keystone model for interactions of management agencies and academic institutions.

and evaluation criteria, and achieve common goals of experimental management and scientific inference. Both management and academic institutions should reward individuals for writing successful proposals and publishing useful results and both institutions should thereby select personnel who are successful collaborators.

Are there examples of the keystone model? Case studies of ecosystem management (Gunderson et al. 1995) include a number of situations in which keystones emerged. In the Columbia River basin, the Northwest Power Planning Council was formed as a neutral institution responsible for system-wide decisions and coordination of diverse investigations (Lee 1993; McLain and Lee 1996). The International Joint Commission and the Great Lakes Fisheries Commission functioned as keystones at certain times in the management of the Great Lakes (Francis and Regier 1995).

Keystones may be short-lived, however. Westley's (1995) review of interorganizational collaborations implies that successful collaborations are often transient. She also notes advantages of informality:

. . . too much consensus and organization may make the interorganizational system vulnerable. For the actor in such systems, therefore, it is important to resist too much organization and centralization . . . Although the idea that policy should be treated as experiment is a brilliant one, it is difficult to achieve. Scientists should continue to be wary of politicians prepared to turn theory into policy. A healthy tension between the two, a redundancy of efforts and activities may offer the most fertile ground for individuals to manage the process of change. (Westley 1995).

Thus, the most effective keystone organizations may have short lifetimes relative to the constituent institutions and the ecosystems being managed. They are "ad-hocracies," not bureaucracies (Gunderson et al. 1995). For the ecosystem experimenter, this raises a dilemma. Informal collaborations may be the optimum for flexibility and creativity (McLain and Lee 1996), however, stable research arrangements (for years to decades) are essential to learn how ecosystem processes respond to management interventions.

Successful examples of ecosystem experimentation and adaptive ecosystem management show that both are possible. But these successes may be a consequence of happenstance collaborations and the dedicated leadership of a few individuals. In other words, luck seems to be involved. Can we create circumstances that increase the frequency of successful ecosystem experiments and adaptive policy design? If so, we can break a critical bottleneck that limits ecosystem experimentation.

People and Nature

Humans are now the dominant species of Earth's ecosystems. Ecologists can no longer study nature separately from people, and in fact many ecosystem studies do measure human impacts. Humans have appeared as independent variates or covariates in many ecosystem studies (McDonnell and Pickett 1993), including a few ecosystem experiments (Kitchell 1992; McAllister and Peterman 1992; Walters 1993). However, to my knowledge no ecosystem experiments have explicitly included people or institutions as *dependent* variables. When human response has been studied in ecosystem experiments, their behavior has always created surprises for ecologists. Some of the best examples come from fisheries. Hilborn et al. (1995) note that ". . . the biggest failure in natural resource management has been the widespread neglect of the dynamics of the exploiters." They point out that commercial fishing regulations have almost universally failed to control exploitation in the ways expected by scientists and managers.

These experiences point to a need for ecosystem studies, including experiments, that address people and nature as an interacting system (Folke chapter 13). Human activity changes ecosystems, and the altered ecosystems prompt change in human behaviors, economies, and institutions. For example, Likens (1992) describes "leapfrog degradation" of lake districts. As cabin sites develop around lake shores, riparian vegetation is removed, and fishing and water quality decline and consequently, a better view of the lake makes the lake less desirable to view (Kitchell 1992). Development activity then shifts to other, more remote lakes. Such feedbacks are rarely studied by ecologists, yet the effect of a changing environment on human attitudes and behavior toward the environment is surely one of the most potent ecological forces on the planet.

Ecosystem experiments that view people as interactors with nature (i.e. as both dependent and independent variates) will require new collaborations of ecologists and social scientists. It will take concerted effort to build these cross-disciplinary links. Pioneering collaborations of social and natural scientists suggest some exciting lines of inquiry but also point to significant conceptual, methodological, and empirical problems (Costanza 1991; Gunderson et al. 1995; Hanna et al. 1996; Folke chapter 13). We must find ways to improve environmental management using large-scale experiments that integrate the dynamics of humans and ecosystems (Lee 1993; Gunderson et al. 1995). To succeed, we must carefully describe the systemic context of these experiments: conceptual, temporal, and spatial. Theory, long-term studies, and comparisons have provided crucial context for ecosystem experiments to date, and will continue to do so in the future. Building an ecosystem ecology that includes Earth's dominant species as an interactive component, not an external component, will require the tools that have served us well in the past, plus novel and surprising ones that may arise from new cross-disciplinary work.

Acknowledgments. I thank Tom Frost, Jim Kitchell, and Mike Pace for years of dialogue about ecosystem experiments. Opportunities to learn by doing have been provided by the University of Notre Dame Environmental Research Center and the Wisconsin Department of Natural Resources (WDNR). Several colleagues at WDNR, especially Jim Addis, Paul Cunningham, Doug Knauer, Dick Lathrop, and Mike Staggs, have influenced my thinking about ecosystem experiments. Gene Likens offered insights about rationales for ecosystem experiments and Gordon Orians shared his views on environmental impact assessment. Participants in a planning workshop at the National Center for Ecological Analysis and Synthesis provided very useful reviews of an early outline. Helpful comments on the draft came from Nina Caraco, Stuart Fisher, Tom Frost, Bob Hall, Jeff Houser, Jim Karr, Conrad Lamon, Dick Lathrop, Tara Reed, Bill Robertson, Daniel Schindler, Gaius Shaver, and Kathy Webster; this does not imply that they agree with everything I have written. I am grateful for the support of the National Science Foundation, the Andrew W. Mellon Foundation, and the WDNR.

References

Abrahamson, G., A.O. Stuanes, and B. Tvelte, eds. 1994. *Long-term experiments with acid rain in Norwegian forest ecosystems.* Springer-Verlag, New York.

Allen, T.F.H., and T.W. Hoekstra. 1992. *Toward a unified ecology.* Columbia University Press, New York.

Bower, P.M., C.A. Kelly, E.J. Fee, J.A. Shearer, D.R. DeClerq, and D.W. Schindler. 1987. Simultaneous measurement of primary production by whole-lake and bottle radiocarbon additions. *Limnology and Oceanography* 32:299–312.

Carpenter, S.R. 1989. Replication and treatment strength in whole-lake experiments. *Ecology* 70:453–463.

Carpenter, S.R. 1993. Statistical analysis of the ecosystem experiments. Pages 26–42 in S.R. Carpenter and J.F. Kitchell, eds. *The trophic cascade in lakes.* Cambridge University Press, London, England.

Carpenter, S.R. 1996. Microcosm experiments have limited relevance for community and ecosystem ecology. *Ecology* 77:677–680.

Carpenter, S.R., S.W. Chisholm, C.J. Krebs, D.W. Schindler, and R.F. Wright. 1995. Ecosystem experiments. *Science* 269:324–327.

Carpenter, S.R., P. Cunningham, S. Gafny, A. Muñoz del Rio, N. Nibbelink, M. Olson et al. 1995. Responses of bluegill to habitat manipulations: power to detect effects. *North American Journal of Fisheries Management* 15:519–527.

Carpenter, S.R., T.M. Frost, D. Heisey, and T.K. Kratz. 1989. Randomized intervention analysis and the interpretation of whole-ecosystem experiments. *Ecology* 70:1142–1152.

Carpenter, S.R., T.M. Frost, J.F. Kitchell, T.K. Kratz, D.W. Schindler, J. Shearer et al. 1991. Patterns of primary production and herbivory in 25 North American lake ecosystems. Pages 67–96 in J. Cole, G. Lovett, and S. Findlay, eds. *Comparative analyses of ecosystems.* Springer-Verlag, New York.

Carpenter, S.R., and J.F. Kitchell, eds. 1993. *The trophic cascade in lakes.* Cambridge University Press, London, England.

Carpenter, S.R., M. Olson, P. Cunningham, S. Gafny, N. Nibbelink, T. Pellett et al. 1998 (in press). Macrophyte structure and growth of bluegill (*Lepomis macrochirus*): design of a multi-lake experiment. Pages 217–226 in E. Jeppesen, K. Christofferson, and M. Sondergaard, eds. *The structuring role of macrophytes in lakes.* Springer-Verlag, Berlin.

Coale, K.H, K.D. Johnson, S.E. Fitzwater, R.M. Gordon, S. Tanner, F.P. Chavez et al. 1996. A massive phytoplankton bloom induced by an ecosystem-scale iron fertilization experiment in the equatorial Pacific Ocean. *Nature* 383:495–501.

Cole, J., G. Lovett, and S. Findlay, eds. 1991. *Comparative analyses of ecosystems.* Springer-Verlag, New York.

Costanza, R., ed. 1991. *Ecological economics: the science and management of sustainability.* Columbia University Press, New York.

Cottingham, K.L., and S.R. Carpenter. 1998. Scale of aggregation for ecological indicators: phytoplankton responses to whole-lake enrichment. *Ecological Applications*: in press.

Ellison, A.M. 1996. An introduction to Bayesian inference for ecological research and environmental decision-making. *Ecological Applications* 6:1036–1046.

Francis, G.R., and H.A. Regier. 1995. Barriers and bridges to the restoration of the Great Lakes basin ecosystem. Pages 239–291 in L.H. Gunderson, C.S. Holling, and S.S. Light, eds. *Barriers and bridges to the renewal of ecosystems and institutions.* Columbia University Press, New York.

Frost, T.M., D.L. DeAngelis, S.M. Bartell, D.J. Hall, and S.H. Hurlbert. 1988. Scale in the design and interpretation of aquatic community research. Pages 229–260 in S.R. Carpenter, ed. *Complex interactions in lake communities*. Springer-Verlag, New York.

Gelman, A., J.B. Carlin, H.S. Stern, and D.B. Rubin. 1995. *Bayesian data analysis*. Chapman & Hall, New York.

Gibson, D.J., T.R. Seastedt, and J.M. Briggs. 1993. Management practices in tallgrass prairie: large- and small-scale effects on species composition. *Journal of Applied Ecology* 30:247–255.

Gulati, R.D., E.H.R.R. Lammens, M.L. Meijer, and E. van Donk, eds. 1990. *Biomanipulation—tool for water management*. Kluwer, Dordrecht, The Netherlands.

Gunderson, L.H, C.S. Holling, and S.S. Light. 1995. *Barriers and bridges to the renewal of ecosystems and institutions*. Columbia University Press, New York.

Gurevitch, J., and L.V. Hedges. 1993. Meta-analysis: combining the results of independent experiments. Pages 378–398 in S.M. Scheiner and J. Gurevitch, eds. *Design and analysis of ecosystem experiments*. Chapman & Hall, New York.

Hall, R.J., and G.E. Likens. 1981. Chemical flux in an acid-stressed stream. *Nature* 292:329–331.

Hanna, S.S., C. Folke, and K.-G. Maler. 1996. *Rights to nature: ecological, economic, cultural, and political principles for the environment*. Island Press, Washington, DC.

Hasler, A.D., O.M. Brynildson, and W.T. Helm. 1951. Improving conditions for fish in brown-water lakes by alkalization. *Journal of Wildlife Management* 15:347–352.

Hilborn, R. 1992. Can fisheries agencies learn from experience? *Fisheries* 17:6–14.

Hilborn, R., and M. Mangel. 1997. *The ecological detective*. Princeton University Press, Princeton, NJ.

Hilborn, R., C.J. Walters, and D. Ludwig. 1995. Sustainable exploitation of renewable resources. *Annual Review of Ecology and Systematics* 26:45–67.

Holling, C.S., ed. 1978. *Adaptive environmental assessment and management*. John Wiley & Sons, New York.

Holling, C.S. 1995. What barriers? What bridges? Pages 3–36 in L.H. Gunderson, C.S. Holling, and S.S. Light, eds. *Barriers and bridges to the renewal of ecosystems and institutions*. Columbia University Press, New York.

Howson, C., and P. Urbach. 1989. *Scientific reasoning: the Bayesian approach*. Open Court Press, La Salle, IL.

Hurlbert, S.H. 1984. Pseudoreplication and the design of ecological field experiments. *Ecological Monographs* 54:187–211.

Jefferys, W.H., and J.O. Berger. 1992. Ockham's razor and Bayesian analysis. *American Scientist* 80:64–72.

Kitchell, J.F., ed. 1992. *Food web management: a case study of Lake Mendota*. Springer-Verlag, New York.

Krebs, C.J., S. Boutin, R. Boonstra, A.R.E. Sinclair, J.N.M. Smith, M.R.T. Dale et al. 1995. Impact of food and predation on the snowshoe hare cycle. *Science* 269:1112–1115.

Lee, K.N. 1993. *Compass and gyroscope.* Island Press, Washington, DC.

Lee, R.G. 1992. Ecologically effective social organization as a requirement for sustaining watershed ecosystems. Pages 73–90 in R. Naiman, ed. *Watershed management.* Springer-Verlag, New York.

Levin, S.A. 1992. The problem of pattern and scale in ecology. *Ecology* 73:1943–1967.

Likens, G.E. 1985. An experimental approach for the study of ecosystems. *Journal of Ecology* 73:381–396.

Likens, G.E., ed. 1989. *Long-term studies in ecology.* Springer-Verlag, New York.

Likens, G.E. 1992. *The ecosystem approach—its use and abuse.* The Ecology Institute, Oldendorf/Luhe, Germany.

Likens, G.E., F.H. Bormann, R.S. Pierce, and W.A. Reiners. 1978. Recovery of a deforested ecosystem. *Science* 199:492–496.

Lodge, D.M., S.C. Blumenshine, and Y. Vadeboncoeur. 1998. Insights and application of large-scale, long-term ecological observations and experiments. In W.J. Resetarits and J. Bernardo, eds. *The state of experimental ecology: questions, levels and approaches.* Oxford University Press, London, England.

McAllister, M.K., and R.M. Peterman. 1992. Experimental design in the management of fisheries: a review. *North American Journal of Fisheries Management* 12:1–18.

McDonnell, M.J., and S.T.A. Piclcett. 1993. *Humans as components of ecosystems.* Springer-Verlag. New York.

McLain, R.J., and R.G. Lee. 1996. Adaptive management: promises and pitfalls. *Environmental Management* 20:437–448.

Mooney, H.A., E. Medina, D.W. Schindler, E.-D. Schulze, and B.H. Walker. 1991. *Ecosystem experiments.* John Wiley & Sons, New York.

Ojima, D.S., D.S. Schimel, W.J. Parton, and C.E. Owensby. 1994. Long- and short-term effects of fire on nitrogen cycling in tallgrass prairie. *Biogeochemistry* 24:67–84.

Orians, G. 1986. Ecology and environmental problem solving. *Northwest Science* 60:287–292.

Paine, R.T. 1969. A note on trophic complexity and community stability. *American Naturalist* 103:91–93.

Paine, R.T., J.L. Ruesink, A. Sun, E.L. Soulanille, M.J. Wonham, C.D.G. Harley et al. 1996. Trouble on oiled waters: lessons from the *Exxon Valdez* oil spill. *Annual Review of Ecology and Systematics* 27:197–235.

Pickett, S.T.A. 1989. Space-for-time substitution as an alternative to long-term studies. Pages 110–135 in G.E. Likens, ed. *Long-term studies in ecology.* Springer-Verlag, New York.

Pickett, S.T.A., J. Kolasa, and C.G. Jones. 1994. *Ecological understanding.* Academic Press, New York.

Pole, A., M. West, and J. Harrison. 1994. *Applied Bayesian forecasting and time series analysis.* Chapman & Hall, New York.

Rasmussen, P.W., D.M. Heisey, E.V. Nordheim, and T.M. Frost. 1993. Time-series intervention analysis: unreplicated large-scale experiments. Pages 138–158 in S.M. Scheiner and J. Gurevitch, eds. *Design and analysis of ecological experiments.* Chapman & Hall, New York.

Schindler, D.W. 1977. The evolution of phosphorus limitation in lakes. *Science* 195:260–262.

Schindler, D.W. 1992. A view of NAPAP from north of the border. *Ecological Applications* 2:124–130.

Schindler, D.W., T.M. Frost, K.H. Mills, P.S.S. Chang, I.J. Davies, D. Findlay et al. 1991. Comparisons between experimentally- and atmospherically-acidified lakes. *Proceedings of the Royal Society of Edinburgh* 97B:193–226.

Slobodkin, L.B. 1988. Intellectual problems of applied ecology. *BioScience* 38:337–342.

Stewart-Oaten, A., W. Murdoch, and K. Parker. 1986. Environmental impact assessment: "pseudoreplication" in time? *Ecology* 67:929–940.

Tilman, D. 1989. Ecological experimentation: strengths and conceptual problems. Pages 136–157 in G.E. Likens, ed. *Long-term studies in ecology*. Springer-Verlag, New York.

Wallace, J.B., M.R. Whiles, S. Eggert, T.F. Cuffney, G.J. Lucthart, and K. Chung. 1996. Long-term dynamics of coarse particulate organic matter in three Appalachian Mountain streams. *Journal of the North American Benthological Society* 14:217–232.

Walters, C.J. 1986. *Adaptive management of renewable resources*. MacMillan, New York.

Walters, C.J. 1993. Dynamic models and large scale field experiments in environmental impact assessment and management. *Australian Journal of Ecology* 18:53–61.

Walters, C.J., J.S. Collie, and T. Webb. 1988. Experimental design for estimating transient responses to management disturbances. *Canadian Journal of Fisheries and Aquatic Sciences* 45:530–538.

Westley, F. 1995. Governing design: the management of social systems and ecosystems management. Pages 391–429 in L.H. Gunderson, C.S. Holling, and S.S. Light, eds. *Barriers and bridges to the renewal of ecosystems and institutions*. Columbia University Press, New York.

Wright, R.F., E. Lotse, and A. Semb. 1988. Reversibility of acidification shown by whole-lake catchment experiments. *Nature* 334:670–675.

13
Ecosystem Approaches to the Management and Allocation of Critical Resources

CARL FOLKE

Summary

The chapter identifies concepts, approaches, and frontiers that can increase the understanding of the linkages between ecology, society, and the economy for improved ecosystem management. The first part focuses on ecosystem services and ecological footprint analyses as indicators of the dependence by societies on ecosystem processes and functions, with examples from specific economic sectors and ecosystems to global values of ecosystem services. Such social and economic driving forces behind ecosystem degradation as misplaced economic incentives and trade policies are investigated. The critical role of human institutions is elucidated, and insights from linked social-ecological systems for ecosystem management are presented. The chapter identifies the six important research frontiers that treat humans as endogenous factors in ecosystems, which are: (1) the performance of analyses that illuminate and communicate societies' dependence on ecosystem support; (2) the identification and prediction of the effects of social and economic driving forces on ecosystem processes, functions, and services; (3) the study of the interrelations within and between ecosystems in drainage basins, and how they affect and are affected by human activities on various scales and across scales; (4) the building of ecological knowledge through adaptive management, and the combination of practical ecological knowledge of local resource users with scientific knowledge; (5) the design of institutions that work in synergy with ecosystem processes and functions; and (6) the learning about ecosystem management from linked social-ecological systems. Most of these frontiers will require substantive collaboration with social scientists.

Introduction

In this chapter, humans are seen as components of ecosystems, as a part of and not apart from the processes and functions of nature. Traditionally in ecology, humans have been treated as nonexistent or as external to ecosystems. In recent years, however, an increasing number of ecologists have started to emphasize the need explicitly to include human activities in ecosystem research (e.g. Lubchenco et al. 1991; McDonnell and Pickett 1993; Groffman and Likens 1994; Gunderson et al. 1995; Christensen et al. 1996). As stated by O'Neill (1996), the natural world, isolated from large-scale human impacts, exists only in our imaginations. We have already altered the atmosphere, exterminated or geographically shuffled species, and broken migration routes to such an extent that no place on the planet is truly natural. The dynamics of an ecosystem cannot be understood unless the dynamics of the dominant species is understood. *Homo sapiens* is the dominant species and its dynamics are an integral part of every ecosystem we study (Odum 1971; Ehrlich 1994).

In this chapter, the human system is seen as a subsystem of the ecosphere, being dependent for development and survival on the structure and function of ecosystems, including biological diversity. The life-support capacity of ecosystems must be accounted for in any society and economy, but this is seldom the case today. The necessity of functional ecosystems is rarely reflected in the prices of goods and services bought and sold on markets or traded between nations, and is seldom fully taken into account by the institutions that provide the framework for human action, whether political or social. Ecosystem contribution to economic development and growth is "mentally hidden" to many parts of modern society, reflected, for example, in most models of economic growth in which the environmental resource base is neglected (Dasgupta 1997; Cleveland et al. 1997), or in the field of development economics in which the essential role of life-support systems is largely absent (Dasgupta and Mäler 1996). Many societies employ social norms and rules that: (1) bank on future technological fixes; (2) use narrow indicators of welfare; (3) employ worldviews that alienate people from their dependence on life-support by ecosystems; and (4) assume that it is possible to find humanmade substitutes for the loss of ecosystems and the services they generate. Viewing the social and economic subsystem in its proper perspective relative to the ecosphere is crucial for moving human society toward a sustainable relationship with the ecosphere (Berkes and Folke 1994).

Two new ecological journals addressing such issues are in the pipeline, however; the Internet journal *Conservation Ecology* (published by the Ecological Society of America, editor-in-chief C.S. Holling), which started in mid-June 1997, and *Ecosystems* (Springer-Verlag, editors-in-chief Stephen Carpenter and Monica Turner), the first issue of which appeared in February 1998. Both journals encourage submissions of papers that integrate natural and social processes at appropriate scales. Several professional societies addressing the interactions between humans and nature, which are founded on the ecosystem perspective, and which also publish associated scientific journals, have been established in the last few years. They include the International Society for Ecological Economics (ISEE) in

1989, the International Ecological Engineering Society (IEES) in 1993, and the International Society for Ecosystem Health (ISEH) in 1994.

Ecological economics is a transdisciplinary field of study that aims to address critical relationships between ecosystems and socioeconomic systems in order to develop a deeper understanding of the entire system of humans and nature as a basis for effective policies for sustainability (Costanza 1991; Jansson et al. 1994). It focuses particularly on the issues of sustainability facing humanity and the life-supporting ecosystems on which society depends (Odum 1989; Folke 1991a). These problems involve: (1) assessing and ensuring that human activities are ecologically sustainable, from local to global levels; (2) distributing resources and property rights fairly within the present-day generation of humans, between this generation and future generations, and between humans and other species; and (3) efficiently allocating resources as defined by the first two points in this list, including both marketed and nonmarketed resources.

The reason that a transdisciplinary approach is especially needed now is that the globalization of human activities, population growth, and large-scale movements of people have placed humankind in an era of novel dynamics and interdependence of ecological, social, and economic systems at regional and even planetary scales (Daily and Ehrlich 1992; Holling 1994). These interdependent subsystems have become so interwoven that actions taken locally may generate regional and global effects, witnessed in, for example, climate change (Houghton et al. 1996) and the evolution of new diseases (McMichael et al. 1996). This trend is reinforced by increasing trade and opportunities for economic growth that can cause environmental effects that spread over larger and larger geographical areas (Ekins et al. 1994; Andersson et al. 1995).

The rapid expansion of human actions on Earth (Turner et al. 1990) is altering the capacity of ecosystems to generate a continuous flow of natural resources and ecosystem services. This capacity is becoming an increasingly limiting factor for social and economic development (Costanza and Daly 1992). As the human dimension grows relative to its supporting ecosystems, this may affect the dynamics of both, and the dynamics may become increasingly unpredictable and potentially discontinuous as the jointly determined ecological-social-economic system draws closer to critical biophysical thresholds at regional and global levels (Common and Perrings 1992; Arrow et al. 1995; Turner et al. 1997). It is not sufficient in this novel situation simply to describe the effects on ecosystems caused by humans. It is also not sufficient for ecology to treat humans as external or nonexistent. Integrated modes of inquiry are needed for understanding, and understanding (but not complete explanation) is needed to form policies.

How do we as ecologists analyze such critical interdependencies and communicate the necessity of ecosystem support for a prosperous development of human society, to the social sciences and to management and policy? In this chapter, some concepts, approaches, and frontiers will be identified that can increase the understanding and improve management of critical ecological-social-economic linkages. In particular, the need to foster the use of ecological knowledge and information in practical management, environmental problem-solving, and in society in general will be emphasized.

In the first part of this chapter, I will focus on two approaches. The first is the identification, quantification, modeling, and valuation of ecosystem services, and the second is the quantification of ecological "footprints" as one indicator of dependence by societies on ecosystem support. Examples will be provided from specific economic sectors and particular ecosystems, to large-scale drainage basins and aggregate global values of ecosystem services.

The second part of this chapter discusses human behavior behind ecosystem degradation, focusing on such indirect driving forces as misplaced economic incentives, trade policies, and other institutional failures. The critical roles that institutions play in ecosystem management are examined, in particular in relation to property rights and their significance for ecosystem management.

In the third section, the need to foster ecological knowledge through adaptive management and through insights drawn from linked social-ecological systems is emphasized. In the conclusion, I synthesize the frontier issues discussed in this chapter.

The Identification, Quantification, Modeling, and Valuation of Ecosystem Services

Ecosystems are essential in global biogeochemical cycles and the water cycle, and produce the bulk of renewable resources and ecosystem services on which the well-being of human society rests. In economics, land, labor, and capital have traditionally been the three main inputs to economic development—the three factors of production. The ecosystem may also be seen as an essential "factor of production" for social and economic development. Economics is concerned with allocating scarce resources among alternative uses in the most efficient way (e.g. Pearce and Turner 1990; Common 1996). During the industrial era, the availability of land (generally interpreted as agricultural land) was not considered a limiting factor. Economic development was thought to be constrained by labor productivity and capital [albeit supported by cheap fossil energy (Hall et al. 1986)], and consequently land, an abundant resource, was basically dropped from economic-growth theory. Ecological economists have tried to bring land back into the picture by using the concept of natural capital for this purpose (e.g. Costanza and Daly 1992; Jansson et al. 1994).

Natural capital generates to human society a flow of nonrenewable resources, renewable resources, and ecosystem services. Ecosystems produce renewable resources and services. The concept of ecosystem services (also referred to as ecological services, environmental services or functions) provides a mental link to social scientists and policy-makers, and has been useful for communicating the dependence of society on the work of ecosystems (Odum 1975; Westman 1977; Ehrlich and Mooney 1983; Folke 1991b; de Groot 1992; Perrings et al. 1992; Myers 1996; Baskin 1997; Daily 1997). Ecosystem services include maintenance of the composition of the atmosphere, amelioration of climate, flood controls, drinking water supply, waste assimilation, recycling of nutrients, generation of soils, pollination of crops, predation on pest insects, provision of food and medicine, maintenance of species and a vast genetic library, and also preservation of the scenery of the landscape, recreational sites, which have both aesthetic and amenity values (Ehrlich and Ehrlich 1981; Folke 1991a; de Groot 1992; Baskin 1997; Daily 1997).

Biological diversity plays an important role in generating and sustaining the flow of renewable resources and ecosystem services (Tilman and Downing 1994; Naeem et al. 1994 Tilman chapter 19). Biodiversity also contributes to the capacity of ecosystems to absorb disturbance, including that which is caused by humans. Biodiversity aids in maintaining an ecosystem's resiliency in the face of change (Holling et al. 1995; Folke et al. 1996a) and therefore sustains social and economic development (Perrings et al. 1992). The capacity to absorb disturbance that biological diversity provides can be viewed as "natural insurance capital" for securing the generation of essential ecosystem services under present-day and future environmental conditions (Barbier et al. 1994).

An ecosystem perspective stresses that there cannot be any production of renewable resources and ecosystem services without an ecosystem to "produce" them. For example, a fish in the sea is not just a resource in the water. The fish is a part of the ecosystem in which it lives, in which it is produced, and the interactions that produce and sustain the fish are inherently complex. Ecosystems are multifunctional. A forest produces timber but also assimilates nutrients and carbon dioxide. As an example, the diversity of natural resources and services generated by a mangrove ecosystem is shown in Table 13.1.

The human use and misuse of resources and ecosystem services greatly affect the flows of other resources and ecosystem services. Degradation and simplification of ecosystems may not only reduce these flows but may also push the system toward a threshold and then thrust it into another stability domain (Ludwig et al. 1997). Moreover, impacts on one ecosystem may spill over into another, and consequently affect the functioning of the latter system and the services it generates. The human use of an ecosystem service is dependent on the existence, operation, and maintenance of a multifunctional ecosystem linked to other multifunctional ecosystems; it also is dependent on energy, biogeochemical and hydrological flows, and other human uses and misuses affecting those flows (Figure 13.1). Because

TABLE 13.1. Resources and ecosystem services generated by mangrove ecosystems.

Resources
Firewood
Timber
Boat-building and other construction materials
Fish and shellfish
Fodder and green manure for agriculture
Various household items and products for textile and leather production

Ecosystem services
Prevention of coastal erosion and soil deposition
Trapping of nutrients, waste, and pollutants
Provision of spawning, nursery, and forage areas for numerous fish and shellfish
Provision of fish-attraction shelters
Food, shelter, and sanctuary for birds and mammals
Maintenance of water quality of such surrounding coastal ecosystems as coral reefs

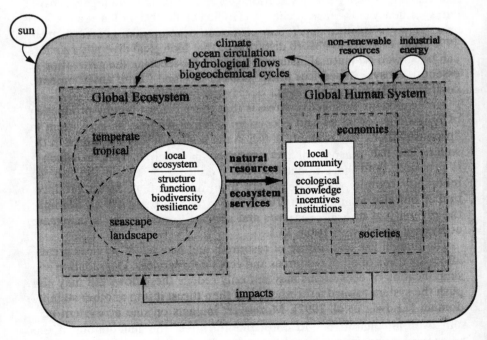

FIGURE 13.1. The combined system of humans and nature. Ecosystems sustain human societies with a flow of essential natural resources and ecosystem services. The rapid expansion of the human system has turned the capacity of ecosystems to generate these flows into an increasingly limiting factor for social and economic development.

of the scale of human activity, such cross-scale interactions, both temporal and spatial, will increasingly affect the flow of essential ecosystem services and support upon which human welfare depends (Levin 1992; Holling 1994).

According to economic theory, the total value of the environment has a very different origin than that which is biophysical, however. It is founded on the preferences of individual human beings. The approach to valuation is to try to reveal consumer preferences and express them as how much people are willing to pay for various aspects of the environment. The total economic value has usually been divided into use and nonuse values (see e.g. Krutilla 1967; Pearce and Turner 1990). Use values refer to the direct and indirect use of the environment, some of which have a market price, as with fish, grain, and timber (in a market in "perfect" competition, the price of a commodity is thought to reflect consumers' preferences for that commodity). Nonetheless, many resources and most ecosystem services do not have a market price. There are also so-called nonuse values, as in bequest values and existence values, which reflect how much people are willing to pay to preserve ecosystems for future generations, or preserve them in their own right (Krutilla and Fisher 1975). Nonuse values are not related to the structure and function of ecosystems.

Several studies have tried to place monetary values on wetland ecosystems (reviewed in Gren and Söderqvist 1994). These studies have tried to explicitly value such ecosystem services as flood and storm protection, nitrogen purification, and water buffering in monetary terms (Thibodeau and Ostro 1981; Faber 1987; Gren 1995). Attempts have also been made to estimate the life-support value of the entire wetland ecosystem (Gosselink et al. 1974; Costanza et al. 1989).

Folke (1991b) showed that it takes expensive and environmentally degrading fossil fuels to try to replace the loss of wetland services with technical substitutes (Table 13.2). The annual cost of the substitutes was estimated from $400,000 to 1,085,000 (U.S. dollars, 1989), and despite those costs the substitutes do not even fully replace lost ecosystem services. Technological development should therefore focus on tools and techniques that maintain the "unpriced" support of an ecosystem, instead of replacing this support with fossil fuel-dependent technologies after it has been destroyed (Bormann 1976). One solution is to apply the technology that restore ecosystems, for example, wetlands for drinking-water supply, recreation, and as filters for nitrogen to coastal waters (e.g. Mitsch and Jörgensen 1989). Indeed, wetland restoration for nitrogen abatement is a technology that has been increasingly applied and that seems to be a cost-effective solution that also generates other ecosystem services (Leonardsson 1994; Gren 1995).

My research group is in the process of quantifying the nitrogen-retention capacity of wetlands in the large-scale drainage basins of the Baltic Sea, the

320   C. Folke

TABLE 13.2. Ecosystem services of a Swedish wetland, and the cost of technologies (in 1989 Swedish money (SEK), and in fossil fuel equivalents, TJ (10^{12}J)) to replace lost services following wetland exploitation.[1]

Ecosystem services	Replacement Technologies	Monetary Costs 1000 SEK	Energy Costs TJ
Maintaining drinking-water quality	Water quality controls	12–40	0.020–0.180
	Water purification plant	32	0.180–0.230
	Nitrogen and saltwater Filtering	0–460	0–3.290
	Silos for manure from domestic animals	370–950	1.160–5.405
Maintaining drinking-water quantity	Water transports		
	—humans	0–500	0–4.050
	—domestic animals	0–335	0–2.715
	Pipe line to distant source	0–990	0–5.570
Maintaining surface water level	Dams for irrigation	57–205	0.200–1.167
	Pumping water to dams	11–36	0.025–0.225
	Irrigation pipes and machines	58–184	0.230–1.045
Maintaining ground water level	Well drilling	33–53	0.080–0.300
Moderation of waterflows	Regulating wire	10–18	0.060–0.200
	Pumping water to stream	7–11	0.015–0.070
Peat accumulation	Redraining and clear-cutting of ditches and stream	50–56	0.200–0.330
		880–1428	7.650–20.275
	Martificial fertilizers		
Processing sewage and other waste	Mechanical sewage treatment and storing	625–750	1.360–4.265
	Sewage transport	63–126	0.165–1.020
	Sewage treatment plant	3–200	0.005–1.170
Filter to coastal waters	Nitrogen reduction in sewage treatment plant	40–45	0.085–0.255
Sustaining trout populations	Hatchery raised trout and farmed salmon	210–250	0.720–1.670
Sustaining wetland dependent flora or fauna	Work by NGOs to protect endangered species	68–314	0.120–1.965
Total		2530–6985	12–55

1. Modified from Folke 1991b.

role of parks in cities and other urban areas as filters for car emissions, the value of birds for pest control, and of coral reefs for fisheries, recreational diving, and wind and wave protection. A recent book, *Nature's Services* (Daily 1997) attempts to quantify and value such ecosystem services as soil generation, pollination, food stability, pest control, wastewater treatment, biodiversity and the genetic library, in addition to freshwater support and services generated by forests, grasslands, marine systems, and the South African fynbos ecosystem.

The fynbos ecosystem study provides a very interesting example of ecological knowledge translated directly into policy and action. In contrast to many other economic valuation studies, this was founded on ecological understanding of the ecosystem. The fynbos is the predominant vegetation of the Cape Floristic Region supporting 8,500 plant species, of which 68% are endemic. Alien invasive plants (including shrubs and trees from other fire-prone, Mediterranean-climate ecosystems) are the major human-induced threat to fynbos biodiversity and ecosystem services (Cowling 1992). They eliminate native plant diversity and reduce water production from the fynbos ecosystem by 30% to 80%. Without management, pristine watersheds will have alien plant cover of between 80% to 100% after 100 years (Le Maitre et al. 1996). The major ecosystem services provided by the fynbos ecosystem are wildflower harvesting, biodiversity storage for various uses including floricultural varieties, water runoff providing water for human consumption, hiker and ecotourist visitation, and aesthetic as well as cultural and existence values (Cowling et al. 1997). The value of the services of a pristine fynbos system reached more than $82 million (U.S. dollars) annually and was derived from the development of a dynamic simulation model that integrated ecological and economic data and processes (Higgins et al. 1977). Clearing alien species amounted to only 0.6% to 4.8% of the value to society of the ecosystem services supplied by fynbos watersheds. Alternative technical solutions to water delivery would cost 1.8 to 6.7 times more than fynbos catchment management (van Wilgen et al. 1996). The analysis was presented to the South African Minister of Water Affairs and Forestry, who decided to invest 100 million Rand (more than $20 million U.S. dollars) annually to clear the fynbos of invading alien plants, starting in 1995. The "alien plant removal from watersheds project" is envisaged to run for twenty years, will generate thousands of jobs, involve numerous training programs that will communicate ecological knowledge, and safeguard many thousands of plant and animal species (Cowling, personal communication).

A popular approach to environmental valuation, particularly among U.S. economists, is the contingent valuation method (CVM). Using CVM, people are asked, generally through a questionnaire, about their willingness to pay for various aspects of nature. It is comparable to a market investigation of the environment and was used, for example, to evaluate the effects of the *Exxon-Valdez* oil spill in Alaska (Carson et al. 1992; Arrow et al. 1993). Another common valuation technique is the travel-cost method, in which the money people spend on visiting an ecosystem is used as an estimate of how much they value that system. It has been applied to evaluating the recreational value of nature reserves and marine parks (Dixon and

Sherman 1990). For a description of these and other economic valuation techniques see Freeman (1994).

Through extensive literature, Costanza et al. (1997) identified seventeen different specific ecosystem services and collected or calculated economic values for sixteen different biomes to arrive at a minimum estimate of the global value of the world's ecosystem services. Their controversial study, which has generated a lot of debate, found that the majority of the values of services that were identified in the literature concern nutrient retention, waste treatment, disturbance regulation, and gas regulation. At present, the majority of these values are not reflected in the prices of goods and services bought and sold on markets or traded between nations. The division by biome is about 62% from marine systems, which are largely coastal systems, and about 38% from terrestrial systems, which are mainly forests and wetlands. This does not imply that other ecosystems or services are of less importance. It merely reflects that only certain ecosystem services have been studied to date, and reveals that there are large gaps in the coverage of ecosystem services. Their estimate of the global total value was approximately $32 trillion (U.S. dollars) per year, which is 1.8 times the present-day global Gross Domestic Product (Costanza et al. 1997).

The monetary values presented in these studies are directly or indirectly derived from consumer preferences. Human preferences are not necessarily related to the importance of multifunctional ecosystems for human welfare. Often humans are not aware of their indirect uses of critical resources and ecosystem services, and their dependence on ecosystem support. And even if they are aware, they may not value it. Human preferences are often decoupled from biophysical realities.

Most ecosystem scientists would presumably argue that ecosystem services and support are essential for society irrespective of whether or not they are perceived as important by humans. There are many ecosystem functions and services that meet the criteria of having economic value (they contribute to well-being and are scarce) but for which humans have not yet developed preferences. This can be the result of ignorance of the contribution of ecosystems to human well-being, or because cultural or social mores tend to preclude preference formation. Therefore, valuation of the environment should go beyond a simple aggregation of preferences to encompass the social and cultural dimension in which individual preferences act, as well as the ecological preconditions for social and economic development (Costanza and Folke 1997). As will be discussed later, institutions are critical in this context because they provide the framework, which are the norms and rules for individuals.

Integrated modeling approaches that directly address the functional value of ecosystem services by looking at the long-term, spatial, and dynamic linkages between ecosystems and socioeconomic systems are in progress (Bockstael et al. 1995; Cowling et al. 1997; Higgins et al. 1977). Such modeling may allow evaluation of the indirect effects over long-time horizons of present-day policy options. These effects are almost always ignored in partial analyses.

There is a wide gap and scope for innovative approaches that more fully capture the essential role and value of biodiversity and ecosystems for social and economic development. The ecosystem approach has an important contribution to make at this frontier.

Ecological Footprints: A Spatial Indicator of Ecosystem Support

Borgström (1967) used the term "ghost acreage" to reflect the area of agricultural land required for human food consumption. Using the concept of energy density (i.e. the amount of energy consumed per unit of area per year), Odum (1989) estimated that cities and industrial areas occupy 6% of the continental United States, but when their ecosystem shadow area is included they appropriate about 35%. Jansson and Zucchetto (1978) and Folke (1988) showed that offshore fisheries and intensive aquaculture require extensive areas of marine ecosystem support. Consequently, Rees and Wackernagel (1994) introduced the concept of the "ecological footprint," to reflect the ecosystem area necessary to sustain present-day levels of resource consumption and waste discharge by a given human population (Wackernagel and Rees 1996).

The appropriate ecosystem area, or the ecological footprint, is an indicator that attempts to capture the ecosystem support that is required to generate a flow of resources and ecosystem services. The footprint concept illuminates the "hidden" requirements for ecosystem support, and places the scale of human activities within an ecosystem framework. It also demonstrates that human activities, which at first glance may seem separate from nature, would not function without ecosystem support. The work of ecosystems is hidden because people and policy seldom perceive it, but it is nevertheless real. Using the concept of the ecological footprint, we have analyzed the area of coastal and marine ecosystems that is required to sustain the yields of fish, shrimp, shellfish, and seaweed from various fisheries and aquaculture (reviewed in Folke et al. 1998). We have also analyzed the ecological footprints of cities and large-scale drainage basins (Folke et al. 1996b; Folke et al. 1997a; Jansson et al. 1998).

Larsson et al. (1994) estimated the spatial ecosystem support that is required to produce food inputs, nursery areas, and clean water for semi-intensive shrimp farms in Colombia. Ecosystem support areas were estimated for six different systems of which four were marine and coastal ecosystems, indicating that a semi-intensive shrimp farm needs an ecological footprint that is thirty-five to 190 times larger than the area of the farm. Fish farming in cages depends on marine ecosystem areas as large as 10,000 to 50,000 times the area of the cages for producing the food, and 100 to 200 times for processing parts of its waste (Folke et al. 1998). Hence, to perceive modern aquaculture as it really is and manage its problems, human thoughts and actions must be expanded far beyond the site of the farm (Figure 13.2).

The dependence of modern aquaculture on resources and services from marine ecosystems makes it difficult to believe that yields from a rapidly growing aquaculture industry, dominated by intensive monocultures, will replace the decline in fisheries harvest (e.g. FAO 1995). To the contrary, exploitation of fish to support aquacultured species with feed may further reduce commercially important fish stocks on which fisheries are based. There may be a risk that fish used for human consumption (e.g. in lesser developed countries) will be exploited for fish meal that is exported to intensive aquaculture production in economically and materially developed countries (Folke and Kautsky 1989; Folke et al. 1998). An ecosystem approach that addresses the relationships between fisheries and aquaculture is needed and it is in this context that the concept of ecological footprints is useful. Such analyses can provide important information about ecological constraints that may otherwise be overlooked in ecosystem management.

City inhabitants require productive ecosystems to produce the food, water, and renewable resources that are consumed inside the city. They also

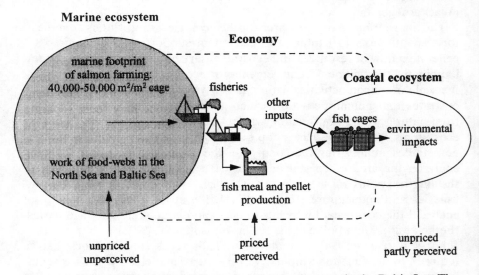

FIGURE 13.2. Atlantic salmon (*Salmo salar*) farming in cages in the Baltic Sea. The salmon farming industry uses huge marine ecosystems for production of farmed salmon and coastal ecosystems for waste release. This ecosystem appropriation is not paid for by the industry. The necessity of fundamental ecosystem support is seldom perceived in decision-making and policy.

depend on ecosystems to provide clean air and process waste. We estimated that the twenty-two million inhabitants of the largest cities in the Baltic Sea drainage basin (about 26% of the region's population), require an area from forest, agricultural, and marine ecosystems for their present consumption of wood, papers, fiber and food that is about 200 times the area of the cities themselves (Folke et al. 1996b). The region's forests, inland bodies of water, and wetlands sequester Carbon dioxide emissions from the cities. About 40% of the excretory release of nitrogen from city inhabitants remain in the wastewater released from sewage treatment plants. There is a potential to process this wastewater in wetlands, and agricultural land in the region has the potential to assimilate the excretory release of phosphorous by the city inhabitants that is found in sewage sludge.

We estimated that the Baltic cities' total appropriation of terrestrial and marine ecosystems in order to secure these natural resources and ecosystems services is at least 600 to 1200 times the area of the cities themselves, or $60,000\,m^2$ to $120,000\,m^2$ for the average citizen. The actual area of the twenty-nine cities corresponds to 0.1% of the area of the whole Baltic Sea drainage basin, but when their "hidden demand" for ecosystem support is taken in to account their appropriation of ecosystem-produced resources and services requires an area corresponding to as much as 75% to 1.5 times the whole Baltic Sea drainage basin (Folke et al. 1997a) (Figure 13.3).

The same resources and ecosystem services are appropriated by the entire population of eighty-five million people living within the Baltic Sea drainage basin. At present, they require ecosystem support corresponding to an area 3.2 to six times larger than the whole drainage basin (Jansson et al. 1998). These estimates are conservative, however, because only the spatial capacity of some ecosystems to produce some renewable resources and ecosystem services used by people have been quantified. Nevertheless, they indicate that the well-being of people in cities and from the fourteen nations in the drainage basin not only depends on their own ecosystems, but also on the productivity of ecosystems of other nations. Hence, economies are dependent not only on imported goods from other countries but also on the support capacity of other nations' ecosystems for the production of these goods. Trade means that people in one country become indirectly dependent on their access to natural resources and ecosystem services outside the boundaries of their own country. It should therefore be in their self-interest to contribute to the conservation of functional diversity and resilience also in other regions of the world (Andersson et al. 1995).

The present-day practice of the ecological footprint method provides a static measure—a snapshot of the ecosystem area required to produce natural resources and ecosystem services. In its present form the footprint does not capture the dynamic nature of change in ecosystems, in human demand, in technology, and so on. The footprint results presented here are

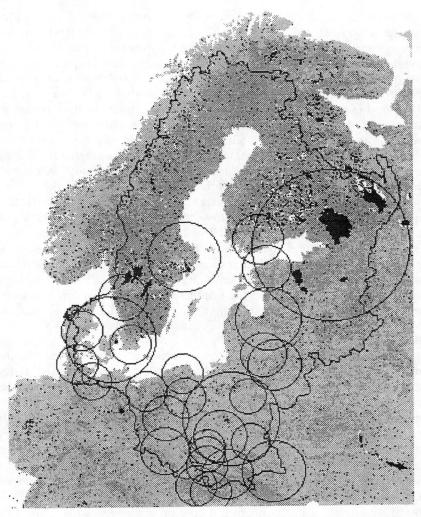

FIGURE 13.3. The ecological footprint of the largest cities in the Baltic Sea drainage basin. The area of the "hidden ecosystem demand" is illustrated by the circles around each city. The cities do not appropriate the actual area of the circles, but they demand this area. Because of trade, the appropriation may take place elsewhere on Earth (adapted from Cities et al. 1997a).

only a reflection of underlying ecological, social, and economic conditions at a given point in time. The footprint concept could be developed into an analytical tool that may capture dynamic changes in complex ecosystems in relation to different human activities and technologies. For example, the spatial extent and configuration of a footprint may change, for example with technology, the development of new markets, or trade opportunities. Those human actions that degrade the life-support capacity of ecosystems and use it in a wasteful manner would demand a much larger footprint than those that are more in tune with the processes and functions of ecosystems. Furthermore, there are those human activities that will leave an imprint for a long time, and even cause irreversible changes, although the footprints of other activities may disappear more rapidly. Land and water uses that reduce the diversity of the landscape, and with it ecological communities, populations and species, are apt to produce less resilient ecosystems. The ecosystem approach has an important role to play in determining the change and dynamics of ecosystem support under different human activities.

Because ecosystems are multifunctional, several human activities may exploit, abuse, and yet also depend on the same footprint. Double counting has to be avoided. Thus far, we have only looked at the ecosystem support required to generate a particular resource or service. The interconnectedness of resources and services—how they are linked across spatial and temporal scales and to a diversity of human activities—is an important frontier that needs to be addressed and that requires ecological knowledge. For example, when we estimated the water consumption by humans in relation to annual rainfall, we found that only 3% to 6% was appropriated by humans in cities and industries. The conventional conclusion would be that there is plenty of water in the Baltic Sea drainage basin. But when we analyzed the amount of water that is needed to produce the resources and ecosystem services that humans consume, the result indicates that 45% to 75% of the annual precipitation is appropriated by humans directly and indirectly (Jansson et al. 1998).

The complex issue of joint products of ecosystems must be addressed and it requires knowledge of the internal relationships of ecosystems, analyzed for example through network analysis (Wulff et al. 1989). An important frontier in ecosystem science is to increase understanding of, for example patch dynamics (Pickett and White 1985) and cross-scale interactions (Levin 1992) within and between ecosystems, and how they relate to human appropriation of resources and ecosystem services (Costanza et al. 1993). For instance, if a decision is made to expand an urban area, how will that affect the surrounding ecosystems, their functional diversity, their buffer capacity, their generation of ecosystem services for other human activities? A landscape ecology or preferably a watershed or drainage basin perspective is needed, in which humans and ecosystems are analyzed as one system (Naiman 1992). Such management has been attempted, for example by the Murray Darling Basin Commission in Australia, an intergovernmental agency that was developed for active regional-scale multiple-use management. In the large-scale Murray Darling River system there are major resource allocation issues in relation to water scarcity, urbanization, tourism, primary industry, and biological diversity that need to be approached from a large-scale drainage basin perspective.

What is happening on land also must be linked to aquatic systems. For example, to secure the ecosystem services generated by a coral reef, it is not sufficient to look at only the effects of fishing, diving, or coral blasting. The impacts caused by mangrove destruction and forest clear-cutting upstream have to be accounted for, because such land-use changes will cause leakage of organic matter and nutrients into the coastal waters surrounding coral reefs. An ecosystem approach must also take into account other effects on coral reefs caused by such human activities on land as waste discharges of nutrients and toxic substances from industries, infrastructure development, agriculture, urban areas, other land-use changes, and a growing human population. These are proximate driving forces of humans behind ecosystem deterioration, but to deal with these causes, the indirect driving forces have to be much more explicitly addressed than has been to date.

Driving Forces of Humans Behind Ecosystem Degradation

It is crucial to address the underlying causes—the economic, institutional, social, and cultural factors that direct human activities—because those causes must be changed in order to reduce and ultimately halt the misuse or overuse of ecosystems. The underlying causes include:

- The structure of property rights and other institutions;
- Macro-economic, trade, and other governmental policies;
- Economic and legal incentives;
- The behavior of financial markets;
- Causes behind population pressure;
- Transfer of knowledge and technology;
- Misguided development aid;
- Patterns of production and consumption;
- Power relations in society;
- Level of democracy;
- Worldview;
- Lifestyle;
- Religion;
- Ethics;
- Values.

Several examples these various causes follow.

Results generated by a model of the Costa Rican economy indicate that increased capital and labor taxation, as well as high interest rates, can contribute significantly to deforestation (Persson 1995). For example, in the United Kingdom in the 1980s, tax concessions on reforestation were increased but not for the purchase of land. Investors, therefore, minimized land purchases and located forest plantations on land of low monetary value, for example wetlands, heath, and moorland, thereby depleting "unpriced" wildlife values (Jones and Wibe 1992). In the Brazilian Amazon rainforest, a title to land could be acquired if the tropical forest was used for a number of years, with logging as a proof of the land being used. Farms containing "unused" forests were taxed at higher rates than those containing pastures or cropland. The real interest rate on loans for agriculture was lower than for other land uses, and agricultural income was almost exempt from taxation (Binswanger 1989).

Rural poverty and environmental deterioration in Colombia is exacerbated by agricultural policies that favor the large farmer through public investment patterns and policies on trade, credit, and taxes (Heath and Binswanger 1996). Despite low population density, a moderate rate of population growth, and declining rural poverty, Belize is presantly following a path of environmental degradation similar to the rest of Central America (Lopez and Scoseria 1996). The reasons seem to be misguided land-allocation policies, in conjunction with trade policies that favor land-intensive commodities and that tax labor-intensive commodities. Commercial interests play an extraordinarily important role in deforestation and ecosystem degradation in Belize. They generate fast economic growth, but at the expense of a degraded environment (Lopez and Scoseria 1996).

Large profits linked to trade opportunities provide a strong incentive for the widespread clear-cutting of mangroves in tropical coastal waters to make room for a new technology—shrimp farming in ponds (Wilks 1995). Their destruction results in loss of several ecosystem services (see Table 13.1). For example, because of the loss of mangroves fish reproduction decreases, and may thereby cause a reduction in fish catches, not only for subsistence economies and domestic fisheries, but also for other countries exploiting the fish stocks (Primavera 1993). Market prices of farmed shrimp do not cover the costs of deteriorated coastal and marine ecosystems caused by the farming, nor the social impacts generated (Bailey 1988). The contribution of such shrimp farming to economic growth is based on liquidation of environmental assets. Many politicians measure success in term of economic growth but it is well-known that growth in itself is not a good indicator. It is the content of growth that matters (Arrow et al. 1995). Policies that directly or indirectly support ecosystem destruction serve as subsidies from society to deplete resources, ecosystem services, and support.

The underlying causes must be made clear and redirected if society is to move toward sustainability. Evidently, this requires not only monetary incentives, but also the design of institutions that protect the capacity of ecosystems to generate services, maintain biodiversity, and absorb disturbance (Gunderson et al. 1995; Hanna et al. 1996). Ecosystem scientists have an important role to play in fostering knowledge about the effects on ecosystems of different policies, whether monetary, legal or cultural, and the impact this will have in terms of loss of functional diversity, resilience, and loss of essential natural resources and ecosystem services.

Institutions and Ecosystem Management

Institutions are the norms and rules, the humanly devised constraints that shape human interaction (North 1990, 1994). Institutions and organizations are not the same things, but an organization can be a subset of an institution. Property-rights regimes—the structure of rights to resources and the rules under which those rights are exercised—are critical institutions. They are the mechanisms people use to control their use of the environment and their behavior toward each other (Bromley 1991). They link society to nature, and have the potential to coordinate nature and human systems in a complementary way for both ecological and human long-term objectives (Hanna et al. 1996).

In the past, proposals were often made to transform resources and ecosystems into private property in efforts to remedy environmental problems. As understanding of common-property regimes and combined state or common-property regimes has increased, it has become clear that no single type of property-rights regime can be prescribed as a remedy for problems of resource overuse and environmental degradation. In some contexts collective, decentralized regimes are a more appropriate structures for management of the natural environment than private-property regimes.

The issue is not whether there should be private, communal, or state-owned rights to the natural environment. The issue is how the bundles of different rights (i.e. private to communal to national to international; local to regional to global) relate to each other, relate people to each other, and relate people to their ecosystems. Types of property-rights regimes are arrayed along a spectrum from open access to private property, with an almost infinite variety in their components (Table 13.3).

The lessons from the literature on common-property resources indicate that local-level institutions learn and develop the capability to respond to environmental feedbacks faster than centralized agencies do (Feeny et al. 1990; Ostrom 1990, Bromley 1992). There are design principles derived from studies of long-enduring institutions, which have, at least to some extent, been successful in managing resources in a sustainable fashion (Becker and Ostrom 1995; Ostrom 1990). The design principles include:

- Clearly defined boundaries for the use of a forest or of groundwater, as well as clearly defined individuals or households with rights to harvest the resources;
- Rules specifying the amount of harvest by users related to local conditions and to rules requiring labor, materials, or money inputs;

TABLE 13.3. The four major types of property-rights regimes.[1]

Regime type	Owner	Owner rights	Owner duties
Private property	Individual	Socially acceptable uses; control of access	Avoidance of socially unacceptable uses
Common property	Collective	Exclusion of nonowners	Maintenance; constrain rates of use
State property	Citizens	Determine rules	Maintain social objectives
Open access (no property)	None	Capture	None

1. The basic functions of ecosystem management—coordinating users, enforcing rules, and adapting to changing environmental conditions—cannot be fostered without a specified system of property rights. In reality, there is a bundle of property-rights regimes operating at the same time and across scales.
2. Adapted from Hanna et al. 1996.

- Collective-choice arrangements;
- Monitoring of resource conditions and user behavior;
- Graduated sanctions when rules are violated;
- Conflict-resolution mechanisms;
- Long-term tenure rights to the resource and rights of users to devise their own institutions without being undermined by governmental authorities;
- for resources that are parts of larger systems, appropriation, provision, monitoring, enforcement, conflict resolution, and governance activities must be organized in multiple layers of nested enterprises.

The institutions of conventional resource management have been successful in producing yields and economic growth in the short term, but have not been very successful in managing the feedbacks between ecological and social systems for resilience and sustainability (Gunderson et al. 1995). In building centralized bureaucracies for environmental management over the years, it has been assumed that resource management can be scaled up, which in reality does not seem to be the case. However, all centralized management institutions cannot be replaced by community-level institutions. That would be to assume that resource management could be scaled down. The evidence from various case studies support the proposition by Holling et al. (1997) that environmental and renewable resource issues tend to be neither small-scale nor large-scale but rather cross-scale in both space and time. It follows, therefore, that the problems have to be tackled simultaneously at several levels. Thus, the power of centralized management agencies should be redistributed and balanced, not eliminated.

Clearly, conventional approaches will not suffice to cope with a spectrum of potentially catastrophic and irreversible environmental problems.

According to Levin et al. (in press) these problems are characterized by unpredictability and, surprise is to be expected); the potential importance of thresholds and domain shifts; the difficulty of detecting change early enough to allow effective solutions, or to develop scientific consensus on a time scale rapid enough to allow effective solution; the probability that the signal of change, even when detected, will be displaced in space and time from the source, so that motivation for action is small. Conventional market mechanisms are inadequate to handle these problems; response systems and institutions that are flexible and adaptive are needed.

Ecological Knowledge, Adaptive Management, and Social Mechanisms for Improved Ecosystem Management

Integrative ecosystem management, as suggested by Carpenter (Chapter 12) and Walters (Chapter 11) can proceed by a design that simultaneously allows for tests of different management policies and emphasizes learning by doing. Termed adaptive management, this approach treats policies as hypotheses and management as experiments from which managers can learn (Holling 1978; Walters 1986). Adaptive management effectively breaks down the barrier between research and management. As it proceeds in a stepwise fashion, responding to changes and guided by feedback from the ecosystem, it allows for institutional learning (Gunderson et al. 1995).

The volume *Linking Ecological and Social Systems: Managing Practices and Social Mechanisms for Building Resilience* (Berkes and Folke 1998) seeks to integrate two streams of resource-management thought that differ from the conventional approach. The first involves rethinking resource-management science in a world of complex systems with nonlinear relationships, thresholds, uncertainty, and surprise, using systems approaches and adaptive management. The second involves rethinking resource-management social science by focusing on institutions, largely property-rights institutions and in particular common-property systems.

The combination of those two streams of resource management have resulted in new insights concerning social-ecological linkages that contribute to building resilience. These insights were generated from a number of case studies from a variety of ecosystems in different parts of the world, and from both traditional and contemporary societies.

First, management practices based on ecological knowledge were identified (Table 13.4). The practices ranged from monitoring and managing specific resources, to ecologically sophisticated practices that respond to and manage disturbance and build resilience across scales. Such management practices include protection of certain species, knowledge of the

TABLE 13.4. Ecologically adapted management practices and social mechanisms for resilience and sustainability.

1. Management practices based on ecological knowledge
    Monitoring change in ecosystems and in resource abundance
    Total protection of certain species
    Protection of vulnerable stages in the life-history of species
    Protection of specific habitats
    Temporal restrictions of harvest
    Multiple species and integrated management
    Resource rotation
    Management of succession
    Management of landscape patchiness
    Watershed management
    Managing ecological processes at multiple scales
    Responding to and managing pulses and surprises
    Nurturing sources of renewal

2. Social mechanisms behind management practices
    a) Generation, accumulation, and transmission of ecological knowledge
        Reinterpreting signals for learning
        Revival of local knowledge
        Knowledge carriers e.g. folklore
        Integration of knowledge
        Intergenerational transmission of knowledge
        Geographical transfer of knowledge

    b) Structure and dynamics of institutions
        Role of stewards or wise people
        Community assessments
        Cross-scale institutions
        Taboos and regulations
        Social and cultural sanctions
        Coping mechanisms; short-term responses to surprises
        Ability to reorganize under changing circumstances
        Incipient institutions

    c) Mechanisms for cultural internalization
        Rituals, ceremonies, and other traditions
        Coding or scripts as a cultural blueprint

    d) Worldview and cultural values
        Sharing, generosity, reciprocity, redistribution, respect, patience, humility

1. Adapted from Folke et al. 1998b.

stages in their life-history and habitats; multiple species and multiple-scale management of ecosystem processes; temporal restriction of harvest; resource switching and rotation; management of landscape patchiness and different phases of ecosystem development; and whole-watershed management. The first five management practices identified in Table 13.4 also exist in conventional resource management but they often lack effective regulations, enforcements, and social sanctions. The other eight practices are rare

C. Folke

or do not exist in modern society, but are very sophisticated from an ecosystem-management perspective. For example, Chisasibi Cree hunters from the James Bay area in Canada, rotate trapping areas on a four-year cycle to allow populations of beaver to recover, seem to manage fish on a five-to ten-year scale, and caribou on an eighty to 100-year scale (Berkes 1998). Succession is managed in shifting cultivation systems, as with the milpa of tropical Mexico, where agriculture is a sequential cropping of crops and noncrops (Alcorn and Toledo 1998). The small-scale movements of the Sahelian herders of nothern Africa are designed to mimic the variability and unpredictability of the landscape patchiness (Niamir-Fuller 1998). In ancient Hawaii, entire river valleys were managed as integrated farming systems, from the upland forest all the way to the coral reef (Costa-Pierce 1987). The Gitksan people of British Columbia are concerned not only with the production of fiber over several square kilometers, but also with the maintenance of ecological processes involving soil bacteria at the spatial scale of a few square meters (Pinkerton 1998). Range reserves of African herders provide a "savings bank" of forage that serves as a buffer to both disturbance and surprise (Niamir-Fuller 1998). Sacred groves in India absorb disturbance by serving as firebreaks for cultivated areas and villages (Gadgil et al. 1998). Some nomads behave much as a disturbance regime, following the migratory cycles of the herbivores from one area to another, and thus contribute to the capacity of the semiarid grasslands of Africa to function under a wide range of climatic conditions (Niamir-Fuller 1998).

Many of these aforementioned management practices are in stark contrast to conventional resource management (Holling et al. 1998; Folke et al. 1998b). The forest is not viewed as a storehouse for timber, or oceans as standing stocks of fish populations. The resource is regarded as a part of a complex ecosystem. These unconventional management practices monitor and interpret the dynamics of complex ecosystems to secure a flow of resources and ecosystem services and are linked to a diversity of social mechanisms and institutions. The sequence of social mechanisms was organized as a hierarchy that proceeds from ecological knowledge to worldviews. Institutions, in the sense of "rules in use", provide the means by which societies can act on their ecological knowledge and use it to produce a livelihood from resources and services in their environment. Both knowledge and institutions require mechanisms for cultural internalization, so that learning can be encoded and remembered by the social group. Worldview or cosmology gives shape to cultural values, ethics, and basic norms and rules of a society.

A few examples of social mechanisms include the hunter's guild of the Yoruba in Nigeria, which functions as a knowledge carrier to maintain ancient traditions and indigenous ecological knowledge (Warren and Pinkston 1998). Reef and lagoon fishery management in Oceania show pervasive spatial diffusion of ecological knowledge, inferred through striking similarities in the management system across island groups (Johannes 1978). Various kinds of taboos are ecologically functional and have the potential to build resilience in ecosystems (Colding and Folke 1997; Gadgil et al. 1998). Many resources are not managed by numbers, but through the social conduct (Acheson et al. 1998). Rituals help people remember the rules and interpret signals from the environment appropriately (Chapin 1991). There are incipient institutions that are activated following an intensification of resource use or decrease in resource supply. One example is the "sleeping territoriality" of some Pacific Islands, which serves a blueprint of property rights that is activated when fisheries resources are becoming scarce (Hviding, personal communication).

Presumably, the social mechanisms of Table 13.4 represent only a tiny fraction of existing human-environmental adaptations. These practices and mechanisms provide a reservoir of active adaptations in the real world, which may be of universal importance in designing for sustainability (Gadgil et al. 1993; Berkes et al. 1995). Several of them prevent the buildup of large-scale crisis. They allow disturbance to enter at a lower level and they build resilience, in contrast to resource management in large parts of contemporary society in which disturbance is blocked out (e.g. Regier and Baskerville 1986; Ludwig et al. 1993; Finlayson and McCay 1998). Under these institutional frameworks, disturbance does not accumulate and return to challenge the existence of the whole social-ecological system (Holling et al. 1998).

Such adaptive management is a search for a more sustainable relationship with life-supporting ecosystems, and is a response to resource scarcity and management failure (Holling et al. 1998). By responding to and managing feedbacks from the ecosystem instead of blocking them out, adaptive management has the potential to avoid ecological thresholds at scales that threaten the existence of social and economic activities.

The social-ecological practices, mechanisms, and principles identified in Berkes and Folke (1998), have the potential to improve conventional resource management by providing:

- Insights for designing adaptive resource management systems that flow with nature;
- Novel approaches to forestry, agriculture, fisheries, aquaculture and freshwater management;
- Lessons for developing systems of social sanctions and successful implementation and enforcement of sustainable practices;
- Means to avoid surprises caused by conventional resource management;
- Experiences in managing fluctuations and disturbance;
- the building of resilience for sustainability.

However, for the social-ecological system to persist, the integrity of locally adapted systems in which the practices and mechanisms are embedded needs to be protected (but not isolated) from such external driving forces as misguided macroeconomic policies or trade opportunities that lead to unsustainable resource exploitation. Such support can be provided by umbrella institutions (Alcorn and Toledo 1998), or through nested sets of institutions (Ostrom 1990; Hanna et al. 1996). Such nested systems of governance exist, for example, in the comanagement process in Maine's soft shell clam fishery for the sharing of rights and responsibilities between the State of Maine and the local community (Hanna 1998). The Great Barrier Reef Marine Park is an example of multiple use of a huge ecosystem, where more than sixty interest groups, governmental agencies, Aboriginal and Torres Strait Islanders communities collaborate with the Great Barrier Reef Marine Park Authority for long-term sustainable management (Zann 1995).

There is a need for social mechanisms by which information from the environment may be received, processed, and interpreted to build resilience of the linked social-ecological system. More attention in regard to these social mechanisms is required among ecosystem scientists concerned with management and restoration. How can we contribute to the generation, accumulation, and transmission of ecological knowledge in society, provide incentives for monitoring and managing ecosystem change, stimulate the development of institutions that respond to environmental feedbacks and that safeguard the capacity of ecosystems to generate essential resources and ecosystem services? We may have something to learn from institutions that have not undermined their existence by degrading their ecological life-support system, thereby losing ecological and institutional resilience. A major task for modern society is to find similar ways of responding to changes in ecosystems. At present, there is a pervasive lack of social mechanisms for managing changing environmental conditions. Holling (1986) observed that institutions, as with ecosystems, can become "brittle" over time, and resource crises can result in release and reorganization. These observations can lead to new empirical, theoretical, and policy-relevant work on linkages between social and ecological systems, and perhaps an answer to the question of just what produces adaptive capacity in linked social-ecological systems.

Conclusion

The global community needs to be able to handle new types of problems that are fundamentally cross-scale, transsectoral, transcultural, and transdisciplinary, which calls for innovative research approaches, social institutions, and technologies. Innovative research aimed at articulating the methods and mechanisms by which human populations can strike a dynamic balance between social and economic development and maintenance of the productive capacity of ecosystems, including biodiversity is critically needed. The research should be integrated rather than divorced from policy and management processes.

Much ecological research has been presented in a form accessible mainly to ecologists and other natural scientists, and not to social scientists or policy-makers. Ecologists increasingly need to ask questions that are linked to social and economic issues, and communicate findings from ecosystem research in a fashion that is accessible and understandable also for people who are neither ecologists nor scientists.

In this chapter, the emphasis has been that people and nature are one system, and that human societies should be treated as a part of and not apart from ecosystems. Two approaches have been reviewed—ecosystem services and ecological footprint analysis—which have proven effective in communicating to policy-makers that human societies are dependent on ecosystems. I have stressed the necessity of addressing the major social and economic driving forces behind ecosystem degradation, and the critical role of institutions for the environment, and have provided insight for ecosystem management drawn from integrated social-ecological systems. The following six frontiers arise from this chapter.

1. Illuminate "the hidden" ecosystem support. Ecological considerations will receive low priority if people and policy do not know or understand why they should care about ecosystems. It is necessary to illuminate through comprehensive examples that ecosystem support is a precondition for social and economic development. It will affect the mindset, the mental models of reality. The quantification and valuation of ecosystem services and ecological footprints are useful in this context.

2. Study the interrelations within and between ecosystems preferably at the landscape level or in whole drainage basins. We need to understand more thoroughly the mosaic of ecosystems in drainage basins, and how they affect and are affected by human activities and on various scales and across scales. The issue of joint product and multifunctionality in ecosystems should be addressed. For example, what are the relationships between the coverage of mangrove vegetation and harvest of fish and shellfish generated through mangrove ecosystems? Flows and feedbacks between ecosystems in drainage basins should also be analyzed. For example, what will happen with the services from a wetland if an adjacent forest is clear-cut? What will the impacts be on coral reefs in coastal zones when there is deforestation for oil palm production and mangrove clear-cutting for shrimp production upstream in the watershed? What will happen with the Okawango delta in Botswana if Namibia (upstream in the drainage basin) decides to build dams to secure its water flows? There is a large amount of ecological information available, and many methods and approaches exist. They need to be reshaped and synthesized to provide important information for improved decision-making in society. There are no simple answers to such

questions, given the complexity of self-organized systems, change and disturbance, true uncertainty and surprise, but to address them is, in my opinion, a critical frontier, and one that must treat humans as endogenous factors, that is, as a major influence in ecosystems.

3. Ecosystem scientists could also contribute to identifying and predicting the effects of social and economic driving forces on ecosystem processes, functions, and services. For example, what are the positive and negative effects on various ecosystems if a carbon tax is implemented, or if trade is liberalized?

4. Build ecological knowledge through adaptive management is another frontier (see chapters 11 and 12 by Walters and Carpenter respectively). Make use of already existing "practical" ecological knowledge of such local users as fishermen, farmers, and local communities. Combine this knowledge with scientific knowledge as is done, for example in the Maine clam fisheries (Hanna 1998). If this would have been done in the New Foundland cod fisheries the codfish collapse might have been avoided.

5. How can institutions be designed to work in synergy with ecosystem processes and functions? Conservation, ecosystem restoration, and technology based on ecological understanding will never work by themselves. To be successfully implemented, they need to be linked to an institutional framework. There are many studies on the structure and function of institutions. There are many studies on the structure and function of ecosystems. There are few studies on linked ecosystems and institutions. This is a major reason why the International Human Dimension Program (IHDP) of the International Geosphere-Biosphere Program (IGBP) has initiated research on what is called the "problem of fit" between institutions and ecosystems.

6. Therefore, there is scope to learn from linked social-ecological systems, as illustrated in this chapter. Are there practices and social mechanisms that manage ecosystems in a knowledgeable fashion from which we can learn? This is a wide-open frontier for analysis and synthesis.

Most of these frontiers will be difficult to address by ecosystem scientists alone. They will require substantive collaboration with social scientists.

Acknowledgments. This chapter has benefited from most valuable comments from Mary Barber, Lisa Deutsch, Nils Kautsky, Karin Limburg, Stewart Pickett, and an anonymous reviewer. The work with the chapter was supported by The Pew Charitable Trusts.

References

Acheson, J.M., J.A. Wilson, and R.S. Steneck. 1998. Managing chaotic fisheries. Pages 390–413 in F. Berkes and C. Folke, eds. *Linking social and ecological systems: management practices and social mechanisms for building resilience.* Cambridge University Press, Cambridge, UK.

Alcorn, J.B., and V.M. Toledo. 1998. Resilient resource management in Mexico's forest ecosystems: the contribution of property rights. Pages 216–249 in F. Berkes and C. Folke, eds. *Linking social and ecological systems: management practices and social mechanisms for building resilience.* Cambridge University Press, Cambridge, UK.

Andersson, T., C. Folke, and S. Nyström. 1995. *Trading with the environment: ecology, economics, institutions and policy.* Earthscan, London, England.

Arrow, K., B. Bolin, R. Costanza, P. Dasgupta, C. Folke, C.S. Holling, et al. 1995. Economic growth, carrying capacity, and the environment. *Science* 268:520–521.

Arrow, K., E. Leamer, P. Protney, R. Randner, H. Schuman, and R. Solow. 1993. Report of the NOAA Panel on contingent valuation. *Federal Register* 58:4601–4614.

Bailey, C. 1988. The social consequences of tropical shrimp mariculture development. *Ocean and Shoreline Management* 11: 31–44.

Barbier, E.B., J. Burgess, and C. Folke. 1994. *Paradise lost? The ecological economics of biodiversity.* Earthscan, London, England.

Baskin, Y. 1997. *The work of nature: how the diversity of life sustains us.* Island Press, Washington, DC.

Becker, C.D., and E. Ostrom. 1995. Human ecology and resource sustainability: the importance of institutional diversity. *Annual Review of Ecology and Systematics* 26:113–133.

Berkes, F. 1998. Learning to design resilient resource management: indigenous systems in the Canadian subarctic. Pages 98–128 in F. Berkes and C. Folke, eds. *Linking social and ecological systems: management practices and social mechanisms for building resilience.* Cambridge University Press, Cambridge, UK.

Berkes, F., and C. Folke. 1994. Investing in cultural capital for a sustainable use of natural capital. Pages 128–149 in A.M. Jansson, M. Hammer, C. Folke, and R. Costanza, eds. *Investing in natural capital: the ecological economics approach to sustainability.* Island Press, Washington, DC.

Berkes, F., and C. Folke, eds. 1998. *Linking social and ecological systems: management practices and social mechanisms for building resilience.* Cambridge University Press, Cambridge, UK.

Berkes, F., C. Folke, and M. Gadgil. 1995. Traditional ecological knowledge, biodiversity, resilience and sustainability. Pages 269–287 in C. Perrings, K.-G. Mäler, C. Folke, C.S. Holling, and B.-O. Jansson, eds. *Biodiversity conservation.* Kluwer Academic Publishers, Dordrecht, the Netherlands.

Binswanger, H.P. 1989. *Brazilian policies that encourage deforestation in the Amazon.* Working Paper No. 16. Environment Department, The World Bank, Washington, DC.

Bockstael, N., R. Costanza, I. Strand, W. Boynton, K. Bell, and L. Wainger. 1995. Ecological economic modeling and valuation of ecosystems. *Ecological Economics* 14:143–159.

Borgström, G. 1967. *The hungry planet.* Macmillan, New York.

Bormann, F.H. 1976. An inseparable linkage: conservation of natural ecosystems and conservation of fossil energy. *BioScience* 26:759.

Bromley, D.W. 1991. *Environment and economy: property rights and public policy.* Basil Blackwell, Oxford, UK.

Bromley, D.W. ed. 1992. *Making the commons work: theory, practice, and policy.* ICS Press, San Francisco, CA.

Carson, R.T., R.C. Mitchell, W.M. Hanemann, R.J. Kopp, S. Presser, and P.A. Ruud. 1992. *A contingent valuation study of lost passive use values resulting from the Exxon Valdez oil spill.* Report from the Attorney General of the Sate of Alaska, November 19, 1992.

Chapin, M. 1991. Losing the way of the great father. *New Scientist*, 10 August: 40–44.

Christensen, N.C. et al. 1996. The report of the Ecological Society of America Committee on the scientific basis for ecosystem management. *Ecological Applications* 6:665–691.

Cities, Folke, C., Å. Jansson, Larsson, and R. Costanza. 1997. Ecosystem appropriation by cities. *Ambio* 26:167–172.

Cleveland, C., R. Ayres, B. Castaneda, R. Costanza, H. Daly, C. Folke et al. 1997. Is sustainable growth possible? *Ambio*, in review.

Colding, J., and C. Folke. 1997. The relation between threatened species, their protection, and taboos. *Conservation Ecology* 1: article 6.

Common, M. 1996. *Environmental and resource economics: an introduction, 2nd edition.* Addison Wesley Longman Publishing, New York.

Common, M., and C. Perrings. 1992. Towards an ecological economics of sustainability. *Ecological Economics* 6:7–34.

Costanza, R, ed. 1991. *Ecological economics: the science and management of sustainability.* Columbia University Press, New York.

Costanza, R., and H.E. Daly. 1992. Natural capital and sustainable development. *Conservation Biology* 6:37–46.

Costanza, R., R. d'Arge, R de Groot, S. Farber, M. Grasso, B. Hannon et al. 1997. The value of the world's ecosystem services and natural capital. *Nature* 387:253–260.

Costanza, R., C.S. Farber, and J. Maxwell. 1989. Valuation and management of wetland ecosystems. *Ecological Economics* 1:335–361.

Costanza, R., and C. Folke. 1997. Valuing ecosystem services with efficiency, fairness and sustainability as goals. Pages 49–68 in G. Daily, ed. *Nature's services: societal dependence on natural ecosystems.* Island Press, Washington DC.

Costanza, R., L. Waigner, C. Folke, and K.-G. Mäler. 1993. Modeling complex ecological economic systems: toward an evolutionary dynamic understanding of people and nature. *BioScience* 43:545–555.

Costa-Pierce, B.A., 1987. Aquaculture in ancient Hawaii. *BioScience* 37:320–331.

Cowling, R.M. ed. 1992. *The ecology of fynbos: nutrients, fire and diversity.* Oxford University Press, Capetown, South Africa.

Cowling, R.M., R. Costanza, and S.I. Higgins. 1997. Services supplied by South African fynbos ecosystems. Pages 345–362 in G. Daily, ed. *Nature's services: societal dependence on natural ecosystems.* Island Press, Washington DC.

Daily, G. ed. 1997. *Nature's services: societal dependence on natural ecosystems.* Island Press, Washington DC.

Daily, G., and P.R. Ehrlich. 1992. Population, sustainability, and Earth's carrying capacity. *BioScience* 42:761–771.

Dasgupta, P. 1997. Economics of the environment. *Environment and Development Economics* 1:387–428.

Dasgupta, P., and K.-G. Mäler. 1996. Environmental economics in poor countries:

the current state and a programme for improvement. *Environment and Development Economics* 2:3–7.

de Groot, R.S. 1992. *Functions of nature.* Wolters-Noordhoff, Amsterdam, The Netherlands.

Dixon, J.A., and P.B. Sherman. 1990. *Economics of protected areas.* Island Press, Washington, DC.

Ehrlich, P.R. 1994. Ecological economics and the carrying capacity of the Earth. Pages 38–56 in A.M. Jansson, M. Hammer, C. Folke, and R. Costanza, eds. *Investing in natural capital: the ecological economics approach to sustainability.* Island Press, Washington, DC.

Ehrlich, P.R., and A.E. Ehrlich. 1981. *Extinction: the causes and consequences of the disappearance of species.* Random House, New York.

Ehrlich, P.R., and H.A. Mooney. 1983. Extinction, substitution and ecosystem services. *BioScience* 33:248–254.

Ekins, P., C. Folke, and R. Costanza. 1994. Trade, environment and development: the issues in perspective. *Ecological Economics* 9:1–12.

Faber, S.C. 1987. The value of coastal wetlands for protection of property against hurricane wind damage. *Journal of Environmental Economics and Management* 14:143–151.

FAO. 1995. The state of the world fisheries and aquaculture. *FAO*, Rome 1995: p. 44.

Feeny, D., F. Berkes, B.J. McCay, and J.M. Acheson. 1990. The tragedy of the commons: twenty-two years later. *Human Ecology* 18:1–19.

Finlayson, A.C., and B.J. McCay. 1998. Crossing the threshold of ecosystem resilience: the commercial extinction of northern cod. Pages 311–337 in F. Berkes and C. Folke, eds. *Linking social and ecological systems: management practices and social mechanisms for building resilience.* Cambridge University Press, Cambridge, UK.

Folke, C. 1988. Energy economy of salmon aquaculture in the Baltic Sea. *Environmental Management* 12:525–537.

Folke, C. 1991a. Socioeconomic dependence on the life-supporting environment. Pages 77–94 in C. Folke and T. Kåberger, eds. *Linking the natural environment and the economy: essays from the Eco-Eco Group.* Kluwer Academic Publishers, Dordrecht.

Folke, C. 1991b. The societal value of wetland life-support. Pages 141–171 in C. Folke and T. Kåberger, eds. *Linking the natural environment and the economy: essays from the Eco-Eco Group.* Kluwer Academic Publishers, Dordrecht, the Netherlands.

Folke, C. F. Berkes, and J. Colding. 1997. Ecological practices and social mechanisms for building resilience and sustainability. Pages 414–436 in F. Berkes and C. Folke, eds. *Linking social and ecological systems: management practices and social mechanisms for building resilience.* Cambridge University Press, Cambridge, UK.

Folke, C., C.S. Holling, and C. Perrings. 1996a. Biological diversity, ecosystems, and the human scale. *Ecological Applications* 6:1018–1024.

Folke, C., and N. Kautsky. 1989. The role of ecosystems for a sustainable development of aquaculture. *Ambio* 18:234–243.

Folke, C., N. Kautsky, H. Berg, Å. Jansson, J. Larsson, and M. Troell. 1998. The ecological footprint concept for sustainable seafood production: a review. *Ecological Applications.* 8(1) supplement:63–71.

Folke, C., J. Larsson, and J. Sweitzer. 1996b. Renewable resource appropriation by cities. Pages 201–221 in R. Costanza, and O. Segura, eds. *Getting down to Earth: practical applications of ecological economics.* Island Press, Washington, DC.

Freeman, A.M. 1994. *The measurement of environmental and resource values: theory and methods.* Resources for the Future, Washington DC.

Gadgil, M., F. Berkes, and C. Folke. 1993. Indigenous knowledge for biodiversity conservation. *Ambio* 22:151–156.

Gadgil, M., N.S. Hemam, and B.M. Reddy. 1998. People, refugia and resilience. Pages 30–47 in F. Berkes and C. Folke, eds. *Linking social and ecological systems: management practices and social mechanisms for building resilience.* Cambridge University Press, Cambridge, UK.

Gosselink, J.G., E.P. Odum, and R.M. Pope. 1974. *The value of a tidal marsh.* Publication No. LSU-SG-74–03, Center for Wetland Resources, Louisiana State University, Baton Rouge.

Gren, I.-M., 1995. The value of investing in wetlands for nitrogen abatement. *European Review of Agricultural Economics* 22:157–172.

Gren, I.-M., and T. Söderqvist. 1994. Economic valuation of wetlands: a survey. Beijer Discussion Papers 54. Beijer International Institute of Econlogical Economics, Royal Swedish Academy of Sciences, Stockholm, Sweden.

Groffman, P.M., and G.E. Likens, eds. 1994. *Integrated regional models: interactions between humans and their environment.* Chapman & Hall, New York.

Gunderson, L., C. S. Holling, and S. Light, eds. 1995. *Barriers and bridges to the renewal of ecosystems and institutions.* Columbia University Press, New York.

Hall, C.A.S., Cleveland, C.J., and R. Kaufmann. 1986. *Energy and resource quality: the ecology of the economic process.* John Wiley & Sons, New York.

Hanna, S. 1998. Managing for human and ecological context in the Maine soft shell clam fishery. Pages 190–211 in F. Berkes and C. Folke, eds. *Linking social and ecological systems: management practices and social mechanisms for building resilience.* Cambridge University Press, Cambridge, UK.

Hanna, S., C. Folke, and K.-G. Mäler, eds. 1996. *Rights to nature.* Island Press, Washington, DC.

Heath, J., and H. Binswanger. 1996. Natural resource degradation effects of poverty and population growth are largely policy induced: the case of Colombia. *Environment and Development Economics* 1:65–84.

Higgins, S.I., J.K. Turpie, R. Costanza, R.M. Cowling, D.C. Le Maitre, C. Marais, and G.F. Midgley. 1997. An ecological economic simulation model of mountain fynbos ecosystems: dynamics, valuation and management. *Ecological Economics* 22:155–169.

Holling, C.S. ed. 1978. *Adaptive environmental assessment and management.* John Wiley & Sons, London, England.

Holling, C.S. 1986. The resilience of terrestrial ecosystems: local surprise and global change. Pages 292–317 in W.C. Clark and R.E. Munn, eds. *Sustainable development of the biosphere.* Cambridge University Press, London, England.

Holling, C.S., 1994. An ecologists view of the Malthusian conflict. Pages 79–103 in K. Lindahl-Kiessling, and H. Landberg, eds. *Population, economic development, and the environment.* Oxford University Press, Oxford, UK.

Holling, C.S., F. Berkes, and C. Folke. 1998. Science, sustainability, and resource management. Pages 342–362 in F. Berkes, and C. Folke, editors. *Linking social and ecological systems: management practices and social mechanisms for building resilience.* Cambridge University Press, Cambridge, UK.

Holling, C.S., D.W. Schindler, B.H. Walker, and J. Roughgarden. 1995. Biodiversity in the functioning of ecosystems: an ecological synthesis. Pages 44–83 in C.A. Perrings, K.-G. Mäler, C. Folke, C.S. Holling, and B.-O. Jansson, eds. *Biodiversity loss: economic and ecological issues.* Cambridge University Press, Cambridge, UK.

Houghton, J.T., L.G. Meira Filho, B.A. Callander, N. Harris, A. Kattenberg, and K. Maskell. 1996. *Climate change 1995: the science of climate change.* Cambridge University Press, Cambridge, UK.

Jansson, Å., C. Folke, and J. Rockström. 1998. *Linking water and ecosystem services appropriated by people in the Baltic Sea drainage basin.* Beijer Discussion Papers, Beijer Institute, Royal Swedish Academy of Sciences, Stockholm, Sweden.

Jansson, A.M., M. Hammer, C. Folke, and R. Costanza, eds. 1994. *Investing in natural capital: the ecological economics approach to sustainability.* Island Press, Washington DC.

Jansson, A.M., and J. Zucchetto. 1978. Energy, economic and ecological relationships for Gotland, Sweden: a regional systems study. *Ecological Bulletins* 28.

Johannes, R.E. 1978. Traditional marine conservation methods in Oceania and their demise. *Annual Review of Ecology and Systematics* 9:349–364.

Jones, T., and S. Wibe. 1992. *Forests: market and intervention failures—five case studies.* Earthscan England. London, England.

Krutilla, J.V. 1967. Conservation reconsidered. *American Economic Review* 57: 777–786.

Krutilla, J.V., and A.C. Fisher. 1975. *The economics of natural environment: studies in the valuation of commodity and amenity resources.* John Hopkins Press for Resources for the Future, Inc., Baltimore, MD.

Larsson, J., C. Folke, and N. Kautsky. 1994. Ecological limitations and appropriation of ecosystem support by shrimp farming in Colombia. *Environmental Management* 18:663–676.

Le Maitre, D.C., B.W. van Wilgen, R.A. Chapman, and D.H. McKelly. 1996. Invasive plants and water resources in the western Cape Province, South Africa: modelling the consequences of a lack of management. *Journal of Applied Ecology* 33:161–172.

Leonardsson, L. 1994. *Wetlands as nitrogen sinks: Swedish and international experience, Report 4176.* Swedish Environmental Protection Agency, Stockholm.

Levin, S.A., 1992. The problem of pattern and scale in ecology. *Ecology* 73:1943–1967.

Levin, S.A., S. Barrett, W. Baumol, C. Bliss, B. Bolin, N. Chichinsky, et al. in press. Resilience in natural and socioeconomic systems. Environment and Development Economics.

Lopez, R., and C. Scoseria. 1996. Environmental sustainability and poverty in Belize: a policy paper. *Environment and Development Economics* 1:289–307.

Lubchenco, J. et al. 1991. The sustainable biosphere initiative: an ecological research agenda. *Ecology* 72:371–412.

Ludwig, D., R. Hilborn, and C. Walters. 1993. Uncertainty, resource exploitation and conservation: lessons from history. *Science* 260:1736.

Ludwig, D., B. Walker, and C.S. Holling. 1997. Sustainability, stability and resilience. *Conservation Ecology* 1: article 7.

McDonnell, M.J., and S.T.A. Pickett, eds. 1993. *Humans as components of ecosystems: the ecology of subtle human effects and populated areas.* Springer-Verlag, New York.

344   C. Folke

McMichael, A.J., A. Haines, R. Slooff, and S. Kovats. 1996. *Climate change and human health.* World Health Organization, Geneva, Switzerland.

Mitsch, W.J., and S.E. Jörgensen, eds. 1989. *Ecological engineering: an introduction to ecotechnology.* John Wiley& Sons New York.

Myers, N. 1996. Environmental services of biodiversity. *Proceedings of National Academic Science* 93:2764–2769.

Naeem, S., L.J. Thompson, S.P. Lawler, J.H. Lawton, and R.M. Woodfin. 1994. Declining biodiversity can alter the performance of ecosystems. *Nature* 368:734–737.

Naiman, R.J. ed. 1992. *Watershed management: balancing sustainability and evironmental change.* Springer Verlag, New York.

Niamir-Fuller, M. 1998. The resilience of pastoral herding in Sahelian Africa. Pages 250–284 in F. Berkes and C. Folke, eds. *Linking social and ecological systems: management practices and social mechanisms for building resilience.* Cambridge University Press, Cambridge, UK.

North, D.C. 1990. *Institutions, institutional change and economic performance.* Cambridge University Press, Cambridge, UK.

North, D.C. 1994. Economic performance through time. *American Economic Review* 84: 359–368.

Odum, E.P. 1975. *Ecology: the link between the natural and social sciences, 2nd edition.* Holt-Saunders, New York.

Odum, E.P. 1989. *Ecology and our endangered life-support systems.* Sinauer Associates, Sunderland, MA.

Odum, H.T. 1971. *Environment, power and society.* John Wiley & Sons, New York.

O'Neill, R.V. 1996. Perspectives on economics and ecology. *Ecological Applications* 6:1031–1033.

Ostrom, E. 1990. *Governing the commons: the evolution of institutions for collective actions.* Cambridge University Press, Cambridge, U.K.

Pearce, D.W., and R.K. Turner. 1990. *Economics of natural resources and the environment.* Havester Wheatsheaf, Hemel Hampstead, UK.

Perrings, C., Folke, C., and Mäler, K.-G. 1992. The ecology and economics of biodiversity loss: the research agenda. *Ambio* 21:201–211.

Persson, A. 1995. A dynamic computable general equilibrium model of deforestation in Costa Rica. Pages 215–235 in C. Perrings, K.-G. Mäler, C. Folke, C.S. Holling, and B.-O. Jansson, eds. *Biodiversity conservation.* Kluwer Academic Publishers, Dordrecht, the Netherlands.

Pickett, S.T.A., and P.S. White, eds. 1985. *The ecology of natural disturbance and patch dynamics.* Academic Press, Orlando, FL.

Pinkerton, E. 1998. Integrated management of a temperate montane forest ecosystem through wholistic forestry: a British Columbia example. Pages 363–389 in F. Berkes and C. Folke, eds. *Linking social and ecological systems: management practices and social mechanisms for building resilience.* Cambridge University Press, Cambridge, UK.

Primavera, J.H. 1993. A critical review of shrimp pond culture. *Reviews in Fisheries Science* 1:151–201.

Rees, W.E., and M. Wackernagel. 1994. Ecological footprints and appropriated carrying capacity. Pages 362–390 in A.M. Jansson, M. Hammer, C. Folke, and R. Costanza, eds. *Investing in natural capital: the ecological economics approach to sustainability.* Island Press, Washington, DC.

Regier, H.A., and G.L. Baskerville. 1986. Sustainable redevelopment of regional ecosystems degraded by exploitive development. Pages 75–101 in W.C. Clark and R.E. Munn, eds. *Sustainable development of the biosphere.* Cambridge University Press, Cambridge, UK.

Thibodeau, F.R., and B.D. Ostro. 1981. An economic analysis of wetland protection. *Journal of Environmental Management* 12:19–30.

Tilman, D., and J.A. Downing. 1994. Biodiversity and stability in grasslands. *Nature* 367:363–365.

Turner, B.L., W.C. Clark, and W.C. Kates, eds. 1990. *The Earth as transformed by human action: global and regional changes in the biosphere over the past 300 years.* Cambridge University Press, Cambridge, UK.

Turner, R.K., C. Perrings, and C. Folke. 1997. Ecological economics: paradigm or perspective. Pages 25–49 in J. van den Bergh and J. van den Straaten, eds. *Economy and ecosystems in change: analytical and historical perspectives.* Edward Elgar, London, England.

van Wilgen, B.W., R.M. Cowling, and C.J. Burgers. 1996. Valuation of ecosystem services: a case study from South African fynbos ecossytems. *BioScience* 46:184–1189.

Wackernagel, M., and W. Rees. 1996. *Our ecological footprint: reducing human impact on the Earth.* New Society Publishers, Gabriola Island, Canada.

Walters, C. J. 1986. *Adaptive management of renewable resources.* McGraw Hill, New York.

Warren, D.M., and J. Pinkston. 1998. Indigenous African resource management of a tropical rain forest ecosystem: a case study of the Yoruba of Ara, Nigeria. Pages 158–189 in F. Berkes and C. Folke, eds. *Linking social and ecological systems: management practices and social mechanisms for building resilience.* Cambridge University Press, Cambridge, UK.

Westman, W.E. 1977. How much are nature's services worth? *Science* 197:960–964.

Wilks, A. 1995. Prawns, profit and protein: aquaculture and food production. *Ecologist* 25 (2/3):120–125.

Wulff, F., J.G. Fields, and K.H. Mann, eds. 1989. *Network analysis of marine ecosystems: methods and applications.* Springer-Verlag, Heidelberg, Germany.

Zann, L.P. 1995. *Our sea, our future: major findings of the State of the Marine Environment Report for Australia.* Ocean Rescue 2000 program. Department of the Environment, Sport and Territories, Canberra, Australia.

14
Ecosystems and Problems of Measurement at Large Spatial Scales

Carol A. Wessman and Gregory P. Asner

Summary

Both the structure and function of terrestrial ecosystems have significant impact on atmosphere-biosphere dynamics. Analyzing and predicting their roles in global environmental change present a major challenge to ecosystem scientists. Measurements at landscape, regional, and global scales are both necessary and difficult because they incorporate the full complexity of interacting system components, variations inherent in resource and climate constraints, and the effects of land-cover or land-use change. Our objective in this chapter is to identify the large-scale measurements that will most effectively further our understanding of regional and global ecology. To do this, our discussion expands from present-day measurements of biophysical structure and function to the areas in which new measurements can span our knowledge gaps. Although today's observation strategies are conducted at several scales, they tend to address either structure (e.g. vegetation-type distribution) or function (e.g. trace-gas fluxes) and rarely link the two. They lack integration as a result of several gaps, the most significant of which may be knowledge of dynamic carbon allocation processes and changing land-cover and land-use distributions. Continued development of dynamic global vegetation models is needed to couple ecosystem redistribution with biogeochemistry. However, because allocation is not mechanistically simulated by present-day ecosystem process models, remote sensing of aboveground allocation patterns will be required to constrain simulations for the foreseeable future. We contend that a stronger effort to link the physical structure to the function is critically needed to advance our understanding and predictive capability at large scales, and that quantitative remote sensing will provide important measurements to make those links.

Introduction

Measurements at landscape, regional, and global scales are critical to our understanding of large-scale processes because they incorporate the full complexity of interacting system components, variations inherent in resource and climate constraints, and the effects of land-cover and land-use change. Large-scale measurements are needed to evaluate our models and adjust our perception of reality in a time of rapid environmental change. Top-down approaches will elucidate the patterns of significance (Vitousek 1994), and the integration of large-scale measurements with process studies and modeling efforts (e.g. Hunt et al. 1996; White et al. 1997) will improve our understanding of regional and global dynamics. Our present, limited understanding of large-scale ecological processes results from both the difficulty in evaluating the behavior of complex systems and the technical challenge of making the necessary measurements at the requisite scales.

The measurement problem hinges on the processes that are important at these large scales and those measurable variables that are most strongly linked to these processes. In other words, to identify those measurements that must be and can be made at large scales, we must first define the key processes and the operational variables that assume importance with increasing scale. For example, at the global level, the biosphere has its greatest influence on the Earth system through physiological regulation of sensible and latent heat flux, linkage of the carbon-water-energy system to the metabolism of trace gases, the coupling of carbon and nitrogen cycles, and the albedo of the land surface. At the regional scale, ecosystem-specific controls and feedbacks will significantly influence sources and sinks of energy and matter and determine their effects on global biogeochemical and atmospheric processes. Additionally, types, rates, and intensity of disturbances that are either natural (e.g. fire, flooding) or human (e.g. land use) tend to be biome- or regionally specific (Houghton 1994; Asner et al. 1997).

The challenge for most large-scale measurement strategies lies in the spatial and temporal heterogeneity resulting from interactions among environmental variables that have scale-dependent attributes. A considerable body of literature has developed around ecological scaling (e.g. Allen and Star 1982; O'Neill et al. 1986; Levin 1992; Wessman 1992) and questions of aggregation (see Gardner et al. 1982; King 1991; Rastetter et al. 1992a). Some studies have been particularly powerful in pointing out the non-additive effects caused by the distribution and connectivity of ecosystems on the landscape; for example, landscape heterogeneity can influence regional climate (Pielke and Avissar 1990; Pielke et al. 1997) and urban centers can affect distant ecosystems through atmospheric transport of nitrogen (Townsend et al. 1996; Magill et al. 1997). However, there remains considerable uncertainty in our knowledge of these larger-scale dynamics and approaches to their measurement.

This chapter discusses the problems that impede large-scale measurements and suggests an effective (and familiar) paradigm within which to couch our large-scale questions. The present suite of large-scale observations and measurements in the context of a nested measurement framework is discussed, and the directions that research may take to enhance the linkages between observations with the goal of scaling ecological knowledge are suggested. In particular, we use relationships between structural and functional attributes of ecosystems to define those measurements, large and small, that are required to ask large spatial-scale questions. The structure and function of ecological systems have long been discussed; however, their application to research efforts focused on large-scale ecological dynamics is underutilized and, in our opinion, represents a frontier in ecosystem research.

Structure and Function: An Organizing Paradigm

When discussing the measurement of ecosystem function and structure, we refer to their measurable, physical characteristics. This is the conceptual framework (or paradigm) that we need to establish to distinguish the large-scale ecosystem properties that can be measured. It is through definition of the variables of interest that our challenges and frontiers in large-scale ecology can begin to be identified. This is not to say that the possible emergence of new aggregate properties of the larger landscapes is neglected; in fact, it is this quantification of aggregate properties that is needed. But with an understanding of the physical controls over our measurements, we can be more certain that the composite measurement is not an artifact of "local conditions" and is truly indicative of large-scale dynamics.

In this discussion "function", as is typically understood, refers to those processes that involve the transfer of matter and energy in the operation of an ecological system; for example, carbon dioxide and water exchange between the biosphere and atmosphere. The use of the term "structure" is more rigidly defined than in most ecological discussions. In this chapter, structure is defined to mean the physical aspects of the vegetation composing the land cover. Structure is used to describe the construction of the vegetation canopy, the distribution of its leaf and stem area in vertical and horizontal space, as well as the spatial orientation of morphologically distinct vegetation types (Figure 14.1). Phenology, then, is used to describe the distribution of these structural properties over time. In this context, structure does not mean species composition, as is often considered in ecological literature. To simplify global characterizations of vegetation "structure" or diversity at large scales, the concept of "functional types" is increasingly used to classify or group species with very similar responses to changing environmental conditions (Smith et al. 1997). The term "structure" is distin-

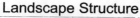

Landscape Structure	Canopy Structure
Distribution of vegetation structural types Canopy dimensions (allometry) Crown Spacing Crown Shading	Leaf Physical Properties Leaf and stem area index Leaf and stem angle distribution Foliage clumping

FIGURE 14.1. Two levels or components of the physical structure or morphology of landscapes. Landscape-level structure includes the factors describing the spatial orientation of vegetation types, their individual canopy dimensions, and shadowing effects between crowns or clusters. Canopy-level structure implies the leaf and stem attributes as well as foliage clumping within the canopy.

guished from even this usage in our discussion of large-scale measurements because neither species composition nor functional groups can be measured directly at large scales (although they may be successfully inferred by their physical characteristics). The aggregation property of our tools (e.g. remote sensing, eddy correlation methods) precludes the separation of species composition as such, and in the case of remote sensing, the aggregate is a result of the physical influence of the vegetation morphology on radiation as it travels through and out of the canopy.

Ecosystem structure emerges from the collective functioning of the components, and from larger-scale constraints imposed by the abiotic environment. This idea has been a central concept among vegetation scientists, but has for the most part been taken as a static property and not a variable of dynamic change (Shugart 1996). Feedbacks between system function and physical structure are central to the dynamics that occur at landscape, regional, and global scales. In a simple canopy example, photosynthetic and carbon allocation processes are strongly constrained by the configuration of the canopy for incoming radiation capture. This feeds back to influence photosynthetic levels and the subsequent level of carbon allocation to above- and belowground biomass. We are becoming increasingly aware of the dramatic influence that changes in landscape structure (e.g. land-cover and land-use change) can have on such functions as carbon storage (Burke et al. 1991) and climatological processes (Pielke and Avissar 1990; Shukla et al. 1990, Fennessy and Xue 1997). However, these relationships are poorly understood in comparison to their smaller-scale counterparts (e.g. landscape structural influences on carbon dynamics vs. canopy structural influences on carbon fixation).

Although both large-scale structure and function are difficult to measure, structure is probably the simpler of the two because it is more easily quantified. Large-scale measurement of function must integrate numerous components and a suite of complex interactions. The interdependence of structure and function requires that both attributes be measured however, if we are to develop sufficient understanding of large-scale system dynamics. Matter and energy fluxes above the canopy and above the landscape are the subject of many ongoing functional studies. Mechanistic understanding of these fluxes is well advanced in simple systems. Interestingly, carbon and nutrient allocation and their phenological controls are less well-understood, as are the resultant structural components and their feedback to these processes (see Zak and Pregitzer chapter 15). As mentioned earlier, other important structural attributes contributing to large-scale processes include the distribution and connectivity of ecosystems and the change in structure through land-cover transitions and alterations in land use. None of these have been sufficiently studied and quantified to determine their feedback on functional processes.

A Nested Measurement Framework

At the present time, measurements are conducted at multiple scales in an effort to understand the mechanisms and products of large-scale biosphere-atmosphere dynamics. In certain respects this is a nested approach, ranging from atmospheric measurements and inverse modeling at continental to global scales, to flux measurements within landscapes that are extended to biomes, and to plot- and chamber-level measurements aimed at identifying mechanistic controls over ecosystem physiological and biogeochemical processes. Simulation models are also constructed and applied at all of these scales. Knowledge of sources of variability across scales is of particular importance to understand and accurately predict ecosystem function in the context of global environmental change. Although present-day observation strategies are conducted at several scales, either structural or functional attributes tend to be addressed and rarely are the two linked. We contend that a stronger effort to make such links is critically needed to advance our understanding and predictive capability at large scales.

Large-Scale Function

Measurements of system function (i.e. energy and material fluxes) are more or less integrative depending on the scale in which they are made, but heterogeneity of contributing variables has significant influence at all scales. Thus, resolving sources of spatial and temporal variation remains one of our critical limitations. Although functional measurements of ecosystems are made in a number of areas (e.g. nitrogen deposition, surface energy balance), the examples provided in this section focus on estimates of carbon dioxide exchange at a range of scales because these measurements are widely made, because the carbon cycle is a critical problem in global change research, and because carbon dioxide flux correlates strongly with many other biogeochemical processes.

Continental to global estimates of carbon dioxide exchange between the atmosphere, land surface, and oceans can be derived from atmospheric chemical measurements coupled with atmospheric transport modeling (Tans et al. 1996). These estimates have been used to test global carbon budgets (e.g. Fung et al. 1987; Tans et al. 1990) and the seasonality of carbon fluxes (e.g. Ciais et al. 1995). Observational methods are based on air samples that are collected continuously at select atmospheric observatories or periodically in "clean air" conditions to avoid the influence of localized sources and sinks. These sites are meant to be both representative of large portions of the globe and to deemphasize regional source or sink heterogeneity by being located mid-ocean, in deserts and on mountaintops. Numerical models of atmospheric transport and mixing are used to drive scenarios of sources and sinks that are consistent with the observations of carbon dioxide, methane, and other species, as well as isotopes of carbon dioxide (e.g. $^{13}C/^{12}C$, $^{18}O/^{16}O$). However, the spatial coverage is still much too sparse to address principal sources and sinks of greenhouse gases, let alone the effects of land-cover or land-use changes on atmospheric trace-gas concentrations at a global level (Enting and Mansbridge 1991; Rayner et al. 1996; Tans et al. 1996).

Micrometeorological techniques integrate carbon dioxide fluxes at scales of tens to hundreds of meters for ground-based instruments, and of tens to hundreds of kilometers for airborne instruments. At the local scale, eddy-flux methods are becoming increasingly common for measuring the exchange of carbon dioxide, as well as water and several trace gases, between the atmosphere, canopy, and soil (Baldocchi et al. 1996; Crawford et al. 1996). These typically tower-based measurements quantify trends in time, but only at discrete sites. A large number of sites representing a wide range of ecosystems are needed to increase the certainty of interannual and intra-annual variation in carbon fluxes (Wofsy et al. 1993; Greco and Baldocchi 1996). Aircraft-mounted systems are implemented to assess the horizontal heterogeneity in carbon fluxes associated with changing cover-type across a landscape (Matson and Harriss 1988; Crawford et al. 1996). Although expanding the extent of the flux studies, aircraft measurements are temporally constrained by limited sampling periods. A combined network of tower sites and aircraft measurements are advantageous to characterize spatial and temporal landscape heterogeneity.

Chamber-based measurements of soil gas fluxes (e.g. carbon dioxide, nitrous oxide, nitric oxide, methane) provide a means to assess the activity of soil microbial communities, fine-scale variation in gas exchange in time and space for a given ecosystem, in addition to a window into general properties of nutrient and organic matter cycling and nutrient limitation (Keller et al. 1986; Matson and Vitousek 1990; Keller et al. 1993; Castro et al. 1995; Townsend et al. 1995). Chambers are a simple, inexpensive method and, if done with a minimum of perturbation of the gas efflux, are particularly valuable for quantifying the soil contribution to gross ecosystem flux at a local level and for model parameterization for larger-scale estimates. Isotope techniques provide additional information on the partitioning of carbon dioxide evolved from soil carbon pools and on rates of soil carbon turnover (Veldkamp 1994; Townsend et al. 1995, 1997). These soil methods effectively measure submeter areas and require high sample numbers to provide sufficient characterization of the naturally high variation in soil properties.

A combination of these measurements of canopy, landscape, and regional trace-gas fluxes are required to improve our understanding of carbon dioxide budgets and for providing data sets to test and parameterize carbon balance models at a range of scales (Baldocchi et al. 1996). Successful multiscale measurement programs obviously require teams of scientists to tackle the multidisciplinary challenges inherent in the methods. Moreover, coordination among sites is imperative for a broad-scale understanding of flux characteristics and ecosystem-specific controlling factors. All of these methods described require significant attention to representativeness for extrapolation purposes and careful accounting of the sources of variation. The highly variable effects of human activities must therefore be incorporated into nested measurement strategies aimed at understanding regional- and global-scale biospheric function.

Large-Scale Structure

Representation of the spatial distribution and biophysical attributes of vegetation at scales ranging from large regions to the entire globe has evolved considerably during the past several decades. The accuracy of global vegetation maps has advanced from models of potential vegetation based on climate patterns to actual measures of vegetation phenology using satellite remote sensing techniques. More detailed mapping and measurements of canopy biophysical characteristics are also evolving, again through the use of remote sensing data.

Holdridge (1947) constructed global maps of potential vegetation based on observed consistencies between vegetation life-forms and such climate factors as precipitation and temperature. Even though these relationships proved useful for characterizing the general patterns of vegetation at large scales, Matthews (1983) and Olson (1983) constructed the first global vegetation maps from actual country-by-country, land-survey and statistics data. This important step resulted in maps of what is essentially potential vegetation with a spatially oriented estimate of land-cover change (e.g. extent of croplands) at one specific point in time. (Consequently, the Matthews global vegetation map has been used extensively in climate and biogeochemical modeling studies (e.g. Schimel et al. 1996).

Remote sensing estimates of land-cover types were considered the next significant contribution to the improvement of global vegetation maps based on the argument that measures of existing land cover are most relevant to global change research (Townsend et al. 1991). These methods have primarily used the reflectance characteristics of vegetation in the visible and near-infrared spectrum to detect coarse-scale differences in vegetation structure. Recently, multitemporal satellite data have been used to classify the land surface; this techniquespis based on the idea that the phenological character (e.g. seasonality) of vegetation is strongly indicative of ecosystem type (Loveland et al. 1991; DeFries et al. 1995). Running et al. (1994b) proposed a classification logic based on vegetation phenological characteristics (inferred from a remotely sensed spectral index) that simplified the global vegetation to six fundamental vegetation classes.

Relationships derived between remotely sensed spectral indices and vegetation biophysical characteristics have also greatly improved the ability to map vegetation types. The most common vegetation index employed for global-scale studies continues to be the normalized difference vegetation index (NDVI). The NDVI is simply the difference between the reflectance characteristics of a surface in the near-infrared (NIR) and the visible spectral regions (VIS), divided by their sum (NIR − VIS/NIR + VIS). This index is not a fundamental ecophysiological variable, although theory and observation support the relationship between NDVI and a number of ecosystem variables, including canopy photosynthetic efficiency, the fraction of absorbed photosynthetically active radiation (fAPAR), stomatal conductance, and leaf area index (LAI) (e.g. Asrar et al. 1984; Tucker and Sellers 1986; Sellers 1987; Myneni and Williams 1994). For example, using an empirical NDVI–fAPAR relationship, along with an efficiency factor to relate fAPAR to aboveground carbon fixation, it is possible to estimate changes in aboveground net primary production (NPP) over time (Field et al. 1995).

However, it is well known that these empirical relationships between the NDVI and canopy biophysical attributes vary with respect to such external factors as background soil color, the presence of surface litter, and the position of the sun and the sensor during image acquisition (Huete 1988; Goward and Huemmrich 1992; van Leeuwen and Huete 1996). Moreover, NDVI/biophysical parameter relationships are often locationally, temporally, and vegetation dependent, and they can require extensive field calibration to achieve a reasonable level of accuracy, which limits their quantitative capability as a large-scale structural measurement.

More recent developments in quantitative remote sensing and radiative transfer (RT) modeling show promise in delivering estimates of key vegetation structural attributes for ecological studies. These newest methods utilize a series of canopy reflectance measurements taken at different viewing and illumination angles, and an inversion of a mechanistic canopy RT model, to simultaneously derive a suite of structural variables from the data. For example, Privette et al. (1994, 1996) and Braswell et al. (1996) estimated LAI and fAPAR of grassland, savanna, and woodland vegetation using the National Oceanic and Atmospheric Administration (NOAA) advanced very high resolution radiometer (AVHRR) satellite. Over a sufficient period of time (e.g. ten days), the AVHRR observes a point on the ground from differing view angles and with slightly differing sun positions in the sky. These multiview angle observations provide increased access to canopy biophysical information because the angular variation in vegetation reflectance is controlled by canopy and landscape-level biophysical characteristics (Goel 1988; Myneni et al. 1995).

These canopy RT inversion methods are promising because they are based on the mechanics of canopy photon transport, and therefore do not require extensive parameterization or field data. They, in fact, are a major improvement over the empirically based methods (e.g. NDVI) because the derived variables do not require prior knowledge of vegetation types; rather they can exploit knowledge (e.g. leaf optical properties) that is generalizable across landscapes (Asner et al. 1998a).

Radar remote sensing of canopy structure has undergone a somewhat parallel evolution to these optical (shortwave) remote sensing developments. Synthetic aperture radar (SAR) instruments emit and receive energy in the microwave region of the spectrum (wavelengths of about 10 mm–1 m). This energy is attenuated and scattered by vegetation canopies depending on three major factors: (1) canopy water content; (2) soil moisture, and (3) the size and orientation of such canopy components as leaves, stems, and trunks.

During the past ten to fifteen years, research using SAR for ecological applications has increased dramatically (Kasischke et al. 1997). The wide variety of SAR applications to ecosystem monitoring has been thoroughly reviewed by Kasischke et al. (1997), and can be summarized into three major categories: (1) land-cover classification; (2) measurement of aboveground biomass; and (3) delineation of flooding or water inundation in forest ecosystems (e.g. Pope et al. 1994; Ranson and Sun 1994; Dobson et al. 1995; Hess et al. 1995). An important point to emphasize is that many SAR studies attempting to relate the radar signal to canopy biomass and structure have relied on extensive field measurements for algorithm development. These efforts, although relatively successful at a given location and at a given time, are limited in their general applicability in a way similar to that of the vegetation index approaches mentioned before.

More recently, and somewhat analogous to the canopy RT methods, radar backscatter modeling approaches have provided both a more mechanistic understanding of the vegetation and soil factors influencing radar signals and a means to directly link canopy structural attributes to radar observations (e.g. Pierce et al. 1994; Saatchi and Moghaddam 1994). These approaches have been carried a step further by linking forest succession models to SAR data using a backscatter model as the interface (Ranson et al. 1997).

Even though all of these remote sensing techniques represent a major improvement over prior vegetation mapping methods, they may be too coarse in spatial resolution when applied at the global scale to adequately represent the land-cover heterogeneity (especially in savannas, shrublands, dry forests, and areas of mixed crops, pastures, and forests). This limitation may be significant for interpretation, extrapolation, and constraint of flux measurements (e.g. the carbon dioxide flux measurements described earlier). At the global scale, both satellite instrument and computational limitations hamper the utility of canopy RT inversion methods, whereas simpler empirical methods, for example the NDVI/biophysical parameter relationships, remain more readily applicable. Thus, at very large spatial scales, we are not only constrained in our ability to resolve key surface detail, but it may not be feasible to measure vegetation structural characteristics with the newest RT methods. These limitations weaken, if not disable, our ability to resolve at a global extent the changes in structure on "human spatial scales" (e.g. approximately sub-kilometer) until those changes are evident at larger spatial scales (e.g. Skole 1994). At regional scales, limitations imposed upon the canopy RT inversion methods may or may not be lifted, depending on the spatial resolution of the sensor. An "optimum" resolution—which can both adequately resolve the surface characteristics of interest and allow for the more robust but computationally expensive approaches—would allow for the most detailed analyses of vegetation structure. Identifying this optimum condition is difficult and may be specific to a given vegetation type or land-cover and land-use change scenario. For example, accurate quantification of deforestation in the Amazon requires a spatial resolution better than 100 m (Skole and Tucker 1993), even though detection of land-cover change in grasslands and savannas requires a spatial resolution no larger than 30 m (Asner et al. 1997; Wessman et al. 1997).

Provided that this issue of resolution and computation can be overcome, these quickly evolving regional-scale remote sensing capabilities provide a hierarchical link between studies of global vegetation-climate interactions and the ecosystem-specific feedback of regions to the global-scale processes. It is probable that remote sensing will continue to stand out as the only efficient method for measuring and monitoring vegetation structure at these large spatial scales.

Linking Function and Structure

Potential links between large-scale function and structure can be approached in two ways: (1) through the direct connection between such remotely sensed biophysical attributes as fAPAR and functional processes including photosynthesis, respiration, and nutrient use; or (2) through the indirect connection provided by ecosystem process models (Figure 14.2). To date, improvements in the ecosystem-process model link have been most broadly demonstrated.

Ecosystem-process models, in one way or another, simulate the dynamics of carbon, water, nitrogen, and sometimes other nutrients (Parton et al. 1987; Rastetter et al. 1992b; Running and Hunt 1993; Field et al. 1995). One of the most important facets of these models is their ability to integrate and test our knowledge of the functional attributes of ecosystems. However, within the model framework, the interaction between structure and function has been largely unidirectional. Typically, these models rely on parameterizations of both above- and belowground allocation to constrain the flow of carbon to various locations in the vegetation (e.g. leaves, woody material, roots). That is, the modeler dictates where assimilated carbon is stored, subject to some modification through the stoichiometric relationships between carbon, water, and nutrients. Many models rely or have relied on vegetation-specific look-up tables to set these carbon allocation constraints. Moreover, most models rely on the vegetation maps described earlier to prescribe the vegetation type, followed by a look-up table of appropriate allocation parameters for each vegetation type.

The important limitations to such an approach are two-fold. First, the ability for vegetation to be redistributed in response to anthropogenic forcing has not been addressed on the time scales most relevant to global

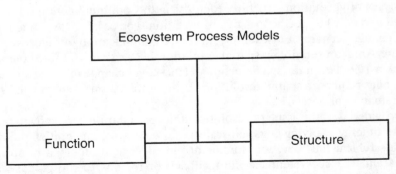

FIGURE 14.2. Ecosystem-process models help to integrate our knowledge of functional processes. However the link between function and structure is unidirectional. Processes are modeled to create structure (although this is not handled mechanistically), even though structure has very limited control on function.

change. For example, even the coupled biogeochemical-biogeographical models (VEMAP Participants 1995; Neilson and Running 1996; Woodward 1996) do not simulate the speed at which herbaceous vegetation has been replaced by woody vegetation resulting from land use or atmospheric change in some biomes (e.g. Archer et al. 1995; Chapin et al. 1996). Second, the characteristics of and potential changes in vertical structure or allocation are not mechanistically simulated in ecosystem-process models, which can be attributed to our insufficient understanding of the links between plant genetics, community and ecosystem ecology, and biogeochemistry. Consequently, the transient response of ecosystems, although simulated with adequate accuracy under conditions of constant vegetation type and allocation pattern, are not modeled well when either the horizontal distribution of vegetation types changes rapidly or when shifts in carbon allocation occur (e.g. caused by land-cover or land-use change or atmospheric change).

Remote sensing is unique, not only in that it allows access to large-scale estimates of vegetation structure, but also that it provides some means to bypass our incomplete understanding of allocation in ecosystem-process models. Recently, some ecosystem-modeling efforts have turned to satellite-derived vegetation maps to provide the vegetation allocation parameters. Field et al. (1995) and Hunt et al. (1996) used the remotely sensed NDVI relationships mentioned earlier to estimate fAPAR and LAI, respectively. These estimates were, in turn, used to constrain processes associated with carbon allocation, and nutrient and water use. These steps in linking remotely sensed structural information to ecosystem-process models have provided a means to circumvent our present-day inability to mechanistically connect principles of allocation to biogeochemistry and ecosystem function.

The Key to Structure: Prospects in Remote Sensing

Ecological remote sensing is evolving at an increasing pace. With the launch of the Earth observing system (EOS) in 1998 and high spatial resolution sensors (e.g. Landsat 7), vegetation remote sensing will undergo a revolution in its capability to provide key ecological information (e.g. Running et al. 1994a; Asner et al. 1998b). These instruments will not only significantly increase the volume and quality of data, but completely new and unique spectral information will become available for any location on Earth. The EOS moderate resolution imaging spectrometer (MODIS) and the multiangle imaging spectroradiometer (MISR) will provide views of the earth surface from multiple angles (Barnsley et al. 1994). Based on early research efforts focused on similar multi-angular and hyperspectral information, there appears to be three areas of ecological remote sensing that will, for the first time, be boosted to an operational level.

The first improvement relates to the difficulty in assessing regional-level changes in land cover. It has been well accepted that increasing the spatial resolution of remote sensing data can have significant effects on estimates of the extent of cover types. For example, estimates of deforestation in the Amazon have been changed by as much as 50% following reanalysis with higher resolution data (Skole and Tucker 1993; Skole 1994). Such improvements in spatial resolution provide a critically needed element to regional ecological-process studies—detection of actual vs. potential land cover.

These regional- and biome-level studies of vegetation dynamics have spawned an entire area of research focused on improving the spatial resolution of remote sensing data. Spectral mixture analysis is a method by which image pixels are separated into components or "endmembers" (Borel and Gerstl 1994; Smith et al. 1994). Depending on the spectral resolution of the data, these endmembers may represent types of land-cover (e.g. forest, pastures, bare soils), types of material (e.g. fresh leaf material, litter, soils), or even land-use characteristics (e.g. heavily grazed, frequently burned) (Roberts et al. 1993; Smith et al. 1994; Wessman et al. 1997). The number and types of endmembers extracted via spectral unmixing are largely dependent on the uniqueness of each "endmember signature," and increasing the information in each signature can have a significant effect on the unmixing results (Smith et al. 1994; Wessman et al. 1997). The advent of MODIS means a significant increase in spectral resolution for regional- and global-scale studies, and these instruments will probably lead to a large improvement in image pixel unmixing. Additionally, the increased accessibility of these data will bring sub-pixel mixture analysis to a much broader range of researchers than is possible now.

The second area of ecological remote sensing to improve during the EOS era is the measurement of canopy biophysical attributes from space. The angular variation of vegetation reflectance with changing sun and sensor-viewing geometry has been termed the bidirectional reflectance distribution function (BRDF) (Figure 14.3). Variation in the BRDF of vegetation results primarily from differences in canopy- and landscape-level structural characteristics, in addition to soil textural attributes (e.g. Goel 1988; Jacquemoud et al. 1992). At the landscape level, both the relative ground cover and spatial distribution of vegetation types, each with differing crown geometry, largely determine the BRDF (Li et al. 1995). However, such canopy-level biophysical characteristics as LAI, leaf angle distribution (LAD), and foliage clumping also play significant roles (Ross 1981; Goel 1988; Chen and Cihlar 1995; Myneni et al. 1995).

Based on this knowledge of the canopy attributes controlling the vegetation BRDF, and the unmatched ability to acquire measurements of this angular reflectance distribution, by the forthcoming MODIS and MISR

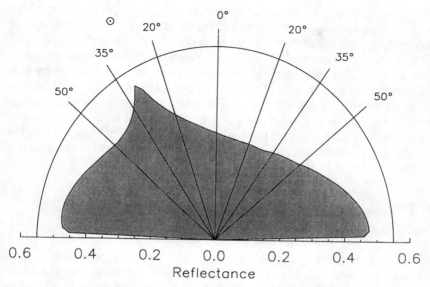

FIGURE 14.3. An example of an angular reflectance distribution along the solar principal plane (the plane in which the sun travels across the sky) collected with a ground-based spectroradiometer during the National Aeronautics and Space Administration First ISLSCP Field Experiment (NASA FIFE) campaign at the Konza Long-Term Ecological Research (LTER) site in Kansas. Notice the strong backscattering effect in the retro-solar direction at 30° left of nadir. This and other characteristics of the BRDF are determined by canopy- and landscape-level structural characteristics such as LAI, foliage clumping, and inter-crown shadowing (see Privette et al. 1994 and Asner et al. 1998a).

(Barnsley et al. 1994), new opportunities for estimating these variables will be available. Canopy RT model inversions offer the most promising approach to retrieval of these vegetation structural characteristics. Because these models mechanistically simulate canopy-level photon interactions between leaves, stems, and soils, inversion of these models can provide reasonable estimates of LAI and fAPAR from multiview angle remote sensing data (Goel and Thompson 1984; Privette et al. 1994; Roujean and Breon 1995; Braswell et al. 1996; Privette et al. 1996; Asner et al. 1998b).

Finally, a continued focus on estimation of such canopy chemical attributes as lignin and nitrogen content, from hyperspectral remote sensing data has resulted in some potential methods for forest canopy biogeochemical analyses (e.g. Wessman et al. 1988; Zagolski et al. 1996; Martin and Aber 1997). Future imaging spectrometer missions will bring this possibility to a spaceborne level, allowing access to forest canopies anywhere on the globe. Prior to the launch of these instruments, canopy chemistry research will continue to be limited to only a few investigations using airborne instrumentation. These investigations have been primarily correlative studies between field-measured chemistry values and high spectral resolution data in closed-canopy forests. Although the relationships between canopy chemistry and reflectance indicate potential for site-specific applications, mechanistically based modeling studies of the influence of foliar chemistry on canopy-scale reflectance are helping to elucidate the relative contributions of leaf and canopy components (Jacquemoud et al. 1995; Fourty and Baret 1997; Asner 1998).

A Remote Sensing Role in Linking Structure and Function

Based on both the history and recent developments in vegetation remote sensing, the potential clearly exists for a greatly expanded role of satellite information in linking structure to function at a variety of large spatial scales (Figure 14.4). For instance, improved quantitative estimates of such key structural variables as vegetation extent and LAI, and such biophysical attributes as fAPAR, may be used as a surrogate for our inconsistent understanding of the links between structure and function. It also provides

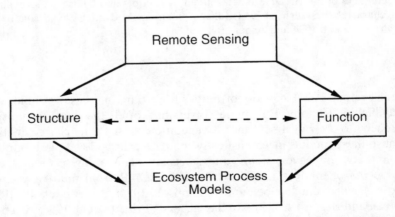

FIGURE 14.4. The role of remote sensing in linking function and physical structure. Ecosystem structure and function will feedback on one another (dashed line), but this is only a unidirectional relationship in ecosystem process models (structure leads to function). Remote sensing can help bridge the functional influence on structure through estimates of aboveground LAI and vegetation type and extent, as well as fAPAR and canopy chemistry.

a means to incorporate relatively fast changes in land cover into the modeling environment. When acquired throughout a growing season, these remote sensing data provide information on large-scale patterns of aboveground allocation. In turn, this information can constrain ecosystem-process model simulations by forcing elemental cycles to "interact" with a measured and changing canopy structure or carbon/nutrient pool (Figure 14.4).

Even though it is difficult to predict the exact contribution of vegetation remote sensing in the coming era, it is very probable that remote sensing methods will continue to stand out as the optimum approach to estimating large-scale structure and land-cover change. It is also probable that remote sensing will provide the only feasible method for moving beyond the limitations presently imposed by our lack of a mechanistic or even empirical understanding of the links between ecological structure and function.

We must note that the use of remote sensing as a link between ecological structure and function at large spatial scales is tempered by two important limitations. First, the quality of an estimate or "retrieval" of such surface characteristics as vegetation cover or LAI is totally dependent on our ability to understand, and to subsequently remove, the effects of the atmosphere in remotely sensed imagery. In land-surface or oceanic remote sensing analyses, the atmosphere becomes a major source of noise in the data (e.g. Kaufman 1989). As we become more quantitative, for example through the use of canopy RT modeling techniques or sub-pixel analyses, issues of atmospheric correction will increase in importance. At present, strong efforts are being made in this area. For instance, atmospheric RT models have been designed and implemented specifically for purposes of remote sensing atmospheric correction (e.g. Vermote et al. 1997). The National Aeronautics and Space Administration's EOS MODIS and MISR instruments will have spectral and angular configurations specifically designed to improve the measurement of such key atmospheric characteristics as aerosol optical depth, atmospheric water vapor, and cirrus cloud cover (King et al. 1992; Running et al. 1994a). These and other advances in atmospheric measurement and correction of data will sharply improve our ability to detect and quantify key surface variables for ecological research and monitoring.

Finally, all of these remote sensing advances, including multi-angle, hyperspectral, and high spatial resolution instruments, mean a concomitant increase in data volume. Ironically, given all of the technical advances in the theory and application of remote sensing for large-scale ecological problems, it remains somewhat unclear as to how we will store, quality check, and disseminate the large volume of data to come with these advances. This is clearly an issue requiring continued and increased attention.

General Limitations to Large-scale Measurements

The continued and carefully planned expansion of a nested measurement strategy for understanding large-scale ecological function, with a concomitant and rapid development of quantitative remote sensing for large-scale structural analyses, will foster a sharp improvement in our ability to scale processes from the landscape to global level. Although we believe that these goals are readily achievable in the next decade, it must also be recognized that there are areas of persistent difficulty that deserve an intense research focus. The combination of nested ecological measurements, remote sensing, and ecosystem-process modeling will be limited without an improved knowledge of: (1) vegetation carbon stocks in relation to land-use change; (2) patterns of belowground biomass; (3) a mechanistic understanding of plant allocation response to climate and biogeochemical processes; and (4) the spatial distribution of soil types and physical soil characteristics.

Vegetation carbon stocks vary significantly with land use, and our present-day knowledge of the extent of these effects is limited. Two categories of questions to ask are: (1) Where, in relation to biome type, climate, and soil characteristics, is land use occurring? What kind of land use takes place, and with what level of frequency and intensity? and (2) What are the structural and functional responses of these biomes, again taking into account the differences in climate and soil characteristics, to these land-use perturbations? What are the transient and long-term effects on carbon stocks? Even though several biomes in the temperate zone have been intensively studied (e.g. temperate grasslands), knowledge gaps persist for most tropical and subtropical biomes.

A good example of land-use change effects on carbon stocks is the increasing percentage of moist tropical forests that are or have been subject to logging and clearing for agriculture and cattle pasture. When abandoned to secondary regrowth, these forests respond differentially based on many factors including the type and duration of land use prior to abandonment and the availability of vegetation growth-forms (e.g. woody arborescent, low stature shrubby, or herbaceous species) to invade the site (Fearnside 1996). Therefore carbon stocks may not change in a consistent, predictable manner during secondary succession depending on the land-use legacy and the composition of the soil seed bank. Fearnside (1996) found that, when secondary regrowth occurred in previously cleared moist tropical forests in the Amazon, biomass was one-tenth that of the original vegetation. From a remote sensing standpoint, it would seem that such large changes in carbon stocks might be detectable from space. However, analyses have consistently shown the detection of secondary forest to be difficult, if not impossible, with optical sensors because the successional canopies grow dense and "appear" similar to primary forest, although the woody (or heavy) component of this secondary canopy is much less than that of the primary forest. Radar remote sensing of forests has been a major research focus, as it has held promise for detecting changes in carbon stocks with land-cover and land-use change (e.g. Yanasse et al. 1997). However, results show that while gross changes in land cover are detectable, quantitative estimates of aboveground biomass remain somewhat elusive (Kasischke et al. 1997).

Both the pattern and processes of above- and belowground carbon allocation continue to stand as an area deserving of intensive research. Although our understanding of aboveground allocation is limited, it still far exceeds that for belowground dynamics (see Zak and Pregitzer, chapter 15). Some recent efforts to compile, synthesize, and analyze a widely dispersed literature on rooting depth and root biomass have yielded useful information for large-scale studies of ecological function (Canadell et al. 1996). However, until a functional understanding of root allocation with respect to biogeochemical processes can be developed, our ability to link structure and function will be hampered.

Finally, spatially oriented information on soil types and characteristics continues to improve through compilations of land-survey information, and efforts to refine these databases should continue. Most important, a minimum set of soil physical attributes required to significantly improve these information sources should be defined (Lathrop et al. 1995). These attributes should not only describe soil characteristics under pristine conditions, but in relation to land-use and land-cover change as well. Some of the soil characteristics needed for ecosystem-modeling efforts include soil texture, bulk density, pH, cation exchange and water-holding capacity, carbon and nutrient pools, in addition to some indication of the magnitude of each major flux pathway (e.g. nitrogen mineralization). This is obviously a large undertaking; however, the need for this information cannot easily be overstated. Without a region-to-region knowledge of these basic soil characteristics, modeling studies of carbon storage (e.g. sources and sinks), the response of ecosystems to nitrogen deposition, and the effects of land-cover and land-use change will continue to be limited in their ability to resolve functional details that scale up and affect global-scale biosphere-atmosphere exchange processes (Asner et al. 1997).

Generalities and Concerns: Present and Future

In conclusion, we stress the need to address ecosystem physical structure and its interaction with ecosystem functioning. The present-day lack of knowledge of the mechanisms behind above- and belowground carbon allocation is a serious limitation to both our understanding and our predictive capabilities of large-scale dynamics. Remote sensing can be used to quantify aboveground structure and allocation patterns, and thereby serve to constrain our ecosystem-process models in a reasonable manner.

However, belowground allocation will remain beyond the reach of any large-scale measurements; our knowledge of these processes will be dependent on experimental studies and subsequent mechanistic modeling. Land-cover and land-use distribution and change are more tractable through remote monitoring; however, the land-use element will require additional knowledge of social systems.

It is very important to recognize that these problems are not endlessly reducible. First, there are physical limits to measurements. If, for example, the spectral reflectance of a species is not distinct, species-level identification using remote sensing will not be possible. In other words, measurement capabilities are set by the phenomena that affect them. Second, further reduction to smaller-scale mechanisms may become divorced from the large-scale (emergent) properties of the aggregate. Better resolution sensors cannot provide all the answers; even denser networks of measurements may be insufficient. Because of the scaled nature of complex systems, finer reduction in measurements will not necessarily be appropriate. We will need to identify acceptable levels of understanding and predictive capabilities within scientific and budgetary reason.

Finally, there is great need to train ecologists in the quantitative skills required of the challenges presented by large-scale ecology. Even though ecosystem science is already a highly interdisciplinary field, it becomes even more so with the increased complexity of large-scale systems. The connections between ecosystem and atmospheric sciences are increasingly apparent and knowledge in both areas is necessary to work at their interface. Coupled with these requirements will be an expanding need for ecologists trained in mathematics, physics, statistics and such cross-disciplinary tools as remote sensing and geographic information systems.

Acknowledgments. We thank Jim Gosz, Carol Johnston, Pamela Matson, and Elaine Matthews for their constructive comments on an early version of this manuscript. Rob Braswell and Alan Townsend shared their clear thinking and substantive insights on a later version. We are grateful for the support of the National Aeronautics and Space Administration.

References

Allen, T.F.H., and T.B. Starr. 1982. *Hierarchy: perspectives for ecological complexity.* University of Chicago Press, Chicago, IL.

Archer, S., D.S. Schimel, and E.A. Holland. 1995. Mechanisms of shrubland expansion: land use, climate or CO_2. *Climatic Change* 29:91–99.

Asner, G.P. 1998. Biophysical and biochemical sources of variability in plant canopy reflectance. *Remote Sensing of Environment* 65:1–20.

Asner, G.P., B.H. Braswell, D.S. Schimel, and C.A. 1998b. Wessman. Ecological research needs from multi-angle remote sensing data. *Remote Sensing of Environment* 63:155–165.

Asner, G.P., T.R. Seastedt, and A.R. Townsend. 1997. The decoupling of terrestrial carbon and nitrogen cycles. *BioScience* 47:226–234.

Asner, G.P., C.A. Wessman, and J.L. Privette. 1997. Unmixing the directional reflectances of AVHRR sub-pixel landcovers. *IEEE Transactions on Geoscience and Remote Sensing* 35:868–885.

Asner, G.P., C.A. Wessman, D.S. Schimel, and S. Archer. 1998a. Variability in leaf and litter optical properties: implications for BRDF models and inversions using AVHRR, MODIS, and MISR. *Remote Sensing of Environment* 63:243–257.

Asrar, G., M. Fuchs, E.T. Kanemasu, and M. Yoshida. 1984. Estimating absorbed photosynthetic radiation and leaf area index from spectral reflectance in wheat. *Agronomy Journal* 76:300–306.

Baldocchi, D., R. Valentini, S. Running, W. Oechel, and R. Dahlman. 1996. Strategies for measuring and modelling CO_2 and water vapour fluxes over terrestrial ecosystems. *Global Change Biology* 2:159–168.

Barnsley, M.J., A.H. Strahler, K.P. Morris, and J.-P. Muller. 1994. Sampling the surface bidirectional reflectance distribution function: 1. Evaluation of current and future satellite sensors. *Remote Sensing Reviews* 8:271–311.

Borel, C.C., and S.A.W. Gerstl. 1994. Nonlinear spectral mixing models for vegetative and soil surfaces. *Remote Sensing of Environment* 47:403–416.

Braswell, B.H., D.S. Schimel, J.L. Privette, B. Moore III, W.J. Emery, E.W. Sulzman, and A.T. Hudak. 1996. Extracting ecological and biophysical information from AVHRR optical data: an integrated algorithm based on inverse modeling. *Journal of Geophysical Research* 101:23335–23345.

Burke, I.C., T.G.F. Kittel, W.K. Lauenroth, P. Snook, C.M. Yonker, and W.J. Parton. 1991. Regional analysis of the central Great Plains. *BioScience* 41(10):685–692.

Canadell, J., R.B. Jackson, and E.D. Schulze. 1996. Maximum rooting depth of vegetation types at the global scale. *Oecologia* 108:583–600.

Castro, M.S., P.A. Steudler, J.M. Melillo, J.D. Aber, and R.D. Bowden. 1995. Factors controlling atmospheric methane consumption by temperate forest soils. *Global Biogeochemical Cycles* 9(1):1–10.

Chapin III, F.S., M.S. Bret-Harte, and H. Zhong. 1996. Plant functional types as predictors of transient responses of arctic vegetation to global change. *Journal of Vegetation Science* 7:347–365.

Chen, J.M., and J. Cihlar. 1995. Plant canopy gap-size analysis theory for improving optical measurements of leaf area index. *Applied Optics* 34:6211–6220.

Ciais, P., P.P. Tans, M.Trolier, J.W.C. White, and R.J. Francey. 1995. A large Northern Hemisphere terrestrial CO_2 sink indicated by the $^{13}C/^{12}C$ ratio of atmospheric CO_2. *Science* 269:1098–1102.

Crawford, T.L., R.J. Dobosy, R.T. McMillen, C.A. Vogel, and B.B. Hicks. 1996. Air-surface exchange measurement in heterogeneous regions: extending tower observations with spatial structure observed from small aircraft. *Global Change Biology* 2:275–285.

DeFries, R.S., C.B. Field, I. Fung, C.O. Justice, S. Los, P.A. Matson, E. Matthews et al. 1995. Mapping the land surface for global atmosphere-biosphere models: toward continuous distributions of vegetation's functional properties. *Journal of Geophysical Research* 100:20867–20882.

Dobson, M.C., F.T. Ulaby, L.E. Pierce et al. 1995. Estimation of forest biomass characteristics in northern Michigan with SIR-C/XSAR data. *IEEE Transactions on Geoscience and Remote Sensing* 33:877–894

Enting, I.G., and J.V. Mansbridge. 1991. Latitudinal distribution of sources and sinks of CO_2: results of an inversion study. *Tellus* 43B:156–170.

Fearnside, P.M. 1996. Amazonian deforestation and global warming: carbon stocks in vegetation replacing Brazil's Amazon forest. *Forest Ecology and Management* 80:21–34.

Fennessy, M.J., and Y. Xue. 1997. Impact of USGS vegetation map on GCM simulations over the United States. *Ecological Applications* 7(1):22–33.

Field, C.B., J.T. Randerson, and C.M. Malmstrom. 1995. Global net primary production: combining ecology and remote sensing. *Remote Sensing of Environment* 51:74–88.

Fourty, T., and F. Baret. 1997. Vegetation water and dry matter contents estimated from top-of-atmosphere reflectance data: a simulation study. *Remote Sensing of Environment* 61:34–45.

Fung, I.Y., C.J. Tucker, and K.C. Prentice. 1987. Application of advanced very high resolution radiometer to study atmosphere-biosphere exchange of CO_2. *Journal of Geophysical Research* 92:2999–3015.

Gardner, R.H., W.G. Cale, and R.V. O'Neill. 1982. Robust analysis of aggregation error. *Ecology* 63(6):1771–1779.

Goel, N.S. 1988. Models of vegetation canopy reflectance and their use in estimation of biophysical parameters from reflectance data. *Remote Sensing Reviews* 4:1–212.

Goel, N.S., and R.L. Thompson. 1984. Inversion of canopy reflectance models for estimating agronomic variables. V. Estimation of leaf area index and average leaf angle using measured canopy reflectances. *Remote Sensing of Environment* 16:69–85.

Goward, S.N., and K.F. Huemmrich. 1992. Vegetation canopy PAR absorptance and the normalized difference vegetation index: an assessment using the SAIL model. *Remote Sensing of Environment* 39:119–140.

Greco, S., and D.D. Baldocchi. 1996. Seasonal variations of CO_2 and water vapour exchange rates over a temperate deciduous forest. *Global Change Biology* 2:183–197.

Hess, L.L., J.M. Melack, S. Filoso, and Y. Wang. 1995. Delineation of inundated area and vegetation along the Amazon floodplain with the SIR-C synthetic aperture radar. *IEEE Transactions on Geoscience and Remote Sensing* 33:896–904.

Holdridge, L.R. 1947. Determination of world plant formations from simple climatic data. *Science* 105:367–368.

Houghton, R.A. 1994. The worldwide extent of land-use change. *BioScience* 44:305–313.

Huete, A.R. 1988. A soil-adjusted vegetation index (SAVI). *Remote Sensing of Environment* 25:295–309.

Hunt, E.R., S.C. Piper, R. Nemani, C.D. Keeling, R.D. Otto, and S.W. Running. 1996. Global net carbon exchange and intra-annual atmospheric CO_2 concentrations predicted by an ecosystem process model and three-dimensional atmospheric transport model. *Global Biogeochemical Cycles* 10:431–456.

Jacquemoud, S., F. Baret, B. Andrieu, F.M. Danson, and K. Jaggard. 1995. Extraction of vegetation biophysical parameters by inversion of the PROSPECT +

SAIL models on sugar beet canopy reflectance data. Application to TM and AVIRIS sensors. *Remote Sensing of Environment* 52:163–172.

Jacquemoud, S., F. Baret, and J.F. Hanocq. 1992. Modeling spectral and bidirectional soil reflectance. *Remote Sensing of Environment* 41:123–132.

Kasischke, E.S., J.M. Melack, and M.C. Dobson. 1997. The use of imaging radars for ecological applications: a review. *Remote Sensing of Environment* 59:141–156.

Kaufman, Y.J. 1989. The atmospheric effect on remote sensing and its correction. Pages 336–428 in G. Asrar, ed. *Theory and applications of optical remote sensing.* John Wiley & Sons, New York.

Keller, M., W.A. Kaplan, and S.C. Wofsy. 1986. Emissions of N_2O, CH_4, and CO_2 from tropical soils. *Journal of Geophysical Research* 91:11791–11802.

Keller, M., E. Veldkamp, A.M. Weitz, and W.A. Reiners. 1993. Effect of pasture age on soil trace-gas emissions from a deforested area of Costa Rica. *Nature* 365:244–246.

King, A.W. 1991. Translating models across scales in the landscape. Pages 479–517 in M.G. Turner, and R.H. Gardner, eds. *Quantitative methods in landscape ecology.* Springer-Verlag, New York.

King, M.D., Y.J. Kaufman, W.P. Menzel, and D. Tanre. 1992. Remote sensing of cloud, aerosol, and water vapor properties from the moderate resolution imaging spectrometer (MODIS). *IEEE Transactions on Geoscience and Remote Sensing* 30:2–27.

Lathrop Jr., R.G., J.D. Aber, and J.A. Bognar. 1995. Spatial variability of digital soil maps and its impact on regional ecosystem modeling. *Ecological Modelling* 82:1–15.

Levin, S.A. 1992. The problem of pattern and scale in ecology. *Ecology* 73(6):1943–1967.

Li, X., A.H. Strahler, and C.E. Woodcock. 1995. A hybrid geometric optical-radiative transfer approach for modeling albedo and directional reflectance of discontinuous canopies. *IEEE Transactions on Geoscience and Remote Sensing* 33:466–480.

Loveland, T.R., J.W. Merchant, D.O. Ohlen, and J.F. Brown. 1991. Development of a land-cover characteristics database for the conterminous U.S. *Photogrammetric Engineering and Remote Sensing* 57:1453–1463.

Magill, A.H., J.D. Aber, J.J. Hendricks, R.D. Bowden, J.M. Melillo, and P. Steudler. 1997. Biogeochemical response of forest ecosystems to simulated chronic nitrogen deposition. *Ecological Applications* 7:402–415.

Martin, M.E., and J.D. Aber. 1997. High spectral resolution remote sensing of forest canopy lignin, nitrogen, and ecosystem processes. Ecological Applications 7(2):431–443.

Matson, P.A., and R.C. Harriss. 1988. Prospects for aircraft-based gas exchange measurements in ecosystem studies. *Ecology* 69(5):1318–1325.

Matson, P.A., and P.M. Vitousek. 1990. Ecosystem approach to a global nitrous oxide budget. *BioScience* 40(9):667–672.

Matthews, E. 1983. Global vegetation and land-use: new high-resolution databases for climate studies. *Journal of Applied Meteorology* 22:474–487.

Myneni, R.B., S. Maggion, J. Iaquinta et al. 1995. Optical remote sensing of vegetation: modeling, caveats, and algorithms. *Remote Sensing of Environment* 51:169–188.

Myneni, R.B., and D.L. Williams. 1994. On the relationship between fAPAR and NDVI. *Remote Sensing of Environment* 49:200–209.

Neilson, R.P., and S.W. Running. 1996. Global dynamic vegtation modelling: coupling biogeochemistry and biogeography models. Pages 451–465 in B. Walker, and W. Steffen, eds. *Global change and terrestrial ecosystems.* International Geosphere-Biosphere Programme Book Series 2. Cambridge University Press, Cambridge, UK.

Olson, J.S., J.A. Watts, and L.J. Allison. 1983. *Carbon in live vegetation of major world ecosystems. ORNL-5862, Environmental Sciences Division Publication No. 1997.* Oak Ridge National Laboratory, Oak Ridge, TN.

O'Neill, R.V., D.L., DeAngelis, J.B. Waide, and T.F.H. Allen. 1986. *A Hierarchical Concept of Ecosystems.* Princeton University Press, Princeton, NJ.

Parton, W.J., D.S. Schimel, C.V. Cole, and D.S. Ojima. 1987. Analysis of factors controlling soil organic matter levels in Great Plains grasslands. *Soil Science Society of America Journal* 51:1173–1179.

Pielke, R.A., T.J. Lee, J.H. Copeland, J.L. Eastman, C.L. Ziegler, and C.A. Finley. 1997. Use of USGS-provided data to improve weather and climate simulations. *Ecological Applications* 7(1):3–21.

Pielke, R.A., and R. Avissar. 1990. Influence of landscape structure on local and regional climate. *Landscape Ecology* 4(2/3):133–155.

Pierce, L.E., K. Sarabandi, and F.T. Ulaby. 1994. Application of an artificial neural network in canopy scattering inversion. *International Journal of Remote Sensing* 15:3263–3270.

Pope, K.O., J.M. Rey-Benayas, and J.F. Paris. 1994. Radar remote sensing of forest and wetland ecosystems in the Central American tropics. *Remote Sensing of Environment* 59:205–219.

Privette, J.L., W.J. Emery, and D.S. Schimel. 1996. Inversion of a vegetation reflectance model with NOAA AVHRR data. *Remote Sensing of Environment* 58:187–200.

Privette, J.L., R.B. Myneni, C.J. Tucker, and W.J. Emery. 1994. Invertibility of a 1-D discrete ordinates canopy reflectance model. *Remote Sensing of Environment* 48:89–105

Ranson, K.J., and G. Sun. 1994. Mapping biomass of a northern forest using multifrequency SAR data. *IEEE Transactions on Geoscience and Remote Sensing* 32:388–396.

Ranson, K.J., G. Sun, J.F. Weishampel, and R.G. Knox. 1997. Forest biomass from combined ecosystem and radar backscatter modeling. *Remote Sensing of Environment* 59:118–133.

Rastetter, E.B., A.W. King, B.J. Cosby, G.M. Hornberger, R.V. O'Neill, and J.E. Hobbie. 1992a. Aggregating fine-scale ecological knowledge to model coarser-scale attributes of ecosystems. *Ecological Applications* 2(1):55–70.

Rastetter, E.B., R.B. McKane, G.R. Shaver, and J.M. Melillo. 1992b. Changes in carbon storage by terrestrial ecosystems: how carbon-to-nitrogen interactions restrict responses to CO_2 and temperature. *Water, Air, Soil Pollution* 64:327–344.

Rayner, P.J., I.G. Enting, and C.M. Trudinger. 1996. Optimizing the CO_2 observing network for constraining sources and sinks. *Tellus* 48B(4):433–456.

Roberts, D.A., M.O. Smith, and J.B. Adams. 1993. Green vegetation, non-photosynthetic vegetation, and soils in AVIRIS data. *Remote Sensing of Environment* 44:255–269.

Ross, J.K. 1981. *The radiation regime and architecture of plant stands.* Kluwer Boston, Inc., Hingham, MA.

Roujean, J., and F. Breon. 1995. Estimating PAR absorbed by vegetation from bidirectional reflectance measurements. *Remote Sensing of Environment* 51:375–384.

Running, S.W., and E.R. Hunt, Jr. 1993. Generalization of a forest ecosystem process model for other biomes, BIOME-BGC, and an application for global-scale models. Pages 141–158 in J.R. Ehleringer, and C. Field, eds. *Scaling physiological processes: leaf to globe.* Academic Press, London, England.

Running, S.W., C.O. Justice, V. Salomonson, D. Hall, J. Barker, Y.J. Kaufmann, A.H. Strahler et al. 1994a. Terrestrial remote sensing science and algorithms planned for EOS/MODIS. *International Journal of Remote Sensing* 15:3587–3620.

Running, S.W., T.R. Loveland, and L.L. Pierce. 1994b. A vegetation classification logic based on remote sensing for use in global biogeochemical models. *Ambio* 23(1):77–81.

Saatchi, S., and M. Moghaddam. 1994. Biomass distribution in boreal forest using SAR imagery. Pages 26–30 in E. Mougin, K.J. Ranson, and J. Smith, eds. *Multispectral and microwave sensing of forestry, hydrology, and natural resources.* Europta Series, SPIE Volume 2314. The International Society of Optical Engineering, Bellingham, Washington.

Schimel, D.S., B.H. Braswell, R. McKeown, D.S. Ojima, W.J. Parton, and W. Pulliam. 1996. Climate and nitrogen controls on the geography and time scales of terrestrial biogeochemical cycling. *Global Biogeochemical Cycles* 10:677–692.

Sellers, P.J. 1987. Canopy reflectance, photosynthesis, and transpiration II: the role of biophysics in the linearity of their interdependence. International Journal of Remote Sensing 6:1335–1372.

Shugart, H.H. 1996. The importance of structure in understanding global change. Pages 117–126 in B. Walker and W. Steffen, eds. *Global Change and Terrestrial Ecosystems.* Cambridge University Press, Cambridge.

Shukla, J., C. Nobre and P. Sellers. 1990. Amazon deforestation and climate change. *Science* 247:1322–1325.

Skole, D.L. 1994. Data on global land-cover change: acquisition, assessment, and analysis. Pages 437–471 in W.B. Meyer and B.L. Turner II, eds. *Changes in Land Use and Land Cover: A Global Perspective.* Cambridge University Press, Cambridge.

Skole, D.L., and C. Tucker. 1993. Tropical deforestation and habitat fragmentation in the Amazon: satellite data from 1978 to 1988. *Science* 260:1905–1910.

Smith, M.O., J.B. Adams, and D.E. Sabol. 1994. Spectral mixture analysis—new strategies for the analysis of multispectral data. Pages 125–144 in J. Hill and J. Megier, eds. *Imaging Spectrometry—A Tool for Environmental Observations.* Kluwer Academic Publishers, The Netherlands.

Tans, P.P., P.S. Bakwin, and D.W. Guenther. 1996. A feasible global carbon cycle observing system: a plan to decipher today's carbon cycle based on observations. *Global Change Biology* 2:309–318.

Tans, P.P., I.Y. Fung, and P.T. Takahasi. 1990. Observational constraints on the global atmospheric CO_2 budget. *Science* 247:1431–1438.

370   C.A. Wessman and G.P. Asner

Townsend, A.R., B.H. Braswell, E.A. Holland, and J.E. Penner. 1996. Spatial and temporal patterns in terrestrial carbon storage due to deposition of anthopogenic nitrogen. *Ecological Applications* 6:806–814.

Townsend, J., C. Justice, W. Li, C. Gurney, and J. McManus. 1991. Global land cover classification by remote sensing: present capabilitites and future possibilities. *Remote Sensing of Environment* 35:243–255.

Townsend, A.R., P.M. Vitousek, and A. Tharpe. 1997. Soil carbon pool structure and temperature sensitivity inferred using CO_2 and $13CO_2$ incubation fluxes from five Hawaiian soils. *Biogeochemistry* 38(1):1–18.

Townsend, A.R., P.M. Vitousek, and S.E. Trumbore. 1995. Soil organic matter dynamics along gradients in temperature and land use on the island of Hawaii. *Ecology* 76(3):721–733.

Tucker, C.J., and P.J. Sellers. 1986. Satellite remote sensing of primary production. *International Journal of Remote Sensing* 7:1395–1416.

van Leeuwen, W.J.D., and A.R. Huete. 1996. Effects of standing litter on the biophysical interpretation of plant canopies with spectral indices. *Remote Sensing of Environment* 55:123–134.

Veldcamp, E. 1994. Organic carbon turnover in three tropical soils under pasture after deforestation. *Soil Science Society of America Journal* 58:175–180.

VEMAP Participants. 1995. Vegetation/ecosystem modeling and analysis project: comparing biogeography and biogeochemistry models in a continental-scale study of terrestrial ecosystem processes. *Global Biogeochemical Cycles* 9:407–420.

Vermote, E.F., D. Tanre, and J.J. Morcrette. 1997. Second simulation of the satellite signal in the solar spectrum, 6S: An overview. *IEEE Transactions in Geoscience and Remote Sensing* 35:675–690.

Vitousek, P.M. 1994. Beyond global warming: ecology and global change. *Ecology* 75(7):1861–1876.

Wessman, C.A. 1992. Spatial scales and global change: bridging the gap from plots to GCM grid cells. *Annual Review of Ecology and Systematics* 23:175–200.

Wessman, C.A., J.D. Aber, D.L. Peterson, and J.M. Melillo. 1988. Remote sensing of canopy chemistry and nitrogen cycling in temperate forest ecosystems. *Nature* 335:154–156.

Wessman, C.A., C.A. Bateson, and T. Benning. 1997. Detecting fire and grazing patterns in the Konza tallgrass prairie using spectral mixture analysis. *Ecological Applications* 7:493–511.

White, M.A., P.E. Thornton, and S.W. Running. 1997. A continental phenology model for monitoring vegetation responses to interannual climatic variability. *Global Biogeochemical Cycles* 11(20):217–234.

Wofsy, S.C., M.L. Gouldin, J.W. Munger, S.M. Fan, P.S. Bakwin, B.C. Daube, S.L. Bassow et al. 1993. Net exchange of CO_2 in a mid-latitude forest. *Science* 260:1314–1317.

Woodward, F.I. 1996. Developing the potential for describing the terrestrial biosphere's response to a changing climate. Pages 511–528 in B. Walker, and W. Steffen, eds. *Global change and terrestrial ecosystems. International geosphere-biosphere programme book series 2.* Cambridge University Press, Cambridge, UK.

Yanasse, C.C.F., S.J.S. Sant'Anna, A.C. Frery, C.D. Renno, J.V. Soares, and A.J. Luckman. 1997. Exploratory study of the relationship between tropical forest regeneration stages and SIR-C L and C data. *Remote Sensing of Environment* 59:180–190.

Zagolski, F., V. Pinel, and R. Joffre. 1996. Forest canopy chemistry with high spectral resolution remote sensing. *International Journal of Remote Sensing* 17:1107–1120.

15
Integration of Ecophysiological and Biogeochemical Approaches to Ecosystem Dynamics

Donald R. Zak and Kurt S. Pregitzer

Summary

Our ability to predict the extent to which climate change will influence the composition, structure, and function of ecosystems is contingent on understanding and integrating the response of organisms across all levels of ecological organization (i.e., physiological, population, community, and ecosystem levels). The integration of ecophysiology and biogeochemistry holds promise for working across levels of ecological organization and for increasing our understanding of ecosystem dynamics. In this chapter, ecophysiology is integrated with biogeochemistry using the C cycle of terrestrial ecosystems as a primary example. The fixation, redistribution, and loss of C from terrestrial ecosystems are largely controlled by the physiological activities of plants and soil microorganisms; however, there are several key gaps in our understanding of plant and microbial ecophysiology that limit our ability to predict the response of the terrestrial C cycle to a changing climate. The most significant gap in our understanding lies belowground and centers on the physiological links among the allocation of C to the production and maintenance of fine roots, the longevity of these structures, and the extent to which the metabolism and longevity of plant roots influence substrate availability for microbial activity in soil. In this chapter, we identify how a physiologically based understanding of fine-root production, maintenance, and longevity can be used to understand ecosystem-level patterns of C allocation. We then explore the extent to which the amount, timing, and biochemistry of root-associated C inputs influence the composition and function of microbial communities in soil. Understanding the ecophysiological links between plant roots and soil microorganisms lies at the heart of understanding the belowground C budget of terrestrial ecosystems.

Introduction

As ecologists, we are increasingly being asked to predict the potential consequences of climate change[1] on the future composition, structure, and function of the Earth's ecosystems. Most often, we conduct relatively short-term, small spatial-scale experiments and extrapolate from these results to predict the response of more complex ecological systems covering large land areas. This is particularly true for forest ecosystems in which most experimentation has focused on the short-term physiological response of juvenile trees with the expectation that insight can be gained into long-term dynamics of entire ecosystems. To accomplish this task, the responses of organisms across all levels of ecological organization must be understood, including physiological, population, community, and ecosystem ecology. There are clear conceptual connections among the physiological response(s) of organisms to environmental variables, competitive abilities that structure populations and communities, and the biotic and abiotic interactions within communities that give rise to ecosystem-level dynamics (Figure 15.1). Nevertheless, integrating over these levels of organization is a daunting task, and it remains an important challenge that must be met to understand the response of ecosystems to a changing environment.

The carbon (C) cycle is an area in which the integration of physiological ecology and biogeochemistry holds particular promise for working across levels of ecological organization and for improving our understanding of ecosystem dynamics. For example, the assimilation and allocation of C within plants, the accumulation of detritus, and the release of C from detritus during decomposition are mediated by the physiological processes of plants, microorganisms, and animals. These processes control the flow of C in both terrestrial and aquatic ecosystems, albeit at different magnitudes in time and space. As such, the direct and indirect physiological responses of organisms to rising atmospheric carbon dioxide (CO_2), altered temperature regimes, and increases in the biological availability of nitrogen (N) have important implications for ecosystem-level cycling and storage of C— a topic that has received considerable attention over the past several years. In terrestrial ecosystems, however, there are fundamental gaps in our knowledge of plant and microbial ecophysiology that limit our ability to predict the response of the C cycle to climate change. The largest gap in our understanding lies belowground and centers on plant-microbial interactions in soil, a frontier in ecosystem science in which ecophysiology and biogeochemistry can be integrated to improve our understanding of ecosystem dynamics.

---

[1] Our use of climate change extends to include rising atmospheric carbon dioxide altered precipitation and temperature regimes, and anthropogenic nitrogen deposition.

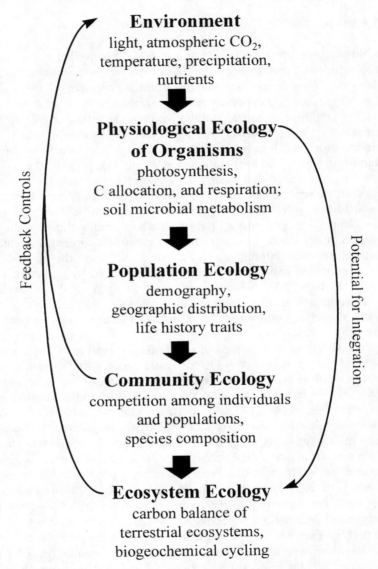

FIGURE 15.1. Linkages among levels of ecological organization. Conceptually, the physiological response(s) of organisms to environmental factor(s) can influence competitive abilities, and those competitive interactions structure populations and communities. It is the biotic and abiotic interactions within communities and ecosystem feedbacks that drive ecosystem processes. We contend that the integration of physiological ecology and biogeochemistry can lead to a better understanding of ecosystem processes. Understanding the physiological ecology of plants and soil microorganisms can bring new insight to ecosystem-level dynamics, especially the belowground C budget of terrestrial ecosystems.

We contend that our ability to predict the extent to which climate change could alter the C budget of terrestrial ecosystems is contingent, in part, on better understanding the physiological response of roots and mycorrhizae to a changing environment and how these responses influence the composition and function of microbial communities in soil. Belowground net primary production (NPP) is a substantial component of the C budget of terrestrial ecosystems, and it is probable that rising atmospheric CO_2, warmer soil temperatures, and enhanced rates of N deposition will alter the production, mortality, and biochemistry of fine roots. Nevertheless, we have an incomplete understanding of the environmental controls that regulate the amount of C that plants allocate to the production and maintenance of fine roots and mycorrhizae. This results from several methodological and conceptual limitations, including our inability to accurately measure the production and mortality of plant roots and to quantify allocation of C to mycorrhizae.

Our limited understanding of fine-root production and mortality also poses obstacles for understanding the physiological activities of soil microorganisms. Because the growth of soil microorganisms is constrained by the amount of energy entering soil from above- and belowground plant production (Smith and Paul 1990), changes in the dynamics of fine roots and mycorrhizae that modify substrate availability for microbial metabolism could alter the composition and function of microbial communities in soil. We do not understand the extent to which environmental variables influence the production, mortality, and biochemistry of plant roots, the very physiological processes that directly influence substrate availability for microbial growth and maintenance. Moreover, the linkage between C inputs from roots, substrate availability, and changes in microbial community composition and function remains mostly unexplored.

Understanding the physiological controls over fine-root dynamics, allocation of C to mycorrhizae, and the influence of soil C availability on microbial activity is fundamental to understanding soil biogeochemistry. First and foremost, we cannot close the belowground C budget of terrestrial ecosystems without understanding the physiological processes that regulate C transformations in the soil. Carbon storage depends on a complex set of processes that transform C from the end products of plant metabolism to stable soil organic matter. Second, the respiration of roots and soil microorganisms is a substantial component of the global C cycle (Raich and Schlesinger 1992), but we are unable to determine their separate contribution to the flux of CO_2 from the soil. And third, because we do not understand the physiological controls on fine-root longevity and biochemistry, we do not fully understand the processes controlling substrate availability for microbial metabolism in soil. This is particularly important because energy inputs to soil (i.e. plant-derived C) directly influence whether inorganic N is assimilated or released from microbial cells, the dynamics of which control the amount of N available for plant uptake. All of these physiological processes have clear implications for the cycling and storage of C within terrestrial ecosystems.

In this chapter, we illustrate how new insights into ecosystem-level processes might be gained by deepening our understanding of the physiological ecology of plants and soil microorganisms. In the sections that follow, physiological processes we integrate with biogeochemistry using the C cycle of terrestrial ecosystems as the primary example. To accomplish this task, key aspects of the terrestrial C cycle are identified in which our knowledge is incomplete and examples of how understanding the physiological ecology of plants and soil microorganisms can aid in filling these gaps are provided. The discussion focuses on the belowground C budget of forest ecosystems, because there is considerable debate and conflicting views regarding patterns of belowground C allocation to fine roots (sensu Hendricks et al. 1993). Presently, it is unclear whether C allocation to fine roots increases or decreases along an increasing gradient of soil N availability. This debate is reviewed and a possible resolution is identified by examining the physiological controls on fine-root longevity (i.e. production, maintenance, mortality). Next, we consider how physiological principles could be used to resolve the separate contribution of plant roots and microorganisms to the flux of CO_2 from soil. Finally, these principles are further built on to identify physiological aspects of fine-root dynamics that control substrate availability for microbial metabolism in soil, thus linking the physiological ecology of plants and microorganisms in a biogeochemical context.

Belowground C Allocation in Terrestrial Ecosystems

Most evidence suggests that a substantial portion of the C fixed by forest ecosystems is allocated to the production of fine roots (<2 mm), the ephemeral nutrient and water absorbing structures of plants (Vogt et al. 1986). Dissimilar to leaves of many temperate trees, which senesce and abscise during a defined time of the year, fine-root production and mortality are continuous processes that occur throughout the entire growing season (Hendrick and Pregitzer 1992). Most often, belowground NPP is estimated by measuring the monthly changes in fine-root and mycorrhizal biomass occurring over an entire year (Keyes and Grier 1981). Other techniques to measure root production include the use of minirhizotrons (Hendrick and Pregitzer 1992, 1993) and several methods that employ an elemental budget (Nadelhoffer et al. 1985; Raich and Nadelhoffer 1989).

Two contrasting views regarding the influence of soil N availability on fine-root dynamics have emerged (Figure 15.2). One hypothesis argues that fine-root longevity (i.e. mortality) does not change along gradients of soil N availability. It also predicts that fine-root production declines in N-rich soil,

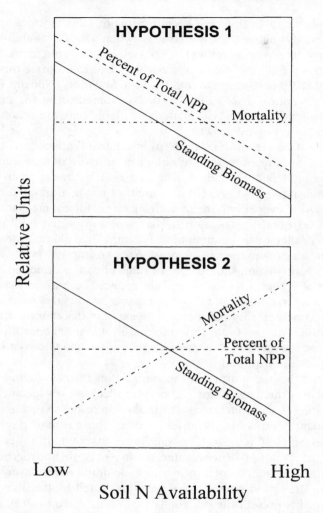

FIGURE 15.2. Two alternative hypotheses describing the relative allocation of NPP (percentage of total NPP) to fine roots in forest ecosystems (after Hendricks et al. 1993).

enabling the plant to allocate proportionately more carbohydrate to leaves and stems (Hypothesis 1 in Figure 15.2). Such a view was developed from the sequential harvest of fine-root biomass and implies that fewer roots are needed to acquire essential soil resources when they are in abundant supply (Keyes and Grier 1981). An alternative hypothesis contends that C allocation to fine roots remains constant along increasing gradients of N availability, but the life span of fine roots declines when soil N availability is relatively high (Hypothesis 2 in Figure 15.2). Therefore, the percentage of total NPP allocated to fine-root growth remains constant, but increased rates of mortality reduce fine-root biomass when soil N availability is high (i.e. fine-root turnover increases). This view of fine-root dynamics was developed from the use of an ecosystem N budget to estimate the relative allocation of NPP (percentage of total NPP) to fine roots (Nadelhoffer et al. 1985). The allocation of N to fine roots is assumed to be the difference between the amount of N assimilated by plants and the amount of N allocated to aboveground growth.

The different ecosystem-level patterns of relative C allocation illustrated in Figure 15.2 are driven by the physiological response of individual plants to environmental factors, and as such can best be understood from an ecophysiological perspective. However, neither of the aforementioned approaches provide insight into the underlying physiological mechanisms that control belowground C allocation in plants. Belowground NPP is fundamentally controlled by the amount of C that individual plants allocate to the construction and maintenance of roots. At a conceptual level, C that is allocated to construction should control rates of root production, whereas the amount of C that is allocated to maintenance (i.e. respiration used for protein repair, ion uptake, and the maintenance of ion gradients) should in part control root mortality and, hence, longevity. In this chapter, the potential relationships among C allocation to fine-root maintenance, those environmental factors that influence maintenance costs, and fine-root longevity are developed and explored.

We contend that the contrasting views regarding the belowground allocation of NPP illustrated by Figure 15.2 simply reflect species-specific patterns of root foraging (i.e. production and mortality) in response to a limiting soil resource. Plant species clearly differ in their aboveground physiological responses to light and temperature, and it is probable that they respond to belowground resources in a manner consistent with aboveground life-history traits (i.e. shade tolerance vs. shade intolerance, evergreen vs. deciduous). The broad physiological traits exhibited by the aboveground portion of plants present the possibility that plants have evolved different mechanisms to forage for soil resources. At the ecosystem level, root demography and its physiological control should influence whether the relative allocation of NPP to roots increases or decreases along gradients of soil N availability.

In the following section, we focus on the physiological processes controlling the demography of fine roots, and this information is used to illustrate how important it is to link physiology with ecosystem-level processes. Alternative demographic patterns of root production (i.e. birth) and mortality (i.e. death) that could produce a relative decline in the standing crop of roots along an increasing gradient of soil N availability are presented. This is followed by a discussion of the physiological control of fine-root maintenance, which in turn should influence fine-root longevity and the demographic response of roots to soil N availability. And last, we build on these principles to resolve the contrasting views regarding C allocation at an ecosystem level.

Plant Carbon Allocation and Fine-Root Demography

In individual plants, there is little doubt that allocation to above- vs. belowground biomass can dramatically change along gradients of soil resource availability, especially N (Tilman 1988). Most often, the relative allocation (percentage of total plant biomass) to leaves increases, whereas the relative allocation to roots declines along increasing gradients of a limiting soil resources. Such a change might indicate that plants allocate less C to root production when soil resources are abundant. Theoretically, relative allocation (percentage of total plant biomass) to roots could decline by several permutations of root production (i.e. birth) and mortality (i.e. death), all of which have very different implications for the absolute amount of C allocated belowground. The following are three different demographic responses that could lower the relative allocation to fine roots along a gradient of increasing N availability:

1. Fine-root production could decline while mortality remains constant.
2. Fine-root production could remain constant while mortality increases.
3. Fine-root production and mortality could both increase, but rates of mortality would increase more rapidly than production.

The first response suggests that fine-root longevity (i.e. mortality) is unchanged by greater soil N availability, but the allocation of C to fine-root production declines. In the second response, increases in fine-root mortality with greater N availability decrease the standing crop of fine roots. In the third, fine-root production and mortality both increase, but a disproportionately greater increase in fine-root mortality gives rise to a relatively smaller fine-root biomass. These demographic responses have very different implications for the amount of photosynthate allocated to fine-root production, wherein C allocation to fine-root production declines in the first response, remains constant in the second, and increases in the third. All three result in a relative decline in allocation to fine roots (percentage of total plant biomass), even if aboveground plant biomass only increases by a small margin. However, the absolute amount of C allocated to fine-root production dramatically changes in each scenario. Clearly, conclusions cannot be drawn regarding above- vs. belowground C allocation in plants (and ecosystems) without understanding the demographic response of fine roots to soil resource availability.

Recent advances in root biology have enabled us to directly observe the birth, growth, and death of individual fine roots. Clear plastic tubes (i.e. minirhizotrons) can be placed in the soil, and a miniaturized video camera can be used to capture images of individual roots from their birth to death

(Hendrick and Pregitzer 1992). The root images can be digitized, processed by computer, and fine-root production, longevity, and mortality can be directly measured (Pregitzer et al. 1993; Eissenstat and Yanai 1997). The results of such a study are shown in Figure 15.3, which illustrates the cumulative change in root production and mortality for *Populus* x

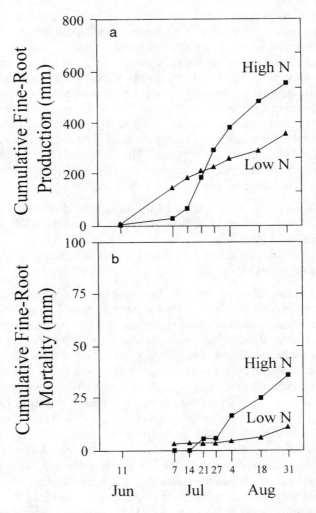

FIGURE 15.3. The influence of soil N availability on fine-root production (a) and mortality (b) of a *Populus* x *euramericana* clone. Note that both fine-root production and mortality increase with greater levels of soil N availability. Although production increases in the high N soil, the standing crop of fine roots is kept constant by rapid rates of fine-root mortality. Such a pattern suggests that the total amount of carbohydrate allocated to fine roots increases in N-rich soil (Pregitzer et al. 1995).

euramericana cv Eugenei growing in experimental soils of low and high N availability (Pregitzer et al. 1995). The aboveground C allocation patterns of this hybrid are particularly well understood, making it an effective model to understand patterns of belowground C allocation (Isebrands and Nelson 1983; Horwath et al. 1994).

Relative to plants in soil with low N availability, the leaf biomass of plants growing in soil with high N availability increased by 35% and fine-root biomass decreased by 25%. The relative decline in fine roots and relative increase in leaves is consistent with traditional views on how root/shoot ratios change along gradients of soil N availability. In the soil with a high level of N, however, fine-root production increased, whereas fine-root longevity decreased (i.e. increase in mortality; Figure 15.3a and b), supporting the idea that more carbohydrate was allocated to fine-root production in N-rich soil (Pregitzer et al. 1995). As a consequence, the amount of C allocated to fine-root production was absolutely greater in soil with high N availabity, even though the relative standing crop of fine roots declined. It could be very misleading about absolute patterns of C allocation, if conclusions are drawn from such relative measures as the root/shoot ratio of the aforementioned plants.

This example illustrates that plant C allocation and fine-root longevity are very responsive to soil N availability. Understanding these belowground plant processes has important implications for the allocation of C to belowground NPP in terrestrial ecosystems, and, hence, the amount of organic substrate entering the soil for microbial metabolism. However, several important questions remain: What are the underlying physiological mechanisms controlling fine-root mortality? How do they respond to changing environmental factors? Does fine-root mortality respond to environmental factors in a similar way for all plant species, or do species-specific responses vary widely to the same change in soil resource availability? The answers to these questions may be related to the amount of C allocated to root maintenance respiration and the extent to which environmental variables influence the sink strength of fine roots for photosynthate.

Fine-Root Longevity and Maintenance Respiration

The amount of C allocated to maintenance is apt to be an important factor influencing fine-root longevity. In some circumstances, the cost of maintaining the metabolic capacity of fine roots over their lifespan can equal or exceed their initial cost of construction (Eissenstat and Yanai 1997) Protein synthesis and replacement, membrane repair, ion uptake, the maintenance of ionic gradients across cell membranes, exudation, and mycorrhizal associations all require the expenditure of C initially fixed through photosynthesis (Ryan 1991; Bloom et al. 1992; Eissenstat and Yanai 1997). Similar to all other enzymatic processes, the rate of maintenance respiration in plant tissues exponentially increases with temperature in a predictable manner

(Ryan 1991). Maintenance respiration also is strongly influenced by tissue N concentration (Ryan 1991), wherein tissues with high N concentrations also have high rates of maintenance respiration. In fact, maintenance respiration in leaves can be predicted from N concentration for a wide range of plant species (Ryan 1991). This relationship arises because a large amount of N in plant tissue resides in proteins, and a large percentage (i.e. 60%) of maintenance respiration is allocated to their repair and replacement (Penning de Vries 1975).

In leaves, there is an important relationship among photosynthetic rate, tissue N concentration, and longevity, wherein leaf longevity is inversely related to metabolic rate (i.e. photosynthesis and dark respiration) and tissue N concentration (Larcher 1969; Chabot and Hicks 1982; Reich et al. 1991, 1992). If a similar relationship occurs in fine roots, then fine-root longevity might be expected to decline as soil N availability increases, as it did in *P.* x *euramericana* (Figure 15.3b). That is, fine roots produced in N-rich soil should have high tissue N (i.e. protein) concentrations, more rapid rates of metabolism, and greater maintenance costs than those produced in N-poor soil. If this was true for *P.* x *euramericana*, then the absolute amount of C allocated to fine-root production and maintenance was greatest in N-rich soil. Unfortunately, few studies have attempted to link fine-root longevity to maintenance costs, and we do not fully understand the degree to which C allocated to fine-root maintenance varies among plant species and with a changing environment.

Developing this understanding is particularly important, because there are widespread efforts to construct mechanistic, environmentally regulated models of forest C and nutrient cycles (Agren et al. 1991; Landsberg et al. 1991), of which fine roots are often an important component. Many ecosystem models predict altered C and nutrient allocation to roots as climate, water, and N availability change (Aber et al. 1991; Bonan 1991; Ewel and Gholz 1991; Rastetter et al. 1991, Running and Gower 1991), and most models adjust root respiration as soil temperature increases under seasonal or climate warming scenarios. The influence of soil N availability on tissue maintenance respiration has only recently been incorporated into ecosystem-level C budgets (Ryan et al. 1996), but their link to root longevity is still not considered. Moreover, there are few reports in the literature of how fine-root longevity responds to soil temperature (Hendrick and Pregitzer 1993) or N availability at the ecosystem level.

Notwithstanding these limitations, it is becoming clear that temperature and soil N availability are two environmental factors that could greatly influence the maintenance costs of fine roots (Burton et al. 1996; Zogg et al. 1996), their longevity, and the cycling of C in forest ecosystems. Using a latitudinal gradient of *Acer saccharum*-dominated forests, Hendrick and Pregitzer (1993) observed that fine-root mortality significantly increased at warmer soil temperatures. If root longevity decreases (i.e. mortality increases) as soil temperature rises, then the amount of C allocated to producing and maintaining a given biomass of fine roots will also increase, a response that could significantly influence the C balance of forest ecosystems.

Maintenance respiration, measured in excised fine roots of *A. saccharum*, is highly responsive to temperature; rates increase by a factor of 2.7 for a 10°C increase in temperature (Burton et al. 1996; Zogg et al. 1996). In combination, these observations support the idea that soil temperature may be linked to belowground C allocation in terrestrial ecosystems, because temperature affects the maintenance cost of fine roots and consequently, their longevity. This relationship illustrates how physiological processes can be directly scaled to ecosystem dynamics. It therefore becomes important to: (1) understand the inter- and intraspecific variation in the response of fine-root production, maintenance and mortality to temperature; and (2) incorporate this understanding into models of ecosystem C cycling.

Given the physiological link between tissue N concentration and the maintenance cost of plant tissues (Ryan 1991), natural gradients of soil N availability and anthropogenic additions also have the potential to affect the production, maintenance, and mortality of fine roots. Understanding the physiological link between soil N availability, belowground C allocation, and fine-root longevity lies at the heart of resolving whether relative allocation to belowground NPP increases or decreases with greater soil N availability (Figure 15.2; Hendricks et al. 1993). It is probable that fine-root tissue N concentration and maintenance respiration increase with greater soil N availability in most plant species. However, there are several possible mechanisms whereby fine-root longevity could be influenced by high tissue N concentrations and maintenance costs. For example, if plants allocate a fixed amount of C to the maintenance of fine roots, then greater maintenance costs associated with high tissue N concentrations (the result of greater soil N availability) should increase fine-root mortality and lower longevity. Conversely, if plants continually subsidize the maintenance of fine roots, then fine-root mortality may be little affected by soil N availability. These patterns of allocation to maintenance would be even more exaggerated if plants assimilate nitrate (NO_3^-) in their roots, because this process is fueled by carbohydrate (Pate and Layzell 1990; Smirnoff and Stewart 1985).

The aforementioned mechanisms represent very different patterns of root foraging in response to soil N availability, and they directly reflect the contrasting ecosystem-level patterns of belowground C allocation illustrated in Figure 15.2. They also pose several important questions that must be answered if C allocation at an ecosystem level is to be understood. How do temperature and soil N availability influence the sink strength of fine roots for photosynthate? Do plants allocate fixed amounts of C to fine-root maintenance, or are maintenance costs subsidized by a continued allocation of C to fine roots? Do plant species differ in the degree to which they allocate C for fine-root maintenance? Do increases in anthropogenic N deposition increase the maintenance cost of roots and diminish their longevity?

In *A. saccharum*, a shade-tolerant, late-successional tree species, there are clear relationships among soil N availability, fine-root N concentration, and fine-root respiration (Burton et al. 1996; Zogg et al. 1996). We have observed that net N mineralization (Figure 15.4a) is significantly related to the N concentration of fine roots in a series of *A. saccharum*-dominated forests in Michigan (Figure 15.4b), wherein tissue N concentrations are greatest in stands with rapid rates of annual net N mineralization. In turn, the highest rates of fine-root respiration (Figure 15.4c) occur in roots with high tissue N concentrations. In these forests, soil temperature accounts for a substantial portion of the seasonal variation in root respiration within a particular stand, and annual net N mineralization accounts for variation among stands (Zogg et al. 1996). In fact, fine root respiration can be predicted using the following expression:

$$R_{fr} = 0.0622 N_{fr} e^{0.101(T)} \quad R^2 = 0.91$$

where R_{fr} is fine-root respiration (nmol O_2 g^{-1} s^{-1}), N_{fr} is the tissue N concentration of fine roots (g kg^{-1}), and T is temperature in °C (Burton et al. 1996). Given that fine-root maintenance costs exponentially increase with temperature and linearly increase with tissue N concentration, it might be predicted that fine-root mortality will increase in N-rich soil if *A. saccharum* allocated a fixed amount of photosynthate to fine-root maintenance. Alternatively, if fine-root maintenance is continually subsidized by photosynthate in N-rich soil, then fine-root mortality could be unchanged or even decline.

Some preliminary results illustrate the survivorship of *A. saccharum* roots produced midway during the growing season (Figure 15.5). Notice that root cohorts from N-rich stands B and C consistently had the highest survivorship, indicating that fine-root mortality was relatively lower compared to that in N-poor stands A and C (Figure 15.5). Also notice that fine-root survivorship is positively related to net N mineralization, fine-root tissue N concentration, and fine-root respiration among stands (see Figures 15.4 and 15.5). Taken together, these results suggest that greater soil N availability increased fine-root tissue N concentration, respiration, and fine-root longevity. In *A. saccharum*, metabolically active fine roots growing in N-rich soil appear to be maintained as sinks for photosynthate to a greater extent than those growing in N-poor soil. This pattern supports the idea that the amount of C allocated to the production and maintenance of fine roots increases with increased soil N availability, at least in *A. saccharum*.

The pattern of root survivorship, or longevity, in *A. saccharum* is in direct contrast to that of *P.* x *euramericana*, a rapidly growing, shade-intolerant

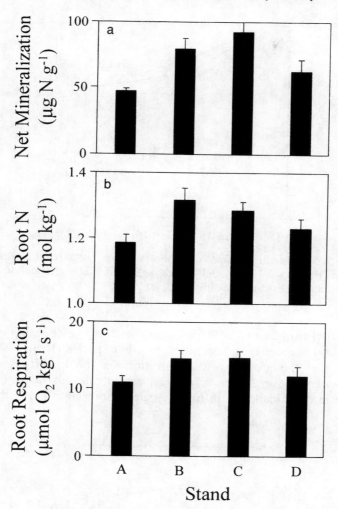

FIGURE 15.4. Net N mineralization during the growing season (a), root N concentration (b), and fine-root respiration (c) in four *Acer saccharum*-dominated forests (A, B, C, D) in Michigan. Values are the mean of twenty-four observations in each stand and the bars represent one standard error. Net N mineralization was measured monthly using buried bags, root N concentration was quantified with a CE Elantech CHN analyzer, and fine-root respiration was measured using a Hansatech O_2 electrode (see Burton et al. 1996 and Zogg et al. 1996 for details). Fine-root respiration was measured at 24 °C.

tree (Figure 15.6). In this hybrid, soil N availability increased fine-root mortality (Figure 15.4) to the extent that survivorship significantly declined (Figure 15.6). The fine-root N concentration of *P.* x *euramericana* increased in soil with high N availability, but we did not measure rates of fine-root

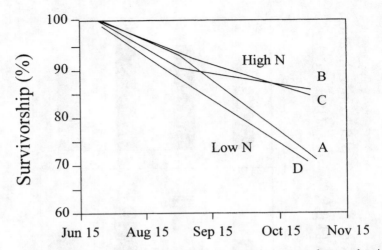

FIGURE 15.5. Survivorship of fine-root cohorts in four *A. saccharum*-dominated forests that differ in soil N availability. Sites A, B, C, and D correspond to the sites in Figure 15.6 (A.J. Burton, unpublished data).

respiration (Pregitzer et al. 1995). Given the strong relationship between tissue N and maintenance respiration in other species (Ryan 1991), it is probable that the greater N concentrations in the fine roots of *P.* x *euramericana* produced more rapid respiration rates. A decline in root longevity in combination with higher maintenance costs imply that *P.* x

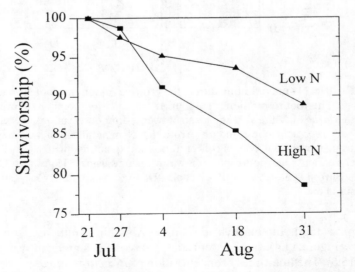

FIGURE 15.6. Survivorship of fine-root cohorts of *Populus* x *euramericana* growing in experimental soils of low and high N availability (Pregitzer et al. 1995).

euramericana significantly increased C allocation to fine roots as soil N availability increased (Figures 15.3 and 15.6). Such a response suggests that either a fixed amount of photosynthate allocated to fine-roots maintenance was metabolized more rapidly with greater soil N availability, or that the sink strength of fine roots for photosynthate diminished with greater soil N availability. Nevertheless, the very different responses of fine-root mortality to soil N availability that were observed in *A. saccharum* and *P. x euramericana* suggest that maintenance costs alone do not control patterns of fine-root longevity. Understanding the extent to which soil N availability or other environmental variables influences the sink strength of fine roots for photosynthate may be a key to understanding fine-root longevity, and, hence, belowground C allocation in terrestrial ecosystems.

The contrasting belowground responses of *A. saccharum* and *P. x euramericana* to soil N availability demonstrate why it is important to understand the underlying physiological mechanisms that drive ecosystem processes. Plants exhibit a wide array of morphological (i.e. crown architecture, deciduous vs. evergreen) and physiological (i.e. shade tolerance vs. shade intolerance) adaptations to capture light, and it is probable that they also display a similar range of belowground adaptations to forage for water and nutrients (Fitter 1987; Fitter and Stickland 1991). Although we often seek unifying principles in the study of nature, plants have evolved dramatically different physiological mechanisms to fix C and acquire soil resources. To some extent, the debate on whether the proportion of NPP allocated belowground increases or decreases with soil N availability is artificial—both patterns are apt to occur in nature.

Ecosystem-level patterns of C allocation along gradients of soil N availability should be a direct expression of the C allocation patterns of individual plants and the extent to which they produce and maintain fine roots. The extent to which soil N availability modifies belowground C allocation and root longevity in plants of contrasting life-history traits are just beginning to be understood. Clearly, we need to better understand the morphological and physiological mechanisms that plants use to capture soil resources if belowground allocation of C at an ecosystem level is to be predicted. This remains an important frontier, which can be broken by integrating physiological ecology with biogeochemistry.

Respiration and the Belowground C Budget of Terrestrial Ecosystems

The present inability to partition soil respiration into the separate contribution of plant roots and soil microorganisms poses an additional constraint on the accuracy of estimates of belowground C budgets. This is particularly true for ecosystem C budgets constructed using small chamber and micrometeorological techniques (Botkin et al. 1970; Woodwell and Botkin

1970; Wofsy et al. 1993). Most often, respiration from plant roots and soil microorganisms is estimated as a fixed proportion of soil respiration, but there are ways in which root or microbial respiration could respond to environmental variables that would alter this assumed fixed relationship. The solution to this problem is complex, because of the interdependent relationship between microbial respiration and C allocated to roots and mycorrhizae. It is clear that soil temperature and N availability exert an important control over the respiration of fine roots, and that soil temperature has an equivalent influence on microbial respiration. Because microbial respiration in soil is also controlled by substrate availability, and because root litter is a substantial input of organic substrates to soil (Vogt et al. 1986; Hendrick and Pregitzer 1992), patterns of microbial respiration should be driven, in part, by fine-root production, mortality, and biochemistry (i.e. substrate quality).

Contrasting patterns of fine-root production, maintenance, and mortality suggest that fine root and microbial respiration could change in ways that alter their contribution to soil respiration. As soil N availability increases, fine-root respiration should increase, but differences in root longevity among species provide different amounts of organic substrate for microbial metabolism. For example, if fine-root longevity declines along a gradient of increasing N availability (Figure 15.6) then substrate availability for microbial metabolism should increase. Conversely, if fine-root longevity increases (Figure 15.5) with soil N availability, then substrate availability for microbial metabolism should decline (i.e. other things being equal). Given the dependence of microbial metabolism on fine-root dynamics, it appears unrewarding to exclude roots from soil in an attempt to gain insight into microbial respiration and its contribution to the flux of CO_2 from soil.

Developing a better understanding of the physiological controls over root biomass and respiration is one approach to determine the contribution of plants to soil respiration. The amount of CO_2 that plant roots release from soil should be the product of their respiration rate $(g\,CO_2\,kg^{-1}\,h^{-1})$ and biomass $(kg\,m^{-2})$. Many observations suggest that the fine-root biomass declines as soil N availability increases (Alexander and Fairley 1983; Keyes and Grier 1981; Nadelhoffer et al. 1985; Vogt et al. 1983). However, as discussed before, the respiration rate of fine roots increases with greater soil N availability, because higher tissue N concentrations produce greater rates of maintenance respiration. These two opposing trends suggest that the total flux of CO_2 from fine roots (biomass $\times$ respiration rate), as well as the contribution of plant roots to soil respiration, could change with increasing soil N availability, depending on how rapidly fine-root biomass declined and respiration rate increased. Clearly, the biomass and respiration of coarse roots would need to be considered in order to estimate the total flux of CO_2 from roots.

The relationship between fine-root biomass and respiration rate for *A. saccharum*-dominated forests that differ in soil N availability is illustrated in

Figure 15.7 (Zogg 1996). Fine roots growing in soil with high N availability had high tissue N concentrations and consequently more rapid rates of respiration. At the same time, fine-root biomass declined with increasing soil N availability, such that fine-root biomass was lowest in stands with the highest N availability. As a result of these contrasting responses, the respiration of fine roots (respiration rate × biomass) was a relatively substantial and constant proportion of soil respiration (Figure 15.8). This relationship is unexplored in other plants and ecosystems, and it is probably an important factor contributing to variation in the contribution of fine roots to soil respiration.

Estimating the amount of CO_2 released from microbial respiration in soil is somewhat more complicated for several reasons. First, the amount, timing, and quality of root litter inputs to soil (as discussed above) is not understood, and there is evidence to suggest that microbial respiration responds to temperature in ways that are not presently considered (MacDonald et al. 1995). Nevertheless, our understanding of microbial respiration from a physiological perspective is important at an ecosystem level, because the amount of C stored in soil is a significant component of the global C cycle, and its release to the atmosphere is mediated by microbial metabolism. Consequently, the extent to which microbial activity in soil

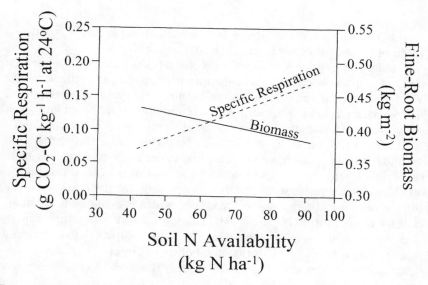

FIGURE 15.7. Fine-root respiration rate and biomass response to soil N availability in four *Acer saccharum*-dominated forests in Michigan (Zogg 1996). Soil N availability is the total net N mineralized from May to November 1994. Fine-root respiration rates were measured using an Hansetech O_2 electrode, and rates of O_2 consumption were converted to CO_2 production using a respiratory quotient of 0.8.

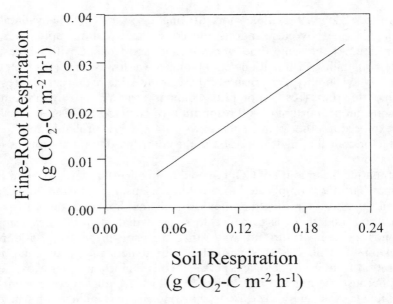

FIGURE 15.8. The relationship between root respiration and soil respiration in four forest stands dominated by *Acer saccharum*.

responds to environmental factors has important implications for the C balance of terrestrial ecosystems and of the entire Earth.

Microbial respiration in soil conforms to first-order kinetics and can be estimated with knowledge of: (1) substrate pool size; and (2) the response of the first-order rate constant (k) to soil temperature and water potential. This approach has been incorporated into computer models simulating soil organic matter dynamics (e.g. Parton et al. 1988; van Veen et al. 1984), wherein seasonal patterns of temperature, soil water potential, and substrate inputs have been used to scale first-order kinetics to an ecosystem level. Traditionally, the temperature dependence of microbial respiration is modeled as an increase in the first-order rate constant, whereas the substrate pool metabolized by soil microorganisms is assumed to be unaffected by temperature (Figure 15.9a). However, the temperature dependence of microbial respiration appears to be more complex than commonly assumed and probably involves the metabolism of larger substrate pools as soil temperature increases (Figure 15.9b; see MacDonald et al. 1995 and Zogg et al. 1997 for details).

In laboratory experiments, microbial respiration clearly does not fit the assumption of one labile substrate pool and a rate constant that exponentially increases with temperature (compare Figure 15.9a with Figure 15.10). These observations suggest that changes in substrate use with increasing soil temperature have a more substantial influence on microbial

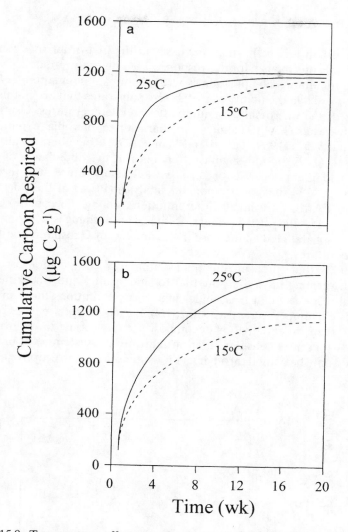

FIGURE 15.9. Temperature effects on the first-order kinetic model of microbial respiration: $Y = A_o(1 - e^{-kt})$, in which Y is the cumulative respired C ($\mu g\ C\ g^{-1}$) produced over time t (week), A_o is the amount of substrate C ($\mu g\ C\ g^{-1}$) present at the start of the experiment, and k is the rate constant (week^{-1}). In panel a, temperature-dependent changes in k are reflected in a change in the rate at which the amount of cumulative C respired over time approaches a finite value, the pool size of substrate (A_o, bold line). For example, these hypothetical data depict how an increase in soil temperature from 15 to 25 °C leads to an approximate doubling of the rate constant ($k_{15°C} = 0.18\ wk^{-1}$, $k_{25°C} = 0.36\ wk^{-1}$), resulting in a more rapid depletion of what is assumed to be a relatively static pool of substrate ($A_o = 1,200\ ug\ C\ g^{-1}$) available for microbial respiration. Alternatively, in panel b, temperature could have a relatively smaller effect on the rate constant ($k = 0.18\ wk^{-1}$), but might result in an increase in the pool size of substrate respired ($A_{o15°C} = 1,200\ ug\ C\ g^{-1}$, $A_{o25°C} = 1,600\ ug\ C\ g^{-1}$) caused by some temperature-dependent constraint on microbial utilization of substrate (Zogg et al. 1997).

respiration than relatively small increases in the first-order rate constant. One plausible mechanism for this response is a temperature-induced shift in microbial community composition, wherein dominant populations at higher temperatures have the ability to metabolize substrates that are not used by members of the microbial community at lower temperatures. Using fatty acid methyl ester (FAME) analysis to characterize microbial communities (Vestal and White 1989; Tunlid and White 1992; Ringelberg et al. 1994), Zogg et al. (1997) found significant changes in the abundance of gram-positive and gram-negative bacteria as soil temperature was increased from 5 to 25°C. Principal components analysis (PCA) of the thirty most abundant FAMEs revealed differentiation among microbial communities at 5, 15, and 25°C (Figure 15.11). These changes in community composition paralleled an increase in the amount of C respired at different soil temperatures.

It appears that shifts in microbial community composition associated with soil warming have the potential to significantly alter the kinetics of microbial respiration, at least under laboratory conditions that simulate a range of field temperatures (i.e. 5 to 25 °C). If such a response takes place in the field, then organic matter decomposition may occur through mechanisms not presently considered, an insight into ecosystem-level processes that could only be gained through an understanding of the ecophysiology of

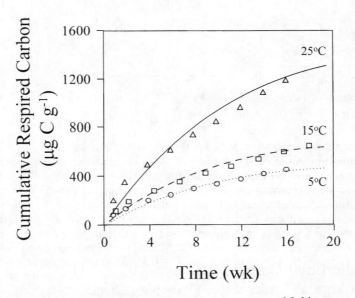

FIGURE 15.10. Microbial respiration measured over a range of field temperatures in the laboratory (see Zogg et al. 1997 for details). The following first-order models can be used to predict microbial respiration at different temperatures: i) C respired at 5°C = 515 $(1 - e^{-0.120t})$; C respired at 15°C = 742$(1 - e^{-0.107t})$; C respired at 25°C = 1,607 $(1 - e^{-0.015t})$.

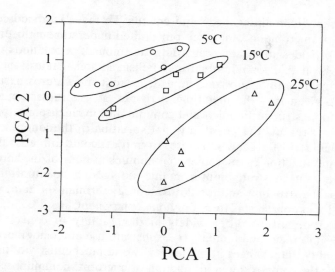

FIGURE 15.11. The PCA of microbial communities at 5, 15, and 25°C in a forest soil incubated in the laboratory (Zogg et al. 1997). In general, monoenoic, unsaturated FAMEs indicative of gram-negative soil bacteria were heavily weighted on the first PCA axis, whereas saturated FAMEs indicative of gram-positive bacteria were heavily weighted on the second PCA axis.

soil microorganisms. The extent to which microbial community composition and function control such ecosystem-level processes as organic matter decomposition and N mineralization, represents a new frontier and an opportunity to use ecophysiology to deepen our understanding of biogeochemistry.

Although physiological processes of plant roots and soil microorganisms control the flux of CO_2 from soil, defining their individual contribution still remains a considerable challenge. Resolving the contribution of roots and microorganisms to the flux of CO_2 from soil appears to be tied to developing our understanding of belowground substrate inputs (i.e. fine-root demography and biochemistry) and how environmental factors control C metabolism by microbial communities. Until this task is accomplished, we will be faced with uncertainty in our estimates of the belowground C budget of terrestrial ecosystems.

Soil Microbial Communities and Fine-Root Dynamics: Physiological Links with Ecosystem-Level Implications

Microbial growth and maintenance is commonly thought to be limited by the amount of organic substrate entering the soil from plant litter production (Smith and Paul 1990), wherein annual litter inputs are only sufficient to meet microbial maintenance requirements. This idea has become a central tenet of microbial ecology, but our limited understanding of fine-root dynamics provides uncertainty regarding the amount, timing, and quality of substrates available for microbial metabolism in soil. Understanding the type and magnitude of substrate input from root dynamics is particularly important, given that the metabolic activities of live roots are a strong selective force structuring microbial communities in rhizosphere soil. This observation presents the possibility that variation in the amount, timing, and biochemistry of organic substrates from roots could influence the composition and function of microbial communities in non-rhizosphere soil as well. Changes in substrate input from fine roots also have important implications for the structure and the flow of energy through soil food webs, of which soil microorganisms form only one component.

Soil microorganisms differ widely in their ability to convert organic substrates into microbial biomass (i.e. substrate-use efficiency), presenting the possibility that changes in the longevity and biochemistry of fine roots could alter the composition and function of microbial communities in soil. Bacterial substrate-use efficiencies range from 20% to 40% (Chahal and Wagner 1965; Hernandez and Johnson 1967; Behera and Wagner 1974; Elliott et al. 1983), indicating that these microorganisms return 60% to 80% of substrate C to the atmosphere as CO_2. Substrate-use efficiencies of fungi are substantially higher (40% to 75%; Harley 1971; Griffin 1972; Ingham 1981) than those of soil bacteria. Differences in substrate-use efficiency between soil bacteria and fungi suggest that changes in fine-root litter could alter the function and composition of microbial communities in soil, particularly if the biochemistry of fine roots responds to changing environmental factors.

Several lines of evidence suggest that the biochemistry of live, fine roots can change seasonally (Nambiar 1987), with increasing soil N availability (Burton et al. 1996) and with elevated levels of atmospheric CO_2 (Pregitzer et al. 1995). Live, fine roots, for example, display marked seasonal variation in starch concentrations, but concentrations of N and phosphorus(P) are relatively stable (Nambiar 1987). Also, very fine roots (<0.5 mm) have higher N concentrations and lower starch concentrations than larger-diameter fine roots (0.5 to 5 mm; Nambiar 1987). Although the processes of root senescence are not well understood, it is possible that fine roots may not retranslocate substantial quantities of nutrients prior to death (Nambiar 1987). If N is not retranslocated from fine roots, then the positive relationship between soil N availability and fine-root N concentration suggests that the biochemistry of dead roots could change with soil N availability.

If soil N availability alters the amount of N in dead, fine roots, then this has important implications for the balance between N mineralization and immobilization. When plant litter enters the soil, it is colonized by microorganisms, and the high energy-yielding constituents (i.e. organic acids, simple sugars, carbohydrates) are preferentially used for microbial biosynthesis. Because most plant litter is a C-rich and N-poor substrate for microbial growth and maintenance (i.e. its C/N is higher than that of microbial biomass), N must be assimilated from soil solution or directly from plant litter to form new N-containing compounds within microbial cells. If the N concentration of fine roots parallels an increase in soil N availability, then N may be more rapidly mineralized from dead, fine roots with high N concentrations. This scenario presents the possibility of a positive feedback between root litter produced in N-rich soil and the process of net N mineralization.

However, it also is important to understand the extent to which soil N availability modifies the organic constituents of root litter, because organic compounds produced by roots serve as substrate for microbial growth and maintenance. Microbial metabolism of simple carbohydrates, organic acids, and other labile substrates provides energy for the biosynthesis of proteins, creating a need for inorganic N. Recalcitrant substrates (i.e. lignified cell-wall constituents) provide a lower energy yield for microbial biosynthesis and therefore create a lower demand for inorganic N. As such, it is both the energy yield of organic compounds and their N concentration that control whether soil microorganisms release or assimilate N during organic matter decomposition. Nonetheless, relatively little is understood about the biochemistry of root-associated (i.e. dead roots, cell sloughing, exudation, mycorrhizae) organic matter inputs to soil. What are the chemical constituents of dead, fine roots, and how do they change among plant species? What are the contributions of exudation, cell sloughing, and mycorrhizae to the amount and types of organic compounds in soil, and how do they influence the demand for N during microbial biosynthesis? Answering these questions will be important, because litter biochemistry (i.e. organic compounds and N concentration) in combination with microbial substrate-use efficiency provide a direct physiological link between patterns of root and microbial activity in soil. This relationship has important implications for the ecosystem-level cycling of C and N.

Although there are obvious connections between the belowground physiological activities of plants and soil microorganisms, relatively few studies have attempted to link fine-root dynamics with the composition and function of microbial communities in soil (Zak et al. 1996; Zogg et al. 1997). Linking compositional changes with functional responses in microbial communities has been difficult, but many molecular tools hold promise for accomplishing this task. Recently, we used FAME analysis to investigate changes in microbial community composition in response to fine-root growth under experimental atmospheric CO_2 and soil N availability treatments. We modified soil N availability by varying the proportion of A horizon (high organic matter) and C horizon (no organic matter) of a native forest soil in Michigan (see Pregitzer et al. 1995 for details). Our high-N-availability (all A horizon) and low-N-availability (4:1 C horizon to A horizon) had the same source of organic matter and therefore the same microbial inoculum at the beginning of our experiment. We grew *Populus tremuloides* in these experimental soils for two years at ambient and elevated levels of CO_2 (Kubiske et al. 1998).

In this experiment, soil N availability significantly increased the production and mortality of fine roots (Figure 15.12), suggesting that the amount of organic substrate entering soil from roots increased with soil N

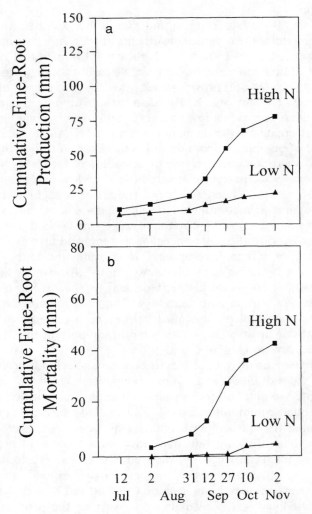

FIGURE 15.12. The production (panel a) and mortality (panel b) of *Populus tremuloides* growing in experimental soils of high and low N availability (M.E. Kubiske and K.S. Pregitzer, unpublished data).

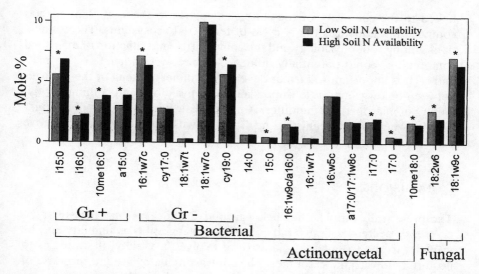

FIGURE 15.13. The FAME profile of microbial communities beneath *Populus tremuloides* growing in soil of high and low N availability (C.J. Mikan, unpublished data). Soil N availability was modified by mixing different proportions of A horizon material that was rich in organic matter with C horizon material that was poor in organic matter (see Pregitzer et al. 1995 for details), providing the same inoculum of soil microorganisms but in different amounts. After two years of growth, the microbial communities in soil with high and low N availability differed significantly, the probable result of changes in the biochemistry of root litter.

availability. At each level of soil N availability, plants grown under elevated levels of atmospheric CO_2 increased the rates of fine-root production and mortality (data not shown). It appears that changes in root litter production (and probably chemistry) in response to soil N availability had a significant influence on the microbial community composition as well, wherein FAMEs indicative of gram-positive soil bacteria significantly increased in the soil with high N availability (Figure 15.13). In contrast to this increase, actinomycetal and fungal FAMEs significantly declined under the same conditions (Figure 15.13; C.J. Mikan unpublished data). Changes in fine-root dynamics related to atmospheric CO_2 did not alter microbial community composition (data not shown; C.J. Mikan, unpublished data).

If the N concentration of fine roots increased with soil N availability in this experiment, then gross and net rates of N mineralization (per unit of root litter input) should be most rapid in the soil with high N availability. Such a functional change would correspond to the increase in gram-positive bacteria and the decline in actinomycetes and fungi in the high N availability treatment of Figure 15.13. These potential relationships suggest that the composition and function of the microbial community could be directly linked to the chemistry of root litter, analogous to the way plant species composition influences the N cycle of terrestrial ecosystems. The connections among root physiology and the composition and function of microbial communities could potentially influence the ecosystem-level cycling of C and N. It is important to further develop our understanding of the linkages between belowground litter inputs and the physiology and community dynamics of microbial communities. A mechanistic understanding of the linkages between microorganisms and C inputs will lead to greater predictive power at the ecosystem level.

Conclusions

The present debate whether belowground C allocation in ecosystems increases or decreases along gradients of increasing soil N availability may be artificial to the extent that plants have evolved a variety of strategies to acquire belowground resources. The contrasting belowground response of plants with different life-history traits suggests that individual plants may increase or decrease the absolute allocation of C to fine roots as soil N availability increases, even though the relative allocation of C to fine-root biomass (percentage of total biomass in fine roots) declines with greater soil N availability. A current challenge is to further develop the understanding of how environmental variables control C allocation in individual plants and to use this information to understand the above- and belowground partitioning of NPP in terrestrial ecosystems.

Uncertainty regarding the contribution of plant roots and soil microorganisms to the flux of CO_2 from soil may be resolved, in part, by understanding the physiological controls over plant and microbial respiration. There is a potential tradeoff between fine-root biomass and rates of respiration (per unit of root) along a gradient of soil N availability that could influence the relative contribution of plants to soil respiration. However, little is understood regarding the extent to which fine-root biomass and respiration rate change along gradients of N availability for most plant species.

Determining the contribution of soil microorganisms to soil respiration is a difficult task. First-order kinetics have traditionally been used to scale microbial respiration to an ecosystem level. However, it appears that microbial communities metabolize different substrate pools as soil temperatures vary over a range of field values. As a consequence, changes in substrate pools may have a larger influence on microbial respiration, compared to relatively smaller increases in rate constants (k). Evidence suggests that compositional shifts in microbial communities with increasing temperature alter the kinetics of microbial respiration in a way not considered in any present-day simulation models. An important challenge is to link the composition and function of microbial communities in soil to better understand the processes controlling the cycling of C and N within terrestrial ecosystems.

Although microbial growth in soil is thought to be limited by substrate inputs from plant production, the limited understanding of fine-root and mycorrhizal dynamics provides a substantial degree of uncertainty regarding the amount, timing, and quality of substrates available for microbial metabolism. Changes in root production and biochemistry, associated with changes in soil N availability, could feedback on the composition and function of microbial communities, which in turn could influence ecosystem-level cycling of C and N. Such changes also have important implications for the structure and flow of energy through soil food webs. Nevertheless, few studies have specifically focused on linking belowground plant dynamics with the composition and function of microbial communities in soil. Molecular techniques for identifying soil microorganisms in combination with the use of stable isotopes (e.g. ^{13}C and ^{15}N) provide the opportunity to ask species-specific questions regarding the flow of C and N through the microbial community in soil. This area appears to hold promise for integrating physiological ecology with biogeochemistry.

We have illustrated several linkages between ecophysiology and biogeochemistry using examples from our own research. There are many other important frontiers in ecosystem science that link physiology with ecosystem processes. Such linkages occur in both terrestrial and aquatic ecosystems, because the physiological activities of microorganisms, plants, and animals control the ecosystem-level cycling of C and many associated nutrients. In terrestrial ecosystems, soil is a tremendous reservoir of biological diversity and an important C cycle component. Transformations of soil C lie at the heart of understanding nutrient availability and NPP in many terrestrial ecosystems. We have only scratched the surface of the soil when it comes to understanding how plants are linked to microbial physiology and community dynamics. Because soil microorganisms account for many important transformations of soil organic matter, a much more mechanistic understanding of how microbial physiology influences C and nutrient cycles is truly a frontier in ecosystem science.

Acknowledgments. We gratefully acknowledge Andy Burton, Mark Kubiske, Carl Mikan, David Rothstein and Greg Zogg for their willingness to share their thoughts and unpublished data; it greatly contributed to this synthesis. Patrick Bolen, Laurie Drinkwater, Stuart Findley, Peter Groffman, Paul Hendricks, Clive Jones, George Kling, Mike Pace, Bill Schlesinger, and Bill Sobczak provided comments that substantially improved this manuscript; we sincerely thank each of them. We also acknowledge support from the National Science Foundation (DEB 9496197, DEB 9629842, BIR 9413407, and E24101), the Department of Energy's Program for Ecosystem Research (DE-FG02-93ER61666), and National Center for Global Environmental Change (21507-0195).

References

Aber, J.D., J.M. Melillo, K.J. Nadelhoffer, J. Pastor, and R. Boone. 1991. Factors controlling nitrogen cycling and nitrogen saturation in northern temperate forests. *Ecological Applications* 1:118–138.

Agren, G.I., R.M. McMurtrie, W.J. Parton, J. Pastor, and H.H. Shugart. 1991. State-of-the-art of models of production-decomposition linkages in conifer and grassland ecosystems. *Ecological Applications* 1:118–138.

Alexander, I.J., and R.I. Fairley. 1983. Effects of N fertilization on populations of fine roots and mycorrhizas in spruce humus. *Plant and Soil* 71:49–53.

Behera, B., and G.H. Wagner. 1974. Microbial growth rates in glucose amended soil. *Soil Science Society of America Proceedings* 38:591–594.

Bloom, A.J., S.S. Sukrapanna, and R.L. Warner. 1992. Root respiration associated with ammonium and nitrate absorption and assimilation by barley. *Plant Physiology* 99:1294–1301.

Bonan, G.B. 1991. Atmosphere-biosphere exchange of carbon dioxide in boreal forests. *Journal of Geophysical Research* 96:7301–7312.

Botkin, D.B., G.M. Woodwell, and N. Tempel. 1970. Forest productivity estimated from carbon dioxide uptake. *Ecology* 51:1057–1060.

Burton, A.J., K.S. Pregitzer, G.P. Zogg, and D.R. Zak. 1996. Latitudinal variation in sugar maple fine root respiration. *Canadian Journal of Forest Research* 26:1761–1768.

Chabot, B.F., and D.J. Hicks. 1982. The ecology of leaf life spans. *Annual Review of Ecology and Systematics* 11:233–257.

Chahal, K.S., and G.H. Wagner. 1965. Decomposition of organic matter in Sanborn field soils amended with ^{14}C glucose. *Soil Science* 100:96–103.

Eissenstat, D.M., and R.D. Yanai. 1997. The ecology of root lifespan. *Advances in Ecological Research* 27:1–60.

Elliott, E.T., C.V. Cole, B.C. Fairbanks, L.E. Woods, R. Bryant, and D.C. Coleman. 1983. Short-term bacterial growth, nutrient uptake, and ATP turnover in sterilized, inoculated and C-amended soil: the influence of N availability. *Soil Biology and Biochemistry* 15:85–91.

Ewel, D.C., and H.L. Gholz. 1991. A simulation model of the role of belowground dynamics in a Florida pine plantation. *Forest Science* 37:397–438.

Fitter, A.H. 1987. An architectural approach to the comparative ecology of plant root systems. *New Phytologist* 106 (supp.):61–77.

Fitter, A.H., and T.R. Stickland. 1991. Architectural analysis of plant root systems 2. Influence of nutrient supply on architecture in contrasting plant species. *New Phytologist* 118:383–389.

Griffin, P.M. 1972. *Ecology of soil fungi.* Syracuse University Press, Syracuse, New York.

Harley, J.L. 1971. Fungi in ecosystems. *Journal of Animal Ecology* 41:1–16.

Hernandez, E., and M.J. Johnson. 1967. Energy supply and cell yield in aerobically grown microorganisms. *Journal of Bacteriology* 94:996–1001.

Hendrick, R.L., and K.S. Pregitzer. 1992. The demography of fine roots in a northern hardwood forest. *Ecology* 73:1094–1104.

Hendrick, R.L., and K.S. Pregitzer. 1993. Patterns of fine root mortality in two sugar maple forests. *Nature* 361:59–61.

Hendricks, J.J., K.J. Nadelhoffer, and J.D. Aber. 1993. Assessing the role of fine roots in carbon and nitrogen cycling. *Trends in Ecology and Evolution* 8:174–178.

Horwath, W.R., K.S. Pregitzer, and E.A. Paul. 1994. ^{14}C allocation in tree-soil systems. *Tree Physiology* 14:1163–1176.

Ingham, E.R. 1981. *Use of fluorescein diacetate for assessing functional fungal biomass in soil.* Ph. D. dissertation. Colorado State University, Fort Collins.

Isebrands, J.G., and N.D. Nelson. 1983. Distribution of $[^{14}C]$-labeled photosynthates within intensively cultured Populus clones during the establishment year. *Physiologia Plantarum* 59:9–18.

Keyes, M.R., and C.C. Grier. 1981. Above- and below-ground net production in 40-year-old Douglas-fir stands on low and high productivity sites. *Canadian Journal of Forest Research* 11:599–605.

Kubiske, M.E., K.S. Pregitzer, C.J. Mikan, D.R. Zak, J.L. Masiasz, and J.A. Teeri. 1998. Populus tremuloides photosynthesis and crown architecture in response to elevated CO_2 and soil N availability. *Oecologia* in press.

Landsberg, J.J., M.R. Kaufmann, D. Binkley, J. Isebrands, and G.G. Jarvis. 1991. Evaluating progress toward closed forest models based on fluxes of carbon, water and nutrients. *Tree Physiology* 9:1–15.

Larcher, W. 1969. The effect of environmental and physiological variables on the carbon dioxide gas exchange of trees. *Photosynthetica* 3:167–198.

MacDonald, N.W., D.R. Zak, and K.S. Pregitzer. 1995. Temperature effects on the kinetics of microbial respiration and the net mineralization of N and S. *Soil Science Society of America Journal* 59:233–240.

Nadelhoffer, N.J., J.D. Aber, and J.M. Melillo. 1985. Fine roots, net primary production, and soil nitrogen availability: a new hypothesis. *Ecology* 66:1377–1390.

Nambiar, E.K.S. 1987. Do nutrients retranslocate from fine roots? *Canadian Journal of Forest Research* 17:913–918.

Parton, W.J., C.V. Cole, J.W.B. Stewart, D.S. Ojima, and D.S. Schimel. 1988. Simulating regional patterns of soil C, N, and P dynamics in the US central grasslands region. *Biogeochemistry* 2:3–27.

Pate, J.S., and D.B. Layzell. 1990. Energetics and biological costs of nitrogen assimilation. Pages 1–42 in B.J. Mifflin and P.J. Lea, eds. *The biochemistry of plants: a comprehensive treatise.* Academic Press, CA.

Penning de Vries, F.W.T. 1975. The cost of maintenance processes in plant cells. *Annals of Botany* 39:77–92.

Pregitzer, K.S., R.L. Hendrick, and R. Fogel. 1993. The demography of fine roots in response to patches of water and nitrogen. *New Phytologist* 125:575–580.

Pregitzer, K.S., D.R. Zak, P.S. Curtis, M.E. Kubiske, J.A. Teeri, and C.S. Vogel. 1995. Atmospheric CO_2, soil nitrogen, and turnover of fine roots. *New Phytologist* 129:579–585.

Raich, J.W., and K.J. Nadelhoffer. 1989. Belowground carbon allocation in forest ecosystems: global trends. *Ecology* 70:1346–1354.

Raich, J.W., and W.H. Schlesinger. 1992. Global carbon dioxide flux in soil respiration and its relationship to vegetation and climate. *Tellus* 44B:81–99.

Rastetter, E.B., M.G. Ryan, G.R. Shaver, J.M. Melillo, K.J. Nadelhoffer, J.E. Hobbie, et al. 1991. A general biogeochemical model describing the response of

402   D.R. Zak and K.S. Pregitzer

the C and N cycles in terrestrial ecosystems to changes in CO_2, climate and N deposition. *Tree Physiology* 9:101–126.

Reich, P.B., C. Uhl, M.B. Walters, and D.S. Ellsworth. 1991. Leaf life-span as a determinant of leaf structure and function among 23 species in Amazonian forest communities. *Oecologia* 86:16–24.

Reich, P.B., M.B. Walters, and D.S. Ellsworth. 1992. Leaf life-span in relation to leaf, plant and stand characteristics among diverse ecosystems. *Ecological Monographs* 62:365–392.

Ringelberg, D., G.T. Townsend, K.A. DeWeerd, J.M. Suflita, and D.C. White. 1994. Detection of the anaerobic dechlorinating microorganism *Desulfomonile tiedjei* in environmental matrices by its signature lipopolysaccharide branched-long-chain hydroxy fatty acids. *FEMS Microbial Ecology* 14:9–18.

Running, S.W., and S.T. Gower. 1991. FOREST-BGC, A general model of forest ecosystem processes for regional applications. II. Dynamics of carbon and nitrogen budgets. *Tree Physiology* 9:147–160.

Ryan, M.G. 1991. Effects of climate change on plant respiration. *Ecological Applications* 1:157–167.

Ryan, M.G., R.M. Hubbard, S. Pongracic, R.J. Raison, and R.E. McMurtrie. 1996. Foliage, fine-root, woody-tissue and stand respiration in *Pinus radiata* in relation to nitrogen status. *Tree Physiology* 16:333–343.

Smirnoff, N., and G.R. Stewart. 1985. Nitrate assimilation and translocation in higher plants: comparative physiological and ecological consequences. *Physiologia Plantarum* 64:133–140.

Smith, J.L., and E.A. Paul. 1990. The significance of soil microbial biomass estimations. Pages 357–393 in J. Bollag and G. Stotzky, eds. *Soil biochemistry*. Merkel Deker, New York.

Tilman, G.D. 1988. *Plant strategies and the dynamics and structure of plant communities*. Princeton University Press, Princeton, NJ.

Tunlid, A., and D.C. White. 1992. Biochemical analysis of biomass, community structure, nutrient status, and metabolic activity of microbial communities in soil. Pages 229–262 in J. Bollag and G. Stotzky, eds. *Soil biochemistry*. Mercel Dekker, New York.

van Veen, J.A., J.N. Ladd, and M.J. Frissel. 1984. Modeling C and N turnover through the microbial biomass in soil. *Plant and Soil* 75:257–274.

Vestal, J.R., and D.C. White. 1989. Lipid analysis in microbial ecology. *BioScience* 39:535–541.

Vogt, K.A., C.C. Grier, and D.J. Vogt. 1986. Production, turnover, and nutrient dynamics of above- and belowground detritus in world forests. *Advances in Ecological Research* 15:303–377.

Vogt, K.A., E.E. Moore, D.J. Vogt, M.L. Redlin, and R.L. Edmonds. 1983. Conifer fine root and mycorrhizal root biomass within the forest floor of Douglas-fir stands of different ages and site productivities. *Canadian Journal of Forest Research* 13:429–427.

Woodwell, G.M., and D.B. Botkin. 1970. Metabolism of terrestrial ecosystems by gas exchange techniques: the Brookhaven approach. Pages 73–85 in D. Reichle, ed. *Analysis of temperate forest ecosystems*. Springer-Verlag, New York.

Wofsy, S.P., M.L. Goulden, J.W. Munger, S.-M. Fan, P.S. Bakwin, B.C. Daube, et al. 1993. Net exchange of CO_2 in a mid-latitude forest. *Science* 260:1314–1317.

Zak, D.R., D. Ringelberg, K.S. Pregitzer, D.L. Randlett, D.W. White, and P.S. Curtis. 1996. Soil microbial communities beneath *Populus grandidentata* Michx. growing at elevated atmospheric CO_2. *Ecological Applications* 6:257–262.

Zogg, G.P. 1996. *Belowground carbon dynamics in northern hardwood forests*: factors controlling respiration from roots and soil microorganisms. Ph.D. dissertation. University of Michigan, Ann Arbor.

Zogg, G.P., D.R. Zak, A.J. Burton, and K.S. Pregitzer. 1996. Fine root respiration in northern hardwood forests in relation to temperature and nitrogen availability. *Tree Physiology* 16:719–725.

Zogg, G.P., D.R. Zak, D.B. Ringelberg, N.W. MacDonald, K.S. Pregitzer, and D.C. White. 1997. Compositional and functional shifts in microbial communities related to soil warming. *Soil Science Society of America Journal* 61:475–481.

16
Simulation Modeling in Ecosystem Science

WILLIAM K. LAUENROTH, CHARLES D. CANHAM, ANN P. KINZIG,
KAREN A. POIANI, W. MICHAEL KEMP, and STEVEN W. RUNNING

Summary

Ecosystem science has lost its past close connection to simulation modeling. At the inception of the field, simulation was viewed as a crucial technology to provide organization and integration. The small role that simulation modeling played in the 1997 Cary Conference, which was organized to highlight the past successes and future challenges in ecosystem science, was the impetus for this chapter. The key underlying assumption of our analysis is that simulation modeling represents one of the most powerful tools available to ecosystem scientists. This is particularly true with respect to the role that the scientific community is expecting ecosystem science to play in the analysis of problems associated with global change. Our objective for this chapter was to address the following four questions about the role of simulation modeling in ecosystem science:

1. What role could or should simulation modeling play in ecosystem science?
2. How has the relationship between ecosystem science and simulation modeling evolved over the past few decades, and what are some examples of successes in that relationship?
3. How has the use of simulation models in ecosystem science been limited, and what factors have contributed to those limitations?
4. How might simulation models contribute to advances in ecosystem science in the future?

Introduction

The 1997 Cary Conference was convened to assess the present state of progress in ecosystem science, and to explore potential frontiers in the field. Given the important role that simulation modeling has played in ecosystem science in the past, it seems reasonable to have expected that such a conference would include explicit consideration of the role of simulation modeling. However, both the overall agenda and the individual talks were notable for the relative lack of consideration of simulation modeling. Our concern is that the conference agenda reflects the often wide and troubling schism that has developed between simulation modeling and the rest of the field. That concern was the stimulus to write this brief chapter, the purpose of which is to raise issues concerning the role of simulation modeling in ecosystem science, and the apparent gulf between modeling and empirical research.

In the spirit of the focus of the conference on "Successes, Limitations, and Frontiers," we address several questions in this chapter. First, what role could or should simulation modeling play in ecosystem science? How has the relationship between ecosystem science and simulation modeling evolved over the past few decades, and what are some examples of successes in that relationship? How has the use of simulation models in ecosystem science been limited, and what factors have contributed to those limitations? Finally, how might simulation models contribute to advances in ecosystem science in the future?

The Role of Simulation Modeling in Ecosystem Science

Before turning to a history and some examples of successes and limitations, we first address our view of the potential contributions of ecosystem-level simulation models to ecosystem science. In this brief analysis four such contributions are highlighted.

Development of Simulation Models

The development of simulation models contributes to conceptualizations of the structure and function of ecosystems, allows a common framework and language for cross-system comparisons, and facilitates recognition of key patterns and processes. Development first requires a clear statement of the objectives. Models are abstractions and simplifications of the ecosystems they represent (Innis 1979). To be useful, they must include the key components of the ecosystem as conceived by the research team. Box and arrow diagrams have been a common and useful way for ecosystem scientists to formalize their conceptualization of the structure and function of an ecosystem (Innis 1979; Pearlstine et al. 1985; Sasser et al. 1996). Constructing such a diagram forces the researcher or team to identify the components of interest (i.e. the boxes) and the relationships among them (i.e. the arrows). Moreover, this type of visualization is accessible to scientists of different perspectives and expertise in ways that computer codes or systems of equations frequently are not, because they allow a common "language" with which competing or complementary conceptualizations can be discussed. Finally, the broader, qualitative relationships among boxes and arrows in one system can be compared to those in another to facilitate identification of general or universal patterns across different ecosystems (Burke et al. 1990; Lauenroth et al. 1993, Taylor et al. 1994).

Uses of Simulation Models

Simulation can be used to generate testable hypotheses, inform the design of complex and potentially expensive ecosystem experiments, and identify potential mechanisms governing observed patterns and dynamics. Ecosystem-level experiments, however, are often time-consuming, expensive, and in some cases not possible (Lee et al. 1992; Liu et al. 1995; Carpenter chapter 12). Generation of hypotheses and selection of suitable experimental designs can be a formidable task. Moreover, results from large-scale and complex ecosystem experiments are difficult to interpret. Use of simulation models can facilitate both experimental design and interpretation of results. As soon as a model has been developed, sensitivity and uncertainty analyses can be used to identify components or relationships that appear to play the most significant roles in governing ecosystem structure and dynamics or contribute the largest amount of variability to the output (DeAngelis and Cushman 1990; Parton et al. 1994; Poiani et al. 1995). Experiments can be conducted on models to identify potential outcomes of proposed manipulations, and to design experiments that can distinguish most effectively between two or more competing hypotheses (Sirois et al. 1994). Finally, simulation models can be employed after data collection to test competing hypotheses concerning cause-and-effect relationships or the mechanistic underpinnings of observed phenomena, and identify the most probable or plausible candidates.

Frameworks

Simulation models provide a framework around which the huge accumulation of empirical observations collected over the past few decades by ecosystem scientists can be organized. One of the consequences of the significant growth in ecosystem science over the past three decades is that the field is awash in data (e.g. Bildstein and Brisbin 1990). In many cases, we lack the conceptual and operational frameworks that provide answers to questions about relationships among variables within or across data sets and regions. Simulation models provide an important tool for organizing these data (and our thoughts), examining the relationships that emerge from these data, and allowing identification of within- and across-system patterns in the data and emergent relationships (Shugart 1990; Parton et al. 1994).

Tools

Simulation models provide a tool with which knowledge obtained in specific experiments can be extrapolated to broader spatial and temporal scales. Ecosystem experiments are limited in spatial and temporal scope. Simulation models are the most powerful tool available to extrapolate experimental results across space and time (Lauenroth et al. 1993; Parton et al. 1994; Poiani et al. 1996). This facility is becoming increasingly important as the scientific community strives to understand and predict the consequences of global-level changes in climate and land use; this feature is discussed in more detail in the "Frontiers" section.

We end this section by noting that there is a two-way relationship between simulation modeling and empiricism in ecosystem science. Although simulation models can aid in experimental design and interpretation of experimental data, the reverse process is even more important. Ecosystem simulation models require prior knowledge and data for the system of interest. After a model has been built, additional data are required to test the model. Does the model produce results that are consistent with field observations? If not, where does the model fail most dramatically? How must it be modified so that it does produce results that are consistent with data? Simulation models will continue to be powerful tools in ecosystem science only if they are integrated into the processes producing empirical observations.

The History of Simulation Modeling in Ecosystem Science: Origins and Successes

Ecosystem science and ecosystem simulation modeling share a common inception. The U.S. International Biological Program (IBP)—initiated in the 1960s and funded by the National Science Foundation—provides a simple and convenient marker for their emergence. Even though the origins of ecosystem science and ecosystem simulation modeling date from well before IBP (Shugart 1997), and both certainly draw on research and disciplines with much older histories, the IBP projects brought ecosystem research and simulation modeling to the attention of a large segment of the scientific communities in ecology and resource management, and led to the training of many new practitioners.

The IBP projects were oriented explicitly toward ecosystem science, and it was conceived that simulation models would play an integral role in the success of those studies (Golley 1993). Many of the groups that began using simulation models during the IBP projects have continued to do so (Golley 1993). To a large extent, the most prominent groups or individuals using simulation models in ecosystem-level research can be traced directly to an

IBP project, or are only one generation removed (i.e. were trained by scientists involved with the original IBP models).

We suggest that the failure of the original IBP models to reach the lofty objectives that were set out for them has contributed to the present-day gulf between modelers and empiricists, and therefore has acted as a limitation on the use of modeling in ecosystem science. However, there are many well-defined cases in which simulation models have made significant contributions to the advancement of ecosystem science during the past thirty years. Three examples of such contributions are presented in the next sections.

Conceptualization and Cross-System Comparisons

One of the key contributions that simulation models can make to ecosystem science is to provide a conceptual framework that can be used to compare results from a variety of ecosystems. The Century model is an excellent example of such a framework for the biogeochemistry of grasslands (Parton et al. 1987). This framework consists of two relatively simple ideas that have proven to be extremely powerful in their explanatory value and fruitful in terms of stimulating interest in testing the framework. The first idea is that the amount of net primary production (NPP) is determined by water availability and is modified primarily by nitrogen availability and secondarily by temperature. The second idea is the soil organic matter (carbon and nitrogen) can be represented by the following three kinetically defined pools: (1) active (turnover time of months to a few years); (2) slow (turnover time of twenty to fifty years); and (3) passive (turnover time of 400 to 2000 years). The Century framework has been so influential that essentially any new biogeochemical results for grasslands must be explained and justified in terms of Century. Recently, Century has been used to compare the responses of thirty-one temperate and tropical grasslands from six continents to climate change (Parton et al. 1995, 1996). Because of the uncertainty associated with predictions of changes in climate, the key value of these simulations lies in the comparisons of the biogeochemical dynamics of the thirty-one grasslands using a common conceptual framework. Such comparisons allow the researchers at any one of the sites to place the dynamics of their grassland into a worldwide context. This is exactly the perspective that is most often missing in the presentation of results from a single ecosystem and is at least part of the reason why the knowledge bank of our science is dominated by data fragments. Simulation models are one of the key tools that can contribute to the integration and synthesis called for by Likens (chapter 10).

Design, Interpretation, and Analysis of Experiments

A widespread trend of declining abundance of seagrasses and related submersed vascular plants, which has been observed in shallow coastal environments worldwide (Short and Wyllie-Echeverria 1996), was the focus for a multi-investigator ecosystem study in Chesapeake Bay (Orth and Moore 1983). Simulation modeling played a central role in several phases of this investigation, including: (1) developing a coordinated research plan; (2) designing and extrapolating from controlled experiments to natural ecosystems; (3) integrating diverse data to determine causes of the seagrass decline; and (4) computing the ecological and economic consequences of this habitat loss (Kemp et al. 1980, 1994). In this case, alternative hypotheses regarding the cause of submersed plant loss in this estuary were tested via simulation studies that integrated information from diverse field and laboratory experiments (Kemp et al. 1983), revealing the dominant role of eutrophication (Duarte 1995). Models were calibrated with data from mesocosm experiments (Twilley et al. 1985; Goldsborough and Kemp 1988; Neundorfer and Kemp 1993) and simulation scenarios were used to extrapolate results to conditions in the open estuary (Kemp et al. 1995). Modeling analyses (Kemp et al. 1984) provided an integrated view of the relative importance of seagrass communities as habitats for fish (Lubbers et al. 1989) and as sites for enhanced sedimentation (Ward et al. 1984) and biogeochemical cycling (Caffrey and Kemp 1992). Again, simulation studies were useful in translating observed ecological relationships into a framework for estimating the socioeconomic consequences of changing habitat conditions (Boynton et al. 1982; Kahn and Kemp 1985). Ecosystem simulation studies continue to assist in establishment of monitoring indices for assessing trends in seagrass habitat conditions and for devising effective strategies for plant restoration (Madden and Kemp 1996).

Synthesis, Sensitivity Analysis and Extrapolation to Broader Scales

Prediction of the effects of global climate change on the biosphere represents one of the major frontiers (and most daunting challenges) in ecosystem science. Even a prediction as outwardly simple as the probability of a state change in biome type (e.g. tundra to boreal forest) under a given climate change scenario could conceivably involve consideration of processes ranging from plant ecophysiology to herbivory, nutrient cycling, and disturbance dynamics. Predictions at scales that would be relevant for dynamic incorporation in Global Climate Models (GCM) models would further require that processes be understood across a much wider range of environments and time scales than typically represented in the empirical literature. Starfield and Chapin (1996) presented a novel approach to this issue for prediction of transient changes in arctic and boreal vegetation under changing climate and land management scenarios. Starfield and Chapin's model attempts to predict the occupancy of large (i.e. $25\,km^2$) patches of present-day Alaskan tundra by one of four different ecosystem types (i.e. upland tundra, conifer forest, broad-leaved deciduous forest, and grassland). The model uses a set of relationships and decision rules unique to each of the four ecosystem states to specify the probability of transition to a different state. These transitions are governed by climate, the composition of neighboring patches (to incorporate seed source effects), and land-management policies that affect fire frequency, logging, and herbivore (e.g. moose) activity. Both calibration and validation of the model are problematic, because many of the key relationships are only understood qualitatively, and the authors attempt only a perfunctory validation by comparing steady-state predictions to extant patterns of vegetation distribution under different climate regimes. The model's greatest strength lies in its use by researchers attempting to integrate transient dynamics in the physical environment (i.e. climate change and associated effects on fire regimes) with human land-use and management decisions—specifically fire suppression, logging, and predator control, which all have surprisingly large effects on the model predictions. In this sense, the model is best viewed as a thought experiment, in which sensitivity analysis allows the comparison of the relative importance of different processes, given the specified model structure and present parameters.

Limitations to the Use of Simulation Modeling in Ecosystem Science

Ecosystem research in general has continued to develop and grow since the IBP era, and ecosystem science has become an accepted and often distinct discipline within many academic research communities, through the incorporation of faculty in biology, ecology, or natural-resource science departments. Most large universities with active research programs in ecology have one or more ecosystem scientists. Additionally, a number of institutes have arisen in the past three decades that have become important centers of ecosystem research.

In contrast to these impressive advances for ecosystem science in general, the progress of simulation modeling in the field has been more restricted. Compared to individuals with an empirical or theoretical orientation, ecosystem-level simulation modelers are much less prevalent in academic settings. Many researchers and teams that have a significant involvement in simulation modeling are located at institutes and centers, and many of these scientists are largely or wholly supported on "soft money" research contracts and grants.

In spite of obvious successes in the use of simulation models in ecosystem science, and some good examples of tight integration between modeling and empirical research in this field (Coffin and Lauenroth 1989, 1990; Coffin et al. 1993), the gulf between modelers and empiricists remains. We suspect that the gulf is growing as the development of ecosystem models for use in global change research outstrips the field's ability to keep pace with empirical research at the appropriate spatial and temporal scales. Why have empirical and simulation approaches for understanding ecosystems become largely decoupled over the past few decades? A full answer to this question requires a more detailed and scholarly analysis than is possible here. Nevertheless, we offer a few observations and hypotheses from within the discipline. One possible explanation for the limited advance of simulation modeling since the IBP is the perceived failure of simulation modeling in that initial endeavor. Many members of the ecological research community believe that the BIP models failed to produce the whole-system descriptions and the testable hypotheses that had been promised (Golley 1993). Unrealistic claims for the benefits of simulation modeling without significant proof of success have contributed to skepticism or neglect. A "feedback" effect in graduate training also provides a partial explanation: An initial neglect of simulation modeling in academic settings contributes to a general deficit in practitioners of simulation modeling who could be tapped to fill future faculty positions. Finally, the development and maintenance of ecosystem simulation models can be a time-consuming process, often requiring several years and many researchers. This long-term, "team" approach more often is more easily conducted in centers and institutes than in many academic departments, which place heavy emphasis on individual research.

These are only some of the possible explanations for the lack of use of simulation models in much of ecosystem research. Regardless of the ultimate explanation or explanations, we assert that the gulf between the two is detrimental to both; the small role played by models at a conference devoted to illuminating the most important advances in the field of ecosystem science is evidence enough of a worrisome rift.

Frontiers for Simulation Modeling in Ecosystem Science

One of the most compelling problems facing the world today is the potential for global-level changes in climate, biomes, and biodiversity as a result of human activity. Improved predictions of the course of climate change and the response of the biosphere to regional or global stresses will require coupling climate system models with models of the biosphere over large spatial scales and long temporal scales. For instance, most GCMs presently utilize very simple (and in some cases incorrect) distributions and dynamics of the Earth's biomes. Ecologists have data and knowledge that could better inform efforts to incorporate realistic ecosystem responses in GCMs, but such incorporation will mean translating ecological information to spatial scales on which GCMs operate—typically thousands of square kilometers. There is also much interest in understanding how human-driven changes in temperature and precipitation will affect ecosystems over the next twenty-five to 250 years (VEMAP Participants 1995), but such an understanding requires temporal information of ecosystem response that extends beyond that of the typical experiment. Simulation models of ecosystems must be used to extrapolate ecological information to relevant spatial and temporal scales if ecosystem science is to make a contribution to understanding the causes and consequences of global change. An extension of ecosystem models to the study of global change is already occurring. Nonetheless, it is in this area of regional and global biospheric response to climate change, and to the role of the biosphere in determining the course of climate change, that the most challenging frontiers for ecosystem science in general and simulation modeling in particular lie.

Ecosystem science is concerned explicitly with understanding the structure and function of ecological systems. One key approach for understanding ecological systems is simulation modeling. At the inception of ecosystem science, simulation modeling was an integral part of the methods used to investigate ecosystems. Over the past thirty years, however, ecosystem science has lost its close connection to simulation modeling and, in our opinion, both efforts are suffering as a result. Any transition of ecosystem science from its present-day state to a mature science will require a rediscovery of the crucial role simulation modeling can and must play in its future.

References

Bildstein K.L., and I.L. Brisbin, Jr. 1990. Lands for long-term research in conservation biology. *Conservation Biology* 4:301–308.

Boynton, W.R., W.M. Kemp, et al. 1982. The decline of submerged macrophyte systems in Chesapeake Bay: modeling the eco/energetic and socio/economic implications. Pages 441–454 in W.J. Mitsch, R. Bosserman, and J. Klopatik, eds. *Energy & ecological model.* Elsevier, Amsterdam, The Netherlands.

Burke, I.C., D.S. Schimel, C.M. Yonker, W.J. Parton, L.A. Joyce, and W.K. Lauenroth. 1990. Regional modeling of grassland biogeochemistry using GIS. *Landscape Ecology* 4:45–54.

Caffrey, J., and W.M. Kemp. 1992. Influence of the submersed plant, *Potamogeton perfoliatus* L., on N cycling in estuarine sediments: use of 15N techniques. *Limnology and Oceanography* 37:1483–1495.

Coffin, D.P., and W.K. Lauenroth. 1989. Disturbances and gap dynamics in a semiarid grassland: a landscape level approach. *Landscape Ecology* 3:19–27.

Coffin, D.P., and W.K. Lauenroth. 1990. A gap dynamics simulation model of succession in a semiarid grassland. *Ecological Modelling* 49:229–266.

Coffin, D.P., W.K. Lauenroth, and I.C. Burke. 1993. Spatial dynamics in the recovery of shortgrass steppe ecosystems. Pages 75–108 in R. Gardner, ed. *Spatial processes in ecological systems.* American Mathematical Society, Providence Rhode Island.

DeAngelis, D.L., and R.M. Cushman. 1990. Potential application of models in forecasting the effects of climate changes on fisheries. *Transactions of the American Fisheries Society* 119:224–239.

Duarte, C. 1995. Submerged aquatic vegetation in relation to different nutrient regimes. *Ophelia* 41:87–112.

Goldsborough, W.G., and W.M. Kemp. 1988. Light response and adaptation for the submersed macrophyte, *Potamogeton perfoliatus*: implications for survival in turbid tidal waters. *Ecology* 69:1775–1786.

Golley, F.B. 1993. *A history of the ecosystem concept in ecology*. Yale University Press, New Haven CT.

Innis, G.S. 1979. A spiral approach to ecosystem simulation, I. Pages 211–386 in G.S. Innis and R.V. O'Neill, eds. *Systems analysis in ecosystems. Statistical Ecology Series Vol. 9*. International Cooperative Publishing House, Burtonsville, MD.

Kahn, J.R., and W.M. Kemp. 1985. Economic losses associated with the degradation of an ecosystem: The case of submerged aquatic vegetation in Chesapeake Bay. *Journal of Environmental Economics and Management* 12:246–263.

Kemp, W.M., W.R. Boynton, and A.J. Hermann. 1994. Simulation models of an estuarine macrophyte ecosystem. Pages 262–278 in B. Patten and S.E. Jørgensen eds. *Complex ecology*. Prentice Hall, Englewood Cliffs, NJ.

Kemp, W.M., W.R. Boynton, and A.J. Hermann. 1995. Ecosystem modeling and energy analysis of submerged aquatic vegetation in Chesapeake Bay. Pages 28–41 in C.A.S. Hall, ed. *Maximum power*. University Press of Colorado, Niwot.

Kemp, W.M., W.R. Boynton, J.C. Stevenson, R.R. Twilley, and J.C. Means. 1983. The decline of submerged vascular plants in upper Chesapeake Bay summary of results concerning possible causes. *Marine Technology Society Journal* 17:78–89.

Kemp, W.M., W.R. Boynton, R.R. Twilley, J.C. Stevenson, and L.G. Ward. 1984. Influences of submersed vascular plants on ecological processes in upper Chesapeake Bay. Pages 367–394 in V.S. Kennedy ed. *The estuary as a filter*. Academic Press, New York.

Kemp, W.M., M.L. Lewis, J.J. Cunningham, J.C. Stevenson, and W.R. Boynton. 1980. Microcosms, macrophytes and hierarchies: environmental research in the Chesapeake Bay. Pages 911–939 in J. Giesy, ed. *Microcosm research in ecology*. NTIS, Springfield, VA.

Lauenroth, W.K., D.L. Urban, D.P. Coffin, W.J. Parton, H.H. Shugart, T.B. Kirchner, et al. 1993. Modeling vegetation structure-ecosystem process interactions across sites and ecosystems. *Ecological Modelling* 67:49–80.

Lee, J.K., R.A. Park, and P.W. Mausel. 1992. Application of geoprocessing and simulation modeling to estimate impacts of sea level rise on the northeast coast of Florida. *Photogrammetric Engineering & Remote Sensing* 58:1579–1586.

Liu, J., J.B. Dunning, Jr., and H.R. Pulliam. 1995. Potential effects of a forest management plan on Bachman's sparrows (*Aimophila aestivalis*): linking a spatially explicit model with GIS. *Conservation Biology* 9:62–75.

Lubbers, L., W. Boynton, and W.M. Kemp. 1989. Variations in structure of estuarine fish communities in relation to abundance of submersed vascular plants. *Marine Ecology Progress Series* 65:1–14.

Madden, C., and W.M. Kemp. 1996. Ecosystem model of estuarine submersed plant community: calibration and simulation of eutrophication responses. *Estuaries* 19(2B):457–474.

414   W.K. Lauenroth, et al.

Neundorfer, J.V, and W.M. Kemp. 1993. Nitrogen versus phosphorus enrichment of brackish waters: response of *Potomogeton perfoliatus* and its associated algal communities. *Marine Ecology Progress Series* 94:71–82.

Orth, R., and K. Moore. 1983. Chesapeake Bay: an unprecedented decline in submerged aquatic vegetation. *Science* 222:51–53.

Parton, W.J., M.B. Coughenour, J.M.O. Scurlock, D.S. Ojima, T.G. Gilmanov, R.J. Scholes, et al. 1996. Global grassland ecosystem modelling: development and test of ecosystem models for grassland systems. Pages 229–269 in A.I. Breymeyer, D.O. Hall, J.M. Melillo, and G.I. Ågren, eds. *Global change: effects on coniferous forests and grasslands.* SCOPE 56, John Wiley & Sons, New York.

Parton, W.J., D.S. Ojima, and D.S. Schimel. 1994. Environmental change in grasslands: assessment using models. *Climate Change* 28:111–141.

Parton, W.J., D.S. Schimel, C.V. Cole, and D.S. Ojima. 1987. Analysis of factors controlling soil organic matter levels in Great Plains grasslands. *Soil Science Society of America Journal* 51:1173–1179.

Parton, W.J., J.M.O. Scurlock, D.S. Ojima, D.S. Schimel, D.O. Hall, M.B. Coughenour, et al. 1995. Impact of climate change on grassland production and soil carbon worldwide. *Global Change Biology* 1:13–22.

Pearlstine, L., H. McKellar, and W. Kitchens. 1985. Modelling the impacts of a river diversion on bottomland forest communities in the Santee River floodplain, South Carolina. *Ecological Modelling* 29:283–302.

Poiani, K.A., W.C. Johnson, and T.G. F. Kittel. 1995. Sensitivity of a prairie wetland to increased temperature and seasonal precipitation changes. *Water Resources Bulletin* 31:283–294.

Poiani, K.A., W.C. Johnson, G.A. Swanson, and T.C. Winter. 1996. Climate change and northern prairie wetlands: simulations of long-term dynamics. *Limnology and Oceanography* 41:871–881.

Sasser, C.E., J.G. Gosselink, E.M. Swenson, C.M. Swarzenski, and N.C. Leibowitz. 1996. Vegetation, substrate and hydrology in floating marshes in Mississippi river delta plain wetlands, USA. *Vegetatio* 122:129–142.

Sirois, L., G.B. Bonan, and H.H. Shugart. 1994. Development of a simulation model of the forest-tundra transition zone of northeastern Canada. *Canadian Journal of Forest Research* 24:697–706.

Short, F., and S. Wyllie-Echeverria. 1996. Natural and human-induced disturbance of seagrasses. *Environmental Conservation* 23:17–27.

Shugart, H.H. 1990. Using ecosystem models to assess potential consequences of global climatic change. *Trends in Ecology and Evolution* 5:303–307.

Shugart, H.H. 1997. *Terrestrial ecosystems in a changing world.* Cambridge University Press, Cambridge, U.K.

Starfield, A.M., and F.S. Chapin. 1996. Model of transient changes in arctic and boreal vegetation in response to climate and land use change. *Ecological Applications* 6:842–864.

Taylor Jr., G.E., D.W. Johnson, and C.P. Andersen. 1994. Air pollution and forest ecosystems: a regional to global perspective. *Ecological Applications* 4:662–689.

Twilley, R.R., W.M. Kemp, K.W. Staver, J.C. Stevenson, and W.R. Boynton. 1985. Nutrient enrichment of estuarine submersed vascular plant communities: I. Algal growth and effects on production of plants and associated communities. *Marine Ecology Progress Series* 23:179–191.

VEMAP Participants. 1995. Vegetation/ecosystem modeling and analysis project: comparing biogeography and biogeochemistry models in a continental-scale study of terrestrial ecosystem responses to climate change and CO_2 doubling. *Global Biogeochemical Cycles* 9:407–437.

Ward, L., W.M. Kemp, and W. Boynton. 1984. The influence of water depth and submerged vascular plants on suspended particulates in a shallow estuarine embayment. *Marine Geology* 59:85–103.

17
Understanding Effects of Multiple Stressors: Ideas and Challenges

DENISE L. BREITBURG, JAMES W. BAXTER, COLLEEN A. HATFIELD,
ROBERT W. HOWARTH, CLIVE G. JONES, GARY M. LOVETT,
and CATHLEEN WIGAND

Summary

Predicting and understanding the effects of multiple stressors is one of the most important challenges presently facing ecologists. Human activities expose ecological systems to a wide range of stressors, whose direct, indirect, and interactive effects can vary depending on system, species, and stressor characteristics. Understanding how multiple stressors affect natural systems will improve our ability to manage and protect these systems, as well as contribute to the understanding of fundamental ecological principles. However, a concerted effort is needed to explore this issue through experiments, modeling, and sampling conducted at a range of spatial and temporal scales, and in ways that take advantage of management-initiated as well as unintentional changes in human-influenced systems.

Introduction

Ecosystem structure and function are dependent on responses to natural and anthropogenic stressors that can influence the physiology and behavior of organisms, ecological interactions within assemblages, and such system-level processes as biogeochemical cycles. Because of both the explosion of the human population (approaching six billion by the year 2000; Flavin 1997), and the social, economic, and political factors that determine how humans interact with the natural environment, most ecosystems are directly or indirectly affected by human activities. Given the range of human activities and the propagation of their effects, it is probable that most human-influenced systems are exposed to multiple stressors of anthropogenic origin, rather than to only a single discrete anthropogenic effect. In addition, both the percentage of systems exposed to multiple stressors and the number of stressors to which most systems are exposed should increase as the effects of global climate change are overlaid on existing processes.

Predicting and understanding the effects of multiple perturbations on ecosystems is one of the most pressing problems facing ecologists. The challenge is important both to the development of management and restoration practices, and to progress in improving our understanding of basic ecological processes. Although multiple stressors have received little specific attention, there have been many ecological studies that simultaneously address several natural or anthropogenic factors influencing the system under study. In this chapter, some key features of the effects of multiple stressors on ecosystems are discussed, emphasizing particularly the relationship between ecosystem structure and the effects of multiple stressors, and the potential for interactions among individual stressors. Important approaches and challenges for addressing the issue of multiple stressors are also explored. Although natural and anthropogenic stressors share many similarities and both are extremely important, in this chapter, the focus is primarily on stressors that are directly or indirectly anthropogenic in origin.

Characteristics of ecosystem structure and function, individual species, and stressors will all be important in determining how systems respond to multiple stressors. Although the list of system, species, and stressor characteristics that potentially influence ecosystem responses is vast, there are several issues that seem especially important in moving from a discussion of individual stressors to multiple stressors. Among the characteristics of an ecosystem that will likely influence its response to multiple stressors are the number of functionally similar species, openness of the system, successional processes (and other forms of temporal variability), inherent resistance and resilience, spatial patterns, and both ecological and evolutionary history. The specificity of stressors, the similarity or dissimilarity of their modes of action, their sequence, the potential for interactions among stressors, and the magnitude of stressor effects will also influence the response or output of the perturbed ecosystem.

System and Stressor Characteristics and Multiple Stressor Effects

Functional Similarity and Multiple Stressors with Independent Modes of Action

Species diversity may promote stability at the ecosystem level, in part, because species vary in their tolerance to natural perturbations and anthropogenic stressors (Tilman 1996). This variation provides the potential for complementary responses (i.e. increases in more tolerant species and concomitant declines in others) when a system is exposed to a stress. A single stressor has the potential to reduce the number of functionally similar species that have the capacity for complementary or compensatory responses. As the number of stressors with independent modes of action increases, it is probable that there will be fewer resistant species with the potential to increase in direct response to the altered environment (which could now be more favorable for them), or in response to negative effects of stressors on their predators and competitors. For example, consider system resistance to a suite of toxic substances each with a different mode of action. If the proportion of species with similar functional roles that are resistant to stressor 1 is r_1, to stressor 2 is r_2, and so forth, and the proportion of species with life-history and physiological characteristics conducive to permitting complementary responses is q, then the probability of functional group-level resistance to the suite of stressors is $q*(r_1*r_2*..r_n)$. As this quantity declines below the reciprocal of the number of functionally similar species, the probability of a complementary or compensatory response should become diminishingly small, and system-level change becomes inevitable.

The relationship between the number of functionally similar species and the magnitude of system-level responses also suggests that the effects of multiple stressors should be quite different when the target assemblage or trophic group is characterized by strong dominants or keystone species than when the target assemblage is more diverse. If a keystone or strongly dominant species is targeted, the system-level response may be similar whether the exposure is to single or multiple stressors. In contrast, single and multiple stressors may have quite different effects in diverse systems containing many species with the capacity for complementary responses.

Caribbean coral reefs provide an example of a diverse system in which a single stressor had very little apparent effect, but a second stressor caused major system-wide changes because it targeted the species that had responded in a complementary manner in the presence of the first stressor. Overharvesting of fish in Caribbean coral reefs greatly reduced herbivory by fishes that feed on benthic macroalgae, which potentially outcompete various coral recruits (Hay 1984; Hughes 1994). Harvest pressure on fish may also have reduced fish predation on sea urchins. With the decline of herbivory by fishes, sea urchin (*Diadema antilarum*) grazing became relatively more important and coral abundance and diversity persisted. A disease that greatly reduced densities of the dominant sea urchin over large areas of the Caribbean (Lessios 1988) then altered the relationship between algal production and consumption. As a consequence, macroalgae increased and often outcompeted coral recruits. This shift in competitive abilities of coral vs. macroalgae has led to dramatic declines in the abundance of live coral.

This example of Caribbean coral reefs raises two important questions: (1) How often does a second stressor target the particular species with the potential for a compensatory or complementary response to the initial stressor? and (2) Is it common for the addition of a second stressor to preclude potential compensatory or complementary responses that occur in systems primarily influenced by a single stressor?

Linked Responses and Nonindependent Modes of Action

Sometimes quite different categories of stressors may target the same species because of similarities in their mode of action at the physiological or physical level. When the susceptibilities to two or more stressors are not independent either mechanistically or statistically, the relationship between the level of functional similarity and the effects of single vs. multiple stressors should be less clear. In combination, these stressors may target no wider a range of species than a single stressor alone, thus weakening the link between the number of functionally similar species and stability of processes at the ecosystem-level.

One example of this linkage that may potentially be important in aquatic systems is the tendency for some anionic trace elements to be taken up and incorporated by phytoplankton by the same mechanisms as nutrients. For example, arsenate acts as a chemical analog of phosphate, and chromate acts as an analog of sulfate (Sanders 1979; Reidel 1984, 1985; Sanders and Riedel 1987). The bioavailability and, therefore, the ecological effects of these trace-element anions is dependent on the concentration of their analogs. As a result, the level of anthropogenic nutrient loadings (and the consequent influence on the abundance and limitation of various nutrients) may modify the effects of relatively low levels of trace-metal contamination. In nitrogen-enriched systems in which phosphorus becomes the limiting nutrient, some phytoplankton species readily take up arsenate, with the consequence of inhibition of primary production at nanomolar arsenic concentrations (Sanders 1979; Wangberg and Blanck 1990; Sanders et al. 1994).

In experiments in the Complexity and Stressors in Estuarine Systems (COASTES) mesocosms, located on the Patuxent River, a tributary of the Chesapeake Bay, the same estuarine algal species (a large centric diatom *Rhizosolenia fragilissima*) bloomed and dominated primary production and biomass in nutrient-enrichment treatments, and was also the algal species that was the most susceptible to trace-element additions (Sanders and Breitburg, in review). Because of their opposing influences (i.e. nutrients increase phytoplankton production while elevated trace elements can decrease production or cause mortality) and the relationship between arsenate and phosphate uptake, increased nutrient loadings and increased arsenic loadings have the potential to cancel each other's effects when phosphorous becomes limiting. When both nutrients and trace elements were added to mesocosm assemblages, and the arsenate/phosphate ratio was high, the density of *Rhizosolenia*, and as a result, total primary production, total chlorophyll, bacterial production, and heterotrophic nanoflagellate abundance were similar in controls (ambient Patuxent River water) and treatments to which both stressors were added. Essentially, arsenic prevented *phytoplankton* from blooming in response to increased nutrients at levels of arsenic addition that caused little effect in tanks to which no nutrients were added.

How common is it for one stressor to prevent the response to a second stressor? Another example in which this occurs, but in which very different species are potentially targeted, may come from studies of colored dissolved organic carbon (DOC) and ultraviolet radiation (UV) in freshwater lakes. High DOC, which can be caused by high nutrient loadings, reduces penetration of UV and, consequently, limits its potentially deleterious effects on fish to a thinner layer of surface water (Williamson et al. 1997).

Potential Masking of Stressor Responses by Additional Stressors with Similar Ecological Effects

In contrast to the examples discussed thus far in which one stressor either exacerbates, prevents or reduces the effects of another stressor (i.e. there are strong interactions among stressor effects), successive stressors may have little or no further effect on the system when all stressors cause similar changes. This may occur even when the underlying physiological or physical mechanisms are quite different if similar species are susceptible to each stress. When it is apparent that multiple stressors occur in the environment, it may be quite difficult to determine which stressor has caused the observed changes. Removal of any single stressor may cause no detectable change if other stressors are not simultaneously manipulated.

Toxic chemical stress often leads to alteration of benthic aquatic community structure such that these systems become dominated by opportunistic species (Howarth 1991). Quite similar changes in benthic community structure result from eutrophication caused by excess nutrient inputs. Therefore, when examining ecosystems that are exposed to both elevated nutrients and toxic substances, it can be difficult—if not impossible—to determine which stressor is having the greater effect. Are the effects of eutrophication and toxification additive? Howarth (1991) hypothesized that they are not, and suggested that a benthic system that has already been affected by eutrophication is less sensitive to the effects of a toxic substance because the more sensitive species have already been replaced. Some evidence in support of this proposal is found in experimental studies of benthic communities in two Norwegian fjords: When oil was artificially added to sediments from the two fjords, recovery was faster in the more eutrophied estuary although the eventual diversity was greater in the sediments from the more pristine fjord (Berge et al. 1987; Berge 1990).

Based on such observations, Howarth (1991) proposed a general model of the response of aquatic ecosystems to toxic chemical stress in which more open ecosystems are hypothesized to be more resistant to toxic stress; openness is defined here in terms of a recycling ratio, which is the ratio of uptake of a limiting nutrient by primary producers divided by the external supply to the ecosystem of that nutrient. DeAngelis (1980) has similarly proposed that ecosystems that are more open are also more resilient and will recover more quickly following a stress-related disturbance.

Interactions Among Sequential Stressors: Increasing System Susceptibility to Stress

The sequence of exposure to multiple stressors, and, therefore, the ecological history of a system, may also be extremely important. The potential exists for any single stressor to alter an ecological system in such a way that subsequent stressors have a greater effect on the stressed system than on an unperturbed system. Sequence dependence of stressor effects may occur because of the way an initial stressor affects the physical environment or the outcomes of biotic interactions.

Oak forests in the eastern United States are generally very retentive of added nitrogen, even under the stress of enhanced nitrogen from atmospheric deposition of pollutant nitrogen oxides. Both plot-scale and watershed-scale studies have shown negligible leaching of nitrogen from oak forests under ambient conditions of atmospheric nitrogen deposition in the eastern United States (Aber et al. 1993; Webb et al. 1995). However, much greater leaching losses of nitrogen may occur when a forest has been defoliated by the gypsy moth (*Lymantria dispar*), a lepidopteran that was introduced to the United States from Europe in the 1860s. The effect of defoliation on nitrogen leaching varies with the level of nitrogen deposition. Although defoliation of forests by the gypsymoth can sometimes result in substantial increases in the concentration of nitrate in streamwater (Webb et al. 1995), very little nitrogen is leached (i.e. less than 1 kg. of nitrogen per hectare per year) in response to defoliation in ecosystems receiving lower levels of nitrogen deposition (Lovett and Ruesink 1995). This result implies that, if nitrate leaching is used as a response variable, some forests are quite resistant to the stress of excess nitrogen deposition until exposed to the additional stress of gypsy moth attack. The increased availability of nitrogen from accumulated nitrogen deposition appears to promote nitrification and nitrate loss, which occurs only to a small extent when the defoliation is absent.

The potential importance of the sequence of stressors is also illustrated by the reduced ability of Caribbean coral reefs to recover from hurricane damage under the greatly reduced herbivore densities described earlier. The sequence of finfish and then sea urchin reductions may not have been particularly important to the eventual results of reduced total herbivory. However, when total herbivory on benthic algae (and consequently coral recruitment) was greatly reduced, additional stressors like hurricanes that kill existing coral colonies potentially have more dramatic and longer-lasting effects. Thus, regrowth and recolonization of coral reefs was more rapid and more complete prior to the sea urchin decline (Hughes 1994).

Succession and Temporal Variability

One of the problems that makes the issue of multiple stressors difficult to tackle is the compounding of unpredictability. At any point in time or space, there may be an array of species that will vary in their response to a stressor. The effect of any particular stressor-related change on system-level processes will vary with the connectedness of the affected species to the rest of the system. Both interaction strengths and the diversity of alternate trophic pathways should be important. Consequently, even with a single stressor, predicting system-level responses is often difficult because of the number of contingent processes. Multiple stressors compound this problem.

The temporal changes that characterize dynamic natural systems also increase the unpredictability of multiple stressor effects by overlaying temporal variability on the relevant processes. The same stressors may have quite different effects depending on the successional state of an assemblage because of differences in species composition, as well as system-level processes such as rates of recycling. Similarly, seasonal variation in species composition, reproductive cycles, or process rates may cause variation in both the presence and magnitude of responses to stressors.

Landscape Level Processes

When multiples stressors are expressed at the broader scale of the landscape, the impact may be translated across community and ecosystem boundaries. The extent and severity of the impact varies with the resistance and resilience of each community and ecosystem, as well as the "leakiness" (i.e. how the impact is expressed across system boundaries) of the systems. Consequently, the extent and magnitude of system responses within the landscape can vary, ranging from no apparent effect at the landscape level (although individual components of the landscape may be altered) to a dramatic and far-reaching alteration of the landscape. For example, insect outbreaks often target specific vegetation types, frequently stressing the vegetation beyond recovery. The stressed systems are more vulnerable to subsequent disturbance such as fire, while adjacent vegetation types that are less stressed may be less damaged and thus have a faster recovery potential (Turner 1987; Pickett et al. 1989). As a result, the extent of the multiple stressor effects at landscape scales may be formed by a patchwork of responses, which can often make interpretation difficult.

One of the more pervasive examples of landscape responses to multiple stressors is fragmentation. Deforestation, agriculture, and urban processes have resulted in continued fragmentation of most natural ecosystems. This fragmentation may have dramatic effects on both ecological and evolutionary patterns of species (Gilpin 1991), as well as on community structure and ecosystem dynamics. In many instances, fragmentation may increase habitat susceptibility to additional stresses and alter the adaptive ability of species to respond to environmental change. For example, in forested landscapes in which clear-cutting practices have greatly increased forest fragmentation, newly exposed forest edges experience increases in light, gas flux, and soil temperature, in addition to wind exposure. Eventually many of the trees along the edge die as a result of wind damage or stress, effectively reducing the patch size even further (Murcia 1995).

Fragmentation also changes biotic interactions, potentially isolating species with limited vagility, changing community structure, and altering competitive interactions within and between communities. For example, increased fragmentation has led to a decrease in the interior-to-edge ratio in many forested areas. This decreased interior/edge ratio decreases habitat available to "interior" species, which require relatively large tracts of forested habitat and live only or primarily away from the habitat edge. Decreased interior/edge ratios and reduced distances to the patch interiors have also allowed edge species to have greater access to the interior of fragmented patches, and in some cases have been the source of population declines in the interior species. Such is the case with the red-eyed vireo (*Vireo olivaceus*), which requires relatively large forest patches for breeding habitat. Habitat loss caused by fragmentation has directly contributed to population decline in this species. Further vireo population declines have been attributed to the parasitic brown-headed cowbird (*Molothrus ater*), an edge species that has experienced marked range expansion primarily in response to landscape fragmentation. The decreased interior/edge ratio and reduced distance to forest interior has allowed this edge species to target and parasitize the nests of such interior bird species as the vireo (Gustafson and Crow 1994). This has ultimately led to reduced vireo brood success and exacerbated the population decline associated with habitat loss. Thus, fragmentation (which itself can be caused by multiple stressors) can cause a cascade of stresses and stress responses. In the preceding example, the direct effects of fragmentation include decreased red-eyed vireo populations and invasion of the interior by brown-headed cowbirds. The invasion of cowbirds then acts as a secondary stressor, causing further decreases in vireo populations.

When multiple stressors are pervasive at the landscape level, it is probable that many of the underlying systems have experienced dramatic changes. Disentangling the causal effects of the various stressors at this stage will be challenging and difficult, but is imperative because the scale and magnitude of the effects raise serious concern for potential recovery of many systems.

Linking Population and Ecosystem-Level Processes: Multiple Stressors in an Urban Landscape

The effects of multiple anthropogenic stressors on terrestrial ecological systems are perhaps most evident in urban and suburban landscapes. In the northeast United States, for example, encroachment of urban and suburban areas into forests is increasing (National Research Council 1990). As human development expands into deciduous forests, it brings an increased exposure to a variety of stressors, including atmospheric and water pollutants, altered disturbance regimes and species introductions (National Research Council 1990; Grey and Deneke 1986). The novelty of these stressors and their multiple differential effects on individual species and ecosystem functions requires a substantial advance in our understanding of multiple stressors in urban areas.

Research conducted along a 130 km transect extending from Bronx County in New York to Litchfield County in Connecticut indicates the potential for variation in responses among individual components of a system, and between species and system-level responses. Forests along this urban to rural land-use gradient can be structurally similar, even though forests at the urban end of the gradient are exposed to a multitude of abiotic and biotic anthropogenic stressors (McDonnell and Pickett 1990). Soils at the urban end of the gradient have higher levels of heavy metals than rural soils (White and McDonnell 1988; Pouyat and McDonnell 1991) and are more hydrophobic (White and McDonnell 1988). Nitrogen deposition rates are also higher at the urban end of the gradient (Lovett et al. unpublished data). In addition, forests in the urban area have a higher proportion of non-native species than comparable rural forests (Rudnicky and McDonnell 1989; McDonnell and Pickett 1990; McDonnell and Roy 1996). For example, exotic earthworms are abundant in urban soils but are absent in rural soils (Steinberg et al. 1997).

Consistent with the notion that the multiple stresses associated with urbanization should create depauperate communities in urban environments (Goudie 1990), saprotrophic fungal biomass and microinvertebrate densities are reduced in urban soils (Pouyat et al. 1994). However, despite decreased decomposer biomass, the ecosystem rates of litter decomposition and nitrogen-cycling are, in fact, higher at the urban end of the gradient (Pouyat 1992; McDonnell and Pickett 1993; Pouyat et al. 1997). Although depauperate fungal and microinvertebrate communities may slow nutrient-cycling processes in the urban forests, their reduction does not necessarily translate directly into reduced ecosystem functioning. Instead, exotic earthworms, higher nitrogen deposition, and somewhat warmer soil temperatures at the urban end of the gradient appear to override any reduction in these processes making the prediction of these net effects from population or community level knowledge extremely difficult. It will require an understanding of the particular ecological contingencies of a given ecosystem and a greater understanding of the links between population and community-level phenomena and ecosystem processes.

Gradient analysis may provide a promising analytical approach to understanding the effects of multiple stressors on ecosystem functioning (Whittaker 1967; McDonnell and Pickett 1993) by integrating the complexity of multiple stress effects across the landscape. The gradient approach relies on the assumption that graduated spatial environmental patterns govern the structure and functioning of ecological systems (McDonnell and Pickett 1990). Changes in population, community, or ecosystem variables along the gradient can then be related to the corresponding spatial variation in the environmental variables, with specific statistical techniques dependent upon whether or not environmental variation is ordered sequentially in time or space, and whether single or multiple variables are being monitored. In the case of system responses to multiple stressors, complex, nonlinear gradients are apt to be present and ordination techniques may provide insight into the biotic responses to these gradients (ter Braak and Prentice 1988; Jongman et al. 1995).

Theoretical and Empirical Challenges

Development of a conceptual and empirical framework for multiple-interacting stressors will require a sustained and concentrated effort on the part of ecologists. General theoretical development of the response of whole ecosystems to single stressors is still limited. We cannot yet generalize, for example, under what conditions we would expect stresses to be amplified and travel through multiple components of an ecosystem, as opposed to being dampened and limited to the initially stressed component. Given this state of our understanding of single-stress effects, it is perhaps not surprising that the subject of multiple stressors has not been well-developed either theoretically or empirically. However, delaying a focus on multiple stressors until there is a thorough understanding of the actions of individual stressors is not warranted and would only delay improvements in the understanding and management of natural systems, most of which are now directly or indirectly affected by multiple stressors caused by human activities.

Single-stress experiments have most commonly been conducted on such tractable systems as terrestrial ecosystems dominated by small annual or perennial plants or in small lakes and streams. Examples of such experimental studies include acidification of lakes and streams (Hall and Likens 1981; Schindler et al. 1985), eutrophication of lakes (Schindler 1977), and warming of tundra plants and soils (Harte and Shaw 1995). Single-stress experiments on whole forests are rare because of the complexity of the logistics and engineering involved, except in the case when the stress is easily applied (e.g. nitrogen fertilizer). These whole-ecosystem experiments are critical to perform (see Carpenter, chapter 12, for a discussion of large-scale manipulations), even though logistical and financial constraints often limit replication and require that data are analyzed as a time-series effect or a simple comparison of unreplicated treatment and reference systems.

Multiple-stress experiments are significantly more complicated than single-stress experiments, and are rarely attempted on whole ecosystems in a research framework. The need for many "replicate" ecosystems to complete the more complex experimental designs is a major limiting factor to empirical research of the effects of multiple stressors. For this reason, in the near future it is expected that most experimental examinations of multiple stressors at an ecosystem level will be made in small, easily manipulated natural ecosystems and in experimental systems such as mesocosms. Small-scale experimental work with rigorous control over both the stressors and the biological components of the system to be manipulated should also contribute greatly to understanding multiple-stressor effects.

There is, nonetheless, an urgent need for knowledge about the interactions of multiple stressors on larger, less tractable ecosystems as well. In the near term, progress on these systems most probably will come from novel statistical analyses of complex environmental gradients (e.g. the urban to rural gradient discussed before), from manipulation of single stressors against a spatially or temporally varying background of other stressors, and from fortuitous co-occurrences of stressors, for example, the arrival of an introduced species into a well-studied system already under stress from another factor.

Additionally, one of the most important opportunities for large-scale manipulations that could contribute greatly to understanding multiple-stressor effects could result from stronger partnerships between researchers and managers. In efforts to reduce anthropogenic impacts on natural systems, local and national agencies often sequentially manipulate nutrient loadings, pollutants, and fisheries, game or timber harvests. The implementation of these management-initiated manipulations often varies at temporal and spatial scales amenable to statistical analyses and modeling techniques that have the potential to distinguish individual and combined effects of stressors (e.g. Schmitt and Osenberg 1997). For example, the tidal portion of the Chesapeake Bay system includes dozens of major and minor tributaries that vary in loadings of nutrients and contaminants, as well as the degree to which harvesting and diseases have reduced top-down control of the system by predatory finfish and suspension-feeding bivalves. Management actions have resulted in sequential manipulations of nutrients, with phosphorous reductions preceding nitrogen reductions. The progress of nitrogen-reduction efforts has also varied greatly among Chesapeake Bay tributaries both in the quantitative reduction achieved and in timing. At the same time that nutrient loadings have been manipulated, oyster reefs that are protected from harvest have been constructed in several tributaries, and fisheries regulations were implemented for several years that prohibited harvest of one of the top predators in the Bay, the striped bass (*Morone saxatilis*), and resulted in a resurgence of striped bass populations. The sort of temporal and spatial variation in management-initiated stressor manipulation that has occurred in the Chesapeake Bay system is probably common. Adequate planning and funding of data collection and monitoring may allow these management-led manipulations to provide information on the effects of multiple stressors that is just as useful as researcher-initiated manipulations. Furthermore, even more might be learned if a consideration of the knowledge that could be gained in the understanding of multiple-stressor effects became a more important factor that was used to determine the timing and location of management-initiated stressor manipulations. Changes in the spatial or temporal implementation of stressor manipulation, and adequate funding for monitoring of reference sites, could make a substantial difference in the ability to evaluate whether and why management actions achieve their desired results, and could greatly improve our understanding of how multiple stressors affect natural systems.

Modeling will also be an important component in the development of a more comprehensive understanding of multiple-stressor effects because models can provide a framework for organizing the massive amounts of data generated by complex empirical studies, and can facilitate extrapolations beyond the temporal and spatial confines of particular studies (see Lauenroth et al. Chapter 16). Models, as well as empirical studies, that improve the understanding of the role of indirect effects in food webs will be particularly important to the ability to predict how direct effects of stressors on individual species will be expressed throughout the ecosystem (Abrams et al. 1996).

Whatever the approaches used, the issue of multiple stressors clearly presents a challenge to ecologists that is both timely and important. Preliminary attempts at understanding the relationship between the complexity and organization of ecological systems and stressor effects will often raise more questions than they answer (Box 17.1). Nevertheless, both experimental and theoretical advances on the topic are needed if we are to understand and reduce ecological changes in human-influenced natural systems.

Acknowledgments. This chapter is an outgrowth of discussions of the Multiple Stressors Workgroup at the 1997 Cary Conference. We would like to thank additional workgroup participants, including J. Meyer, S. Tartowski, R. Adams-Manson, J. Karr, J. Bengston, and I. Reche, for a stimulating discussion, and the organizers of the Cary Conference for including this topic and for helpful suggestions on an earlier draft of this manuscript. Support was provided by the Institute of Ecosystem Studies and by NOAA-Coastal Ocean Program (to D. Breitburg).

Box 17.1

As is typical of any emerging ecological issue, there are more questions than answers. Some key questions we have identified are listed here. These are not meant to represent an exhaustive list, but instead are intended to stimulate thought on what we consider to be an important issue:

Are effects of single and multiple stressors fundamentally different? Is it more difficult for a system to return to its predisturbed state when disturbed by multiple stressors than by a single stressor? Do the cumulative effects of multiple stressors have a greater effect than does a similar level of perturbation by a single factor?

Contingent and dependent effects. Perturbations can presumably change system structure, and system structure is presumably a key component of system response. Does this mean that one perturbation necessarily alters the sensitivity of the system to other perturbations? How will the timing, mixture, and history of perturbations alter the responses to the next set of perturbations?

Functional complementarity. Resistance or resilience to perturbations may be strongly influenced by the extent of functional complementarity within the system. How often and under what conditions will functional complementarities be sufficient to dampen system-level responses to multiple stressors if we consider the diverse effects of many perturbations operating on many species and processes across many different ecosystems with many different structures? Are certain types of functional complementarities more important, or is the functional complementarity of importance entirely dependent on the combination of stressors? What attributes of system structure in addition to functional complementarity determine whether or not the effects of multiple stressors will become dampened (i.e. have reduced or no effects beyond the directly affected component) or amplified (i.e. passed on and increased in other components) throughout a system?

Empiricism vs. theory. What will be the most profitable partnerships between empirical and theoretical approaches to the study of systems under multiple stresses?

Limited outcomes. Any particular system function is generally limited to three kinds of outcomes: an increase, a decrease, or no net change. Does the constrained diversity of outcomes preclude elucidation of underlying causality, or will causality readily map onto outcome?

Box 17.1 *Continued*

Predictability. Does understanding system responses to multiple stressors increase our capacity to predict outcomes in new situations, or is low predictability inherent with multiple stressors because the magnitude and direction of effects of each of the individual stressor can vary for different components of the system?

References

Aber, J.D., A. Magill, R. Boone, J.M. Melillo, P. Steudler, and R. Bowden. 1993. Plant and soil responses to chronic nitrogen additions at the Harvard Forest, Massachusetts. *Ecological Applications* 3:156–166.

Abrams, P., B.A. Menge, G.G. Mittelbach, D. Spiller, and P. Yodzis. 1996. The role of indirect effects in food webs. Pages 371–395 in G.A. Polis and K.O. Winemiller, eds. *Food webs: integration of patterns and dynamics.* Chapman & Hall, New York.

Berge, J.A. 1990. Macrofauna recolonization of subtidal sediments. Experimental studies on defaunated sediment contaminated with crude oil in two Norwegian fjords with unequal eutrophication status. I. Community responses. *Marine Ecology and Progress Series* 66:103–115.

Berge, J.A., R.G. Lichtenhtaler, and F. Orled. 1987. Hydrocarbon depuration and abiotic changes in artificially oil contaminated sediment in the subtidal. *Estuarine, Coastal and Shelf Science* 24:567–583.

DeAngelis, D.L. 1980. Energy flow, nutrient cycling, and ecosystem resilience. *Ecology* 61:764–771.

Flavin, C. 1997. The legacy of Rio. Pages 3–22 in L.R. Brown, C. Flavin, and H. French, eds. *State of the world, 1997, Worldwatch Institute.* W.W. Norton and Company, New York.

Gilpin, M. The genetic effective size of a metapopulation. 1991. *Biological Journal of the Linnean Society* 42:165–172.

Grey, G.W., and F.J. Deneke. 1986. *Urban forestry.* John Wiley & Sons, New York.

Goudie, A. 1990. *The human impact on the natural environment.* MIT Press, Cambridge, MA.

Gustafson, E.J., and T.R. Crow. 1994. Modeling the effects of forest harvesting on landscape structure and the spatial distribution of cowbird brood parasitism. *Landscape Ecology* 9(4):237.

Hall, R.J., and G.E. Likens. 1981. Chemical flux in an acid-stressed stream. *Nature* 292:329–331.

Harte, J., and R. Shaw. 1995. Shifting dominance within a montane vegetation community: results of a climate warming experiment. *Science* 267:876–880.

Hay, M.E. 1984. Patterns of fish and urchin grazing on Caribbean coral reefs: are previous results typical? *Ecology* 65:446–454.

430   D.L. Breitburg, et al.

Howarth, R.W. 1991. Comparative responses of aquatic ecosystems to toxic chemical stress. Pages 169–195 in J. Cole, G. Lovett, and S. Findlay, eds. *Comparative analyses of ecosystems.* Springer-Verlag, New York.

Hughes, T.P. 1994. Catastrophes, phase shifts, and large-scale degradation of a Caribbean coral reef. *Science* 265:1547–1551.

Jongman, R.H.G., C.J.F. ter Braak, and O.F.R. van Tongeren. 1995. *Data analysis in community and landscape ecology.* Cambridge University Press, Cambridge, U.K.

Lessios, H.A. 1988. Mass mortality of *Diadema antillarum* in the Caribbean: what have we learned? *Annual Review of Ecology and Systematics* 19:371–393.

Lovett, G.M., and A.E. Ruesink. 1995. Carbon and nitrogen mineralization from decomposing gypsy moth frass. *Oecologia* 104:133–138.

McDonnell, M.J., and E.A. Roy. 1996. Vegetation dynamics of a remnant hardwoods-hemlock forest in New York City. *Supplement to the Bulletin of the Ecological Society of America* 77:294.

McDonnell, M.J., and S.T.A. Pickett. 1990. The study of ecosystem structure and function along urban-rural gradients: an unexploited opportunity for ecology. *Ecology* 71:1231–1237.

McDonnell, M.J., and S.T.A. Pickett. 1993. *Humans as components of ecosystems.* Springer-Verlag, New York.

Murcia, C. 1995. Edge effects in fragmented forests: implications for conservation. *Trends in Ecology and Evolution* 10:58–62.

National Research Council. 1990. *Forestry research: a mandate for change.* National Academy Press, Washington, DC.

Pickett, S.T.A., J. Kolasa, J.J. Armesto, and S.L. Collins. 1989. The ecological concept of disturbance and its expression at various hierarchical levels. *Oikos* 54:129–136.

Pouyat, R.V. 1992. Soil characteristics and litter dynamics in mixed deciduous forests along an urban-rural gradient. Ph.D. dissertation, Rutgers University, New Brunswick, NJ.

Pouyat, R.V., and M.J. McDonnell. 1991. Heavy metal accumulations in forest soils along an urban-rural gradient in southeastern New York, USA. *Water, Air, and Soil Pollution* 57–58:797–807.

Pouyat, R.V., M.J. McDonnell, and S.T.A. Pickett. 1997. Litter and nitrogen dynamics in oak stands along an urban-rural gradient. *Urban Ecosystems* 1:117–131.

Pouyat, R.V., R.W. Parmelee, and M.M. Carreiro. 1994. Environmental effects on forest soil-invertebrate and fungal densities in oak stands along an urban-rural land use gradient. *Pedobiologia* 38:385–399.

Riedel, G.F. 1984. Influence of salinity and sulfate on the toxicity of chromium (VI) to the estuarine diatom *Thalassiosira pseudonana. Journal of Phycology* 20:496–500.

Riedel, G.F. 1985. The relationship between chromium (VI) uptake, sulfate uptake, and chromium (VI) toxicity in the estuarine diatom, *Thalassiosira pseudonana. Aquatic Toxicology* 7:191–204.

Rudnicky, J.L., and M.J. McDonnell. 1989. Forty-eight years of canopy change in a hardwood-hemlock forest in New York City. *Bulletin of the Torrey Botanical Club* 116:52–64.

Sanders, J.G. 1979. Effects of arsenic speciation and phosphate concentration on arsenic inhibition of *Skeletonema costatum (Bacillariophyceae). Journal of Phycology* 15:424–428.

Sanders, J.G., and G.F. Riedel. 1987. Control of trace element toxicity by phytoplankton. Pages 131–149 in J.A. Saunders, L. Kosak-Channing, and E.E. Conn, eds. *Recent advances in phytochemistry, vol. 21.* Plenum Press, New York.

Sanders, J.G., G.F. Riedel, and R.W. Osman. 1994. Arsenic cycling and impact in estuarine and coastal marine ecosystems. Pages 289–308 in J.O. Nriagu, ed. *Arsenic in the environment, part I: cycling and characterization.* John Wiley & Sons, New York.

Schindler, D.W. 1977. The evolution of phosphorus limitation in lakes. *Science* 195:260–262.

Schindler, D.W., K.H. Mills, D.F. Malley, D.L. Findlay, J.A. Shearer, I.J. Davies, et al. 1985. Long-term ecosystem stress: the effects of years of experimental acidification on a small lake. *Science* 228:1395–1401.

Schmitt, R.J., and C.W. Osenberg. 1997. *Detecting ecological impacts: concepts and applications in coastal habitats.* Academic Press, San Diego, CA.

Steinberg, D.A., R.V. Pouyat, R.W. Parmalee, and P.M. Groffman. 1997. Earthworm abundance and nitrogen mineralization rates along an urban-rural land use gradient. *Soil Biology and Biochemistry* 29:427–430.

ter Braak, C.J.F., and I.C. Prentice. 1988. A theory of gradient analysis. *Advances in Ecological Research* 18:272–327.

Tilman, D. 1996. Biodiversity: population versus ecosystem stability. *Ecology* 77:350–363.

Turner, M.G. 1987. *Landscape heterogeneity and disturbance.* Springer, New York.

Wangberg, S.-A., and H. Blanck. 1990. Arsenate sensitivity in marine periphyton communities established under various nutrient regimes. *Journal of Experimental Marine Biology and Ecology* 139:119–134.

Webb, J.R., B.J. Cosby, F.A. Deviney, K.N. Eshleman, and J.N. Galloway. 1995. Change in the acid-base status of an Appalachian catchment following forest defoliation by the gypsy moth. *Water, Air, and Soil Pollution* 85:535–540.

Williamson, C.E., S.L. Metzgar, P.A. Lovera, and R.E. Moeller. 1997. Solar ultraviolet radiation and the spawning habitat of yellow perch, *Perca flavescens. Ecological Applications* 7:1017–1023.

White, C.S., and M.J. McDonnell. 1988. Nitrogen cycling processes and soil characteristics in an urban versus rural forest. *Biogeochemistry* 5:243–262.

Whittaker, R.H. 1967. Gradient analysis of vegetation. *Biological Reviews* 49:207–264.

18
Within-System Element Cycles, Input-Output Budgets, and Nutrient Limitation

Peter M. Vitousek, Lars O. Hedin, Pamela A. Matson, James H. Fownes, and Jason Neff

Summary

Widely used conceptual models for controls on nutrient cycling and input-outputs budgets of forest ecosystems suggest that: (1) nutrient losses from ecosystems originate in the available nutrient pool in soil; (2) nutrients that limit plant production are retained tightly within those systems; (3) this retention leads to accumulation of the limiting nutrient(s), eventually to the point at which it no longer limits production; and (4) losses of nutrient(s) thereafter should reflect rates of nutrient input, rather than biotic demand. In this chapter, we explore mechanisms that could constrain the accumulation of a limiting nutrient, and therefore could allow nutrient limitation to persist indefinitely. Possible mechanisms include episodic disturbance-related nutrient losses, closed element cycles, and losses of nutrients from sources other than the available inorganic pool of nutrients in soil. For the last mechanism, both a simple and a more complex model are used to show that losses of dissolved organic forms of a nutrient could constrain nutrient accumulation and permit nutrient limitation to persist indefintely. Emissions of nitrogen (N) trace gases produced during nitrification could have a similar effect. To the extent that losses of nutrients by these and related pathways are important, anthropogenic inputs of nutrients (particularly N) could alter forest ecosystems substantially, to an extent greater than standard conceptual models would allow.

Introduction

Ecosystem science has developed a conceptual model that links short-term, largely biological processes of nutrient mineralization and uptake with the longer-term, largely geological and hydrological processes driving inputs and outputs of nutrients. In this chapter, we describe the model as it applies to forest ecosystems, specifically to upland forest watersheds in which precipitation is substantially greater than evapotranspiration. In the model, nutrients (especially organically bound N, phosphorus [P], and sulfur [S]) are transformed to available inorganic forms through microbial mineralization of litter and soil organic matter; these mineralized nutrients can then be taken up by plants and microbes. In the short term, balance between nutrient supply via mineralization and the demand for nutrients by plants and microbes determines whether nutrient limitation constrains biological processes; when supply is insufficient to meet demand, the growth of plants or microbes is reduced. Under these conditions, organisms survive with some combination of reduced growth, more efficient utilization of lower quantities of nutrients, longer retention times of elements within organisms, and allocation of other resources in ways that allow them to compete for the resource(s) in short supply.

This within-ecosystem nutrient cycle takes place in the context of inputs to and outputs from the whole ecosystem. The major pathways of input are rock weathering and atmospheric deposition. Although inputs are often considered to be externally controlled forcing functions, this is not strictly true for most inputs—for example, respired carbon dioxide (CO_2) and organic acids from plants can enhance rates of weathering (Cochran and Berner 1997), and canopy structure can influence rates of atmospheric deposition (Lovett 1994). In contrast to inputs, outputs typically are assumed to originate in the mobile, biologically available pool of inorganic nutrients. In essence, this represents the pool of nutrients that is "left over" after plant and microbial demands for nutrients are met; as a consequence, outputs are inferred to be controlled by processes operating within ecosystems. In the long term, the balance between inputs and outputs of nutrients controls the quantity of nutrients within ecosystems, which in turn determines the supply of nutrients for potential mineralization and uptake. When this quantity is small, nutrient limitation is apt to constrain the growth of plants or microbes.

A simplified view of the link between within-system nutrient cycling and input-output balances is summarized in Figure 18.1. There are many important processes omitted from this figure, including biological N fixation, biochemical mineralization, and plant utilization of dissolved organic nutrients (McGill and Cole 1981; Kielland 1994; Northrup et al. 1995). However, we believe that it can be used to represent a standard approach of ecosystem science, which is expressed in such conceptual models as Vitousek and Reiners (1975), Gorham et al. (1979), and Bormann and Likens (1979). It is also consistent with the underlying conceptual basis for simple models of N saturation in forest ecosystems (Aber 1992). In this chapter: (1) some of the implications of this conventional model for linkages among nutrient limitation, within-system nutrient cycling, and input-output budgets are explored; (2) ways that these linkages might not occur as described by the model in the long term are described and evaluated; and (3) information from both a simple and a more complex model is used to evaluate the importance and

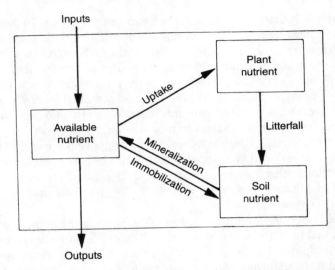

FIGURE 18.1. Simplified diagram of a within-system nutrient cycle. Inputs and outputs go to and from the available nutrient pool in soil, which contains mobile and exchangeable inorganic forms of nutrients; the remainder of each element in soil is in the soil pool, which includes the microbial nutrient pool.

consequences of several processes that are not represented in Figure 18.1, focusing especially on losses of nutrients from sources other than the available inorganic nutrient pool. We are particularly interested in the implications of different—and differently regulated—pathways of N loss. What are the long-term consequences of leaching losses of nitrate vs. dissolved organic N? What are the long-term consequences of N gas emissions produced during nitrification vs. denitrification?

Implications of the Standard Conceptual Model

One important consequence of the conventional model is that there should be a very close connection between nutrient limitation by a particular nutrient and losses of that nutrient from an ecosystem. When the supply of a nutrient limits biological processes (for simplicity, the focus will be on limitations to plant growth), demand for that nutrient should ensure that there will be no surplus remaining in available inorganic pools in the soil (except for a residual that is too dilute to be taken up), and consequently, losses of that nutrient will be very small. Low losses of the limiting nutrient(s), in combination with continued inputs of that nutrient from the atmosphere or weathering, should lead to accumulation of the limiting nutrient within ecosystems. As the total quantity of a nutrient accumulates, mineralization increases the supply of available forms of that nutrient—until, at some point, supply is sufficient to meet potential plant demand. At that time, supply of the nutrient no longer limits plant growth, and losses from the ecosystem can increase to the point at which they balance inputs.

In general, this conventional model suggests that when nutrients limit biological processes within an ecosystem, losses should be small and controlled largely by the demands of organisms. When nutrients are not limiting, losses should be greater, and controlled primarily by inputs. (Losses of elements with insoluble forms, such as P, may also be controlled by the formation of secondary minerals in soil—whether or not they are limiting). Spatial and temporal heterogeneity in nutrient supply or demand complicate this model. For example, losses of putatively limiting nutrients often increase during winter in northern forests, reflecting diminished nutrient uptake by plants (Likens and Bormann 1995). Similarly, as discussed later, the spatial heterogeneity of plant cover in dryland ecosystems may be sufficient to allow losses of limiting nutrients from the spaces between plants (Schlesinger et al. 1990).

When the supply of a nutrient(s) limits plant growth in an ecosystem, both the intensity of limitation and the lack of nutrient losses can be accentuated by a plant-soil-microbial (PSM) positive feedback. This feedback arises because many of the mechanisms by which plants cope with nutrient-poor sites, as well as many of the biotic consequences of low nutrient availability (low tissue nutrient concentrations, long residence times for nutrients within plants, production of recalcitrant carbon compounds) lead to further decreases in rates of nutrient supply. Low tissue concentrations and recalcitrant carbon compounds reduce rates of litter decomposition and increase rates of microbial immobilization of any nutrients that are released (Vitousek 1982), thereby reducing rates of net mineralization. Slow decomposition and mineralization lead to the accumulation of organically bound nutrients in litter and soil, and a further reduction in nutrient availability. This feedback can occur as a result of plasticity within a single species on a range of sites (cf. Miller et al. 1976; Crews et al. 1995) or as a consequence of species replacement (Pastor et al. 1987; Berendse 1993).

The operation of this PSM positive feedback can maintain very low levels of available nutrients within low nutrient systems, can reduce the probability of nutrient losses even during times when plants are less active, and thereby can increase the rate of accumulation of limiting nutrients within ecosystems. Consequently, the contribution of the PSM feedback to low nutrient availability should be a transient phenomenon—as should nutrient limitation itself. It should be important for a few years to a few centuries, until inputs of limiting nutrients from outside the system build up nutrient pools to the point at which even the slow turnover of a large, organically bound nutrient pool in the soil can supply as much of the nutrient as plants require. At that point, nutrient limitation should disappear, nutrient concentrations in plant tissue should increase, and the PSM positive feedback should operate to increase rates of nutrient cycling.

A further implication of the conventional model is that availability of all the major nutrients in ecosystems should equilibrate (eventually) at more or less characteristic ratios, and these ratios should be close to the relative requirement of organisms for the different nutrients. An element in short supply relative to the needs of organisms will be retained (i.e. inputs > outputs) and accumulated; that which is in excess will have high residual concentrations in the available inorganic pool, and therefore, be subject to loss. These characteristic ratios are analogous to the Redfield ratios of marine systems (Redfield 1958), although the situation in terrestrial ecosystems is complicated by the importance of element inputs via weathering, by abiotic as well as biotic pathways of nutrient retention, and by the variety of ways that terrestrial plants can allocate nutrients. At equilibrium, ratios of element concentrations in outputs from ecosystems should reflect their ratios in inputs, regardless of the cycling rate of particular nutrients within ecosystems (Hedin et al. 1995).

A Simple Simulation Model

We developed a straigthtforward simulation model that can be used to illustrate the dynamics and limitations of the "conventional" model discussed here. In our base case, this model resolves three pools of an element—in plants, in organic forms in soil, and in mobile, biologically available, inorganic forms in soil. The transfer of nutrients from the plant pool to organic pool in soil (i.e. litter production) and from the organic pool in soil to the available inorganic pool (i.e. mineralization) are modeled as first-order processes, with annual transfer coefficients of 0.1 and 0.01 respectively. Transfer from the available inorganic pool in soil (hereafter called the *available pool*) to the plant pool (i.e. plant uptake) removes up to ten units of the nutrient (say, $10 \, g/m^{-2}/y^{-1}$); any nutrient remaining in the available pool is then lost via leaching. Removal of $10 \, g/m^{-2}/y^{-1}$ from the available pool represents the rate of nutrient uptake that will support maximum production on the site; the maximum intrinsic growth rate of plants or the supply of light, water, or space set that maximum. In this base case, we set inputs of the nutrient to $0.2 \, g/m^{-2}/y^{-1}$; these inputs are transferred into the available pool. We initialize the within-ecosystem nutrient pool (i.e. plant plus soil) at $550 \, g/m^{-2}$, and run the model for 5,000 years. Simulating long-term nutrient budgets while holding all else constant is not realistic—climate and other boundary conditions are very apt to change in any 5,000 year period—but it is a good way to examine the underlying controls of nutrient accumulation and loss. Results are illustrated in Figure 18.2. When the soil pool contains $<980 \, g/m^{-2}$ (plant plus soil pool $<1,078 \, g/m^{-2}$) of the nutrient, mineralization plus input supplies an amount

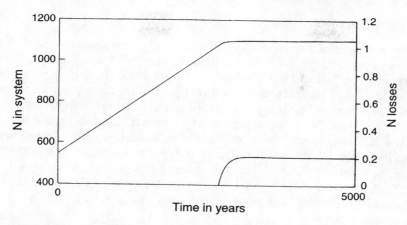

FIGURE 18.2. Simulated changes in the plant plus soil nutrient pools and in losses of dissolved inorganic nutrients over time, using the simple model described in the text. Input of the nutrient is set to 0.2 units.

of the nutrient less than is required to support maximum production. Consequently, nutrient supply limits plant production, losses of the nutrient are zero, and inputs of the nutrient are accumulated within the ecosystem. However, at equilibrium (plant plus soil pool equal to $1,100\,g/m^{-2}$) nutrient supply no longer limits production, and losses are equal to inputs of the nutrient.

Although simple, this simulation model provides a reasonable representation of the standard model discussed earlier. Given that the assumptions underlying the conventional model (especially that ecosystems are open systems with inputs of nutrients, and that losses of limiting nutrients are less than inputs of those nutrients), the expectation that nutrient limitation to primary production should be a transient phenomenon seems robust. Moreover, many important biological processes that are not considered in the model (i.e. biological N fixation that responds positively to N deficiency within the system, mineralization of nutrients by extracellular enzymes) will tend to offset the PSM feedback and reduce the extent or duration of nutrient limitation. Why, then, do observations and experimental studies suggest that nutrient limitation to plant production is widespread in the real world?

Persistent Nutrient Limitation

Three mechanisms that could maintain long-term nutrient limitation in ecosystems are: (1) repeated disturbances that cause substantial losses of nutrients; (2) little or no input of a nutrient; and (3) losses of nutrients from sources other than the available pool in soil.

Disturbance

The significance of disturbance in long-term ecosystem nutrient budgets has been discussed widely (cf. Vitousek and Reiners 1975; Woodmansee 1978; Bormann and Likens 1979; Pickett and White 1985). If disturbances (e.g. harvesting, fire, destructive windstorms) occur frequently enough, and cause sufficiently large losses of nutrients, then the behavior of systems will be characterized by sustained periods of nutrient accumulation, and potentially limitation (Miller 1981), punctuated by brief, intense periods of loss. Moreover, some kinds of disturbance (i.e. pathogen attack, fire) may be more frequent in nutrient-limited ecosystems and possibly accentuated by the PSM feedback (Matson and Waring 1984; Pastor et al. 1987), thereby establishing a longer-term positive feedback that can reinforce nutrient limitation.

We simulated the effects of disturbance by removing all of the nutrients in the plant compartment at 500-year intervals. We did not simulate additional losses of nutrients by other pathways (e.g. leaching or volatilization) following disturbance; these would be small in comparison with removal of the plant nutrient pool in our model, as happens (most often) in the real world. Under these conditions, it takes more than 10,000 years for the system to reach its equilibrium nutrient content (Figure 18.3). Even when it does so, the simulated disturbance regime is intense enough to prevent any losses of the nutrient via leaching during the long phases of nutrient accumulation between disturbances.

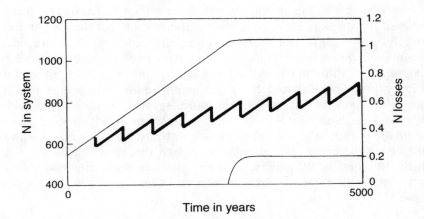

FIGURE 18.3. The effect of simulated disturbance on the plant plus soil nutrient pool and nutrient loss. The thinner lines represent the base case (Figure 18.2); the bold line represents the effect of removing all of the nutrients in the plant pool every 500 years. No losses of dissolved nutrients during the regrowth phase are observed in this simulation, and little would occur even when the plant plus soil pool reached the maximum level it will achieve under these conditions.

We believe that disturbance is an important process that drives nutrient losses over large areas of terrestrial ecosystems, causing many of them to be nutrient limited most of the time. However, its implications already have been widely discussed and are relatively well known, and we will not emphasize the effects of disturbance in this chapter.

Closed Element Cycles

A second mechanism that could lead to persistent nutrient limitation is an element cycle for which inputs from outside the ecosystem are zero. In this case, the lack of nutrient inputs could constrain the accumulation of organic matter in both plants and soils indefinitely; the limiting nutrient could not accumulate within the system and, ultimately, could not reverse limitation.

How realistic is this scenario? Ecosystems are open systems; inputs of nutrients may be low, especially in settings that are modified little by human activity (Likens et al. 1987; Galloway et al. 1996), but they are not zero. An element cycle with zero inputs may be approached most closely for rock-derived elements on old, deeply leached soils, from which weatherable minerals have been nearly or wholly depleted. We are presently working on a 4.1-million-year developmental sequence of sites in the Hawaiian Islands, where primary minerals and weathering inputs from them have essentially disappeared tens of thousands of years into the sequence (Crews et al. 1995; Vitousek et al. 1997). Pools of the major cations in plants and soil decline dramatically across this sequence. If weathering were the only source, the cycles of the major cations would be closed. However, there is atmospheric deposition of marine-derived cations, and indeed isotopic analyses show that most major cations in the older sites are from the ocean rather than from weathering (Kennedy et al. in press). This source of cations is sufficient to maintain forest production in the long term; experimental studies show that cation availability does not limit forest production even in the oldest site (Herbert and Fownes 1995; Vitousek and Farrington 1997). A similiar situation is found in temperate rain forests in southern Chile; isotope analyses demonstrate that most of the major cations in the forests and soils are derived from atmospheric deposition of dilute sea salts (Hedin and Likens 1996; Kennedy et al. unpublished data). Analyses of cation ratios in hydrologic losses from forests in southern Chile and the Pacific Northwest further support the conclusion that atmospheric deposition can be the dominant source of cation inputs to ecosystems (Hedin et al. 1995; Hedin and Hetherington 1996).

Phosphorus may provide a better example of a closed element cycle. Weathering-derived P is retained within systems much longer than cations, but ultimately it is lost or bound up in chemically or physically protected forms (Walker and Syers 1976; Vitousek et al. 1997). Atmospheric inputs of P to ecosystems in humid regions are small, and of uncertain biological availability (Newman 1995). Nevertheless, some P is transported through the atmosphere—dust from distant arid areas is ubiquitous (Swap et al. 1992), and any available P that does reach terrestrial ecosystems is retained very effectively by both biotic and abiotic processes (Wood et al. 1984; Crews et al. 1995). In practice, we believe that low inputs of nutrients, by themselves, cannot sustain nutrient limitation indefinitely.

Losses of Nutrients from Other Pools

A third mechanism that could maintain nutrient limitation in the long term is losses of elements by pathways other than leaching of nutrients from the available pool in soil. For example, Hedin et al. (1995) demonstrated that dissolved organic N (DON) dominates hydrologic losses of N from old-growth forest watersheds in Chile, and that the importance of DON losses from forests (relative to nitrate losses) decreases systematically as anthropogenic N inputs increase. Hedin et al. (1995) speculated that large losses of DON could be sufficient to balance the (very low) inputs of fixed N via atmospheric deposition in remote regions, thereby preventing the accumulation of N within ecosystems, and potentially maintaining N limitation in the long term. There is no direct evidence that losses of DON represent an uncontrollable leak of N (in contrast to losses of leftover N as nitrate)—but that possibility is logical, and consistent with the relative constancy of DON losses from many watershed ecosystems, in contrast to the very wide variation in losses of nitrate (McDowell and Asbury 1994; Hedin et al. 1995). Most unpolluted (and presumably N-limited) temperate forests lose little nitrate or ammonium, but substantial quantities of DON are lost. Moreover, experimental studies suggest that DON losses change much less in response to fertilization and to disturbance than do losses of nitrate (Sollins and McCorison 1981; Currie et al. 1996). Losses of dissolved organic S (DOS) and P (DOP) from forests also occur; DOP in particular may be a dominant pathway of P loss, especially since leaching losses of inorgnic P generally are very small.

Dissolved organic nutrients are not the sole pathway of loss that is independent of the available nutrient pool in soil. Wind and water erosion remove nutrients from ecosystems in both organic and particulate mineral forms (although they can also increase the supply of rock-derived elements by exposing primary minerals to weathering). The movement of animals can also be a pathway of nutrient loss, and (as described later) some pathways of gaseous loss do not depend on the accumulation of leftover nutrients in the available pool.

Here, the consequences of loss by the dissolved organic pathway are explored, for N in particular. We assume that DON comes from a depletable pool that is renewed during microbial processing of soil N; we simulate this pathway by making the loss of DON proportional to N mineralization (setting it to 3% of mineralization). Even this relatively small loss removes enough N to balance N inputs at a pool size that is substantially below the equilibrium level observed in the base case (Figure 18.4). Losses of DON constrain nutrient accumulation within the system, sustain nutrient limitation to plant growth for the long term, and prevent leaching losses of N in inorganic forms.

Some of the multiple pathways of gaseous N losses from ecosystems could largely remove leftover nutrients from the available pool, in the same way that the conventional model characterizes leaching losses (Figure 18.2). Other pathways could remove N in proportion to the rate of mineralization rather than the size of the available pool, in a way analogous to our characterization of dissolved organic losses of nutrients (Figure 18.4). Any definitive classification is arbitrary, but it is reasonable to think of the production of nitric and nitrous oxide (NO and N_2O) during nitrification as being analogous to DON losses; production of these gases occurs in the course of an N transformation that is part of the within-system cycle (cf. Firestone and Davidson 1989; Parton et al. 1996). In contrast, we consider ammonia (NH_3) volatilization and the production of N_2O and N_2 during denitrification to be analogous to nitrate (NO_3) leaching; they represent losses from an accumulated pool of a mobile, available nutrient.

Where gaseous losses of N dominate outputs from ecosystems, the regulation of N loss could be complex. When N availability is very low, nitrification rates also may be low and plants may acquire most N as NH_4 (Vitousek

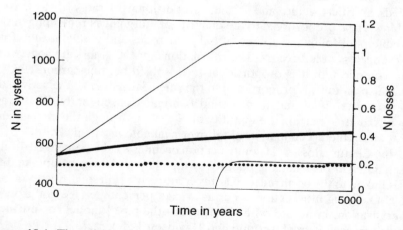

FIGURE 18.4. The effects of losses of a dissolved organic form of nutrient on the plant plus soil nutrient pool and on nutrient loss. The thinner lines represent the base case (Figure 18.2); the bolder lines represent the effect of simulating loss of dissolved organic nutrient by removing a constant fraction of mineralization (set to 0.03) from the system by this pathway. The solid line represents the plant plus soil nutrient pool; the dotted line represents dissolved organic loss. No losses of dissolved inorganic nutrient occur in this simulation.

1982; Tamm 1991); gaseous losses of nitrogen oxides or N_2 would then be low. At higher levels of N availability, more N may cycle through NO_3, and gaseous losses of NO and N_2O during nitrification may become more important (and analogous to DON losses as discussed earlier) (Figure 18.4). However, more recent evidence suggests that rates of nitrification in infertile sytems may have been systematically underestimated in the past (Stark and Hart 1997). In any case, when N availability is so high that NO_3 accumulates in soils, losses via denitrification could be important and analogous to leaching of leftover NO_3 from soils. An additional complication is that denitrification can take place within anaerobic, N-rich microsites within a soil matrix, and denitrification may therefore be more responsive to fine-scale spatial and temporal occurrences of accumulated NO_3 pools than is NO_3 leaching.

A More Complex Model

Although the model underlying Figure 18.2 to 18.4 is simple, the dynamics expressed by this model can be seen in far more complex models as well. This point is illustrated by using the Century model to simulate N pools and losses in a generic tropical ecosystem (Parton et al. 1987, 1993). Century calculates hydrologic losses of both inorganic and organic N, and gaseous N losses by a range of pathways. Losses of NO_3 and DON are controlled in part by the flow of water through soil profiles; this occurs when water flow exceeds a critical value, and is then proportional to rates of flow. NO_3 leaching is also proportional to the quantity of inorganic N remaining in the available pool following uptake of N by microbes and plants; it therefore represents loss of leftover N. The calculations for DON loss are somewhat more complicated; it is proportional to dissolved organic carbon (DOC) loss, which, in turn, is proportional to the rate of turnover of the active soil organic matter pool and to soil sand content. Consequently, DON loss depends in large part on the quantity of N cycling through the system, and not at all on the accumulation of leftover available inorganic N.

Using Century, losses via denitrification occur as a fixed percentage of the N remaining in the available pool following plant and microbe uptake; this loss is linked to the occurrence of leftover inorganic N in soils. Losses of NO and N_2O during nitrification occur as a fixed percentage (1%) of gross N mineralization, in accord with the "hole-in-the-pipe" model for nitrification-driven gas fluxes (Firestone and Davidson 1989).

We ran Century using a tropical forest parameterization in version 4.0 (Metherell et al. 1993), setting precipitation to 2,600 mm/y. Nitrogen losses are grouped into two classes distinguished by their dependence on the accumulation of inorganic N in available pools. NO_3 leaching and denitrification represent what we termed available-pool-dependent losses; DON leaching and trace gases emitted during nitrification represent what we termed transformation-dependent losses. In the first simulation, the full model was run for 5,000 years (following a 1,500-year initialization), and we plotted the ecosystem N pool (plant plus soil) and losses of N by available-pool-dependent and transformation-dependent pathways (Figure 18.5). With these low N inputs, the total quantity of N in the system declined over time, and plant production was limited by N throughout the run. Transformation-dependent losses were nearly three times greater than available-pool-dependent losses, and N-containing gases produced during nitrification represented the single most important pathway of loss.

Our second run was much as the first, except that we turned off the transformation-dependent loss pathways (i.e. DON leaching and gases produced during nitrification). With these N losses removed, the quantity of N within the ecosystem increased over time, as did N losses by available-pool-dependent pathways (Figure 18.5). Plant production tended toward not being limited by N. The results in Figure 18.5 clearly illustrate that in the Century model, N accumulation is constrained and N limitation can be maintained by transformation-dependent losses of N—in a manner wholly analogous to the way simulated DON losses sustain N limitation in the simple model illustrated in Figure 18.4.

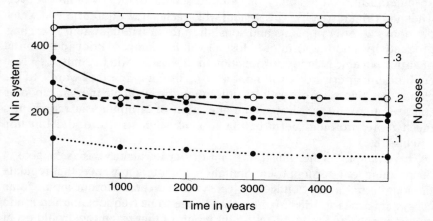

FIGURE 18.5. Patterns of N pools and losses predicted using the Century model, with a tropical forest parameterization. Inputs are set to $0.2\,\mathrm{gm^{-2}y^{-1}}$. Results obtained with the full model, with all output pathways included, are represented with solid symbols; results obtained when transformation-dependent pathways are turned off are represented with hollow symbols. The total N pool in the system is represented with solid lines, losses by available-pool-dependent pathways (NO_3 and NH_4 leaching plus denitrification) with the dotted lines, and losses by transformation-dependent pathways (leaching of dissolved organic N plus N trace-gas emissions during nitrification) with the dashed line.

Anthropogenic N Enrichment, Pathways of Loss, and N Saturation

Anthropogenic N inputs to terrestrial ecosystems can cause a coarse-scale alteration of N availability, productivity, and N losses, in a syndrome termed N saturation (Agren and Bosatta 1988; Aber et al. 1989; Aber 1992). Assume that the "conventional" model outlined in Figure 18.2 captures the essential features of ecosystem response to enhanced inputs of N. In that case, an increase in N inputs from 0.2 to $1 g/m^2/y$ (a moderate anthropogenic enhancement, much less than is observed in northwestern Europe; Berendse et al. 1993) will increase the rate at which N accumulates within aggrading ecosystems; losses of N will occur sooner, and will be five times greater at equilibrium than in the base case (reflecting the enhancement of inputs) (Figure 18.6a). However, the equilibrium quantity of N within the system, and rate of N uptake and, therefore, plant productivity at equilibrium, will not be affected by increased inputs.

In contrast, if losses of DON (or nitrification-dependent gaseous losses) are sufficient to balance background N inputs at some lower quantity of N in the system (as in Figure 18.4), then an anthropogenic increase in N inputs could alter the system more profoundly. If we impose increased inputs (from 0.2 to $1 g/m^{-2}/y^{-1}$) on a system in which DON losses are equal to 3% of mineralization, the ability of DON losses to balance N inputs is overwhelmed (Figure 18.6b). This system will reach a new equilibrium at which the pool size of N in plants and soils, the rate of within-system N cycling, productivity, and the form of N lost all will be changed. Prior to the onset of anthropogenic N inputs, production in the system will be limited by N at equilibrium; afterward N will no longer be limiting. In other words, to the extent that N losses from sources other than the available inorganic pool in soil are important, anthropogenic alteration of N inputs can represent a much more profound perturbation to ecosystem structure and function (Figure 18.6).

This example further illustrates that the various mechanisms capable of maintaining long-term nutrient limitation can interact in ways that regulate the long-term nutrient budgets of ecosystems. When nutrient inputs to an ecosystem are large, disturbance would have to be frequent and intense to maintain nutrient limitation. Nutrient inputs of that magnitude could occur naturally when readily weatherable minerals are present in soil, or when the vegetation is dominated by active biological N fixers; alternatively, they could result from human activity. When nutrient inputs are very low (as in old, deeply leached soils in which weatherable minerals have been depleted, or in remote areas that are not very affected by human activity), even relatively low rates of nutrient loss by dissolved organic pathways, or infrequent disturbance, could be sufficient to maintain nutrient limitation indefinitely.

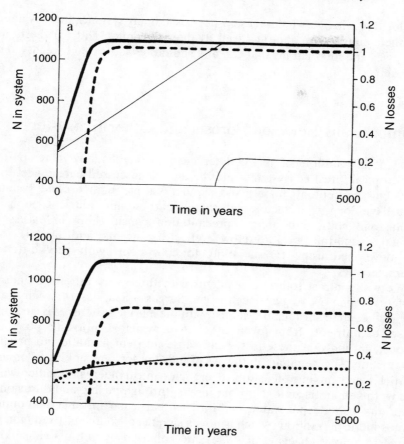

FIGURE 18.6. a. Effects of enhanced nutrient inputs on the plant plus soil nutrient pool and nutrient loss. The thinner lines represent the base case (Figure 18.2); the bold lines represent the effect of increasing inputs from 0.2 to 1.0 units of nutrient per year. The bold solid line represents the plant plus soil nutrient pool; the bold dashed line represents loss of dissolved inorganic nutrient. b. Effects of increased input when the dissolved organic loss pathway is important. Thinner lines represent the base case with an organic loss flux (Figure 18.4); the bolder lines represent the effect of increasing inputs from 0.2 to 1.0 units per year. The bold solid line represents the plant plus soil pool, the bold dashed line represents dissolved inorganic loss, and the bold dotted line represents dissolved organic loss.

Which of these situations best describes the world now? Which described it best in the not-so-distant past, before humans altered the N cycle so profoundly? We can only speculate about these questions—and we speculate that: (1) processes that can maintain nutrient limitation at equilibrium are widespread; (2) human alteration of the N cycle in particular has overwhelmed or will overwhelm these processes in many systems, thereby altering their equilibrium state, as well as their dynamics; and (3) these are processes and interactions that can and should be addressed by ecosystem research.

Comparisons Between Forests and Other Ecosystems

Most of the authors of this chapter have worked mainly in temperate and tropical forest ecosystems, and the models discussed here reflect that fact. Our conceptual models and simulations are built around systems that have long-lived plants, substantial carbon and nutrient pools in plants and soils, and annual precipitation considerably in excess of annual evapotranspiration. How well do concepts and models that are developed using forests apply to ecosystems with very different characteristics?

We will contrast forests ecosystems with three very different kinds of systems—open-ocean gyres such as the Sargasso Sea, desert streams, and dry woodland or shrubland systems. With the marine systems, it is clear that losses of nutrients from forms other than available inorganic pools are important, and that these losses can cause nutrient limitation to primary production. The dominant pathways of nutrient loss are sinking of organic particles and mixing of dissolved organics into colder and saltier water below the euphotic zone. Nutrient inputs, through mixing and diffusion of nutrients from deeper water, and for N in particular through fixation and atmospheric deposition (which is now a significant source of fixed N in the Sargasso Sea; Michaels et al. 1996), are balanced by outputs via sinking; nutrient supply remains limiting to primary production at this steady-state level. Nutrient inputs to surface water (in contrast to recycled nutrients) support "new production" in marine systems, in the same sense that nutrient inputs can support net ecosystem production in aggrading forest ecosystems within our simple model.

Desert streams are shaped by floods, and by the extremely rapid successional dynamics that follow. After flood-induced scouring, developing algal communities rapidly make use of streamwater nitrate, sequestering N in biomass and depleting nitrate in streams, in dynamics analogous to but hundreds of times faster than those described by the conventional model of aggrading forests (Fisher et al. 1982). However, streamwater nitrate concentrations may not increase substantially after the algal communities are fully established. Grimm and Fisher (1986) suggested that streamwater nitrate could remain low if algal nitrate uptake is balanced by losses of dissolved and particulate organic N from the algal mat. This mechanism appears to be analogous to DON losses from forests (Figure 18.4); whether organic losses can maintain nutrient limitation in addition to low streamwater nitrate concentrations is not known.

In comparison to mesic forests, dryland ecosystems are characterized by: (1) infrequent hydrologic losses of nutrients, especially via leaching through the soil; (2) a high degree of spatial heterogeneity in plant use of soil resources; (3) enormous temporal variation, with infrequent rains causing intensive bursts of biological activity; and (4) a lesser degree of biotic control over wind and water erosion. Erosion may represent an important pathway of nutrient loss from dryland systems, removing nutrients predominately from pools other than the available inorganic pool in soil. Alternatively, the spatial and temporal heterogeneity of these systems can uncouple nutrient release from nutrient uptake, allowing substantial losses of nutrients from the available inorganic pool in soil—even for limiting nutrients. This could be especially important for gaseous losses of N. Available nutrients in interplant spaces of dryland ecosystems may remain unutilized for long periods, and accumulate to relatively high concentrations. There is convincing evidence that available inorganic nutrient pools generally increase as one moves to progressively drier sites along moisture gradients (Hogberg 1986; Scholes 1993; Austin and Vitousek 1998). Episodic leaching or gaseous emissions could then remove these pools of available nutrients or redistribute them on the landscape. Within a site, available nutrients or readily mineralizable forms accumulate in soil during dry periods. When rains occur, they can be mobilized or volatilized more rapidly than plants can respond (Davidson et al. 1993). These dynamics differ substantially from those we believe to be important in mesic forest ecosystems.

The Frontier

We suggest that the controls of ecosystem-nutrient dynamics in the long term (i.e. beyond the time scale of Long Term Ecological Research, reaching into centuries and millenia) represent a potentially rewarding area of research in ecosystem ecology. Our analyses in this chapter are mainly heuristic models that allow an exploration of the implications of assumptions about ecosystems; we believe they are useful for thinking about ecosystems and for designing research. Additionally, new tools are becoming available for long-term analyses of the questions and mechanisms that are discussed here. These include minerological analyses of soils and a wide variety of isotopes. There are also complex as well as simple models that can be used as tools for exploration, integration, and analysis, and a growing body of paleoecological information that provides a more direct look at changes in ecosystems over time. Even though studies of long-term nutrient dynamics and their implications may seem esoteric, we believe that there is an excellent chance that they will contribute substantially to understanding, and even to predicting, contemporary human effects on ecosystem structure and functioning.

Acknowledgments. We thank J. Bengtsson, J. Cole, N.B. Grimm, P. Groffman, J. Hobbie, M. Pace, M. Schachak, and S. Tartowski for critical comments on a prior draft of the manuscript, D. Schimel for advice on implementing and interpreting Century, and B. Lilley and C. Nakashima for preparing figures and text for publication. This research was supported by The Andrew W. Mellon Foundation.

References

Aber, J.D. 1992. Nitrogen cycling and nitrogen saturation in temperate forest ecosystems. *Trends in Ecology and Evolution* 7:220–223.

Aber, J.D., K.J. Nadelhoffer, P. Steudler, and J.M. Melillo. 1989. Nitrogen saturation in northern forest ecosystems. *BioScience* 39:378–386.

Agren, G.I., and E. Bosatta. 1988. Nitrogen saturation of terrestrial ecosystems. *Environmental Pollution* 54:185–197.

Austin, A.T., and P.M. Vitousek. 1998. Nutrient dynamics on a precipitation gradient in Hawai'i. *Oecologia* 113:519–529.

Berendse, F., R. Aerts, and R. Bobbink. 1993. Atmospheric nitrogen deposition and its impact on terrestrial ecosystems. Pages 104–121 in C.C. Vos, and P. Opdam eds. *Landscape ecology of a stressed environment.* Chapman & Hall, London England.

Berendse, F. 1993. Ecosystem stability, competition, and nutrient cycling. Pages 409–431 in E.-D. Schulze, and H.A. Mooney eds. *Biodiversity and Ecosystem Function.* Springer-Verlag, Berlin, Germany. (and p. 435)

Bormann, F.H., and G.E. Likens. 1979. *Pattern and processes in a forested ecosystem.* Springer-Verlag, New York.

Cochran, M.F., and R.A. Berner. 1997. Promotion of chemical weathering by higher plants: field observations on Hawaiian basalts. *Chemical Geology* 132:71–85.

Crews, T.E., K. Kitayama, J. Fownes, D. Herbert, D. Mueller-Dombois, R.H. Riley, and P.M. Vitousek. 1995. Changes in soil phosphorus and ecosystem dynamics across a long soil chronosequence in Hawai'i. *Ecology* 76:1407–1424.

Currie, W.S., J.D. Aber, W.H. McDowell, R.D. Boone, and A.H. Magill. 1996. Vertical transport of dissolved organic C and N under long-term N amendments in pine and hardwood forests. *Biogeochemistry* 35:471–505.

Davidson, E.A., P.A. Matson, P.M. Vitousek, R. Riley, K. Dunkin, G. Garcia-Mendez, et al. 1993. Processes regulating soil emissions of NO and N_2O in a seasonally dry tropical forest. *Ecology* 74:130–139.

Firestone, M.K., and E.A. Davidson. 1989. Microbiological basis of NO and N_2O production and consumption in soil. Pages 7–21 in M.O. Andreae and D.S. Schimel, eds. *Exchange of trace gases between terrestrial ecosystems and the atmosphere.* John Wiley & Sons, London.

Fisher, S.G., L.J. Gray, N.B. Grimm, and D.E. Busch. 1982. Temporal succession in a desert stream following flash flooding. *Ecological Monographs* 52:93–110.

Galloway, J.N., W.C. Keene, and G.E. Likens. 1996. Processes controlling the composition of precipitation at a remote southern hemispheric location: Torres del Paine National Park, Chile. *Journal of Geophysical Research* 101:6883–6897.

Gorham, E., P.M. Vitousek, and W.A. Reiners. 1979. The regulation of element budgets over the course of terrestrial ecosystem succession. *Annual Review of Ecology and Systematics* 10:53–84.

Grimm, N.B., and S.G. Fisher. 1986. Nitrogen limitation in a Sonoran Desert stream. *Journal of the North American Benthological Society* 5:2–15.

Hedin, L.O., J.J. Armesto, and A.H. Johnson. 1995. Patterns of nutrient loss from unpolluted, old-growth temperate forests: evaluation of biogeochemical theory. *Ecology* 76:493–509.

Hedin, L.O., and E. Hetherington. 1996. Atmospheric and geologic constraints on the biogeochemistry of North and South American temperate rain forests. Pages 57–74 in R.G. Lawford, P.B. Alaback, and E. Fuentes, eds. *High-Latitude Forests and Associated Ecosystems on the West Coast of the Americas.* Springer-Verlag, New York.

Hedin, L.O., and G.E. Likens. 1996. Atmospheric dust and acid rain. *Scientific American* 275:88–92.

Herbert, D.A., and J.H. Fownes. 1995. Phosphorus limitation of forest leaf area and net primary productivity on a weathered tropical soil. *Biogeochemistry* 29:223–235.

Hogberg, P. 1986. Nitrogen fixation and nutrient relations in savanna woodland trees (Tanzania). *Journal of Applied Ecology* 23:675–688.

Kennedy, M.J., O.A. Chadwick, P.M. Vitousek, L.A. Derry, and D. Hendricks. Replacement of weathering with atmospheric sources of base cations during ecosystem development, Hawaiian Islands. *Geology*, in press.

Kielland, K. 1994. Amino acid adsorption by arctic plants: implications for plant nutrition and nitrogen cycling. *Ecology* 75:2373–2383.

Likens, G.E., and F.H. Bormann. 1995. *Biogeochemistry of a Forested Ecosystem, 2nd edition.* Springer-Verlag, New York.

Likens, G.E., W.C. Keene, J.M. Miller, and J.N. Galloway. 1987. Chemistry of precipitation from a remote, terrestrial site in Australia. *Journal of Geophysical Research* 92:13299–13314.

Lovett, G.M. 1994. Atmospheric deposition of nutrients and pollutants in North America: an ecological perspective. *Ecological Applications* 4:629–650.

Matson, P.A., and R.H. Waring. 1984. Effects of nutrient and light limitation on mountain hemlock: susceptibility to laminated root rot. *Ecology* 65:1517–1524.

McDowell, W.H., and C.E. Asbury. 1994. Export of carbon, nitrogen, and major ions from three tropical montane watersheds. *Limnology and Oceanography* 39:111–125.

McGill, W.B., and C.V. Cole. 1981. Comparative aspects of cycling of organic C, N, S, and P through soil organic matter. *Geoderma* 26:267–286.

Metherell, A.K., L.A. Harding, C.V. Cole, and W.J. Parton. 1993. CENTURY soil organic matter model environment: technical documentation. *Great Plains System Research Unit Technical Report 4*, USDA-ARS, Fort Collins, CO.

Michaels, A.F., D. Olson, J.L. Sarmiento, J.W. Ammerman, K. Fanning, R. Jahnke, et al. 1996. Inputs, losses, and transformations of nitrogen and phosphorus in the pelagic North Atlantic Ocean. *Biogeochemistry* 35:181–226.

Miller, H.G. 1981. Forest fertilization: some guiding concepts. *Forestry* 54:157–167.

Miller, H.G., J.M. Cooper, and J.D. Miller. 1976. Effect of nitrogen supply on nutrients in litterfall and crown leading in a stand of Corsican pine. *Journal of Applied Ecology* 13:233–248.

Newman, E.I. 1995. Phosphorus inputs to terrestrial ecosystems. *Journal of Ecology* 83:713–726.

Northrup, R.R., Z. Yu, R.A. Dahlgren, and K.A. Vogt. 1995. Polyphenol control of nitrogen release from pine litter. *Nature* 377:227–229.

Parton, W.J., A.R. Mosier, D.S. Ojima, D.W. Valentine, D.S. Schimel, K. Weier, et al. 1996. Generalized model for N_2 and N_2O production from nitrification and denitrification. *Global Biogeochemical Cycles* 10:401–412.

Parton, W.J., D.S. Schimel, C.V. Cole, and D.S. Ojima. 1987. Analysis of factors controlling soil organic matter levels in Great Plains grasslands. *Soil Science Society of America Journal* 51:1173–1179.

Parton, W.J., J.M.D. Scurlock, D.S. Ojima, T.G. Gilmanov, R.J. Scholes, D.S. Schimel, et al. 1993. Observations and modelling of biomass and soil organic matter dynamics for the grassland biome worldwide. *Global Biogeochemical Cycles* 7:785–809.

Pastor, J., R.H. Gardner, V.H. Dale, and W.M. Post. 1987. Successional changes in nitrogen availability as a potential factor contributing to spruce decline in boreal North America. *Canadian Journal of Forest Research* 17:1394–1400.

Pickett, S.T.A., and P.S. White. eds. 1985. *The ecology of natural disturbance and patch dynamics*. Academic Press, New York.

Redfield, A.C. 1958. The biological control of chemical factors in the environment. *American Scientist* 46:205–221.

Schlesinger, W.H., J.F. Reynolds, G.L. Cunningham, L.F. Huenneke, W.M. Jarrell, R.A. Virginia, et al. 1990. Biological feedbacks in global desertification. *Science* 247:1043–1048.

Scholes, R.J. 1993. Nutrient cycling in semi-arid grasslands and savannas: its influence on pattern, productivity and stability. *Proceedings of the XVII International Grassland Congress*, 1331–1334.

Sollins, P., and F.M. McCorison. 1981. Nitrogen and carbon solution chemisty of an old-growth coniferous forest watershed before and after cutting. *Water Resources Research* 17:1409–1418.

Stark, J.M., and S.C. Hart. 1997. High rates of nitrification and nitrate turnover in undisturbed coniferous forests. *Nature* 385:61–64.

Swap, R., M. Garstang, S. Greco, R. Talbot, and P. Kallbert. 1992. Saharan dust in the Amazon Basin. *Tellus* 44:133–149.

Tamm, C.O. 1991. *Nitrogen in terrestrial ecosystems*. Springer-Verlag, Berlin, Germany.

Vitousek, P.M. 1982. Nutrient cycling and nutrient use efficiency. *American Naturalist* 119:553–572.

Vitousek, P.M., O.A. Chadwick, T. Crews, J. Fownes, D. Hendricks, and D. Herbert. 1997 Soil and ecosystem development across the Hawaiian Islands. *GSA Today* 7(9):1–8

Vitousek, P.M., and H. Farrington. 1997. Nutrient limitation and soil development: experimental test of a biogeochemical theory. *Biogeochemistry* 37:63–75.

Vitousek, P.M., J.R. Gosz, C.C. Grier, J.M. Melillo, and W.A. Reiners. 1982. A comparative analysis of potential nitrification and nitrate mobility in forest ecosystems. *Ecological Monographs* 52:155–177.

Vitousek, P.M., and W.A. Reiners. 1975. Ecosystem succession and nutrient retention: a hypothesis. *BioScience* 25:376–381.

Walker, T.W., and J.K. Syers. 1976. The fate of phosphorus during pedogensis. *Geoderma* 15:1–19.

Wood, T., F.H. Bormann, and G.K. Voight. 1984. Phosphorus cycling in a northern hardwood forest: biological and chemical control. *Science* 223: 391–393.

Woodmansee, R.G. 1978. Additions and losses of nitrogen in grassland ecosystems. *BioScience* 28:488–453.

19
Species Composition, Species Diversity, and Ecosystem Processes: Understanding the Impacts of Global Change

David Tilman

Summary

Invasions by exotic species, climate change, increased atmospheric CO_2, deposition of N, habitat destruction and fragmentation, predator decimation, and many other anthropogenic perturbations to ecosystems can have both direct impacts on ecosystem processes and indirect effects mediated through changes in ecosystem composition and diversity. The work reviewed in this chapter suggests that the long-term effects of global change will depend on the changes in ecosystem composition and diversity that occur in response to global change. If this is so, the discipline of ecology faces its greatest challenge to date: to discover how to predict the effects of global change on ecosystem composition and diversity and how to predict the joint impacts of all these changes on a variety of ecosystem processes.

Introduction

Humans are having unprecedented impacts on the ecosystems of the earth. Atmospheric concentrations of such greenhouse gases as carbon dioxide and nitrogen dioxide are increasing, leading to predictions of rapid climate change for many parts of the globe (e.g. Raval and Ramanathan 1989; Schneider 1989, 1990, 1993; Sellers et al. 1996). Humans now dominate the global nitrogen (N) cycle via production of N fertilizer, cultivation of legumes as crops, and production of nitrogenous gases via fossil fuel combustion (e.g. Vitousek 1994; Galloway et al. 1995; Vitousek et al. 1997). Humans have already directly appropriated, and degraded, about 43% of the land surface of the Earth (Daily 1995), thus impacting the future utility of this land. This habitat modification, destruction, and fragmentation is leading to a marked increase in the rate of species extinction and to great changes in the species composition and diversity of many ecosystems (e.g. Prescott-Allen and Prescott-Allen 1978; Simberloff 1984;

Groom and Schumaker 1993; Tilman et al. 1994; Moilanen and Hanski 1995). Similarly, humans now use 54% of all accessible freshwater runoff and 26% of total terrestrial evapotranspiration (Postel et al. 1996). Human movements of species within and among major biogeographic realms are allowing exotic species to replace native species in many aquatic and terrestrial ecosystems (e.g. Elton 1958; Fox and Fox 1986; Drake et al. 1989; Vitousek 1990; D'Antonio and Vitousek 1992). Humans also release into the environment a wide array of biologically active organic compounds, including pesticides that may mimic endrocrine hormones (Colburn et al. 1996).

As Vitousek (1994) emphasized, we are certain that human society is causing increased atmospheric CO_2, dominating the global N cycle, appropriating land and water, fragmenting habitats, and causing massive invasions by exotic species. However, we are uncertain of the impacts of these actions. There is an increasingly urgent need to know what the long-term impacts of these human activities, loosely called global change, will be on the composition, structure, functioning, sustainability, and stability of Earth's aquatic and terrestrial ecosystems. The need to understand and accurately predict the impacts of various scenarios of global change, and to find ways to alleviate such impacts, is undoubtedly the greatest challenge facing the discipline of ecology, and one of the greatest needs of society. We will meet this challenge only if we make major advances in our knowledge of the fundamental processes that determine ecosystem structure, dynamics, and functioning.

It seems highly improbable that such understanding will come from simple extrapolations of correlational patterns observed in natural ecosystems. Correlation is not causation. Such extrapolations would have to assume that correlation is causation, that the basic processes structuring ecosystems would not be altered by the forces of global change, and that the responses of presently intact systems to small-scale natural variation in temperature, precipitation, nitrogen availability, and so forth, would also occur even after the major changes in ecosystem species composition that are apt to result from global change.

I suggest that we consider a different approach. I hypothesize that the largest effects of global change may be modifications of the biotic composition, biodiversity, and trophic structure of ecosystems, and that these changes will cause both qualitative and quantitative shifts in a variety of ecosystem processes. To predict these shifts will require knowledge of how the forces of global change influence ecosystem species composition and diversity, and how composition and diversity influence ecosystem processes. Thus, this view suggests that the impacts of particular global change scenarios will depend both on how these modify the structure and composition of an ecosystem, and on how the novel ecosystem consequently created functions. For instance, is it worthwhile to extrapolate from correlational data on the effects of normal climatic variation on a

Minnesota oak forest to predict how a 4 °C increase in temperature might affect it, especially if the long-term impact of such a temperature increase would be to replace the forest with grassland? And, would it be appropriate to use knowledge of native prairies to predict the behavior of grasslands created by such temperature increases if they have novel species compositions and lower species diversity than native prairie? It seems plausible that climate change would create novel grasslands because few native prairie species would be able to invade such a region given the slow rate of migration of such species and the difficulty of migration through the fragmented landscapes that now exist in much of North America. Indeed, the dominant native bunchgrasses of Minnesota prairies often require ten to twenty years to invade an abandoned field that is directly bordered by prairie (Inouye et al. 1987; Tilman 1990, 1994), and another twenty to forty years to attain dominance. How long would be required if they had to move ten to 100 km through an inhospitable and fragmented landscape?

I raise these questions because much work suggests that the species and functional group composition and diversity of ecosystems may be an important determinant of ecosystem processes, which are formally defined later in this chapter. The forces of human-caused global change directly impact ecosystem functioning by changing the physical parameters that an ecosystem experiences. However, larger, long-term impacts may be through change in species composition and diversity. In some cases, the resultant ecosystems may bear little resemblance to any that exist today. If the rates of ecosystem processes are highly dependent on species composition and diversity, then a major challenge facing us is to understand how composition and diversity will respond to global change, and how the resultant ecosystems will function.

Although ecosystem ecology, evolutionary ecology, population ecology, and community ecology have not developed in isolation from each other, many of the advances in these disciplines have been achieved by making a series of simplifying assumptions about the other three. One of the central assumptions of modern ecosystem ecology, which traces many of its roots to the pioneering work of Lindeman and the Odums, is that the functional composition of an ecosystem is a greater determinant of ecosystem processes than its species composition or diversity. This simplifying assumption began when initially quantifying the flow of energy and the cycling of materials in ecosystems. Both to make this work practical and to emphasize the roles of trophic structure, organisms were grouped by their trophic roles (e.g. primary producers, primary consumers, secondary consumers, decomposers, and so on), sampled, and analyzed as units. This approach has provided many significant insights into the functioning of ecosystems. Indeed, it seems difficult to imagine how an ecosystem could have been treated as an interactive system, complete with feedback, had this simplifying assumption not been made.

The past half-century has been an era of concomitant advances in population and community ecology, which have revealed many of the details of the interactions among species that determine the location of species and their abundance. Work in ecosystem ecology has shown that ecosystem processes often are controlled by a few key species, and that variation in species composition can cause large differences in these ecosystem processes. As important as this, the accelerating effects of humans on ecosystems via destruction, fragmentation, intensive ecosystem management, introduction of exotic species, predator decimation, and domination of nutrient cycling have led to greatly increased public and scientific concern. This brings to the forefront a series of questions that represent a new conceptual frontier for ecosystem ecology. The questions center on that which controls ecosystem species and functional group composition and biodiversity, and on how and why composition and diversity affect ecosystem processes.

The ability of the discipline of ecology to address these issues will influence our impact on environmental policy during a time when humans are having unprecedented, massive, and potentially irreversible effects on the ecosystems of the earth. To address these issues will require a series of fundamental conceptual, theoretical, observational, and experimental breakthroughs. The groundwork for these has been laid during the past two decades, but the intellectual challenge—the greatest our discipline has ever faced—lies before us.

Ecosystem Processes

There are many different ecosystem processes. Among the most fundamental are rates of primary and secondary production, the stores and rates of cycling of carbon (C), N, and other elements in an ecosystem, and the rates of gain and loss of these elements from an ecosystem. Additionally, the stability of each of these, including both its resistance to disturbance and its resilience after disturbance, are important (Pimm 1979, 1984; King and Pimm 1983). Collectively these will be called *ecosystem processes*. (They also are appropriately called "ecosystem functioning," a term that need not imply any notion of teleology.)

In this chapter, I will use a review of the literature and conceptual speculation to explore three related questions: (1) What impacts may global change have on the composition and diversity of ecosystems? (2) Do changes in composition influence ecosystem processes? and, (3) Do changes in species diversity impact ecosystem processes, and why? Although there are not yet answers to these questions, I hope to show that they are critically important questions that are answerable if we dedicate sufficient intellect and effort.

Global Change and the Composition and Diversity of Ecosystems

Although there are uncertainties about the rate and magnitude of climate change caused by the assimilation of greenhouse gases, most global circulation models (GCMs) predict climatic changes that are of greater magnitude than has occurred naturally within the past several thousand years (e.g. Schneider 1993). There is an immense literature speculating on the possible impacts of such changes in climate on species abundance, geographic distributions, and biodiversity (e.g. Webb and Wigley 1985; Davis 1986; Cronin and Schneider 1990; Huntley 1991; Schwartz 1991; Peters and Lovejoy 1992; Chapin et al. 1995). The climatic changes predicted by most GCMs are sufficiently great that large changes in species composition are expected. However, because of limitations on rates of species migrations (e.g. Schwartz 1992) and slow changes in soil C and N, it seems improbable that there will be a simple mapping of the present relationships between climate and composition. Many species that are currently dominant in areas with a given climate and soil may not remain dominant after certain climate changes and may not become the dominants of new areas that approach the physical conditions of their former natural habitats. Rather, other species— probably those that are presently rare but happen to be present in a region and have traits similar to those that would dominate were dispersal not limiting—may become the new dominants of a region. Well-dispersed, weedy species should also increase in abundance (e.g. Nee and May 1992; Tilman et al. 1994). The number of species in these new ecosystems in apt to be much lower because of dispersal limitation. Thus, if the future were to be painted with bold broad brush strokes, the Picasso-like scene would be one of domination by many fewer species than at present, with many habitats dominated by species, perhaps weedy species, that presently may not be dominant in any habitat.

This same qualitative scenario of changes in ecosystem composition and diversity may occur because of global impacts on the N cycle. Human-caused doubling of global N fixation is causing great increases in N deposition in many areas of the world (e.g. Vitousek 1994; Galloway et al. 1995). Numerous fertilization experiments suggest that increased rates of N deposition will lead to the loss of many native species and to dominance by once-rare species, often non-natives. Similarly, habitat fragmentation also leads to the biased extinction of species (e.g. Diamond 1972, 1973; Terborgh 1974; Lovejoy et al. 1984; Laurance 1991; Tilman et al. 1994; Steadman 1995). In general, top carnivores, rare species, and poor dispersers are preferentially lost following habitat fragmentation and destruction.

The loss of species and changes in species, functional groups, and trophic composition may change a variety of ecosystem processes. Such effects are illustrated by the following three case studies. These are a biased subset of a much larger number of possible studies, chosen because of my greater familiarity with them.

Case 1. Impacts of Nitrogen Addition

Nitrogen addition can cause dramatic shifts in plant composition and declines in plant diversity (e.g. Lawes and Gilbert 1880; Thurston 1969; Silvertown 1980; Bobbink et al. 1987; Tilman 1987, 1996; Aerts and Berendse 1988; Huenneke et al. 1990). A long-term experimental study in Minnesota illustrates several of the impacts of high rates of N deposition (Tilman 1987, 1988, 1994, 1996; Tilman and Downing 1994; Wedin and Tilman 1996). Nitrogen is the major soil resource limiting growth of plant species and determining primary productivity (Tilman 1987). In 1982, a series of 207 plots were established to which various amounts of ammonium nitrate were added. The experiment, replicated in four different fields, had nine different nutrient treatments including controls that received no nutrients, plots that received phosphorus (P), potassium (K), calcium (Ca), magnesium (Mg), sulfur (S), and trace metals but no N, and plots that received P, K, Ca, Mg, and trace metals and one of seven different rates of N addition. The rates of N addition were 0, 1, 2, 3.4, 5.4, 9.5, 17, and $27 \, \text{g m}^{-2} \text{yr}^{-1}$ of N added in two equal applications. Present rates of N deposition (wet and dry fall) in the region are about $0.8 \, \text{g m}^{-2} \text{yr}^{-1}$, whereas rates in the Ohio valley are about 2 to $4 \, \text{g m}^{-2} \text{yr}^{-1}$, and rates in parts of Europe and 10 to $15 \, \text{g m}^{-2} \text{yr}^{-1}$. The rates of experimental addition span these, and include some higher rates that may give results indicative of long-term effects of deposition at lower rates.

Composition and Diversity

This experiment has shown that N addition led to major changes in the plant species composition (Figure 19.1), plant functional group composition, insect species composition, and mycorrhizal fungal species composition (Johnson 1993) of these grasslands. In particular, higher rates of N addition led to the loss of native plant species, especially C-4 grasses, and to greatly increased abundances of non-native plant species, especially C-3 grasses. At the highest rates of N addition, all plots converged on dominance by *Agropyron repens*, a European perennial C-3 grass and a dominant agricultural weed. Plant species diversity declined with increasing rates of N addition, with a detectable decline in plant species richness even for addition of $1 \, \text{g m}^{-2} \text{yr}^{-1}$, and with species richness declining to 35% of its original value at the highest rate of N addition (Figure 19.2).

458   D. Tilman

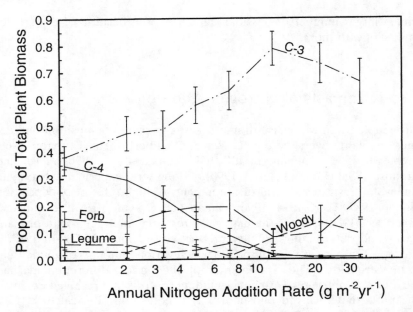

FIGURE 19.1. Results from the Cedar Creek N addition experiments show the effects of different rates of N addition on relative abundance of major plant functional groups, with C-3 being non-native cool-season grasses, C-4 being native warm-season grasses, legume being N-fixing species, forb being all nonlegume broad-leaved herbaceous species, and woody being all plants with perennial aboveground stems. Figure modified from Tilman (1996). Results are for 1994, the thirteenth year of N addition to these four fields.

Composition and Nitrogen Dynamics

Ecosystem processes in this grassland were impacted by these changes in species composition and diversity. In particular, the shifts in species composition caused by N addition led to sharp declines in litter and root C/N ratios (Wedin and Tilman 1996). Because high C/N ratios tend to cause slow rates of decomposition and net N immobilization, and low C/N ratios tend to cause high rates of decomposition and net mineralization of nitrogen, the shifts in plant species composition caused by N addition could greatly impact N cycling in these grasslands. Wedin and Tilman (1996) found just such impacts. Soil nitrate (NO_3) concentrations increased by over an order of magnitude and in situ N mineralization rates increased four- to six-fold. The change in soil NO_3, however, was highly dependent on the C/N ratio of above- and belowground biomass (Wedin and Tilman 1996; Figure 19.3a). Only plots with low C/N ratios in plant biomass had high concentrations of soil nitrate. These were the plots that either initially were dominated by C-3 grasses or that became dominated by C-3 grasses after N addition. Be-

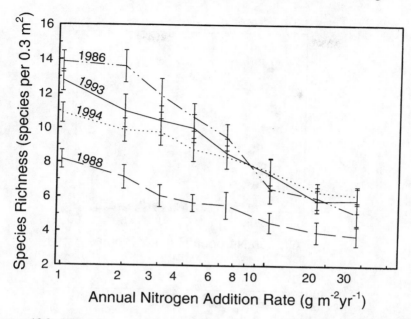

FIGURE 19.2. Dependence of plant species richness (number of plant species per 0.3 m² clipped strip) on the rate of N addition and on year for the Cedar Creek N addition plots. Figure modified from Tilman (1996).

cause of the high motility of NO_3 through the soil, it might be expected that plots with high concentrations of soil NO_3 would also lose more of their total N. This occurred (Figure 19.3b; Wedin and Tilman 1996) and consequently, much of the N added at high rates was lost from these plots. This loss of N occurred whenever the rates of N addition were sufficient to cause the replacement of C-4 grasses by C-3 grasses, which occurred within about twelve years for plots receiving about $5\,g\,m^{-2}yr^{-1}$ of N.

Composition and Carbon Storage

Changes in plant species composition caused by N addition also had a significant impact on C storage in these ecosystems. It has been hypothesized that atmospheric N deposition may increase plant growth, and thus terrestrial C storage, by a sufficient amount to explain the "missing C" issue (Peterson and Melillo 1985; Schindler and Bayley 1993; Hudson et al. 1994; Townsend et al. 1996). It was found that net C storage in the plots depended on plant species composition. In areas initially dominated by C-4 grasses, about 20 to 25 g of C was stored for every g of N added, when N was added at a sufficiently low rate to minimize shifts in species

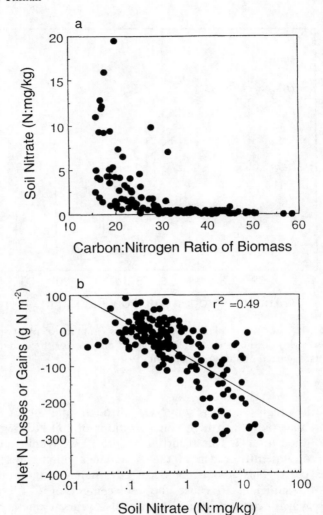

FIGURE 19.3. a. Dependence of soil nitrate on plant biomass C/N ratio in the Cedar Creek N fertilization plots of the three successional grassland fields. b. Net N loss or gain in these plots versus amount of soil nitrate. Figures modified from Wedin and Tilman (1996).

composition. However, there was no net C storage at these rates of N addition in areas initially dominated by C-3 grasses. At higher rates of N addition, the rates of C storage converged in both types of areas on a low rate of C storage (Figure 19.4; Wedin and Tilman 1996). This corresponded to having their species composition converge on dominance by non-native C-3 grasses.

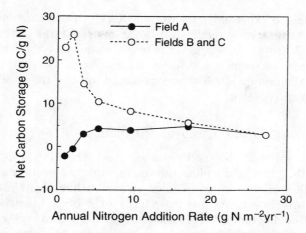

FIGURE 19.4. Effect of the rate of N addition in the Cedar Creek N fertilization plots on the net amount of carbon stored per unit of N added. Note that the fields converged on similar storage when their compositions converged on dominance by C-3 grasses at high rates of N addition. Figure modified from Wedin and Tilman (1996).

Composition, Diversity, and Stability

These same plots have provided insights into the relationships between biodiversity, species and functional group composition, and stability. In 1987 and 1988, this region experienced the third-worst drought of its 150-year recorded meteorological history. On average, this drought caused peak aboveground living standing crop (henceforth, *total plant biomass*) to decrease to less than half of its former level. However, the actual magnitude of the change in total plant biomass depended on the plant species richness of the plots. On average, plots containing one or two plant species had a decrease in total plant biomass about one-fifth to one-tenth of their predrought levels, whereas plots with from fifteen to twenty-six species fell to about one-half of their predrought levels. This dependence of drought resistance on plant species diversity remained highly significant even when a large number of potentially confounding variables were statistically controlled in a series of multiple regressions (Tilman and Downing 1994; Tilman 1996). Interestingly, of the twenty potentially confounding variables entered into a multiple regression, those that were retained as significant in a backward elimination procedure were mainly measures of changes in plant species composition or plant functional group composition. The rate of N addition, which caused plant composition to change, was not retained, but the biomasses of *Agropyron repens*, the dominant C-3 grass of plots receiving high rates of N addition, of *Schizachyrium scoparium*, the dominant C-4 grass of control plots, and of forbs, legumes, and woody plants were all retained. This indicates that changes in their abundance better explained changes in drought resistance than did the actual rate of N addition. In total, this suggests that the stability of this ecosystem is dependent both on its species richness and on its composition, and that the effect of N addition on stability was mediated through the effects of N on composition and diversity.

Composition, Diversity, and Variability

A similar pattern was observed when year-to-year variation in total plant biomass within individual plots during nondrought years was analyzed (Tilman 1996). Higher levels of plant species richness within a plot was associated with lower year-to-year variation in plant biomass, and this dependence held in multiple regression analyses in which a series of potentially confounding variables were controlled. Nine of eighteen candidate variables were retained in backward elimination multiple regressions as significant, with five of these being measures of plant compositional differences among plots (biomasses of *Agropyron repens*, *Schizachyrium scoparium*, *Poa pratensis*, C-3 grasses, and legumes). As before, both diversity and composition were significant controllers of stability in these plots, but the rate of N addition was not.

Case 2. Effects of Species Composition on Nutrient Cycling

The rate of nutrient mineralization, which is an important determinant of ecosystem productivity, depends on the species composition of an ecosystem. For instance, Pastor et al. (1984) found that never-logged forest stands on Blackhawk Island in the Wisconsin River, differed in their in situ rates of N mineralization in ways that corresponded with their species composition. The highest rates of N mineralization occurred in a stand dominated by sugar maple; the lowest in a stand dominated by red pine. Annual above ground production increased with the rate of N mineralization. These differences in N mineralization corresponded with differences in the quality of the litter produced by the tree species that dominated these eight stands, suggesting that species composition controlled N mineralization. A direct test of this possibility was provided by a well-replicated experiment in which five grass species were planted into initially identical soils (Wedin and Tilman 1990). In situ rates of N mineralization diverged through time, with marked differences corresponding to interspecific differences in litter quality (Wedin and Pastor 1993; Wedin et al. 1995). Similarly, the invasion of Hawaii by an exotic legume greatly changed N dynamics (Vitousek and Walker 1989; Walker and Vitousek 1991), as have other plant invasions

(Vitousek 1990). In total, changes in species and functional group composition are apt to impact the rates of supply of limiting nutrients, which, in turn, could feedback on composition.

Case 3. Trophic Structure

Ecosystems differ in their trophic structure, which can be measured as the proportion of the total living biomass that occurs on each trophic level. In both terrestrial and aquatic ecosystems, this can be highly dependent on the presence or absence of one or a few predator species (e.g. Paine 1969; Oksanen et al. 1981; Carpenter and Kitchell 1988, 1993; Leibold 1989; Hairston and Hairston 1993). For instance, at the time of Neolithic settlement, most of the British Isles was covered by dense, productive forests, but much of this was converted within Neolithic times into grassland (Tansley 1949). Similar shifts occurred in Europe (Ellenberg 1988). Tansley, Ellenberg, and others contend that this dramatic shift was caused more by large mammal grazing and by predator decimation by Neolithic people, than by direct clearing of forests. The high densities of goats, sheep, cattle, and other herbivores resulting from human pastoral practices increased seed and seedling mortality on trees and slowed tree recruitment, causing grasslands to replace forests. Predator decimation is a hallmark of human expansion, and is effecting trophic structure in other areas, as in portions of the United States where geese and deer densities are unusually high. Similarly, the presence or absence of some species of predatory fish can greatly impact the trophic structure of lakes (Carpenter and Kitchell 1988, 1993). For instance, in lakes in which large *Daphnia* are the major herbivore on planktonic algae, the absence of a top predator can shift a lake toward high algal abundances, whereas the presence of a top predator can greatly decrease algal biomass. Early work by Darwin (1859), in addition to work by Paine (1966, 1988) in the intertidal have also clearly demonstrated the effects of predators on the trophic structure of food webs.

Changes in trophic structure could impact many ecosystem processes. Consider the impacts of deer on the savannas of Cedar Creek Natural History Area in Minnesota. For a variety of reasons, deer populations have increased greatly at Cedar Creek and in its vicinity since settlement. In 1982, a series of deer exclosures was established at Cedar Creek (Inouye et al. 1994). These exclosures have shown some surprising effects of deer herbivory on primary productivity and soil total N (Ritchie and Tilman 1995; Ritchie et al. 1997; Knops et al. unpublished manuscript). Inside of deer exclosures, peak plant standing crop of oak savanna has increased during the past fifteen years, whereas it has remained about constant outside the deer exclosures (Figure 19.5). This increase in plant biomass corresponds with an increase in total soil N in the plots from which deer were excluded. Why, though, did deer exclosure cause soil total N to increase?

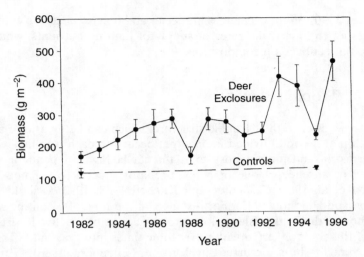

FIGURE 19.5. Effect of deer browsing (broken line) and of deer exclosure (solid line) on total plant biomass of otherwise unmanipulated plots in a field of native oak savanna. High biomass of plots without deer is caused by legumes, especially *Lathyrus*, which became much more abundant when deer were excluded. Figure modified from Knops et al. (in review).

The cause was a dramatic shift in plant community composition resulting from reduced deer herbivory. Deer feed preferentially on the fast-growing legume *Lathyrus venosus*, which has high tissue N levels. After twelve years, the deer exclosures had caused *Lathyrus* abundance to increase six- to eight-fold. Especially in years with cool, moist springs, *Lathyrus venosus* dominated the deer exclosures. By late summer, when the peak standing crop was measured, *Lathyrus* often had died and entered the decomposition loop. The long-term net effect of deer exclosure was the increased primary productivity caused by the increased rates of N fixation associated with a change in plant species composition.

Separating the Effects of Composition and Diversity

The following are but a few of the numerous examples that illustrate that changes in both species composition and diversity that result from global change have the potential to have major impacts on ecosystem processes. These changes in composition and diversity may be a major cause of shifts in ecosystem processes. On one level, it seems impossible for species composition not to matter for ecosystem processes. After all, ecosystem processes are driven by the lives, deaths, and interactions of organisms. One of the tenets of population ecology is that no two species are identical, and that it is the differences among species that allow coexistence in multispecies ecosystems. However, does this imply that ecosystem processes must depend on both the identity and abundance of every species in an ecosystem? Surely not. There may be many species of viruses, bacteria, fungi, vascular plants, insects, and so forth, that are sufficiently similar to other members of their group that they could be substituted for each other with no discernible impact on ecosystem processes (Lawton and Brown 1993; Vitousek and Hooper 1993). Additionally, most species are rare, and, with the exception of keystone species (e.g. Paine 1966; Power et al. 1996), the loss or addition of most rare species may have little detectable effect on ecosystem processes. Those species that should matter, and the reason they should matter for ecosystem processes, will depend on the ecosystem process of interest.

Although it has long been clear that species composition influences ecosystem processes, it has been less clear whether ecosystem processes also would depend on biodiversity, and how the effects of composition can be separated from those of biodiversity. Recent papers suggest that ecosystem processes may depend on biodiversity (Ewel et al. 1991; McNaughton 1993; Vitousek and Hooper 1993; Naeem et al. 1994, 1995; Tilman et al. 1996). Specifically, primary productivity or plant standing crop became larger with increasing diversity, as did nutrient use, and nutrient leaching losses decreased with increasing diversity in these studies.

To experimentally determine the effects of species diversity on ecosystem processes, it is necessary to ensure that all species are represented in an unbiased manner at all levels of diversity. This requires that there be a defined species pool, and that various species combinations be drawn from this pool. For a small species pool, it might be possible to choose all possible combinations of species taken one, two, three, four, and so forth, at a time. This would mean that a given species will compose the same proportion of the total species mix used at each level of diversity. When this occurs, tests for the effects of diversity on ecosystem processes merely involve comparing average responses among the different levels of diversity. This design can be expanded to determine the effects of species or functional group composition by having replicate plots for each unique combination of species or functional groups. Analysis of variation (ANOVA) or regression then could be used to separate the effects of diversity and composition. This same design can be used in a slightly modified form when the species pool is too large to allow use of all possible species combinations. In this case, the species compositions of plots with a given level of diversity would be determined by random draws of that number of species from the species pool, with each plot being a separate random draw. To minimize effects caused by species compositional differences among various sets of random draws, it is necessary to have a relatively large number of plots, for example twenty to forty, at each level of diversity. Even within this design it is possible to distinguish between effects of diversity and composition by having replicates of each particular randomly drawn species combination. By having many random species combinations, the mean response among replicate ecosystems at a given level of diversity becomes independent of particular species combinations, and differences among means measure the effect of diversity. The variance among the various species combinations at a given diversity level measures the effects of alternative species compositions. This is the design used in the Cedar Creek biodiversity experiments (Tilman et al. 1996), which have shown that both functional composition and functional diversity are significant determinants of plant community biomass and other ecosystem processes (Tilman et al. 1997).

It is also possible to impose this experimental design on various models of interspecific interactions to determine how these interactions depend on diversity (Tilman et al. 1997). The resulting theories of the effects of biodiversity on ecosystem processes predict that plant standing crop should be an increasing function of diversity, and that nutrient leaching loss should decrease with diversity. The predicted effects of diversity are the logical outcome of a few simple processes and assumptions. It is assumed that species differ in their abilities to compete for one or more limiting resources. In the simplest case, in which all species compete for the same limiting nutrient, increased diversity leads to increased productivity simply because better, more productive species are more apt to be present at higher levels of diversity. The highest productivity occurs when the single best species is present, but the probability of this occurring increases with diversity. Thus, this effect is a probabilistic effect that only requires that species compete and that they differ in productivity. When species compete for two limiting resources, or when there are two or more niche axes along which species are differentiated, increased diversity leads to increased productivity because of species complementarities, that is, on average, increased diversity leads to a fuller coverage of niche space, allowing a more complete utilization of resources and, therefore, higher productivity. At any given level of diversity, different combinations of species can lead to different productivities or levels of nutrient utilization and loss. The differences among these different species combinations are a measure of the effects of species composition, whereas the differences in mean responses among different levels of diversity measure the effect of diversity. Within the three models in Tilman et al. (1997), of the total variance in productivity, from about one-third to two-thirds was explained by diversity, with the remainder explained by composition. Thus, even in these simple models, both composition and diversity are major determinants of ecosystem processes.

The Challenge

These examples illustrate the challenge before us. There is not yet adequate understanding of what controls species diversity, species composition, or trophic structure, nor how these influence ecosystem processes and vice versa, and yet such understanding is essential if the impacts of present-day and future alternative trajectories of global change are to be predicted.

In other words, we do not yet know that which explains the broad patterns of the natural world. What allows thirty species of phytoplankton to coexist in a few liters of lake water, or 300 species of tropical trees to coexist in a hectare, or 4,500 species of insects to coexist in a few square kilometers at Cedar Creek? What causes latitudinal gradients in diversity? What controls C, N, P and other element cycles on local, continental, and global scales? What causes changes in composition, diversity, life-history, and life-forms along continental gradients in productivity, elevation, or disturbance? What causes the worldwide convergence of traits within a given type of biome?

If a step is taken back from the Earth and the broad, general, repeatable patterns that occur are viewed, most of these have not yet been rigorously described or explained (Lawton 1996). The problem is not that we lack explanations for these, it is that we have too many explanations, and these are conceptual outlines rather than rigorous predictive theory. What is lacking are rigorous tests tied to mechanistic theory that allow us to distinguish among competing hypotheses and establish workable theory. If the actual forces structuring natural ecosystems are not understood, how are we to understand how these ecosystems will change in response to the forces of global change? We need to develop and test simple, mechanistic theories that can explain what we see, and then use these as our tools to predict that which various futures may hold.

This is not just an academic goal. If we fail to achieve this level of knowledge, we will have missed the most important contribution that our discipline could have given to society. We live in an era of unprecedented human impacts of the Earth's ecosystems. Ecology must develop into a predictive science if we are to realistically and honestly inform society of the future consequences of the present patterns of human environmental behavior. The more rigorous and well-tested the conceptual basis for ecology becomes, the greater will be our ability to predict the alternative futures that lie before the Earth and how they depend on the actions that society takes.

How can we achieve as ambitious a goal as this? In addition to using existing resources at an optimum level, we must articulate society's need for major new research programs directed at a predictive understanding of the elements of global change. Such programs should attract and train the next generation of ecologists, a generation dedicated to fundamental advances in ecological understanding and the application of this knowledge to the environmental issues that face the globe. The long-term viability and sustainability of our society depends on advances in ecological knowledge as fundamental as those that led to the dawn of the nuclear age. These can be achieved only by appropriate investments.

References

Aerts, R., and F. Berendse. 1988. The effect of increased nutrient availability on vegetation dynamics in wet healthlands. *Vegetatio* 76:63–69.

Bobbink, R., H.J. During, J. Schreurs, J. Willems, and R. Zielman. 1987. Effects of selective clipping and mowing time on species diversity in chalk grassland. *Folia Geobotanica et Phytotaxonomica* 22:363–376.

Carpenter, S.R., and J.F. Kitchell. 1988. Consumer control of lake productivity. *BioScience* 38:764–769.

Carpenter, S.R., and J.F. Kitchell. eds. 1993. *The trophic cascade in lakes.* Cambridge University Press, Cambridge, UK.

Chapin, III, F.S., G.R. Shaver, A.E. Giblin, K.J. Nadelhoffer, and J.A. Laundre. 1995. Responses of arctic tundra to experimental and observed changes in climate. *Ecology* 76:694–711.

Cronin, T.M., and C.E. Schneider. 1990. Climatic influences on species: evidence from the fossil record. *Trends in Ecology and Evolution* 5(9):275–279.

Daily, G.C. 1995. Restoring value to the world's degraded lands. *Science* 269:350–354.

D'Antonio, C.M., and P.M. Vitousek. 1992. Biological invasions by exotic grasses, the grass/fire cycle, and global change. *Annual Review of Ecology and Systematics* 23:63–87.

Darwin, C. 1859. *The origin of species by means of natural selection.* Murray, London.

Davis, M.B. 1986. Climatic instability, time lags, and community disequilibrium. Pages 169–284 in J. Diamond, and T. Case, eds., *Community ecology.* Harper and Row, Inc., New York.

Diamond, J.M. 1972. Biogeographic kinetics: estimation of relaxation times for avifaunas of Southwest Pacific Islands. *Proceedings of the National Academy of Science* 69:3199–3203.

Diamond, J.M. 1973. Distributional ecology of New Guinea birds. *Science* 179:759–769.

Drake, J.A., H.A. Mooney, F. di Castri, R.H. Groves, F.J. Kruger, M. Rejmánek and M. Williamson, editors. 1989. *Biological invasions. A global perspective.* Wiley, New York.

Ellenberg, H. 1988. *Vegetation ecology of central europe, 4th edition.* Cambridge University Press, Cambridge, UK.

Elton, C.S. 1958. *The ecology of invasions by animals and plants.* Methuen & Co. Ltd., London, England.

Ewel, J.J., M.J. Mazzarino, and C.W. Berish. 1991. Tropical soil fertility changes under monocultures and successional communities of different structure. *Ecological Applications* 1:289–302.

Fox, M.D., and B.J. Fox. 1986. The susceptibility of natural communities to invasion. Pages 57–66 in R.H. Groves, and J.J. Burdon, eds. *Ecology of biological invasions: an Australian perspective.* Australian Academy of Science, Canberra.

Galloway, J.N., W.H. Schlesinger, H. Levy II, A. Michaels, and J.L. Schnoor. 1995. Nitrogen fixation: atmospheric enhancement—environmental response. *Global Biogeochemical Cycles* 9:235–252.

Groom, M.J., and N. Schumaker. 1993. Evaluating landscape change: patterns of worldwide deforestation and local fragmentation. Pages 24–44 in P.M. Kareiva,

J.G. Kingsolver, and R.B. Huey, eds. *Biotic interactions and global change.* Sinauer Associates Inc., Sunderland, MA.

Hairston, Jr. N.G., and N.G. Hairston. 1993. Cause-effect relationships in energy flow, trophic structure, and interspecific interactions. *American Naturalist* 142: 379–411.

Hudson, R.J., S.A. Gherini, and R.A. Glodstein. 1994. Modelling the global carbon cycle: nitrogen fertilization of the terrestrial biosphere and the "missing" CO_2 sink. *Global Biogeochemical Cycles* 8:307–333.

Huenneke, L.F., S.P. Hamburg, R. Koide, H.A. Mooney, and P.M. Vitousek. 1990. Effects of soil resources on plant invasion and community structure in California serpentine grassland. *Ecology* 71:478–491.

Huntley, B. 1991. How plants respond to climate change: migration rates, individualism and the consequences for plant communities. *Annals of Botany 67* (Supplement 1):15–22.

Inouye, R.S., T.D. Allison, and N. Johnson. 1994. Old-field succession on a Minnesota sand plain: effects of deer and other factors on invasion by trees. *Bulletin of the Torrey Botanical Club* 121(3):266–276.

Inouye, R.S., N.J. Huntly, D. Tilman, J.R. Tester, M.A. Stillwell, and K.C. Zinnel. 1987. Old field succession on a Minnesota sand plain. *Ecology* 68:12–26.

Johnson, N.C. 1993. Can fertilization of soil select less mutualistic mycorrhizae? *Ecological Applications* 3:749–757.

King, A.W., and S.L. Pimm. 1983. Complexity, diversity, and stability: a reconciliation of theoretical and empirical results. *American Naturalist* 122:229–239.

Laurance, W.F. 1991. Ecological correlates of extinction proneness in Australian tropical rain forest mammals. *Conservation Biology* 5:79–89.

Lawes, J.B., and J.H. Gilbert. 1880. Agricultural, botanical and chemical results of experiments on the mixed herbage of permanent grassland, conducted for more than twenty years in succession on the same land. Part I. The agricultural results. *Philosophical Transactions of the Royal Society* (B) 192:139–210.

Lawton, J.H. 1996. Patterns in ecology. *Oikos* 75:145–147.

Lawton, J.H. and V.K. Brown. 1993. Redundancy in ecosystems. Pages 255–270 in E.-D. Schulze, and H.A. Mooney, eds. *Biodiversity and ecosystem function.* Springer-Verlag, Berlin, Germany.

Leibold, M.A. 1989. Resource edibility and the effects of predators and productivity on the outcome of trophic interactions. *American Naturalist* 134:922–949.

Lovejoy, T.E., J.M. Rankin, R.O. Bierregaard Jr., K.S. Brown Jr., L.H. Emmons, and M.E. van der Voort. 1984. Ecosystem decay of Amazon forest remnants. Pages 295–325 in M.H. Nitecki, ed. *Extinctions.* University of Chicago Press, Chicago, Il.

McNaughton, S.J. 1993. Biodiversity and function of grazing ecosystems. Pages 361–383 in E.-D. Schulze, and H.A. Mooney, eds. *Biodiversity and ecosystem function.* Springer-Verlag, Berlin, Germany.

Moilanen, A., and I. Hanski. 1995. Habitat destruction and the coexistence of competitors in a spatially realistic metapopulation model. *Journal of Animal Ecology* 64:141–144.

Naeem, S., L.J. Thompson, S.P. Lawler, J.H. Lawton, and R.M. Woodfin. 1994. Declining biodiversity can alter the performance of ecosystems. *Nature* 368:734–737.

Naeem, S., L.J. Thompson, S.P. Lawler, J.H. Lawton, and R.M. Woodfin. 1995. Empirical evidence that declining species diversity may alter the performance of terrestrial ecosystems. *Philosophical Transactions of the Royal Society of London B* 347:249–262.

Nee, S., and R.M. May. 1992. Dynamics of metapopulations: habitat destruction and competitive coexistence. *Journal of Animal Ecology* 61:37–40.

Oksanen, L., S.D. Fretwell, J. Arrud, and P. Niemela. 1981. Exploitation ecosystems in gradients of primary productivity. *American Naturalist* 118:240–261.

Paine, R.T. 1996. Food web complexity and species diversity. *American Naturalist* 100:65–75.

Paine, R.T. 1969. A note on trophic complexity and community stability. *American Naturalist* 103:91–93.

Paine, R.T. 1988. Food webs: road maps of interactions or grist for theoretical development? *Ecology* 69:1648–1654.

Pastor, J., J.D. Aber, C.A. McClaugherty, and J.M. Melillo. 1984. Aboveground production and N and P cycling along a nitrogen mineralization gradient on Blackhawk Island, Wisconsin. *Ecology* 65:256–268.

Peters, R.L., and T.E. Lovejoy. 1992. *Global warming and biological diversity*. Yale University Press, New Haven, CT.

Peterson, B.J. and J.M. Melillo. 1985. The potential storage of carbon caused by eutrophication of the biosphere. *Tellus* 37B:117–127.

Pimm, S.L. 1979. Complexity and stability: another look at MacArthur's original hypothesis. *Oikos* 33:351–357.

Pimm, S.L. 1984. The complexity and stability of ecosystems. *Nature* 307:321–326.

Postel, S.L., G.C. Daily, and P.R. Ehrlich. 1996. Human appropriation of renewable fresh water. *Science* 271:785–788.

Power, M.E., D. Tilman, J. Estes, B.A. Menge, W.J. Bond, L.S. Mills, et al. 1996. Challenges in the quest for keystones. *BioScience* 46:609–620.

Prescott-Allen, R., and C. Prescott-Allen. 1978. *Sourcebook for a world conservation strategy: threatened vertebrates*. International Union for Conservation of Nature and Natural Resources, Gland, Switzerland.

Raval, A., and V. Ramanathan. 1989. Observational determination of the greenhouse effect. *Nature* 342:758.

Ritchie, M.E., and D. Tilman. 1995. Responses of legumes to herbivores and nutrients during succession on a nitrogen-poor soil. *Ecology* 76:2648–2655.

Ritchie, M.E., D. Tilman, and J. Knops. 1997. Herbivore effects on plant and nitrogen dynamic in oak savanna. *Ecology*, 79:165–177.

Schindler, D.W., and S.E. Bayley. 1993. The biosphere as an increasing sink for atmospheric carbon: estimates from increasing nitrogen deposition. *Global Biogeochemical Cycles* 7:717–734.

Schneider, S. 1989. The changing global climate. *Scientific American* 261:38–47.

Schneider, S.H. 1990. *Global warming: are we entering the greenhouse century?* Vintage Books, New York.

Schnieder, S.H. 1993. Scenarios of global warming. Pages 9–23 in P.M. Kareiva, J.G. Kingsolver, and R.B. Huey, eds. *Biotic interactions and global change*. Sinauer Associates Inc., sunderland. MA.

Schwartz, M.W. 1991. Potential effects of global climate change on the biodiversity of plants. *The Forestry Chronicle* 68:462–471.

Schwartz, M.W. 1992. Modeling effects of habitat fragmentation on the ability of trees to respond to climatic warming. *Biodiversity and Conservation* 2:51–61.

Sellers, P.J., L. Bounoua, G.J. Collatz, D.A. Randall, D.A. Dazlich, S.O. Los, et al. 1996. Comparison of radiative and physiological effects of doubled atmospheric CO_2 on climate. *Science* 271:1402–1406.

Silvertown, J.W. 1980. The dynamics of a grassland ecosystem: botanical equilibrium in the Park Grass experiment. *Journal of Applied Ecology* 17:491–504.

Simberloff, D. 1984. Mass extinction and the destruction of moist tropical forests. *Zhurnal Obshchei Biologii* 45:767–778.

Steadman, D.W. 1995. Prehistoric extinctions of Pacific island birds: biodiversity meets zooarchaeology. *Science* 267:1123–1131.

Tansley, A.G. 1949. *The British Isles and their vegetation.* Cambridge Press, Cambridge, UK.

Terborgh, J. 1974. Preservation of natural diversity: the problem of extinction prone species. *BioScience* 24:715–722.

Thurston, J. 1969. The effect of liming and fertilizers on the botanical composition of permanent grassland, and on the yield of hay. Pages 3–10 in I. Rorison, ed. *Ecological aspects of the mineral nutrition of plants.* Blackwell Scientific, Oxford, UK.

Tilman, D. 1987. Secondary succession and the pattern of plant dominance along experimental nitrogen gradients. *Ecological Monographs* 57(3):189–214.

Tilman, D. 1988. *Plant strategies and the dynamics and structure of plant communities.* Monographs in Population Biology, Princeton University Press, Princeton, NJ.

Tilman, D. 1990. Constraints and tradeoffs: toward a predictive theory of competition and succession. *Oikos* 58:3–15.

Tilman, D. 1994. Competition and biodiversity in spatially structured habitats. *Ecology* 75:2–16.

Tilman, D. 1996. Biodiversity: population versus ecosystem stability. *Ecology* 77(3):350–363.

Tilman, D., and J.A. Downing. 1994. Biodiversity and stability in grasslands. *Nature* 367:363–365.

Tilman, D., J. Knops, D. Wedin, P. Reich, M. Ritchie, and E. Siemann. The influence of Functional diversity and composition on ecosystem processes. *Science* 277:1300–1302.

Tilman, D., C.L. Lehman, and K.T. Thomson. 1997. Plant diversity and ecosystem productivity: theoretical considerations. *Proceedings of the National Academy of Science* 94:1857–1861.

Tilman, D., R.M. May, C.L. Lehman, and M.A. Nowak. 1994. Habitat destruction and the extinction debt. *Nature* 371:65–66.

Tilman, D., D. Wedin, and J. Knops. 1996. Productivity and sustainability influenced by biodiversity in grassland ecosystem. *Nature* 379:718–720.

Townsend, A.R., B.H. Braswell, E.A. Holland, and J.E. Penner. 1996. Spatial and temporal patterns in terrestrial carbon storage due to deposition of fossil fuel nitrogen. *Ecological Applications* 6(3):806–814.

Vitousek, P.M. 1990. Biological invasions and ecosystem processes: towards an integration of population biology and ecosystem studies. *Oikos* 57:7–13.

Vitousek, P.M. 1994. Beyond global warming: ecology and global change. *Ecology* 75:1861–1876.

472   D. Tilman

Vitousek, P.M., J.D. Aber, R.W. Howarth, G.E. Likens, P.A. Matson, D.W. Schindler, et al. 1997. Human alteration of the global nitrogen cycle: sources and consequences. *Ecological Applications* 7:737–750.

Vitousek, P.M., and D.U. Hooper. 1993. Biological diversity and terrestrial ecosystem biogeochemistry. Pages 3–14 in E.-D. Schulze, and H.A. Mooney, eds. *Biodiversity and ecosystem function.* Springer-Verlag, Berlin, Germany.

Vitousek, P.M., and L.R. Walker. 1989. Biological invasion by *Myrica faya* in Hawai'i: plant demography, nitrogen fixation, ecosystem effects. *Ecological Monographs* 59(3):247–265.

Walker, L.R., and P.M. Vitousek. 1991. An invader alters germination and growth of a native dominant tree in Hawai'i. *Ecology* 72(4):1449–1455.

Webb III, T., and T.M.L. Wigley. 1985. What past climates can indicate about a warmer world. Pages 239–257 in M.C. MacCracken, and F.M. Luther, eds *Projecting the climatic effects of increasing carbon dioxide.* United States Department of Energy, Washington, DC.

Wedin, D., and J. Pastor. 1993. Nitrogen mineralization dynamics in grass monocultures. *Oecologia* 96:186–192.

Wedin, D., and D. Tilman. 1990. Species effects on nitrogen cycling: a test with perennial grasses. *Oecologia* 84:433–441.

Wedin, D.A., and D. Tilman. 1996. Influence of nitrogen loading and species composition on the carbon balance of grasslands. *Science* 274:1720–1723.

Wedin, D.A., L.L. Tieszen, B. Dewey, and J. Pastor. 1995. Carbon isotope dynamics during grass decomposition and soil organic matter formation. *Ecology* 76:1383–1392.

20
Synthesis: What Kind of a Discipline Is This Anyhow?

PETER M. GROFFMAN and MICHAEL L. PACE

Introduction

We need to evaluate the results of the 1997 Cary Conference and the chapters in this book in terms of: (1) our initial ideas about the factors driving success, limitations, and frontiers in ecosystem ecology, and (2) other lessons and ideas that emerged from the presentations and discussions at the conference and in this book. Coming into the conference, we had two "big questions" driving our analysis of successes, limitations, and frontiers. The first question addressed the hypothesis that ecosystem ecology is somewhat unique as a discipline because advances in this field have been driven by the emergence of new environmental problems that could not be addressed with existing conceptual and practical tools. The second question arose from the perception that ecosystem ecology suffers from a lack of cohesion, focus, and justification relative to many other disciplines. Determining whether this perception is true, and evaluating the factors behind these problems was one of our major goals.

Is Ecosystem Ecology a Problem-Driven Discipline?

Over the last two years, we have discussed the idea that the emergence of new environmental problems that could not be addressed with existing conceptual or practical tools has been a major driver of progress in ecosystem science in many forums; among the staff at the Institute of Ecosystem Studies (IES), within our Cary Conference steering committee, at a workshop at the National Center for Ecological Analysis and Synthesis (NCEAS), at a symposium on "Synthesis in Ecology" at NCEAS, and at the 1997 Cary Conference itself. Most ecologists agree that environmental problems have played at least some role in driving fundamental advances in ecosystem ecology. But, as in so many aspects of science, the story is not completely clear-cut. We suggest that environmental problems have acted as fundamental drivers of advances within the science of ecology, fostering the emergence of ecosystem and landscape ecology, and driving the integration of ecology with physical sciences (in Earth System science) and social sciences (in integrated assessment) (Figure 20.1). As depicted in Figure 20.1, environmental problems have acted as drivers, in addition to more traditionally considered drivers of progress in science, for our purposes grouped as "ideas" produced by "dreamers and thinkers" (see Burke et al., chapter 7).

Historical analysis is complex and, especially in the hands on non-historians, can be quite subjective (Tjossem 1997). Moreover, analysis of the fundamental drivers of a discipline is complicated by a lack of generally agreed-upon criteria for determining just "what a fundamental driver is." There is particular difficulty in separating the "roots" of an advance (e.g. Forbes' original study of a lake ecosystem in the 1880s) from the events that propelled an area into scientific and public prominence (e.g. the eutrophication of Lake Washington, see Smith, chapter 2). Analysis of our "success stories" suggests a common pattern, however, which is: (1) an environmental problem is identified, often by a scientist; (2) public concern develops; (3) the political process directs funding toward scientific analysis and solution of the problem; and (4) the resulting scientific research effort leads to major advances, both basic and applied, in the discipline. Our analysis suggests that ecosystem ecology, as with medical science, reacts to the emergence of new problems, often in the glare of public attention, with well-publicized successes, failures and controversies. Also, as in medical science, we detect new problems by being out in the field, interacting with our subjects (e.g. the discovery of acid rain or forest decline is similar to identification of AIDS or Lyme Disease). If, indeed, ecosystem ecology is in some ways very similar to medical science, there are several important implications for our evaluation of successes, limitations, and frontiers in ecosystem ecology.

First, if we are inherently a problem-solving discipline, we need to recognize this as a fundamental component in the evaluation of success. Thus, we have defined success as a body of work that has: (1) increased our basic science understanding of the structure and function of ecosystems, and (2) contributed to the solution of an environmental problem. After much discussion over the last two years, we suggest that it is useful to partition the second criterion, that is, the practical or applied criterion, into two components, which are: (1) Have we generated the scientific understanding necessary to solve a problem? and (2) Has this knowledge been used by society? Among the success stories presented at the 1997 Cary Conference, all have been successes to some degree by the basic science criterion, and most have been successes by the first component of the applied science criterion. However, there is wide variation in the nature and extent of success in implementation among our different success cases.

The success case studies presented at the 1997 Cary Conference illustrate very readily the different criteria for success. The riparian forest (Lowrance

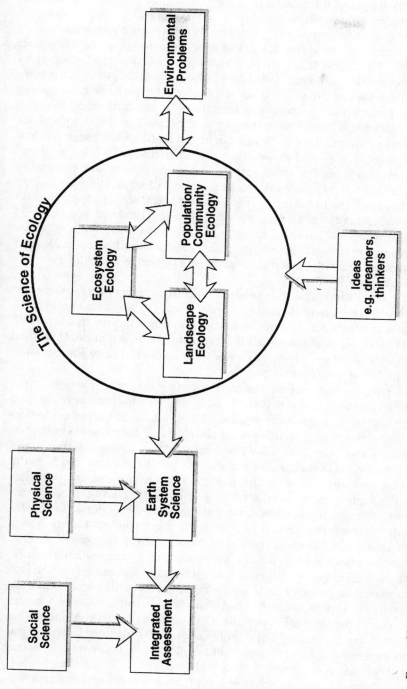

FIGURE 20.1. Applied (environmental problems) and basic (ideas) "drivers" of advances within the science of ecology and of the interface between ecology and physical and social sciences.

chapter 5) and lake eutrophication (Smith chapter 2) success cases are our most complete stories. These chapters describe research efforts that greatly increased our understanding of ecosystem structure and function as well as sophsticated programs for implementation of ecological understanding on a wide scale. The forest (Dale chapter 3), wetland (Zedler et al. chapter 4) and atmospheric chemistry (Weathers and Lovett chapter 8) chapters clearly illustrate cases in which bodies of research have produced advances in understanding the structure and function of ecosystems in addition to specific knowledge useful in managing ecosystems for specific purposes (e.g. wood products, wildlife, water quality), but this knowledge has not been implemented as well as we hope, that is, there is still a lot of "bad" forestry, wetlands are still being lost at an impressive rate, and air quality continues to deteriorate in many areas in a variety of ways. The agroecology (Robertson and Paul chapter 6), restoration (MacMahon chapter 9), and biogeochemical cycles at regional and global scales (Burke et al. chapter 7) cases are "successes that are waiting to happen." These are areas of active basic research in ecosystem ecology, poised to produce a body of ecological understanding that will be useful for managing critical present-day environmental problems.

A second question raised by the idea that ecosystem ecology is inherently a problem-solving discipline relates to advocacy. Do we need to make the value of our discipline obvious to science funding agencies, managers of natural resources, politicians, and the general public? We argue that if we are, in part, a problem-solving discipline, we need to advocate for the value of our discipline with these groups.

A third question arises from the lack of success in our "implementation" criteria. Whose fault is it that forests, lakes, and wetlands are not managed using the latest ecological knowledge? Does a lack of use of ecological information constitute a fundamental failure for our discipline? As Osvaldo Sala asked during the 1997 Cary Conference, "Should our salaries be dependent on the pH of the lake we are managing?" We answer, "No." On the one hand, lack of success in the implementation and use of ecological knowledge highlights the need for stronger interaction between ecosystem scientists, management, political institutions, and the general public. Indeed, this interaction emerged as one of the great frontiers for ecosystem ecology in our discussions at the Cary Conference. On the other hand, we cannot be responsible for societal decision-making processes that are influenced by a wide range of factors. Again, an analogy with medicine is useful. Is the medical community to blame because people still smoke? Medical researchers have demonstrated convincingly that smoking causes health problems and the medical community has been very active in publicizing their results and in reaching out to political, policy, and education organizations to disseminate their message. Yet people still smoke. And people still clear-cut forests on steep slopes, and drain wetlands . . .

Although most of the scientists we have talked to agree that environmental problems have played a role in driving advances in ecosystem science, there is certainly not universal agreement with the idea that ecosystem ecology is inherently a problem-driven, problem-solving discipline. Many would argue that our discipline is similar to other basic sciences (e.g. physics) in which advances are driven by the emergence of new ideas (e.g. "dreamers and thinkers," Burke et al. chapter 7) or new tools rather than such problem-solving disciplines as medicine. Many would argue that environmental problems have been a hindrance to our discipline, distracting effort from truly exciting questions, causing scientists to "chase after" topics determined by public fears and superstitions and generally leading to a lack of focus, predictive power, and cohesion (Edwards 1995). Attention to environmental problems (undue or not) by ecosystem ecologists may also be at the root of the ongoing divisions between ecosystem and population-level ecology, which tends to be driven more by theory and ideas. As depicted in Figure 20.1, it is probable that many factors contribute to advances in ecology at the same time.

As is true for so many topics in science, the question, "Is ecosystem ecology a problem-driven discipline?" has no one simple answer. Nevertheless, we suggest that environmental problems have played an important role, and that recognition of this role has important implications for the way that we design, conduct, interpret, evaluate, and advocate for our science.

Does Ecosystem Science Suffer from a Lack of Cohesion, Focus, and Justification?

We cannot deny that ours is a young discipline, fraught with controversy. Yes, our discipline suffers from a lack of cohesion, focus, and justification. But what can we learn about these problems from our analysis of successes, limitations, and frontiers?

We Must Claim a Core Body of Work for Ecosystem Science

Two factors, the young age and multidisciplinary nature of our discipline make it difficult to define the core of ecosystem ecology. For example, nutrient cycling is one of the key parameters generally considered to be within the domain of ecosystem ecology. Yet, much nutrient-cycling research is based on the idea of "limiting nutrients" as defined by Liebig, a 19th-century chemist and agronomist. Was Liebig an ecosystem ecologist? Certainly not, but he did ecosystem ecology research. We argue that we can define (and have done so) the bounds of ecosystem ecology (e.g. analysis of the flow of matter and energy through ecosystems, biogeochemical cycling, succession, the mass-balance concept) in a disciplinary sense (Likens 1992), and then claim all work that falls within those bounds.

The "Style" of Ecosystem Research Creates Problems with Cohesion, Focus, and Justification

Ecosystem-scale research, which often involves multiple investigators, large amounts of funding, and long-term studies of complex and highly uncertain phenomena is problematic for the discipline given the structure of academic, natural-resource management, science funding, and political institutions. The structure of academic departments favors single-investigator research efforts, with rapid production of results and scientific publications. Natural-resource management agencies and political institutions are poorly equipped to accept the complexity and uncertainty in ecosystem science and management. Science funding agencies are reluctant to commit large amounts of money to long-term studies (although there have been some notable exceptions, discussed later) and more important, these commitments can be divisive and alienating within the discipline itself. These statements have been made before, in many places. What have we learned about solving these problems from our analysis of successes, limitations, and frontiers? First, we need to be clear about the successes in our discipline, and the importance of the "style" of our research for achieving those successes. Second, we need to advocate for the value of our discipline with science funding agencies, managers of natural-resource agencies, political institutions, and the general public. It is important to note that the ecosystem "approach" has been extraordinarily useful serving as a focus to bring together scientists with diverse expertise to solve a wide range of complex environmental problems (Likens 1992).

What Does the Future Hold for Ecosystem Ecology?

When we began this analysis of successes, limitations, and frontiers in ecosystem ecology, we were criticized for an imbalance in topics, that is, why were there only two chapters on limitations, while there were eight on successes and eight on frontiers. In the final analysis, we have talked much more about limitations, about what we have not achieved, than what we have accomplished. Nearly all of our "success stories" describe only partial success, and end with ringing calls for more and better work on these topics. And what do our "frontiers" represent if not the tasks that we have failed to complete in the past? We suggest the themes presented in the next section will be important in ecosystem ecology over the next ten to twenty years.

Explicit Recognition and Incorporation of Human Factors in Ecosystem Studies

The above phrase has been said and written many, many times over the last ten years. The first Cary Conference, held in 1985, listed this topic as a frontier. The fourth Cary Conference, held in 1991, focused on "Humans as Components of Ecosystems" (McDonnell and Pickett 1993). But we cannot deny that the lack of accomplishment in this area is staggering. As discussed before, nearly all of our success cases fail to some extent by the "implementation" criteria, that is, has ecological knowledge been used in the solution of environmental problems or the management of natural resources?

One of our limitations chapters (Walters chapter 11) clearly articulates scientific problems that have inhibited the use of ecological information in practical application, and one of our frontiers chapters (Folke chapter 13) highlights the conceptual challenges and problems to integrating ecology with social sciences. These analyses suggest that "progress is at hand." We now have structures (e.g. adaptive management) that foster better design of experiments and make better use of information produced from these experiments in decision-making. Science funding agencies have explicitly recognized the need for integration of ecological, physical, and social sciences in their requests for proposals on "Methods and Models in Integrated Assessment" (MMIA), "Water and Watersheds" and "Urban Long-Term Ecological Research (LTER)." In the model for progress in ecosystem science described above; i.e. the identification of a problem, societal concern, funding, and progress, we are hopefully just beginning the "progress phase."

There appear to be two somewhat distinct roles for ecosystem ecologists in the incorporation of human factors in environmental science. First, we need to help society achieve its goals for ecosystem management, that is, we need to make recommendations to achieve ecosystem services, functions, outputs desired by humans (Pickett et al. 1997). Second, we need to educate and inform society so that its goals and values incorporate ecological knowledge. Efforts to articulate the wide range of functions and services that ecosystems provide, and to foster ecological literacy in society are important challenges to our discipline. Tim Wirth, the U.S. Undersecretary of State for Global Affairs, made a forceful case for such efforts in his remarks at the 1997 Cary Conference.

Complex Problems, Simplifying Concepts

The problem of deriving and expressing simple laws and concepts to describe the behavior of complex ecosystems has always dogged ecosystem ecology, from the earliest comprehensive models of the IBP, to the theoretical abstractions of H.T. Odum, to the debate about diversity and stability in the 1970s, to the debate about diversity and stability in the 1990s to the debate about diversity and stability . . . Brown (1997) argues that "there are some very general ecological laws still awaiting discovery" and that greater effort must be made to discover these laws. Articulating these laws will require some simplification and efficient expression of complexity. Walters (chapter 11) suggests that scientists' inability to simplify complex issues has greatly inhibited the use of scientific information by managers of natural resources. Likens (chapter 10) observes that for some problems, it may not be possible to find simple explanations and solutions for complex problems. What is needed in these cases is to address and express complexity in a straightforward and clear manner. Zak and Pregitzer (chapter 15) show how the search for simplicity in the face of complexity can lead to serious error.

The struggle between complexity and simplicity will endure in ecosystem ecology. It remains to be seen whether the key challenges in our field, both basic and applied, will be met in a better manner by a more clear articulation of complexity, or by clever simplification of this complexity. This is a clear, general frontier for the discipline.

New Experiments and Better Articulation of the Unique Style of Ecosystem Research

We have argued in this chapter that ecosystem research has a unique style, characterized by multidisciplinary approaches and long-term studies. Several of our frontiers chapters suggest that the future will see further development of this research style and that this expansion will help achieve many of the goals that have eluded us in the past. Carpenter (chapter 12) argues for a new series of long-term experiments that focus on the interface between basic and applied research. Tilman (chapter 19) suggests that a major new research initiative is needed to address the challenge of global change. The initiative would forge a strong link between species characteristics, community change, and ecosystem function, hopefully acting as a unifying force among the subdivisions of ecology (Jones and Lawton 1994), in addition to amplifying the interactions between ecology and physical and social sciences. Other frontiers chapters emphasize the importance of maintaining the multidisciplinary basis of ecosystem ecology (Zak and Pregitzer chapter 15; Vitousek et al. chapter 18; Folke chapter 13) and the application of a variety of modeling and measurement tools (Vitousek et al. chapter 18; Lauenroth et al. chapter 16; Wessman and Asner chapter 14; Breitburg et al. chapter 17) to address complex problems. Eventhough these frontiers highlight areas that we have not yet addressed, they offer a ringing endorsement of the ecosystem "approach" and "style" and suggest that the future will bring further definition and strength to our discipline in both basic and applied contexts.

The Best and the Brightest

A clear challenge to science in general, and to ecosystem ecology in particular, is to continue to attract bright students to our discipline. But how do we define "the best and the brightest"? In discussions at the 1997 Cary Conference, concern was expressed that our students lack the disciplinary depth and quantitative skills necessary to really solve many problems. Yet, the need for depth must to be balanced with the need for breadth—an ability to conceptualize the wide range of factors that contribute to ecosystem studies and environmental problems. The ability to integrate multiple factors is the hallmark of ecosystem science and must not be lost. Ultimately, we suggest that advances in any field are driven largely by the availability of individuals with passionate interest and determination to solve the puzzles that nature presents. The challenge to ecosystem ecologists today is to present our puzzles in a compelling way, and to demonstrate that our discipline possesses the necessary conceptual and practical tools to solve them.

References

Brown, J.H. 1997. *An ecological perspective on the challenge of complexity. Ecoessay series number 1*. National Center for Ecological Analysis and Synthesis, Santa Barbara, CA.

Edwards, P. 1995. Ecological progress the meet the challenge of environmental change. *Trends in Ecology and Evolution* 10:261.

Jones, C.G., and J.H. Lawton. eds. 1994. *Linking species and ecosystems*. Chapman & Hall, New York.

Likens, G.E. 1992. *The ecosystem approach: its use and abuse*. Ecology Institute, Oldendorf/Luhe, Germany.

McDonnell, M.J., and S.T.A. Pickett. eds. 1993. *Humans as components of ecosystems: the ecology of subtle human effects and populated areas*. Springer-Verlag, New York.

Pickett, S.T.A., R.S. Ostfeld, M. Shachak, and G.E. Likens, eds. 1997. *Enhancing the ecological basis of conservation*. Chapman & Hall, New York.

Tjossem, S.F. 1997. *Reply to James H. Brown's "An ecological perspective on the challenge of complexity." EcoEssay Series Number 1*. National Center for Ecological Analysis and Synthesis, Santa Barbara, CA.

Index